Soil Biology

Volume 56

Series Editor

Ajit Varma, Amity Institute of Microbial Technology, Amity University, Noida, Uttar Pradesh, India

The Springer series *Soil Biology* publishes topical volumes in the fields of microbiology, environmental sciences, plant sciences, biotechnology, biochemistry, microbial ecology, mycology and agricultural sciences.

Special emphasis is placed on methodological chapters or volumes. This includes coverage of new molecular techniques relevant to soil biology research or to monitoring and assessing soil quality as well as advanced biotechnological applications. Leading international authorities with a background in academia, industry or government will contribute to the series as authors or editors.

Key Topics: microbial-plant interactions; microbial communities; root symbiosis, mycorrhiza; rhizosphere environment; soil fauna, e.g. termites etc.; biochemical processes, soil proteins, enzymes, nucleic acids; degradation of biomaterials, decomposition, nutrient cycles; soil genesis, mineralization; bioremediation of contaminated sites; biotechnological applications of soil microorganisms.

More information about this series at http://www.springer.com/series/5138

Bhoopander Giri • Ajit Varma
Editors

Microorganisms in Saline Environments: Strategies and Functions

Springer

Editors
Bhoopander Giri
Department of Botany
Swami Shraddhanand College
University of Delhi
New Delhi, India

Ajit Varma
Amity Institute of Microbial Technology
Amity University
Noida, Uttar Pradesh, India

ISSN 1613-3382 ISSN 2196-4831 (electronic)
Soil Biology
ISBN 978-3-030-18974-7 ISBN 978-3-030-18975-4 (eBook)
https://doi.org/10.1007/978-3-030-18975-4

This Springer imprint is published by the registered company Springer Nature Switzerland AG.
The registered company address is: Gewerbestrasse 11, 6330 Cham, Switzerland

Foreword

In the world, there are millions of hectares of land affected by excessive salt accumulation and more are becoming prone to salinity due to the accumulation of salts, making them increasingly unsuitable for agriculture. Indeed, it has been estimated that 25% of irrigated lands are currently affected by excessive salinity (Ali et al. 2016). Soil salinity decreases plant productivity through osmotic and ion-specific stresses, altering plant capabilities for water uptake and causing ion toxicity. Microorganisms are indispensable agents for the right performance of the soil ecosystem and largely responsible for organic matter decomposition, formation of soil aggregates, the promotion/suppression of plant growth, and nutrient cycling (Guo et al. 2019). In this line, plants usually interact with soil microorganisms that make them more efficient in coping with environmental stresses, including salinity. However, many well-known plant growth-promoting microorganisms that operate well in unstressed environments are unable to maintain their efficiency when applied under unfavourable conditions. Hence, understanding the adaptation made by microbes from saline environments can potentially lead to new biotechnological applications in agriculture. Thus, there is an increasing need to understand saline environments, the microbial diversity found in these saline systems, as well as the role of these microorganisms in improving plant salt tolerance, as a strategy to guarantee food production for an evenly increasing world population (Ali et al. 2016). Moreover, the use of beneficial microorganisms native to saline environments and adapted to salinity could be a critical issue for success in recovering saline areas either in natural environments or in salinized agricultural lands. Special attention should be paid to plant growth-promoting bacteria (PGPB) and soil fungi establishing mycorrhizal symbioses with roots of most higher plants, since both can help the host plants to overcome the limiting effects of soil salinity on plant growth and yield production.

The book *Microorganisms in Saline Environments: Strategies and Functions* represents a comprehensive review dealing with salt impact on soil microbial diversity and plant fitness, plant–microbe interactions in saline environments, and different strategies for reclamation of saline soils. Thus, Chap. 1 presents an overview on the problem of soil and water salinization around the world, providing a description of salinity and its causes, global status, and effects of salinity on plant growth and productivity, environment, and economy of salt-affected areas. Chapter 2 focuses on the occurrence and distribution of fungi in saline habitats, with a special emphasis on marine fungi and the fact that these fungi are more ubiquitous and diverse than previously considered. Chapters 3 and 4 describe the current knowledge on the microbial ecology and diversity of halophilic microorganisms, including an overview of their biotechnological, industrial, and environmental applications, leading to the production of new bioactive molecules such as enzymes, antibiotics, osmolytes, and polymers. Thus, these chapters emphasize the importance of saline environments in the discovery of bioactive metabolites and new therapeutic compounds. On the other hand, Chap. 5 summarizes the diversity, distribution, and function of microbes in the extensive Indian Mangroves, covering an area of about 4921 km^2. Chapter 6 refers to the biotechnological potential of halophilic and other PGPR microorganisms, focusing on the role of exopolysaccharides and biofilm formation in the mitigation of salinity stress for plant growth. Subsequently, several chapters are devoted to the interactions of plant and microbes in saline environments. Chapter 7 describes the impact of salinity on growth and development of aquatic fungi. Chapters 8, 9, and 10 offer an overview of the cellular mechanisms of plant salt tolerance and the effects of salinity on plant physiological processes, including information on how microorganisms can help plants to improve their fitness and survival in these environments. Thus, osmotic adjustment, ion homeostasis, antioxidant defence systems, maintenance of membrane integrity, transport system, photosynthetic processes, and stress sensing and signalling are described. Chapter 11 describes how the use of beneficial microorganisms (especially plant growth-promoting bacteria and arbuscular mycorrhizal fungi) improves plant salt tolerance and thereby plant growth when applied alone or in combination with native Pseudomonads. Chapters 12, 14, and 18 summarize strategies used for the isolation of saline-adapted bacteria beneficial for crop production under salinity condition, including Rhizobia and phosphorous-solubilizing bacteria. These chapters also describe different mechanisms used by PGPB to stimulate plant growth under saline conditions (synthesis of phytohormones and siderophores, phosphate solubilization, production of ACC deaminase, etc.). Chapters 13 and 16 focus on the various mechanisms used by arbuscular mycorrhizal fungi to improve plant salt tolerance (improved osmotic adjustment, ionic homeostasis, antioxidant defence, etc.). Chapters 15 and 17 present an overview of the physiological, biochemical, and molecular basis of plant responses to salinity and salt tolerance in plants. Finally, two chapters are devoted to strategies for mitigation and reclamation of saline soils. Thus, Chap. 19 provides interesting information on the methods developed to desalinize the soils, such as leaching, growing halophytes, subsurface drainage, and biomimicry. It also identifies the challenges in the application of these techniques. Chapter 20 provides information on the use of biochar, as it enhances physicochemical properties and reduces soil loss.

The different subjects included in this book provide a thorough and detailed picture of the updated knowledge on the diversity of microorganisms from saline environments and on their functions in assisting plants to overcome the limitations of salinity on plant growth, development, and productivity.

Departamento de Microbiología del Suelo y Sistemas Simbióticos,
Estación Experimental del Zaidín (CSIC),
Granada, Spain

Juan Manuel Ruiz-Lozano

References

Ali I, Prasongsuk S, Akbar A, Aslam M, Lotrakul P, Punnapayak H, Rakshit SK (2016) Hypersaline habitats and halophilic microorganisms. Maejo Int J Sci Technol 10:330–345

Guo H, Hu Z, Zhang H, Hou Z, Min W (2019) Soil microbial metabolic activity and community structure in drip-irrigated calcareous soil as affected by irrigation water salinity. Water Air Soil Pollut 230:44–56

Preface

On this planet, microorganisms occupy diverse habitats due to their ability to grow and survive under both favourable and adverse conditions. Indeed, several microorganisms have been able to adapt and occupy saline environment. Such microbes exhibit their potential to pleasantly grow and survive under saline conditions. On the other hand, only about 2% of plants species (halophytes) are able to survive satisfactorily under saline conditions. The rest of the plant species (glycophytes) do not demonstrate satisfactory growth and yield under saline conditions; however, they could achieve this in association with certain microbes prevalent in the saline environment. Excessive concentration of soluble salts (salinity) in soils or irrigation water adversely affects the growth and productivity of plants. This problem exacerbates more in the arid and semiarid areas due to low precipitation and high evaporation rates.

Toxic effects of specific ions prevailing in the saline environment disrupt the structure of enzymes and other macromolecules, damage cell organelles, disrupt general metabolic activities, inhibit protein synthesis, and induce ions deficiency in plants. In fact, salinity creates nutrient imbalance and physiological drought conditions as it decreases differential nutrient uptake and their translocation to different plant tissues, thereby making plant weak and unproductive. Further, the salinization of arable land is increasing steadily in many parts of the world and is expected to accelerate devastating global effects. Estimation indicates that if the current trend of soil salinization continues, it would lead to 30% land loss within the next 25 years and up to 50% by the middle of the twenty-first century. Rapid industrialization and urbanization in several countries is, however, gradually becoming a major source of water pollution, especially of rivers. Poor irrigation and management practices and policies for using river water in the agriculture fields have been one of the main causes for inducing secondary salinization of soils in several regions. To meet the growing demands of increasing population, it is of paramount importance to know about salt-stress tolerance potentials of microorganisms prevailing in different habitats and their responses to crop plants under saline conditions and to also understand the nature of damage caused by salinity to agricultural crops.

The book *Microorganisms in Saline Environments: Strategies and Functions* comprises four main parts. Part I outlines the definition of salinity, its genesis and impacts, and microbial diversity in saline habitats. Part II deals with impact of salinity on microbial and plant life/health. Part III highlights plant–microbe interactions in saline environments, and Part IV describes strategies for reclamation of saline soils. Attempts have been made to bring together the major aspects of microbial ecology of saline environments for the benefit of academicians and scientists. We hope that this large volume would generate interest and ideas among young researchers that will bring forward new approaches and methodology leading to further advances in our understanding of microbial ecology and interactions under saline conditions and their implications for sustainable agriculture. With a fortune of information, we hope that this volume would serve as a valuable resource for researchers, academicians, environmentalists, and students in the broad field of microbiology and agriculture.

We highly appreciate cooperation and support provided by all the contributors/ authors for preparing very precious and authoritative chapters. Indeed, we are thankful to all our fellow colleagues for their generous support and help to prepare and compile this massive volume. We extend our sincere thanks to Springer officials, especially William F Curtis, Eric Schmitt, Sabine Schwarz, Isabel Ullmann, Beate Siek, and Anand Venkatachalam, for their relentless support and cooperation in accomplishing this massive volume. We specially thank all our near and dear, especially our families for the coherent support and encouragement.

New Delhi, India Bhoopander Giri
Noida, Uttar Pradesh, India Ajit Varma

Contents

Contributors

Mazhar M. F. Abdalla Faculty of Agriculture, Agronomy Department, Cairo University, Giza, Egypt

Magdi T. Abdelhamid Botany Department, National Research Centre, Giza, Egypt

Sunder Singh Arya Department of Botany, Maharshi Dayanand University, Rohtak, India

Ram Avtar Department of Genetics and Plant Breeding, CCS Haryana Agricultural University, Hisar, India

Rajib Bandopadhyay UGC-Center of Advanced Study, Department of Botany, The University of Burdwan, Burdwan, West Bengal, India

Aparna Banerjee Vicerrectoría de Investigación y Postgrado, Centro de Biotecnología de los Recursos Naturales, Universidad Católica del Maule, Talca, Chile

Karima Bencherif Faculté des sciences de la Nature et de la vie, Université de Djelfa, BP, Djelfa, Algeria

Unité de Chimie Environnementale et Interactions sur le Vivant (UCEIV), Université du Littoral Côte d'Opale, Calais cedex, France

Rachna Bhateria Department of Environmental Sciences, Maharshi Dayanand University, Rohtak, Haryana, India

Adèle L. Bunbury-Blanchette Department of Biology, Acadia University, Wolfville, NS, Canada

Bishwanath Chakraborty Department of Biological Sciences, Aliah University, Kolkata, West Bengal, India

Usha Chakraborty Department of Botany, Plant Biochemistry Laboratory, University of North Bengal, Siliguri, West Bengal, India

Sara Cuadros-Orellana Facultad de Ciencias Agrarias y Forestales, Centro de Biotecnología de los Recursos Naturales, Universidad Católica del Maule, Talca, Chile

Yolande Dalpé Agriculture et agroalimentaire Canada, Centre de recherche et développement d'Ottawa, Ottawa, Ontario, Canada

Darwish S. Darwish Faculty of Agriculture, Agronomy Department, Cairo University, Giza, Egypt

Bansal Deepak JBM Group, Gurugram, Haryana, India

Sarita Devi Department of Botany and Plant Physiology, CCS Haryana Agricultural University, Hisar, India

Zhou-Yan Dong State Key Laboratory of Biocontrol and Guangdong Key Laboratory of Plant Resources, School of Life Sciences, Sun Yat-Sen University, Guangzhou, PR China

Ashutosh Dubey Department of Biochemistry, College of Basic Sciences and Humanities, G. B. Pant University of Agriculture and Technology, Pantnagar, India

Nabti Elhafid Faculté des Sciences de la Nature et de la Vie, Laboratoire de Maitrise des Energies Renouvelables, Université de Bejaia, Bejaia, Algeria

Raafat R. El-Masry Botany Department, National Research Centre, Giza, Egypt

Heikham Evelin Department of Botany, Rajiv Gandhi University, Itanagar, India

Abhrajyoti Ghosh Department of Biochemistry, Bose Institute, Kolkata, West Bengal, India

Shubhrima Ghosh Enzyme and Microbial Biochemistry Laboratory, Department of Chemistry, Indian Institute of Technology Delhi, New Delhi, India

Bhoopander Giri Department of Botany, Swami Shraddhanand College, University of Delhi, Delhi, India

Anissa Lounès Hadj-Sahraoui Unité de Chimie Environnementale et Interactions sur le Vivant (UCEIV), Université du Littoral Côte d'Opale, Calais cedex, France

Shyamalina Haldar Department of Microbiology, Goa University, Taleigão, India

Minakshi Jattan Department of Genetics and Plant Breeding, CCS Haryana Agricultural University, Hisar, India

Sunil Kumar Khare Enzyme and Microbial Biochemistry Laboratory, Department of Chemistry, Indian Institute of Technology Delhi, New Delhi, India

Bala Kiran Discipline of Biosciences & Biomedical Engineering, Indian Institute of Technology, Indore, India

Sumit Kumar Enzyme and Microbial Biochemistry Laboratory, Department of Chemistry, Indian Institute of Technology Delhi, New Delhi, India

Nisha Kumari Department of Biochemistry, CCS Haryana Agricultural University, Hisar, India

Bensidhoum Leila Faculté des Sciences de la Nature et de la Vie, Laboratoire de Maitrise des Energies Renouvelables, Université de Bejaia, Bejaia, Algeria

Wen-Jun Li State Key Laboratory of Biocontrol and Guangdong Key Laboratory of Plant Resources, School of Life Sciences, Sun Yat-Sen University, Guangzhou, PR China

Key Laboratory of Biogeography and Bioresource in Arid Land, Xinjiang Institute of Ecology and Geography, Chinese Academy of Sciences, Ürűmqi, PR China

Kamla Malik Department of Microbiology, CCS Haryana Agricultural University, Hisar, India

Mohamed Magdy F. Mansour Faculty of Science, Department of Botany, Ain Shams University, Cairo, Egypt

Min Xiao State Key Laboratory of Biocontrol and Guangdong Key Laboratory of Plant Resources, School of Life Sciences, Sun Yat-Sen University, Guangzhou, PR China

Sharma Mona Department of Environmental Science & Engineering, Guru Jambheshwar University of Science & Technology, Hisar, Haryana, India

Thirumala Mothe Microbial Ecology Laboratory, Department of Applied Biosciences, Mahatma Gandhi University, Nalgonda, Telangana, India

Shayantan Mukherji Department of Biochemistry, Bose Institute, Kolkata, West Bengal, India

Rani Nisha Department of Biosciences, Himachal Pradesh University, Shimla, India

Shinya Oba Faculty of Applied Biological Sciences, Plant Production Control Lab, Gifu University, Gifu, Japan

Okon Godwin Okon Department of Botany, Akwa Ibom State University, Ikot Akpaden, Akwa Ibom, Nigeria

Ragab Ragab Centre for Ecology and Hydrology (CEH), Wallingford, UK

Babita Rani Department of Biochemistry, CCS Haryana Agricultural University, Hisar, India

Manik Prabhu Narsing Rao State Key Laboratory of Biocontrol and Guangdong Key Laboratory of Plant Resources, School of Life Sciences, Sun Yat-Sen University, Guangzhou, PR China

Swarnendu Roy Department of Botany, Plant Biochemistry Laboratory, University of North Bengal, Siliguri, West Bengal, India

Anissa Lounès-Hadj Sahraoui Faculté des sciences de la Nature et de la vie, Université de Djelfa, Djelfa, Algeria

Unité de Chimie Environnementale et Interactions sur le Vivant (UCEIV), Université du Littoral Côte d'Opale, Calais, France

Karima Hamid A. Salama Faculty of Science, Department of Botany, Ain Shams University, Cairo, Egypt

Shrabana Sarkar UGC-Center of Advanced Study, Department of Botany, The University of Burdwan, Burdwan, West Bengal, India

Vinay Shankar Department of Botany, Deshbandhu College, University of Delhi, New Delhi, India

Priyanka Srivastava Department of Botany, Sri Guru Tegh Bahadur Khalsa College, University of Delhi, Delhi, India

Vishnuvardhan Reddy Sultanpuram Microbial Ecology Laboratory, Department of Applied Biosciences, Mahatma Gandhi University, Nalgonda, Telangana, India

Sushil Department of Chemistry, CCS Haryana Agricultural University, Hisar, India

Patricia Velez Department of Botany, Universidad Nacional Autónoma de México, Mexico City, Mexico

A. K. Verma Department of Biochemistry, College of Basic Sciences and Humanities, G. B. Pant University of Agriculture and Technology, Pantnagar, India

Allison K. Walker Department of Biology, Acadia University, Wolfville, NS, Canada

Qiang-Sheng Wu College of Horticulture and Gardening, Yangtze University, Jingzhou, Hubei, China

Mohammad Javad Zarea Faculty of Agriculture, Department of Agronomy and Plant Breeding, Ilam University, Ilam, Iran

Part I
Genesis and Impact of Salinity on the Diversity and Distribution of Microorganisms in Saline Habitats

Chapter 1
Salinity: An Overview

Priyanka Srivastava, Qiang-Sheng Wu, and Bhoopander Giri

Abstract Excessive concentration of soluble salts in soils adversely affects agricultural lands and crops and subsequently the livelihood of people all around the world. More than 100 countries are facing the problem of soil salinity along with salinization of groundwater. Irrigation of agricultural crops with saline water indeed increases the concentration of soluble salts in soil, thereby reducing the productivity of crop plants. Excessive salts primarily disrupt the cellular osmotic balance by lowering the water potential inside cells. The salts like chlorides and sulphates of sodium, calcium and magnesium along with sodium carbonate and sodium bicarbonate prevailed in saline soils negatively affect plant growth and productivity as they change the osmotic balance between plant roots and soil and interfere with physiological and metabolic processes of plant. Several techniques have been developed for the reclamation and management of saline and sodic soils, which involve changes in the farming system irrigation, cropping pattern and use of salt-resistant varieties as well. The efficient and balanced use of fertilizers or more likely the utilization of biofertilizers could be a better option which can serve as a cost-effective technique for resource conservation. This chapter provides brief description of salinity and its causes, global status and effects of salinity on plant growth and productivity, environment and the economy of salt-affected areas. Indeed various techniques are in practice, and there is an urgent need of more site-specific studies for the development of strategies to reclaim saline soils. Under changing climatic conditions, this problem is predicted to increase further and may cover many more areas that necessitate to find out more realistic solutions of this problem. This chapter aims to draw attention of researchers to better understand reclamation and management technologies for sustainable solutions to curtailing the problem of salinization and efficient utilization of saline soils.

P. Srivastava
Department of Botany, Sri Guru Tegh Bahadur Khalsa College, University of Delhi, Delhi, India

Q.-S. Wu
College of Horticulture and Gardening, Yangtze University, Jingzhou, Hubei, China

B. Giri (✉)
Department of Botany, Swami Shraddhanand College, University of Delhi, Delhi, India

B. Giri, A. Varma (eds.), *Microorganisms in Saline Environments: Strategies and Functions*, Soil Biology 56, https://doi.org/10.1007/978-3-030-18975-4_1

Keywords Salt-affected soils · Saline soils · Soil reclamation · Soil salinity · Salt stress · Crop production

1.1 Introduction

Salinity is the term used to describe the condition when soluble salts get accumulated in the solution (soil or water) to a level that has a negative impact on the growth and development of crop plants (Childs and Hanks 1975). A soil can be recognized as salt-affected when certain changes appear on the ground surface like appearance of white salt-crust on bare land and development of salty-spots in the soil; permanent or prolonged waterlogging after precipitation; destruction and crumbling of roads; deterioration of ground and surface water, which become unpalatable for livestock and humans; and presence of salt crystals, which are more likely in the case of highly saline soils. It is quite imperative to understand differences between normal and saline soils. A normal soil is having exchangeable sodium percentage (ESP) less than 15.0, pH between 6.5 and 8.5, electrical conductivity (EC) less than 4.0 dS m^{-1} and an equal proportion of anions and cations. However, a saline soil has ESP around 15.0, sodium adsorption ratio (SAR) less than 13.0 (at 25 °C), EC more than 4.0 dS m^{-1} and pH less than 8.5 (Sharma 2014; Mindari et al. 2015). In these soils, mostly calcium and magnesium are the cations and chloride and sulphate are the anions, which occur in the form of soluble salts. Another type of soil can be categorized into sodic soil where EC is generally less than 4.0 dS m^{-1}. Sodic soils lack occurrence of soluble salts, and the exchangeable sodium is more than 15. The pH of such soils generally falls between 8.5 and 10.5. The most prevalent cation of a sodic soil is sodium and carbonate, and bicarbonate is the most complimentary anions of these soils. Therefore, soil with excessive soluble salts is recognized as saline soil, with high content of exchangeable sodium as sodic soil and with excess of both salts and exchangeable sodium as saline-sodic soils. On the basis of EC, water or soil solutions can be differently categorized (Table 1.1).

The process of accumulation of soluble salts in a solution is known as salinization. Salinization may be primary or secondary. In case of primary salinization, salinity develops due to natural processes like physical and chemical weathering from parent materials, deposition of sandstones and imposition of ocean water into coastal areas and rivers followed by evapotranspiration left behind large amount in the soil. Cyclones bring ocean salts (e.g. sodium chloride) to ground, which accumulates in soil due to precipitation (Chaitanya et al. 2014). Seepage from sea to low-lying areas and lack of water in arid zones to leach out the excess salts could be one of the reasons behind salinization. Secondary salinization is a result of anthropogenic activities like deforestation which causes soil movement in both the upper and lower layers (Said-Al and Omer 2011). Application of chemical fertilizers or salt-rich irrigation with lack of proper precipitation or drainage to leach out extra salts triggers salinization of soil and water and makes soil unproductive with

Table 1.1 Categories of saline solutions, their applications and impact on plant growth and yield

Type of solutions	EC ($dS\ m^{-1}$) of SE	Application/type	Effect on plant
Non-saline	<1.0	Drinking and irrigation	Effect is insignificant
Slightly saline	1.0–2.0	Irrigation water	Plant growth and yields may hamper in the case of sensitive crops
Moderately saline	2.0–10.0	Primary drainage water and groundwater	A large number of crop plants growth and yield often get reduced
Highly saline	10.0–20.0	Secondary drainage water and groundwater	In this condition, only salt-tolerant crop plants growth and yield occur up to a satisfactory level
Very strongly saline	20.0–45.0	Very saline groundwater	In this condition, only a few crop plants yield satisfactorily; such plants exhibit strong tolerance to salt stress

As per International System, the unit (SI unit) of conductivity is 'siemens', which is denoted by a symbol 'S' per metre, while the equivalent non-SI unit is 'mho' where 1.0 mho is equal to 1.0 siemens. The mmhos/cm can be read as dS/m without any numerical change. Conductivity 1.0 S cm^{-1} (1 mho/cm) is equal to 1000 mS/cm (1000 mmhos/cm), and 1.0 mS/cm^{-1} (1 mmho/cm) is equal to 1.0 dS/m = 1000 mS/cm (1000 micromhos/cm). *EC* electrical conductivity, *SE* saturation extract, $dS\ m^{-1}$ deciSiemens per metre at 25 °C, g^{-1} grams per litre

accumulation of salts after evaporation in the root zone of plant (Blaylock 1994; Homaee et al. 2002). Waterlogging and inadequate drainage also accumulate salts in soil and degrade its quality. Salinization is continuously increasing in new areas due to poor irrigation practices (Patel et al. 2011).

Salinity largely affects physical properties of soil by binding the fine soil particles into aggregates by the process called flocculation. Though it is beneficial for soil aeration and root growth, the higher salinity in such a situation further adversely affects plant growth and yield (Abu-Sharar et al. 1987). Calcium and magnesium keeps soil flocculated, while high amount of sodium causes soil dispersion. Salinity converts soil into hard cement-like stuff causing low infiltration with reduced hydraulic conductivity (HC) and development of surface crusting (Abu-Sharar et al. 1987). The World Soil Resources Reports (IUSS Working Group WRB 2007) has considered reference value 15.0 dS m^{-1} of EC as a higher minimum thresholds criterion for defining the saline horizon, while the Soil Taxonomy (Soil Survey Staff 2010) fixes the threshold criterion at EC 30 dS m^{-1}. In fact, availability of water and mineral nutrients for plants and microbes could depend on the concentration of salts present in the soil solution. A high concentration of soluble salts in soil has been found to often influence soil organic matter, content and stability also (Mavi et al. 2012). Saline soils contain low organic matter and mineral nutrient as the latter gets immobilized reaching beyond the approach of plant roots. Further, these soils have increased toxicities of boron, aluminium and carbonates which affect both number and diversity of microbial community (Rengasamy 2002; Rietz and Haynes

2003; Imadi et al. 2016). With the increasing level of salinity, the concentration of cadmium, metal, soluble chloride, sulphate and bicarbonate ions increase, which negatively affects soil composition, while the concentration of magnesium and calcium ions decreases in the soil (Cao et al. 2012; Hashem et al. 2014).

1.2 Status of Saline Areas Globally

Salt-stressed soils adversely affect the livelihood security of people in more than 100 countries. Around 831 million ha (Mha) of the land area, all over the world, is salt-affected. Figure 1.1 depicts salt-affected area of the world. Out of that 397 Mha (47.8%) are saline and 434 Mha (52.2%) are sodic (Sharma and Chaudhari 2012; Sharma 2014). Out of the total cultivated world's land, about 77 Mha of productive land is under the influence of salinity or sodicity (Selvakumar et al. 2014; Arora 2017; Miransari 2017), which is predicted to be further increasing in many parts of the world due to extensive land utilization at global level, especially in arid and semiarid areas. These areas suffer from the low precipitation and high rate of evaporation (Rengasamy 2006, 2010; Selvakumar et al. 2014). According to the Food and Agriculture Organization of the United Nations (FAO 2002), soil salinization is reducing about 1–2% of the area every year, which otherwise could be used for agriculture; this problem is more aggravated in the area with semiarid and arid conditions (FAO 2002). One of the reasons for salinity in countries prevailing arid and semiarid conditions is overuse of groundwater for irrigation (Koohafkan 2012). The FAO of the United Nations (1988) estimated that approximately 50% area of 250 Mha of irrigated land around the world has already gone under salinization and soil saturation problems, whereas 10 Mha area is abandoned annually due to

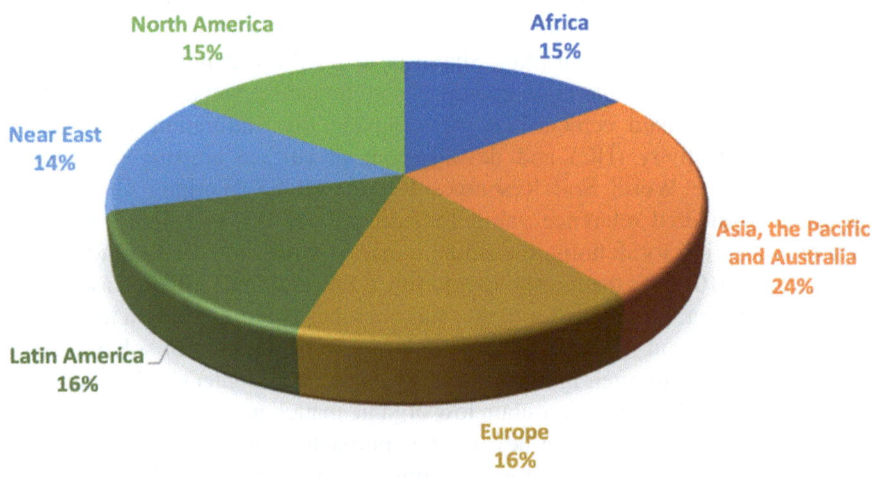

Fig. 1.1 Distribution of salt affected soils in the world (Mha)

salinization. The United Nations Environment Program estimated that about 20% of agricultural land and 50% of cropland in the world is salt-affected (Flowers and Yeo 1995).

As far as regional status of salt-affected soils is concerned, Asia, the Pacific and Australia together have largest salt-affected areas (30%). In these regions, approximately 195 Mha of land is occupied with salinity. Salinity had a large impact on the subsistence and income of people of those countries where agriculture is the main source of economy of agrarian population and a large portion of agriculture production is mainly depends on irrigation. According to United Nations University, Institute for Water, Environment and Health, approximately 2000 ha of productive soil gets degraded daily due to the problem of excessive salts across 75 countries of arid and semiarid regions. The salt-induced soil degradation is a result of poor drainage system and accumulation of salts in the rhizosphere, which is prevalent in the rhizosphere of North and Central Asia along with South America. Nevertheless, coastal areas are also exposed to the problems of high salt concentrations due to salts coming with ocean currents (Hemapriya et al. 2010).

In India alone, about 9.38 Mha of soil is salt-affected either due to salinity, sodicity or both, of this around 5.5 Mha is salinized naturally (Jaleel et al. 2008). Salt-affected soils are increasing across several major states of this country. The maximum salt-affected area occurs in the states like Gujrat (2.23 Mha), Uttar Pradesh (1.37 Mha), Maharashtra (0.61 Mha), West Bengal (0.44 Mha) and Rajasthan (0.38 Mha) (Fig. 1.2). The area in these states accounts for approximately 80% of total salt-affected land in India. Amongst them, Gujarat is the most severely affected state due to salinity as it is covering 1.68 Mha of land followed by West Bengal, which occupy 0.44 Mha of land and by the Rajasthan (0.19 Mha) and Maharashtra (0.18 Mha). As stated above, salt-affected land in Uttar Pradesh is about

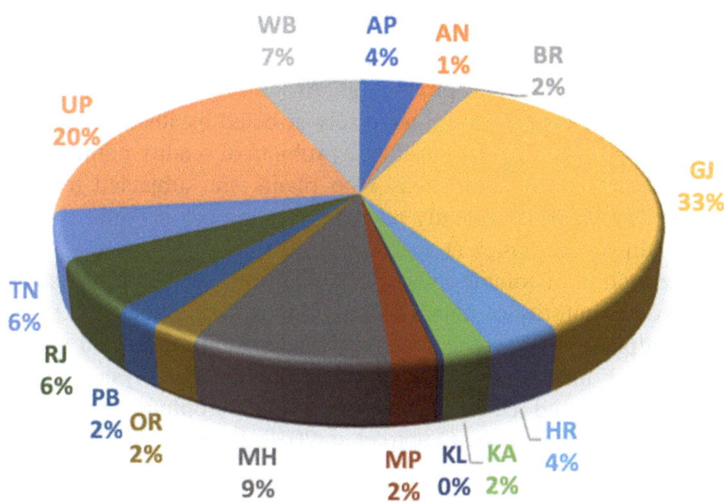

Fig. 1.2 State wise salt affected area (ha) in India

1.37 Mha, of which 0.02 Mha is covered with salinity and 1.35 Mha is facing the problem of sodicity. Amongst all states of India, it is Uttar Pradesh where sodicity has affected more productive lands than salinity. However, the states like Andaman and Nicobar Islands, Kerala, Orissa and West Bengal are having soil salinity as a major issue rather than sodicity (Singh 2018). Actually, 25% of the groundwater resources in India are saline and brackish, and this problem is more prevalent in the states like Haryana and Rajasthan which holds 62 and 84% of poor quality groundwaters, respectively (Sharma 2014). Therefore, these states are largely suffering from the problems of irrigation water quality and regular use of such water for agriculture is provoking the problem of salinity and sodicity. The lack of proper drainage system has also been found as one of the major causes for developing salinity in canal commands in India. It is projected that India would acquire 11.7 Mha area affected by salinity and sodicity by 2025 (Sharma and Chaudhari 2012).

1.3 Effects of Salinity on Agriculture

Any unfavourable condition (external factor or exposure to any substance) which is negatively influencing the development, productivity and reproductive behaviour of plants can be considered as stress (Rhodes and Nadolska-Orczyk 2001). In fact, plant stress can be categorized into two categories; (1) abiotic or natural stresses (like cold, drought, salt, heavy metals as well) and (2) anthropogenic or biotic stresses (like pathogens, herbivory) (Mordecai 2011). Natural or abiotic stresses are of serious concern, particularly with crop plants as these are responsible for severe loss due to their negative effects on every stage/aspects of plant life such as morphological, physiological and biochemical and metabolism processes (Akbarimoghaddam et al. 2011; Fraire-Velázquez et al. 2011). Amongst various abiotic stresses, soil salinity has become a matter of great concern (Lobell and Field 2007; Miransari 2017), as it adversely decreased agricultural production. Around the globe, arid and semiarid regions and coastal areas have been adversely affected by the salinity (Hemapriya et al. 2010). Arid and semiarid regions face a problem of scantly rainfall and lowered precipitation; therefore, in these areas crop plants are subjected to grow under irrigation practice, which is basically mismanaged and has become a major source of secondary salinization (Glick et al. 2007).

High concentration of soluble salts affects plant life even from the stage of seed germination to the maturity of plants (Akbarimoghaddam et al. 2011). These salts adversely affect plant photosynthetic activity by damaging thylakoid membranes of chloroplast, trigger inactivation of ETS (electron transport system) and thereby photophosphorylation and reduce the concentration chlorophyll and other photosynthetic pigments like carotenoids (Omoto et al. 2010; Evelin et al. 2012; Mittal et al. 2012). They also have a negative impact on the turgor pressure, particularly in the leaves which insist on changing the stomatal conductance, making the stomata closed and reducing the rate of transpiration (Chaves et al. 2002, 2008; Silva et al.

2011; Hniličková et al. 2017). Under saline conditions, the increased accumulation of abscisic acid has been observed by a number of researchers (Melcher et al. 2009; Zhang et al. 2006; Wang et al. 2012), and this seems to be a major cause of closing of stomata in the plants growing under saline conditions. Increased accumulation of ABA could influence the depolarization of stomata guard cells and production of hydrogen peroxide. Excessive production of hydrogen peroxide tends to influence production of reactive oxygen species (ROS), which triggers oxidative stress in plants (Chinnusamy et al. 2006a, b). ROS damage plant cell as well as affect plant health (Foyer and Noctor 2005).

High level of soluble salts in the soil also attributes to the promotion of osmotic stress, which on the other hand results in the reduction of water and nutrient uptake and thereby could lead to cell death in certain cases (Munns 2002). In fact, legumes and cereal crops are very sensitive to salinity. In these plants, salinity stress decreased the concentration of photosynthetic pigments like chlorophylls, thereby affecting plant yield (Sairam et al. 2002). Wheat and rice cultivars have been found to be very sensitive to salt stress and are most affected in coastal areas (Sultana et al. 2001; Akbarimoghaddam et al. 2011). It has been observed that in the case of legumes, salinity often affects nitrogen fixation process by interfering with the formation of root nodules (Rao et al. 2002; Ramana et al. 2012). Nutritional imbalance due to excessive salt attributes to chlorosis and senescence and sometimes even boosts more vegetative than reproductive growth, and as a consequence reduced crop yield is obtained (Ali et al. 2004; Munns and Tester 2008). Moreover, salinity declines the reproductive capacity of plant by interfering with enzyme activity, mitosis, DNA and RNA synthesis (Seckin et al. 2009; Tabur and Demir 2010; Javid et al. 2011).

Further, at high salt concentration, a competition could exist between Na^+ and K^+ ions for occupying entry sites in the root plasma membrane for ingression into the symplast that affects K^+-dependent metabolic processes and also disrupts the integrity of root membranes (Grattan and Grieve 1999; Evelin et al. 2009). Potassium acts as a cofactor for many enzymes. It is also required for binding tRNA to ribosomes during protein synthesis. But, excessive concentration of Na^+ ions could replace K^+ ions, which attributes to conformational changes in the proteins, and any change in the protein structure may adversely affect both biochemical and metabolic processes of plants. Besides, high concentration of Na^+ ions could result in the physiological drought and ion toxicity to plants (Zhu 2002; Evelin et al. 2009).

1.4 Effects of Salinity on Environments

Salinization of fresh water bodies like rivers and streams directly affects the quality of drinking and irrigation water due to chloride present in it, while the presence of sodium and magnesium sulphates has adverse effects on other organisms (husbandry). Salinity indeed affects livelihood of both aquatic and terrestrial ecosystems. It often results in declining biodiversity through habitat fragmentation, loss of habitat and loss of intolerant species. Flocculation in water bodies due to salinity could allow more favourable conditions for the growth of harmful algae and hence modifies the functioning and ecological health of aquatic ecosystems (Smyth and Elliott 2016). Loss of soil surface cover due to salinity makes soil prone to soil erosion, consequently reducing the capacity of soil to absorb the rainfall, which results in surface run-off and flooding (Hu and Schmidhalter 2002). Salinity affects the life of both rural and urban communities; however, major effect is on rural communities due to heavy loss of agriculture or business and high cost of its restructuring. Salinity deteriorates physical structure of soil in such a way that water permeability and soil aeration are reduced up to a large extent. The problem of infrastructure destruction due to salinity causes economic loss to a country, for instance, damage to concrete, brick structure, roads, bridges, pipelines and buildings due to the deposition of salt-rich water. The damage needs extra cost for repairing and maintenance.

1.5 Reclamation of Saline Soil

Reclamation includes the removal of extra soluble salts from the soil, in particular from the root zone. Researchers may refer to Chaps. 19 and 20 of this book to study in detail the different reclamation methods either in the practice or can be practiced. However, the common practices adopted for saline soil reclamation are discussed in brief in this chapter. These include, the mechanical removal of the salts accumulated on the soil surface (scraping); however, this method gives a temporary relief but not a solution for long-term removal of salts. Other practice incudes washing of surface soil to remove accumulated salt crust by flushing water over the surface; however, only a small portion of salts get flushed by this method. The process of drainage includes the removal of salts from the soil along with the excess water. It can be achieved ether by surface drainage or sub-surface drainage. In surface drainage, the removal of water from the surface of impermeable or lowland soil takes place, whereas in sub-surface drainage practice ditches are made for the removal of gravitational water. Removing salts from the root zone soil with good quality of water can be achieved by leaching method using approaches like continuous ponding, intermittent ponding and sprinkling (Hanson et al. 2006). It is a most effective and common strategy for the reclamation of saline habitats. Organic amendments are also considered as an effective method for reclamation of saline habitats. The decomposition of organic substances improves soil health by dissolving insoluble salts, maintaining pH and improving soil permeability (Matosic et al. 2018). For instance, livestock manure, biosolids and crop residues are good source of organic produce to reclaim saline soils (Larney and Angers 2012). The other organic produce for the purpose are green manure like sunnhemps (*Crotalaria juncea*), cowpea (*Vigna sinensis*), jantar (*Sesbania aculeata*), etc., which increase biological activity, provide nutrients to soil and act as solubilizing agent of calcium along with neutralizing the pH of soil (Vijayaragavan and Ramachandran 1989).

Farmyard manure and molasses obtained from *Sesbania aculeata* leaves are best for improving rice crop productivity in saline soil (Dutta et al. 2015).

Planting some salt-tolerant species in saline soil (afforestation) can help in improving soil permeability and adding carbonic acid to soil through the root activity. Their organic waste decomposition and release of acids improve mobilization of calcium to replace the sodium from salt-affected soil (Wang et al. 2014). Some of the plant species suitable for the purpose are Babool (*Acacia arabica*), Farash (*Tamarix articulata*), Palash (*Butea monosperma*) and Bada Peelu (*Salvadora oleoides*), as they can tolerate up to a certain range of salinity (Singh 2014). Chemical amendment for decreasing exchangeable sodium concentration from saline soils has been practiced by the application of gypsum (0.59 mm fineness). This method has been proved beneficial for supplying calcium by replacing sodium (Sharma et al. 1998). It helps in removing excess sodium ions from root zone and also for improving soil permeability (Matosic et al. 2018).

The use of microbes for reclamation of saline soils could be an eco-friendly, cheaper and effective approach, as the halophilic microbes have ability to remove salt from saline soils (Arora et al. 2013; Bhuva et al. 2013). These microorganisms could colonize rhizosphere and provide benefits to plant through various mechanisms (Hayat et al. 2010). Certain halophilic bacteria have been found capable enough to alleviate salt stress and improve growth and yield of agricultural crops growing in the saline soils (Nia et al. 2012; Ramadoss et al. 2013; Arora et al. 2014). Indeed, salt-tolerant plant growth-promoting rhizobacteria (PGPRs) such as *Azospirillum*, *Acetobacter*, *Bacillus* and *Pseudomonas* assist in restoring plant health and enhanced crop productivity through various changes in the rhizosphere (Rajput et al. 2013; Vivekanandan et al. 2015; Arora et al. 2016). Giri et al. (2003) demonstrated that inoculation of *Acacia auriculiformis* with arbuscular mycorrhizal fungi (AMF), viz. *Glomus fasciculatum* and *Glomus macrocarpum*, significantly modulated the electrical conductivity and pH of rhizosphere soil. The alteration in the rhizosphere soil property could be due to direct uptake, translocation and transfer of ions by hyphae of arbuscular mycorrhizas to partner plants and also substantiate the fact that these fungi play an important role in making ionic balance, particularly under salt stress conditions. Moreover, AMF have been found to improve growth and productivity of several plants, and under saline conditions, however, their response varies with different levels of salinity (Cantrell and Linderman 2001; Abdel Latef and Chaoxing 2011; Aroca et al. 2013; Evelin et al. 2013).

The production of ethylene often increases under stress conditions. It has been observed that ethylene regulated plant growth by reducing root and shoot growth; however, PGPRs like *Azospirillum*, *Acetobacter*, *Bacillus* and *Pseudomonas* have been found to help in removing ethylene impact under salt stress by increasing the production of enzyme, 1-aminocyclopropane-1-carboxylate deaminase (ACC deaminase) (Glick et al. 2007; Govindasamy et al. 2008). In addition, PGPRs also produce phytohormones like IAA, gibberellins and cytokinin. The latter increases root growth and root surface area and thereby enhances nutrient uptake of plants under salt stress conditions (Egamberdieva and Kucharova 2009; Kang et al. 2012). These bacteria also promote phosphate-solubilizing activity in the rhizosphere and thereby enhance crop production by facilitating crop plants for more phosphate uptake (Bhattacharyya and Jha 2012; Rajput et al. 2013). PGPRs confer plant's tolerance to salt stress. Inoculation of tomato and *Arabidopsis* with *Enterobacter* species enhanced production of ACC deaminase, IAA, proline and activity of reactive oxygen species, which helped in mitigating stress impact through upregulation of conserved salinity responsive factors and thus attributed to enhanced plant growth (Kim et al. 2014). Studies with many other economical plants like tomato, canola, bean and lettuce growing in salt-stressed soils had increased growth when these plants were inoculated with PGPRs (Yildirim and Taylor 2005; Barassi et al. 2006).

1.6 Prevention and Management of Soil Salinity

Frequent irrigation maintains the moisture level of soil, thus preventing the concentration of salts and to an extent minimizing its adverse effects on plant (Chen et al. 2015; Alkhamisi et al. 2016). Sprinkler irrigation is the best method for uniform irrigation using minimal amount of water (Meiri et al. 1992; Liu et al. 2011). Root zone drying methodology and drip irrigation methods are better for sustainable use of water (Shrivastava and Kumar 2015). In this method, water is continuously supplied at a point source very near to root zone for maintaining low saline condition near root. It is also proved good for maintaining increased productivity of plant (Goldberg et al. 1976). To prevent rising water tables and movement of salts to soil surface, some deep-rooted plants are needed to grow with annual plants so that they can utilize unused water (Manchanda and Garg 2008).

Plant waste like rice straw can be applied on the soil surface to protect soil from run off, weed infestation and drying due to evaporation. This method is known as Mulching that not only improves soil properties but maintains soil moisture and enhances plant yield by facilitating plants for better nutrient uptake (Fang et al. 2011; Patil et al. 2013). Mulching protects soil from direct sunlight and reduces evapotranspiration. Soil moisture is an important abiotic factor, which is largely affected by salinity; however, mulching helps in retaining moisture in the soil; therefore, mulching is found to be a suitable technique to prevent salinization of soil (Mia et al. 2015; Abd El-Mageed et al. 2016).

Nevertheless, there are many salt-tolerant plants known as halophytes, which have tendency to nicely withstand salt stress conditions naturally. Moreover, attention is also paid to develop salt-tolerant plant species which are more susceptible to high concentration of salts (Khatun and Flowers 1995). Since salinity is a multigenic trait that makes it difficult to understand the overall mechanism and to identify all the genes involved. Indeed, the development of transgenic varieties may be a successful approach to cope up with the problem of salinity up to certain extent. On the other hand, fertigation approach (application of fertilizers with irrigation water) can also be applied to manage the concentration of nutrient level in soil. The intensive use of fertilizers attributes to the increased concentration/level nutrients in soil because their concentrations increase the actual requirement of plant (Grattan et al. 2002;

Machado et al. 2014). Therefore, frequent and adequate amount of fertilizers with fertigation would be a better option to improve the efficacy of fertilizers. This approach could help the plant for getting enough nutrients as well as in utilizing them properly. In fact, fertigation with nitric and sulfuric acid is found to be an efficient process to removing salinity in arid regions (Machado and Serralheiro 2017).

1.7 Conclusions

Increasing concentration of soluble salts has become a serious problem for agricultural crop production. Various methods for the reclamation of saline soils are in practice along with various management strategies to control this problem. In addition to certain approaches mentioned as well in this manuscript, attention must be paid for applying controlled fertilization and proper irrigation management as they directly influence soil properties, and preference should be given for biological fertilizers/manures as they influence microbial community, which help in conserving and sustaining soil fertility. It is important to provide sustainable substitutes of management practices along with many more type of natural amendments for the soil to control the concentration of salts and to restore soil health and fertility. Development of new salt-resistant varieties of value crops is required with proper field trials in saline environment to figure out their success rate and to provide a better substitute for utilizing salt-affected soils.

References

Abd El-Mageed TA, Semida WM, Abd El-Wahed MH (2016) Effect of mulching on plant water status, soil salinity and yield of squash under summer-fall deficit irrigation in salt affected soil. Agric Water Manag 173:1–12

Abdel Latef AA, Chaoxing H (2011) Effect of arbuscular mycorrhizal fungi on growth, mineral nutrition, antioxidant enzymes activity and fruit yield of tomato grown under salinity stress. Sci Hortic 127:228–233

Abu-Sharar TM, Bingham FT, Rhoades JD (1987) Stability of soil aggregates as affected by electrolyte concentration and composition. Soil Sci Soc Am J 51:309–314

Akbarimoghaddam H, Galavi M, Ghanbari A, Panjehkeh N (2011) Salinity effects on seed germination and seedling growth of bread wheat cultivars. Trakia J Sci 9(1):43–50

Ali Y, Aslam Z, Ashraf MY, Tahir GR (2004) Effect of salinity on chlorophyll concentration, leaf area, yield and yield components of rice genotypes grown under saline environment. Int J Environ Sci Technol 1(3):221–225

Alkhamisi S, Al-Wardy M, Ahmed M, Prathapar S (2016) Impact of reclaimed water irrigation on soil salinity, hydraulic conductivity, cation exchange capacity and macro-nutrients. J Agric Mar Sci 21:8–18

Aroca R, Ruiz-Lozano JM, Zamarreno AM, Paz JA, García-Mina JM, Pozo MJ et al (2013) Arbuscular mycorrhizal symbiosis influences strigolactone production under salinity and alleviates salt stress in lettuce plants. J Plant Physiol 170:47–55

Arora S (2017) Diagnostic properties and constraints of salt-affected soils. In: Arora S et al (eds) Bioremediation of salt affected soils: an Indian perspective. Springer, Cham, pp 41–52
Arora S, Trivedi R, Rao GG (2013) Bioremediation of coastal and inland salt affected soils using halophyte plants and halophilic soil microbes. In: CSSRI Annual Report 2012–13. CSSRI, Karnal, India, pp 94–100
Arora S, Patel P, Vanza M, Rao GG (2014) Isolation and characterization of endophytic bacteria colonizing halophyte and other salt tolerant plant species from Coastal Gujarat. Afr J Microbiol Res 8(17):1779–1788
Arora S, Singh YP, Vanza M, Sahni D (2016) Bio-remediation of saline and sodic soils through halophilic bacteria to enhance agricultural production. J Soil Water Conserv 15(4):302–305
Barassi CA, Ayrault G, Creus CM, Sueldo RJ, Sobrero MT (2006) Seed inoculation with Azospirillum mitigates NaCl effects on lettuce. Sci Hortic 109:8–14
Bhattacharyya PN, Jha DK (2012) Plant growth-promoting rhizobacteria (PGPR): emergence in agriculture. World J Microbiol Biotechnol 28(4):1327–1350
Bhuva C, Arora S, Rao GG (2013) Efficacy of halophilic microbes for salt removal from coastal saline soils. In: National seminar with the theme "Microbes and Human Welfare", Bharathidasan University, Tiruchirappalli, India
Blaylock AD (1994) Soil salinity, salt tolerance and growth potential of horticultural and landscape plants. Co-operative Extension Services, University of Wyoming, Department of Plant, Soil and Insect Sciences, College of Agriculture, Laramie, Wyoming
Cantrell IC, Linderman RG (2001) Preinoculation of lettuce and onion with VA mycorrhizal fungi reduces deleterious effects of soil salinity. Plant Soil 233:269–281
Cao M, Xin P, Jin G, Li L (2012) A field study on groundwater dynamics in a salt marsh—Chongming Dongtan wetland. Ecol Eng 40:61–69
Chaitanya V, Rama Krishna C, Ramana V, Beebi SK (2014) Salinity stress and sustainable agriculture-a review. Agric Rev 35(1):34–41
Chaves MM, Pereira JS, Maroco J, Rodrigues ML, Ricardo CPP, Osorio ML, Carvalho I, Faria T, Pinheiro C (2002) How plants cope with water stress in the field? Photosynthesis and growth. Ann Bot 89:907–916
Chaves M, Flexas J, Pinheiro C (2008) Photosynthesis under drought and salt stress: regulation mechanisms from whole plant to cell. Ann Bot 103(4):551–560
Chen W, Lu S, Pan N, Wang Y, Wu L (2015) Impact of reclaimed water irrigation on soil health in urban green areas. Chemosphere 119:654–661
Childs SW, Hanks RJ (1975) Model of soil salinity effects on crop growth. Soil Sci Soc Am J 39:617–622
Chinnusamy V, Zhu J, Zhu JK (2006a) Gene regulation during cold acclimation in plants. Physiol Plant 126(1):52–61
Chinnusamy V, Zhu J, Zhu JK (2006b) Salt stress signalling and mechanisms of plant salt tolerance. Genet Eng 27:141–177
Dutta T, Rahman MM, Bhuiya SU, Kader MA (2015) Use of organic amendment for amelioration of salinity stress in transplanted aman rice cv. BRRI dhan41. Int J Natl Soc Sci 2(5):82–94
Egamberdieva D, Kucharova Z (2009) Selection for root colonizing bacteria stimulating wheat growth in saline soils. Biol Fertil Soils 45:563–571
Evelin H, Kapoor R, Giri B (2009) Arbuscular mycorrhizal fungi in alleviation of salt stress: a review. Ann Bot 104:1263–1280
Evelin H, Giri B, Kapoor R (2012) Contribution of *Glomus intraradices* inoculation to nutrient acquisition and mitigation of ionic imbalance in NaCl-stressed *Trigonella foenum-graecum*. Mycorrhiza 22:203–217
Evelin H, Giri B, Kapoor R (2013) Ultrastructural evidence for AMF mediated salt stress mitigation in *Trigonella foenum-graecum*. Mycorrhiza 23(1):71–86
Fang S, Xie B, Liu D, Liu J (2011) Effects of mulching materials on nitrogen mineralization, nitrogen availability and poplar growth on degraded agricultural soil. New For 41:147–162

FAO (1988) Soil map of the world. Revised legend, FAO–UNESCO–ISRIC. World Soil Resources Report No. 60. FAO, Rome
FAO (2002) Gender and sustainable development in drylands: an analysis of field experiences. Original and complete version of the present document. FAO, Rome
Flowers TJ, Yeo AR (1995) Breeding for salinity resistance in crop plants where next. Aust J Plant Physiol 22:875–884
Foyer CH, Noctor G (2005) Oxidant and antioxidant signalling in plants: a re-evaluation of the concept of oxidative stress in a physiological context. Plant Cell Environ 28:1056–1071
Fraire-Velázquez S, Rodríguez-Guerra R, Sánchez-Calde-rón L (2011) Abiotic and biotic stress response crosstalk in plants. In: Shanker AK, Venkateswarlu B (eds) Abiotic stress response in plants: physiological, biochemical and genetic perspectives. In Tech, Rijeka, pp 3–26
Giri B, Kapoor R, Mukerji KG (2003) Influence of arbuscular mycorrhizal fungi and salinity on growth, biomass and mineral nutrition of *Acacia auriculiformis*. Biol Fertil Soils 38:170–175
Glick BR, Todorovic B, Czarny J, Cheng Z, Duan J, Mcconkey B (2007) Promotion of plant growth by bacterial ACC deaminase. Crit Rev Plant Sci 26:227–242
Goldberg D, Gornat B, Rimon D (1976) Drip irrigation principles design and agricultural practices. Drip Irrigation Scientific Publications, p 295
Govindasamy V, Murugeasn S, Gaikwad K, Annapurna K (2008) Isolation and characterization of ACC deaminase gene from two plants growth-promoting rhizobacteria. Curr Microbiol 57:312–317
Grattan SR, Grieve CM (1999) Mineral nutrient acquisition and response by plants grown in saline environments. In: Pessarakli M (ed) Handbook of plant and crop stress. Marcel Dekker, New York, pp 203–229
Grattan SR, Grieve CM, Poss JA, Robinson PH, Suarez DL, Benes SE (2002) Reuse of saline-sodic drainage water for irrigation in California: evaluation of potential forages. In: Proceedings of the 17th World Congress on Soil Salinity, Bangkok, Thailand, pp 110–120
Hanson BR, Grattan SR, Fulton A (2006) Agricultural salinity and drainage, revised ed. Division of Agriculture and Natural Resources Publication 3375. University of California
Hashem A, Abd Allah EF, Alqarawi AA, Al-Didamony G, Al-Whibi M, Egamberdieva D, Ahmad P (2014) Alleviation of adverse impact of salinity on faba bean (*Vicia faba* L.) by arbuscular mycorrhizal fungi. Pak J Bot 46(6):2003–2013
Hayat R, Ali S, Amara U, Khalid R, Ahmed I (2010) Soil beneficial bacteria and their role in plant growth promotion: a review. Ann Microbiol 60:579–598
Hemapriya R, Sankar K, Dar IA (2010) Geologic and geomorphologic investigation of Gadilam river basin (India). J Environ Res Dev 4(3):750–757
Hniličková H, Hnilička F, Martinková J, Kraus K (2017) Effects of salt stress on water status, photosynthesis and chlorophyll fluorescence of rocket. Plant Soil Environ 63(8):362–367
Homaee M, Feddes RA, Dirksen C (2002) A macroscopic water extraction model for nonuniform transient salinity and water stress. Soil Sci Soc Am J 66:1764–1772
Hu Y, Schmidhalter U (2002) Limitation of salt stress to plant growth. In: Hock B, Elstner CF (eds) Plant toxicology. Marcel Dekker, New York, pp 91–224
Imadi SR, Shah SW, Kazi AG, Azooz MM, Ahmad P (2016) Phytoremediation of saline soils for sustainable agricultural productivity. Plant metal interaction (emerging remediation techniques), pp 455–468
IUSS Working Group WRB (2007) World reference base for soil resources 2006, first update 2007. World Soil Resources Reports No. 103. FAO, Rome
Jaleel CA, Sankar B, Sridharan R, Panneerselvam R (2008) Soil salinity alters growth, chlorophyll content, and secondary metabolite accumulation in *Catharanthus roseus*. Turk J Biol 32:79–83
Javid MG, Sorooshzadeh A, Moradi F, Sanavy SAMM, Allahdadi I (2011) The role of phytohormones in alleviating salt stress in crop plants. Aust J Crop Sci 5:726–734
Kang NY, Cho C, Kim NY, Kim J (2012) Cytokinin receptor-dependent and receptor-independent pathways in the dehydration response of *Arabidopsis thaliana*. J Plant Physiol 169:1382–1391
Khatun S, Flowers TJ (1995) Effects of salinity on seed set in rice. Plant Cell Environ 18:61–67

Kim ST, Kim SG, Agrawal GK, Kikuchi S, Rakwa R (2014) Rice proteomics: a model system for crop improvement and food security. Proteomics 14:593–610

Koohafkan P (2012) Water and cereals in drylands. Food and Agriculture Organization of the United Nations and Earthscan, Rome

Larney FJ, Angers DA (2012) The role of organic amendments in soil reclamation: a review. Can J Soil Sci 92:19–38

Liu H, Yu L, Yu L, Wang X, Huang G (2011) Responses of winter wheat (*Triticum aestivum* L.) evapotranspiration and yield to sprinkler irrigation regimes. Agric Water Manag 98(4):483–492

Lobell DB, Field CB (2007) Global scale climate–crop yield relationships and the impacts of recent warming. Environ Res Lett 2(1):1–7

Machado RM, Serralheiro RP (2017) Soil salinity: effect on vegetable crop growth. Management practices to prevent and mitigate soil salinization. Horticulture 3:30

Machado RM, Bryla DR, Vargas O (2014) Effects of salinity induced by ammonium sulfate fertilizer on root and shoot growth of highbush blueberry. Acta Hortic 1017:407–414

Manchanda G, Garg N (2008) Salinity and its effects on the functional biology of legumes. Acta Physiol Plant 30(5):595–618

Matosic S, Birkás M, Vukadinovic V, Kisic I, Bogunovic I (2018) Tillage, manure and gypsum use in reclamation of saline-sodic soils. Agric Conspec Sci 83(2):131–138

Mavi MS, Marschner P, Chittleborough DJ, Cox JW, Sanderman J (2012) Salinity and sodicity affect soil respiration and dissolved organic matter dynamics differentially in soils varying in texture. Soil Biol Biochem 45:8–13

Meiri A, Frenkel H, Mantell A (1992) Cotton response to water and salinity under sprinkler and drip irrigation. Am Soc Agron 84(1):44–50

Melcher K, Ng LM, Zhou XE, Soon FF, Xu Y, Suino-Powell KM, Park SY et al (2009) A gate-latch-lock mechanism for hormone signalling by abscisic acid receptors. Nature 462:602–608

Mia S, Dijkstra FA, Abuyusuf M, Hossain A (2015) Synergistic effects of biochar and NPK fertilizer on soybean yield in an alkaline soil. Pedosphere 25(5):713–719

Mindari W, Sasongko Edi P, Kusuma Z, Syekhfani N (2015) Characteristics of saline soil and effect of fertilizer application to rice yield. Int J Agric Res 6(1):7–15

Miransari M (2017) Arbuscular mycorrhizal fungi and soil salinity. In: Wu QS (ed) Arbuscular mycorrhizas and stress tolerance of plant. Springer International, Singapore, pp 147–161

Mittal S, Kumari N, Sharma V (2012) Differential response of salt stress on *Brassica juncea*: photosynthetic performance, pigment, proline, D1 and antioxidant enzymes. Plant Physiol Biochem 54:17–26

Mordecai EA (2011) Pathogen impacts on plant communities: unifying theory, concepts, and empirical work. Ecol Monogr 81:429–441

Munns R (2002) Comparative physiology of salt and water stress. Plant Cell Environ 28:239–250

Munns R, Tester M (2008) Mechanisms of salinity tolerance. Annu Rev Plant Biol 59:651–681

Nia SH, Zarea MJ, Rejali F, Varma A (2012) Yield and yield components of wheat as affected by salinity and inoculation with *Azospirillum* strains from saline or non-saline soil. J Saudi Soc Agric Sci 11:113–121

Omoto E, Taniguchi M, Miyake H (2010) Effects of salinity stress on the structure of bundle sheath and mesophyll chloroplasts in NAD-malic enzyme and PCK type C4 plants. Plant Prod Sci 13:169–176

Patel BB, Patel Bharat B, Dave RS (2011) Studies on infiltration of saline–alkali soils of several parts of Mehsana and Patan districts of north Gujarat. J Appl Technol Environ Sanit 1(1):87–92

Patil SS, Kelkar TS, Bhalerao SA (2013) Mulching: a soil and water conservation practice. Res J Agric For Sci 1(3):26–29

Rajput L, Imran A, Mubeen F, Hafeez F (2013) Salt-tolerant PGPR strain *Planococcus rifietoensis* promotes the growth and yield of wheat (*Triticum aestivum* L.) cultivated in saline soil. Pak J Bot 45:1955–1962

Ramadoss D, Lakkineni VK, Bose P, Ali S, Annapurna K (2013) Mitigation of salt stress in wheat seedlings by halotolerant bacteria isolated from saline habitats. Springer Plus 2(6):1–7

Ramana GV, Padhy SP, Chaitanya KV (2012) Differential responses of our soybean (*Glycine max* L.) cultivars to salinity stress. Legume Res 35:185–193
Rao DLN, Giller KE, Yeo AR, Flowers TJ (2002) The effects of salinity and sodicity upon nodulation and nitrogen fixation in chickpea (*Cicer arietinum* L.). Ann Bot 89:563–570
Rengasamy P (2002) Transient salinity and subsoil constraints to dryland farming in Australian sodic soils: an overview. Aust J Exp Agric 42:351–361
Rengasamy P (2006) World salinization with emphasis on Australia. J Exp Bot 57:1017–1023
Rengasamy P (2010) Osmotic and ionic effects of various electrolytes on the growth of wheat. Aust J Soil Res 48:120–124
Rhodes D, Nadolska-Orczyk A (2001) Plant stress physiology. In: Encyclopedia of life sciences, vol 10. Nature Publishing Group
Rietz DN, Haynes RJ (2003) Effects of irrigation-induced salinity and sodicity on soil microbial activity. Soil Biol Biochem 35:845–854
Said-Al A, Omer EA (2011) Medicinal and aromatic plants production under salt stress. A review. Herba Polonica 57:72–87
Sairam RK, Rao KV, Srivastava GC (2002) Differential response of wheat genotypes to long term salinity stress in relation to oxidative stress, antioxidant activity and osmolyte concentration. Plant Sci 163:1037–1046
Seckin B, Sekmen AH, Turkan I (2009) An enhancing effect of exogenous mannitol on the antioxidant enzyme activities in roots of wheat under salt stress. J Plant Growth Regul 28:12–20
Selvakumar G, Kim K, Hu S, Sa T (2014) Effect of salinity on plants and the role of arbuscular mycorrhizal fungi and plant growth-promoting rhizobacteria in alleviation of salt stress. In: Ahmed P, Wani MR (eds) Physiological mechanisms and adaptation strategies in plants under changing environment. Springer Nature, New York, pp 115–144
Sharma DK (2014) Sustainable technologies for crop production under salt-affected soil in India. In: Proceedings of 3rd International Salinity Forum, Session 2. UC Riverside Convention Center, Riverside, CA, USA
Sharma D, Chaudhari SK (2012) Agronomic research in salt affected soils of India: an overview. Ind J Agronomy 57:175–185
Sharma OP, Sharma DN, Minhas PS (1998) Reclamation and management of alkali soils. In: Minhas PS, Sharma OP, Patil SG (eds) 25 Years of research on management of salt affected soils and use of saline water in agriculture. Central Soil Salinity Research Institute, Karnal, pp 64–84
Shrivastava P, Kumar R (2015) Soil salinity: a serious environmental issue and plant growth promoting bacteria as one of the tools for its alleviation. Saudi J Biol Sci 22(2):123–131
Silva EN, Ribeiro RV, Ferreira-Silva SL, Viégas RA, Silveira JAG (2011) Salt stress induced damages on the photosynthesis of physic nut young plants. Sci Agric 68(1):62–68
Singh A (2014) In salt-affected soils agroforestry is a promising option. Indian Farm 63:19–22
Singh G (2018) Climate change and sustainable management of salinity in agriculture. Res Med Eng Sci 6(2):1–7. https://doi.org/10.31031/RMES.2018.06.000635
Smyth K, Elliott M (2016) Effects of changing salinity on the ecology of the marine environment. In: Solan M, Whiteley N (eds) Stressors in the marine environment. Physiological and ecological responses. Societal implications. Oxford University Press, Oxford, pp 161–174
Soil Survey Staff (2010) Keys to soil taxonomy, 11th edn. USDA-NRCS, Washington, DC
Sultana N, Ikeda T, Kashem MA (2001) Effect of foliar spray of nutrient solutions on photosynthesis, dry matter accumulation and yield in seawater-stressed rice. Environ Exp Bot 46:129–140
Tabur S, Demir K (2010) Role of some growth regulators on cytogenetic activity of barley under salt stress. Plant Growth Regul 60:99–104
Vijayaragavan H, Ramachandran TK (1989) Effect of in situ cultivation and incorporation of green manure crops on the yield of coconut. Cocos 7:26–29

Vivekanandan M, Karthik R, Leela A (2015) Improvement of crop productivity in saline soils through application of saline-tolerant rhizosphere bacteria – current perspective. Int J Adv Res 3 (7):1273–1283

Wang H, Zhang MS, Guo R, Shi DC, Liu B, Lin XY, Yang CW (2012) Effects of salt stress on ion balance and nitrogen metabolism of old and young leaves in rice (*Oryza sativa* L.). BMC Plant Biol 12:194

Wang L, Sun X, Li S, Zhang T, Zhang W, Zhai P (2014) Application of organic amendments to a coastal saline soil in north China: effects on soil physical and chemical properties and tree growth. PLoS One 9(2):e89185

Yildirim E, Taylor A (2005) Effect of biological treatments on growth of bean plants under salt stress. Ann Rep Bean Improv Coop 48:176–177

Zhang J, Jia W, Yang J, Ismail AM (2006) Role of ABA in integrating plant responses to drought and salt stresses. Field Crop Res 97(1):111–119

Zhu JK (2002) Salt and drought stress signal transduction in plants. Annu Rev Plant Biol 53:247–273

Chapter 2
Occurrence and Distribution of Fungi in Saline Environments

Adèle L. Bunbury-Blanchette and Allison K. Walker

Abstract Fungi are present in saline environments that can be broadly categorized as either open ocean or intertidal (coastline) habitats. Many ecological niches exist within these environments, in which fungi adapted to marine conditions may act as saprobes, pathogens, or symbionts. These fungi have a substantial impact on ecosystem functions such as nutrient cycling, productivity, life history and demographics of other organisms, and ecosystem resilience. Recent biodiversity research demonstrates that fungi colonize a wide variety of marine substrates and possess unique adaptations for life submerged in or routinely exposed to saline water. The more well-known marine fungi include decomposers of floating wood, mangrove associates, and saltmarsh saprobes and symbionts. Recent research on less easily accessible or previously overlooked habitats like deep sea vents and sand beaches show that marine fungi are more ubiquitous and diverse than previously considered; there are many opportunities for future research to uncover new marine fungal species and to expand our ecological and functional understanding of known species.

Keywords Biodiversity · Fungi · Marine · Open ocean · Temperate ocean · Tropical ocean

2.1 Introduction: Fungi from Open Ocean

Fungi known from open ocean habitats have traditionally been thought of as primarily decomposers of floating wood; however, recent research has highlighted marine fungi from a wider variety of substrata, including algae, animals, marine sediment, sea ice, and within the water column (Burgaud et al. 2015; Gutiérrez et al. 2016; Hassett and Gradinger 2016; Picard 2017; Xu et al. 2018). Known species continue to be associated with more coastal waters rather than open ocean, as fewer

A. L. Bunbury-Blanchette · A. K. Walker (✉)
Department of Biology, Acadia University, Wolfville, NS, Canada
e-mail: allison.walker@acadiau.ca

B. Giri, A. Varma (eds.), *Microorganisms in Saline Environments: Strategies and Functions*, Soil Biology 56, https://doi.org/10.1007/978-3-030-18975-4_2

substrates for fungal growth, woody or otherwise, are present further from shore and sampling becomes more complicated (Kohlmeyer and Kohlmeyer 1979). Ecologically, the degradation of lignocellulose is generally considered to be an important role of fungi in the ocean; marine fungi possess lignocellulolytic enzymes (Pointing et al. 1998). Other vital roles of marine fungi in ocean ecosystems illuminated by modern DNA- and RNA-based methods include decomposition of, pathogenesis of, and/or symbiosis with algae or marine animals (Gutiérrez et al. 2016; Hassett and Gradinger 2016; Schaumann and Weide 1990; Taylor and Cunliffe 2016; Xu et al. 2018). The actions of marine fungi in the open ocean have cascading effects on pelagic food webs and nutrient cycling (Gutiérrez et al. 2016; Hassett and Gradinger 2016; Taylor and Cunliffe 2016).

The largest group of fungi associated with the open ocean is the polyphyletic order Halosphaeriales, with at least 43 genera and 133 species (Hyde et al. 1998; Jones 1995; Spatafora et al. 1998). Members of the Halosphaeriales are found in a variety of aquatic habitats but are most often recovered from submerged wood in the sea (Hyde et al. 1998; Jones 1995). Genera of the *Halosphaeriaceae* share adaptations for reproduction in marine environments, as do other marine fungi (Jones 1995; Walker et al. 2017). In the Halosphaeriales, these adaptations are thin-walled, unitunicate, deliquescing asci, central pseudo-parenchyma in immature ascomata, predominately hyaline and bicelled ascospores, appendaged ascospores, and passive spore release to aid in spore dispersal and attachment (Jones 1994, 1995).

As in all environments, one of the greatest determinants of fungal diversity in the open ocean is available substrate. As demonstrated by the *Halosphaeriales*, some of the most readily accessible substrates in the sea are submerged wood or other plant materials providing lignocellulose as a food source. Among these substrata, differences in tree species of the available submerged wood may influence fungal community, although it is unclear if any wood-inhabiting marine fungi are host specific (Kohlmeyer and Kohlmeyer 1979; Jones 2000). Several incubation studies using test blocks have demonstrated a preference of fungal species for wood from certain species of wood, although fungal diversity from sampled driftwood is greater than from test blocks (Grasso et al. 1985; Miller and Whitney 1981; Petersen and Koch 1997). There is some evidence that marine fungi in the open ocean may also degrade animal material such as the tunicin present in tunicate remains and calcareous substances produced by mollusks, but little research has been done in this area (Kohlmeyer and Kohlmeyer 1979). As woody and other plant substrates originate on land or in intertidal areas, the geographic ranges of the marine fungi that ultimately decompose these substrates in the ocean are limited by the dispersal of the plant materials from their terrestrial habitats. Some frequently isolated species from woody marine substrates are *Paradendryphiella salina*, *Ceriosporopsis halima*, *Cirrenalia* spp., *Halosphaeria* spp., *Humicola alopallonella*, *Lulworthia* spp., and *Zalerion maritima* (Byrne and Jones 1974; Jones et al. 1972; Schaumann 1975; Schneider 1976).

Other factors that impact fungal diversity in the open ocean are water temperature, salinity, pH, and depth of water (coastal vs. deep sea) as well as sampling depth.

Zones of fungal communities can be drawn according to ocean temperatures, although our knowledge is limited by regions that have been sampled; many areas of ocean remain unexplored (Hughes 1974; Kohlmeyer 1983). A general distinction between pantropical and pantemperate marine fungal species is commonly drawn (Hyde et al. 1998).

There is a low diversity of fungi collected from Antarctic waters due to low temperatures, but some species are considered to have Antarctic distributions, e.g., *Spathulospora antarctica*, (Kohlmeyer and Kohlmeyer 1979; Pugh and Jones 1986). In the Arctic ocean, molecular methods show *Chytridiomycota*, seasonally fluctuating with light and algae blooms, and *Basidiomycota*, to be the dominant phyla, while members of the *Ascomycota*, the dominant phylum in driftwood and warmer ocean climates, are less common (Hasset and Gradinger 2016; Rämä et al. 2017). Fungi in cold environments have adaptations such as increased intracellular trehalose and polyol concentrations, unsaturated membrane lipids, secretion of antifreeze proteins, and cold-avoidant spore germination, e.g., the psychrotolerant marine Arctic fungus *Havispora longyearbyenensis* (Pang et al. 2011; Robinson 2001). Meanwhile, many fungi including *Antennospora quadricornuta* are only present in tropical or subtropical waters, while others have temperate distributions, e.g., *Asteromyces cruciatus* (Jones 1993; Kohlmeyer and Kohlmeyer 1979). Some fungi have cosmopolitan distributions, despite the range of conditions that exist in the world's oceans; examples include *Corollospora maritima* and *Halosphaeria appendiculata* (Jones 1993).

Sensitivity to salinity and pH in marine fungi present in the open ocean is variable. Many species demonstrate the "*Phoma* pattern": increased tolerance for salinity with increased temperatures, although not all (Gessner 1976; Ritchie 1957). Laboratory studies suggest that some marine fungi are best adapted for growth and reproduction in the oceanic salinity of waters where they are naturally found, while others can tolerate a wide range of salinities (Jones 2000; Kohlmeyer and Kohlmeyer 1979). A similar conclusion may be drawn for pH conditions (Jones 2000; Kohlmeyer and Kohlmeyer 1979).

Depth of sampling and other correlated factors such as pressure and dissolved oxygen significantly impact marine fungi community composition. Sampling has shown the distinction between pelagic and benthic zone to better explain differences in fungal community than geographic region (Tisthammer et al. 2016). Fungi recovered from depths >500 m are few but have tolerance for high pressure (Kohlmeyer and Kohlmeyer 1979). Fungi so far found at depths of >500 m are *Abyssomyces hydrozoicus*, *Aspergillus ustus*, *Bathyascus vermisporus*, *Graphium* sp., *Oceantis scuticello*, *Allescheriella bathygena*, and *Periconia abyssa* (Kohlmeyer and Kohlmeyer 1979; Raghukumar and Raghukumar 1998). *A. hydrozoicus* are known to live in hydrozoa and likely decompose chitin (Kohlmeyer 1972). Genetic approaches support that while more species remain to be discovered and identified, fungal diversity in the deep sea is limited; the species present are mostly ascomycetes and basidiomycetes with yeast morphotypes (Bass et al. 2007; Le Calvez et al. 2009). Many of these fungal species may be associated with deep-sea hydrothermal animals: members of the genera *Rhodotorula*,

Rhodosporidium, *Candida*, *Debaryomyces*, and *Cryptococcus* have been isolated from decomposing deep-sea fauna (Burgaud et al. 2010). Deep-sea environments are diverse and offer promising opportunities for further research, e.g., deep-sea sediments, the water column, hydrothermal vents, and methane seeps (Burgaud et al. 2015; Walker et al. 2017).

Some marine fungi present in the ocean, rather than being decomposers, are animal pathogens. Species of the genus *Fusarium* cause shell disease in crustaceans and cause infection in hermit crabs and lobsters (Lightner 1988; Smolowitz et al. 1992; Stewart 1984). Other fungi that infect marine animals include infection of coral by *Aspergillus sydowii* and *Scolecobasidium* sp., lesions caused by *Scolecobasidium humicola* in devil stingers; infection of mollusks by *Bathymodiolus brevior*, *Lasiodiplodia* sp., and *Melanopsamma balani*; and infection of tunicates by *A. quadricornuta* (Kohlmeyer and Kohlmeyer 1979; Le Campion-Alsumard et al. 1995; Norton et al. 1994; Raghukumar and Raghukumar 1991; Van Dover et al. 2007; Wada et al. 1995). Many fungus-like organisms including members of the genera *Branchiomyces*, *Dermatocystidium*, *Ichtyophonus*, and *Saprolegnia* are notable pathogens of fish, while true fungi in the genera *Exophiala* and *Ochroconis* also cause disease in fish and other cold-blooded vertebrates (de Hoog et al. 2011; Ramaiah 2006). Although many studies have taken place within the scope of aquaculture, it is likely these pathogens exist in natural open ocean environments as well. There is also overlap of pathogens in coastal environments and those in the open ocean, as species ranges do no always conform to one or the other, and definitions of "ocean" vs "coastal" waters may differ. Higgins (2000) provides a detailed review of fungal pathogens of mammals in marine environments; common diseases in cetaceans and pinnipeds are caused by *Aspergillus* spp., *Zymonema dermatitidis*, *Coccidioides immitis*, *Cryptococcus neoformans*, *Histoplasma capsulatum*, zygomycetes such as *Mucor* spp., and yeasts including members of the genus *Candida*. Many marine animals remain unexplored as potential hosts for fungi; research is continuously finding fungi in new places, acting in sometimes unexpected roles. Fungal colonization occurs in jellyfish by symbiotic fungi of the genera *Aspergillus*, *Cladosporium*, *Purpureocillium*, and *Tilletiopsis* (Yue et al. 2015). Traditional culture methods combined with DNA barcoding have revealed high fungal diversity associated with corals, including eight fungal genera previously unknown from coral hosts (Xu et al. 2018).

A significant minority of marine fungi are algal pathogens, decomposers, or symbionts, although research has been restricted by difficulties in observing these species in laboratory conditions (Kohlmeyer and Kohlmeyer 1979). Several species, primarily members of the Sordariomycetes, Eurotiomycetes, and Dothideomycetes, may infect algae and some permanently submerged sea grasses (Kohlmeyer and Kohlmeyer 1979). Chytrids like *Algochytrops polysiphoniae* and others are pathogens of algae and diatoms and may have significant impacts on nutrient cycling in the ocean (Gutiérrez et al. 2016; Hyde et al. 1998; Taylor and Cunliffe 2016). Species such as *Phoma laminariae* have been recovered from macroalgae species (the genus *Laminaria* in the case of *P. laminariae)* when washed up on shore (Kohlmeyer and Kohlmeyer 1979).

Fungi of *Blodgettia* and *Mycosphaerella* form mycophycobioses, in which they exist as an obligate symbiont with a marine microalga forming the outer structure (Hawksworth 2000; Kohlmeyer and Kohlmeyer 1972). Additionally, marine lichen-forming fungi such as *Wahlenbergiella mucosa* are known from the littoral zone; however, cases of lichens experiencing constant submersion are few and have not been extensively researched (Fletcher 1975a, b; Hawksworth 2000; Henssen and Johns 1974). Fungal species including *Chadefaudia corallinarum* are capable of loose lichen-like associations with algae, and high recovery of the genus *Buellia* from surface waters suggests a role as a marine lichen symbiont (Kohlmeyer and Kohlmeyer 1979; Picard 2017). There are also symbiotic relationships between marine fungi and sea sponges. For example, members of the genus *Malassezia* are recovered from sea sponges at high incidences using molecular methods, although they have never been successfully cultured from the marine environment (Gao et al. 2008; Li and Wang 2009).

Although not fungi, fungus-like organisms such as members of the protist order *Labyrinthula* and oomycetes including *Pontisma lagenidiodes* are also present in the open ocean and are notable as pathogens of marine algae (Hyde et al. 1998). Thraustochytrids and oomycetes are also pathogens of a wide variety of marine animals including mollusks, crustaceans, and fish (Noga 1990; Porter 1986; Stewart 1984).

The concentration of fungal propagules in seawater remains largely unknown; little sampling has occurred. Numbers range from eight spores per L of seawater to 2800+ propagules per 50 mg of sea foam and sediment following a storm event (Fazzani and Jones 1977; Kirk 1983). Water column and marine sediment sampling are underrepresented in marine fungal research. Diversity in these environments seems to be less than when isolating fungi from substrata such as algae or marine wood but is an important aspect of total abundance and diversity of marine fungi and the niches they fill in ocean environments (Picard 2017).

More research is needed to better understand total fungal diversity in the open ocean, as most diversity studies consider sporulating fungi produced from incubation of substrata. Drawbacks to this approach include omission of non-sporulating species and competition between species during the incubation period affecting recorded fungal diversity (Tan et al. 1995). Metatranscriptomic and targeted cell counting approaches are needed (see Andreakis et al. 2015; Taylor and Cunliffe 2016).

2.2 Tropical and Subtropical Intertidal Habitats

2.2.1 Mangrove Forests

A great deal of research regarding the occurrence and distribution of marine fungi in tropical and subtropical environments has focused on mangrove forests. Mangrove forests are unique coastal ecosystems: in broad terms, mangroves are any salt-

Table 2.1 Mangrove genera with existing fungal community research

Family	Genera
Acanthaceae	*Acanthus*
	Avicennia
Arecaceae	*Nypa*
Combretaceae	*Conocarpus*
	Laguncularia
	Lumnitzera
Lythraceae	*Sonneratia*
Primulaceae	*Aegiceras*
Pteridaceae	*Acrostichum*
Rhizophoraceae	*Bruguiera*
	Cericops
	Kandelia
	Rhizophora

tolerant trees or large shrubs with specialized adaptations to live in fluctuating conditions of salinity, anoxia, and wave and tidal action. Approximately 54 plants species are considered "true mangroves"; common genera and others notable in terms of fungal community are listed in Table 2.1 (Hogarth 1999). The combination of conditions experienced in mangrove forests supports the presence of many species and species interactions that do not occur in other ecosystems, including the occurrence of certain marine fungal species. More than 600 fungal species are known from mangrove habitats, of which approximately 200 are considered obligate marine species (Schmit and Shearer 2003; Shearer et al. 2007). Fungi endemic to intertidal mangrove forests generally possess ascospores that lack elaborate appendages but have mucilaginous sheaths and are actively released (Au et al. 1999; Read et al. 1994).

The abundance of plant detritus associated with mangrove forests creates substrata for fungal colonization. Almost all marine fungi known from mangrove habitats are saprobic, causing soft root of woody areas upon damage to the mangrove by another cause. One possible exception is *Cytospora rhizophorae*, which may be parasitic in *Rhizophora* species (Kohlmeyer and Kohlmeyer 1971, 1979). Diversity of fungal community among mangroves varies with factors such as geographic location, mangrove and other plant species present, type(s) of plant substrata, animal and microbial community, climate, and range of abiotic dynamics such as salinity and oxygen availability (Jones 2000). Many marine fungi living in mangrove forests have large geographical ranges and tolerate a wide variety of environmental conditions; therefore, it is unclear how much sampling bias, rather than other factors, has affected certain determinations of species ranges (Shearer et al. 2007). The number of fungal species recovered from mangroves themselves ranges from 7 to 76 according to location and species of mangrove (Hyde et al. 1990b; Kohlmeyer and Volkmann-Kohlmeyer 1993; Tan and Leong 1990).

Members of the mangrove genus *Avicennia* preferentially attract fungal colonizers including the following species: *Leptosphaeria avicenniae*, *Mycosphaerella pneumatophorae*, and *Rhabdospora avicenniae* (Kohlmeyer and Kohlmeyer 1979). The mangrove associate *Acanthus ilicifolius*, an herbaceous perennial, commonly supports the fungal species *Colletotrichum gloeosporioides* and members of the *Acremonium*, *Phoma*, and *Tubercularia* genera (Sadaba 1996; Sadaba et al. 1995). Common fungal colonizers of *Nypa fruticans* are *Astrosphaeriella striatispora*, *Lignincola laevis*, *Linocarpon appendiculatum*, *Linocarpon bipolaris*, *Linocarpon nypae*, *Oxydothis nypae*, and *Trichocladium nypae* (Hyde 1992a; Hyde and Sarma 2006). The host *Aegiceras corniculatum* supports the marine fungi *Lophiostoma* sp. and *Morosphaeria velatispora* as very frequent isolates (Hyde 1990a). *Cirrenalia pseudomacrocephala* and *Lophiostoma acrostichi* are often isolated from *Acrostichum speciosum* (Hyde 1989a).

The most common fungi found on test blocks of *Bruguiera cylindrica* are *Verruculina enalia*, *Lulworthia* sp., *Aigialus parvus*, and *Payosphaeria minuta*; less frequently isolated but also common are *Halosarpheia marina*, *Nais inornata*, *Halorosellinia oceanica*, and *Natantispora lotica* (Leong et al. 1991). The fungal species *Cytospora rhizophorae*, *Lineolata rhizophorae*, *Etheirophora blepharospora*, *Robillarda rhizophorae*, and *Trematosphaeria mangrovis* have a host preference for *Rhizophora* mangrove species (Kohlmeyer and Kohlmeyer 1979). Test blocks of *Rhizophora apiculata*, similarly to *B. cylindrica*, also are commonly colonized by *V. enalia*, *Lulworthia* sp., *A. parvus*, *H. marina*, and *N. inornata* (Leong et al. 1991). Other common colonizers of *R. apiculata* are *Aigialus mangrovis* and *Savoryella longispora* (Hyde 1990a; Leong et al. 1991). In the family Lythraceae, *Sonneratia griffithii* is very commonly colonized by *Halocyphina villosa* (Hyde 1990a).

It is unclear if any mangrove fungi exhibit true host specificity, although some fungi are only known from a single mangrove species or seem to have strong host preferences (Hyde 1990a; Jones 2000; Sarma and Hyde 2001). *V. enalia*, *Antennospora quadricornuta*, *Dactylospora haliotrepha*, *L. australiensis*, *L. laevis*, and *Lulworthia* spp. are more generalist mangrove colonizers and found associated with multiple mangrove species in many geographical locations (Hyde and Jones 1998; Kohlmeyer and Kohlmeyer 1979; Leong et al. 1991). Other species are more restricted by geography, however.

Trematosphaeria mangrovis is known only from West Africa, *L. rhizophorae* occurs only in the American tropics and subtropics, and except for an introduced population in Hawaii, *E. blepharospora* is found only in American and West Atlantic islands (Kohlmeyer and Kohlmeyer 1979). The most common fungi occurring mangroves of Indian Ocean coastlines are *A. quadricornuta*, *D. haliotrepha*, *Eutypa bathurstensis*, *H. villosa*, *H. oceanica*, *H. marina*, *Kallichroma tethys*, *Leptosphaeria australiensis*, *Rimora mangrovei*, *S. grandispora*, *Rhizophila marina*, *S. longispora*, and *V. enalia* (Sarma and Hyde 2001). The most frequently recorded marine fungi associated with Pacific mangroves are *Caryosporella rhizophorae*, *Hydea pygmea*, *D. haliotrepha*, *H. villosa*, *H. marina*, *L. laevis*, *L. appendiculatum*, *Sammeyersia grandispora*, *Lulworthia* spp., *L. acrostichi*, *M. velatospora*, *Phomopsis* spp., *Torpedospora radiata*, and *Trichocladium linderii* (Sarma and Hyde 2001). *D. haliotrepha*, *Halosarpheia fibrosa*, *H. marina*,

Natantispora retorquens, *Halosphaeria appendiculata*, *S. grandispora*, *Saagaromyces glitra*, and *Tunicatispora australiensis* are common marine fungi of Australian mangrove forests (Hyde 1989b; Sarma and Hyde 2001).

Fungi found on different parts of the mangroves themselves may also differ, reflecting differences in the ratio of exposure time to seawater or air and/or a preference for subsistence on the contents of one part of the plant. On young, submerged prop roots of *Rhizophora* species, *Phomopsis mangrovei* and *S. grandispora* are most common, while on older roots at mid-level, *Aigialus grandis*, *Ascocratera* cf. *manglicola*, *Caryosporella rhizophorae*, *Saagaromyces abonnis*, and *Hypophloeda rhizospora* are common (Hyde 1989b, 1990a, b; Hyde and Jones 1998). *A. quadricornuta* is also a low-level mangrove associate (Jones and Tan 1987). At the high-water mark, or on the upper-level of roots, *Anthostomella* spp., *H. villosa*, *H. oceanica*, *K. tethys*, *Massarina* spp., and *Pyrenographa xylographoides* are found (Jones 2000; Jones and Tan 1987; Hyde and Jones 1998).

Saprobes known from only the bark surface of mangroves are the species *E. blepharospora*, *M. pneumatophorae*, and *R. avicenniae* (Kohlmeyer and Kohlmeyer 1979). It is hypothesized that these species may degrade the suberin in bark tissue (Kohlmeyer and Kohlmeyer 1979). Differences in fungal community between parts of the mangrove may be as subtle as in *S. griffithii*, in which common fungal species on aerial roots (*A. grandis* and *M. velatospora*) differ from common species on twigs (*Dyfrolomyces mangrovei* and *S. longispora*) (Hyde et al. 1990a). Similar patterns occur on *R. apiculata*, *Rhizophora mucronata*, and *Avicennia* species, in which fungal community differs between different types of roots, trunk, branches, leaves, and seedlings (Hyde et al. 1990b; Ravikumar and Vittal 1996; Sarma and Vittal 2001). Many marine fungi do colonize the seedlings of mangroves, which aids in their dispersal. Many mangroves produce viviparous seedlings that fall from the tree and are washed away by waves and ocean currents, sometimes hundreds of kilometers, dispersing fungi to new sites (Gunn and Dennis 1976; Kohlmeyer and Kohlmeyer 1971). Obligate marine fungi isolated with prevalence from *Rhizophora mangle* seedlings were *E. blepharospora*, *S. grandispora*, *Lulworthia medusa*, *Cytospora rhizophorae*, and *Zalerion varium* (Newell 1976).

Temperature (tropical vs. subtropical) does not seem to be a primary driver of species geography in mangrove-associated marine fungi beyond a general decrease of diversity in subtropical areas, correlated with fewer mangrove species and habitats being in subtropical vs. tropical climates (Hyde 1990a; Sarma and Hyde 2001).

Fungi living in mangrove forests are often well adapted to a range of potentially extreme salinities. The nature of tides and seasonality in a mangrove forest creates times of seawater inundation and times of higher salinity resulting from evaporation in tidal pools when waters retreat. Different areas of one mangrove forest may simultaneously experience different salinity conditions, resulting in different fungal communities in those areas (Kohlmeyer 1969). Mangrove fungal communities may also change over time as salinity changes, as exemplified by the differences between the dry and wet seasons experienced by *A. ilicifolius*: obligatory marine fungi dominate in the dry season when salinity is high, while facultative marine fungi and terrestrial fungi also colonize in the wet season when salinities drop (Sadaba 1996). There is evidence for similar patterns of change between seasons in other geographical areas, i.e., greater fungal diversity in the wet season, although further research is needed to confirm and explain this phenomenon (Sarma and Hyde 2001). The marine fungal species remain as constants, however, tolerating both the low and high salinity conditions (Hyde 1992b; Sadaba 1996). Because marine fungi in mangrove forests can tolerate a range of salinity conditions, they, along with oomycetes and other fungus-like protists, play an important role in decomposing the large volume of lignocellulose produced by mangroves (Fell and Master 1973; Newell 1996).

All preceding fungal species listed are associated with woody material; fungal communities associated with mangrove leaf litter differ from them. Fungus-like organisms such as members of the genus *Halophytophthora* and thraustochytrids are common in mangrove leaf litter (Nakagiri et al. 1996; Newell 1992). Of the true fungi, terrestrial species persist on leaves after they fall, while marine fungi may not colonize until the second or third week of submergence in saline water (Fell and Master 1973; Kohlmeyer 1969). Some of these marine fungi species are *Cirrenalia macrocephala*, *Dictyosporium pelagicum*, *Corollospora ramulosa*, *Z. varium*, and *Lulworthia* spp. (Fell and Master 1973).

Fungal species associated with mangrove forest plants other than tree or large shrub species are more terrestrial than marine species and include members of *Cephalosporiopsis*, *Colletotrichum*, *Phomopsis*, and possibly *Septoria*, isolated from the grass *Phragmites australis* and the sedge *Schoenoplectus litoralis* (Poon and Hyde 1998a, b; Sridhar et al. 2012).

Finally, it is noted that fungal community succession related to stage of decomposition occurs on mangrove roots and seedlings; thus, time of sampling and number of sampling points in a time course should be considered when interpreting the composition of a fungal mangrove community (Jones 2000; Leong et al. 1991; Newell 1976; Tan et al. 1989). This is in addition to the typical challenges associated with identifying marine fungi: incubation of substrata selects for faster growing and sporulating fungi able to outcompete other species, and methods may not always completely exclude terrestrial or facultative marine fungi (Tan et al. 1995).

2.2.2 *Other Tropical and Subtropical Intertidal Habitats*

While mangrove forests are a common and well-studied habitat type in many tropical and subtropical intertidal areas, other habitats such as sandy and rocky beaches, saltmarshes, seagrass beds, coral reefs, mudflats, estuaries, mollusk banks, and tidal pools also support marine fungal communities. Marine fungi are associated with the root and basal portions of plants in these habitats that experience tidal flooding, such as tropical and subtropical saltmarshes (Gessner and Kohlmeyer 1976; Kohlmeyer and Volkmann-Kohlmeyer 1998, 1999; Poon and Hyde 1998b; Sadaba et al. 1995). Fungi in the roots that experience greater inundation by saltwater are generally obligate marine species, while fungi found on basal aboveground portion of the plant may be facultative marine species (Kohlmeyer and

Kohlmeyer 1979). These fungi may be parasitic or saprobic, and many are host-specific; common dominant saltmarsh plants are *Sporobolus* (formerly *Spartina*), *Distichlis*, *Juncus*, and *Salicornia* species (Kohlmeyer and Kohlmeyer 1979). Most marine fungi associated with saltmarsh plants are found wherever their host grows, in both temperate and subtropical climates (Gessner and Kohlmeyer 1976; Kohlmeyer and Kohlmeyer 1977). Some common examples of fungi found on *Sporobolus* are *Buergenerula spartinae*, *Claviceps purpurea*, *Halobyssothecium obiones*, and *Juncaceicola typharum* (Gessner and Kohlmeyer 1976). Research on pathogenic and symbiotic fungi in saltmarsh environments has intensified as these ecosystems are increasingly threatened by human activities (Walker and Campbell 2010; You et al. 2012).

While fungi and fungus-like organisms also colonize leaf detritus in tropical intertidal areas, many of those species are not obligate marine fungi but rather facultative marine fungi, terrestrial fungi, or fungus-like organisms such as thraustochytrids (Kohlmeyer and Kohlmeyer 1979). However, leaves of the tropical turtle grass *Thalassia testudinum* are decomposed by several common marine fungi including *Corollospora lacera*, *C. maritima*, *L. marinera*, *Lulworthia* sp., and *D. arenaria* (Kohlmeyer and Kohlmeyer 1977; Meyers 1969; Meyers et al. 1965; Opurt et al. 1964). *Lulworthia* species have also been recovered from washed up rhizomes of *T. testudinum* (Kohlmeyer and Kohlmeyer 1977). Some fungi also form lichen or lichen-like symbiotic associations in intertidal habitats (Kohlmeyer and Kohlmeyer 1979). Many of these symbioses are known from temperate climates, but *Blodgettia bornetii* is one example of a fungal species restricted to tropical and subtropical climates by its hosts, members of the genus *Cladophora* (Feldmann 1938; Kohlmeyer and Kohlmeyer 1979).

Some marine fungi are pathogens of animals living in tropical and subtropical areas. Endolithic fungi may become opportunistic pathogens of the lobe coral *Porites lobate*, while sea fan corals are attacked by *Aspergillus sydowii* (Alker et al. 2001; Le Campion-Alsumard et al. 1995). The tropical mangrove crab *Ucides cordatus* suffers from lethargic crab disease caused by two fungal species in the family Herpotrichiellaceae, and spiny lobsters are also affected by several fungal pathogens (Shields 2011; Vicente et al. 2012). Fungal diseases known from mollusks in tropical or subtropical environments include infection of rock oyster and boring clam by several dematiaceous fungal species and infection of tunicates by *A. quadricornuta* (Kohlmeyer and Kohlmeyer 1979; Norton et al. 1994; Raghukumar and Lande 1988). In tropical fish, lesions caused by *Scolecobasidium humicola* occur in devil stingers (Wada et al. 1995). Marine animals with terrestrial stages in their life cycle are also affected by some nonmarine fungi which subsequently affect the marine ecosystem, e.g., *Fusarium solani* infection in sea turtle eggs (Sarmiento-Ramírez et al. 2010). Several other genera of fungus-like organisms, e.g., *Branchiomyces*, *Dermatocystidium*, *Ichtyophonus*, *Achlya*, and *Saprolegnia*, cause disease in a wide variety of fish species (Ramaiah 2006).

While not strictly tropical or subtropical in distribution, arenicolous fungi are worth mentioning here as many have cosmopolitan distributions, e.g., *Corollospora maritima* and *Corollospora trifurcata*, and are often studied in tropical or subtropical areas (Kohlmeyer and Volkmann-Kohlmeyer 1993; Nakagiri and Tokura 1994). These fungi attach themselves to grains of sand by a subiculum and tolerate extreme conditions of exposure and abrasion, high temperatures, and varying salinity (Sundari 1997). Fungi present in sandy beach ecosystems are important decomposers of organic debris; higher temperatures and productivity in warmer climates result in greater numbers and diversity of sand-dwelling fungi, although this biodiversity is threatened by anthropogenic activities (Hughes 1974; Velez et al. 2013, 2015).

2.3 Temperate Intertidal Habitats

Saltmarshes are the dominant intertidal habitat where marine fungi have been researched in temperate climates (e.g., da Luz Calado et al. 2015; Kohlmeyer and Volkmann-Kohlmeyer 2002; Newell et al. 2000; Torzilli et al. 2006; Walker and Campbell 2010). Saltmarshes are areas in the upper intertidal zone that experience regular tidal flooding and are populated by salt-tolerant grasses, shrubs, and herbaceous plants. Temperate saltmarshes are of research interest due to their unique biodiversity and protective effects on shorelines and water quality (Mcowen et al. 2017). Saltmarshes are widely distributed in coastal areas of both the northern and southern hemispheres; the dominant plants throughout this global range are members of the genera *Sporobolus* (encompassing the subsection *Spartina*), *Juncus*, *Distichlis*, *Salicornia*, *Phragmites*, *Festuca*, *Agrostis*, *Scirpus*, and *Limonium* (Adam 1990; Kohlmeyer and Kohlmeyer 1979; Mcowen et al. 2017). Fungi in temperate saltmarshes are often associated with these and other saltmarsh plants as saprotrophs (Kohlmeyer and Kohlmeyer 1979). Obligate marine fungi colonize plant roots, which experience either constant or regular periodic saline conditions (Kohlmeyer and Kohlmeyer 1979). Fungi present in basal aboveground portions of the plants may be obligate or facultative marine fungi, while terrestrial fungi may be found higher up; these species may also be found on plants adjacent to the tidal range that still experience rare flooding and/or salt spray (Kohlmeyer and Kohlmeyer 1979).

Marine fungi commonly isolated from *Sporobolus* and *Juncus*, two of the most common genera of saltmarsh plants, are *Aniptodera*, *Anthostomella*, *Halosphaeria*, *Lulworthia*, *Phaeosphaeria*, and *Leptosphaeria* species (Kohlmeyer and Volkmann-Kohlmeyer 2001, 2002). Many fungi associated with saltmarsh plants exhibit some host specificity, whether to species or higher taxonomic level, although there are also ubiquitous marine species commonly isolated from continuously submerged roots of many plant species (Kohlmeyer and Kohlmeyer 1979; Torzilli et al. 2006; Walker and Campbell 2010). For example, fungal communities isolated from decomposing grasses *Sporobolus alterniflora* and *Distichlis spicata* are more like each other than those of either *Juncus romarianus* or *Sarcocornia perennis*; however, some fungal species are associated with all four plants (Torzilli et al. 2006).

Parasitic fungi are known from *Sporobolus*, such as the marine *Juncaceicola typharum*, and terrestrial fungi like *Claviceps purpurea* and *Puccinia* and *Uromyces* species (Kohlmeyer and Kohlmeyer 1979). Vertical zones of distribution have been characterized for *Sporobolus*, in which marine fungal species like *Buergenerula spartinae*, *Pleospora pelagica*, *Remispora hamata*, *Leptosphaeria* spp., and *Lulworthia* spp. are found in the culms, species from both marine and terrestrial genera are found in leaves, but primarily terrestrial fungal species are found in the inflorescences (Gessner 1977). *B. spartinae*, *Halobyssothecium obiones*, *Mycosphaerella* sp., and *Phaeosphaeria* spp. have been implicated in decomposition of non-lignocellulosic *Sporobolus* detritus, varying seasonally as well as with bacterial community (Benner et al. 1984; Buchan et al. 2003). In contrast, decaying *J. roemerianus* detritus is dominated by the fungus *Massarina ricifera*, which is less commonly reported from *Sporobolus* (Kohlmeyer et al. 1995; Walker and Campbell 2010). Additionally, only saprobic marine fungi (no pathogenic fungi) are known from *Juncus* (Kohlmeyer and Volkmann-Kohlmeyer 2001). Some common fungi known from detritus of the genus *Phragmites* belong to the genera *Halosarpheia* and *Massarina* (Poon and Hyde 1998a, b; Van Ryckegem and Verbeken 2005). Vertical distribution zones have also been classified in *Phragmites australis*, in which marine species occupy lower parts of plant (Van Ryckegem and Verbeken 2005).

As saltmarshes are numerous and cosmopolitan in geographical distribution, many remain to be investigated; discoveries of new marine species associated with saltmarsh plant species are likely. In the late 1990s alone, 48 new fungal species records from *J. roemerianus* were described (Kohlmeyer and Volkmann-Kohlmeyer 2001). Other recent new species records include *Anthostomella spissitecta*, *Camarosporium roumeguerei*, *Coniothyrium obiones*, and others on *Sporobolus maritima* (da Luz Calado et al. 2015). This research also confirms previous knowledge of vertical zonation of fungi on *Sporobolus maritima* (da Luz Calado et al. 2015). Saltmarshes are increasingly threatened by human activities; thus, new research considering fungal communities in these ecosystems is becoming important to conservation efforts. For instance, naturally occurring, established saltmarshes have more diverse fungal communities compared to newer or created saltmarshes, information crucial to saltmarsh restoration efforts (Walker and Campbell 2010).

Some fungi form symbiotic relationships with saltmarsh plants. Fungal endophytes *Pleospora salicorniae* and *Pleospora bjoerlingii* and species of the genus *Stagonospora* colonize *Salicornia perennis* (Petrini and Fisher 1986). Arbuscular mycorrhizal fungi (AMF) possibly augment oxygen uptake in low-oxygen sediments and increase stress resistance to saline environments in saltmarsh plants (Khan and Belik 1995; Rozema et al. 1986). Factors such as species and sex may affect AMF colonization in plants such as *Sporobolus* and *Distichlis* (d'Entremont et al. 2018; Eppley et al. 2009). Marine fungi are present in intertidal waters and sediments, although it can be difficult to determine if isolates represent true marine fungi or terrestrial fungi that have dispersed propagules into marine environments (Kohlmeyer and Kohlmeyer 1979).

Algae are also present in temperate intertidal zones and provide suitable substrata for marine fungi (Kohlmeyer and Kohlmeyer 1979). Macroalgae, or seaweed, including brown algae (e.g., *Ascophyllum*, *Fucus*, *Laminaria*), red algae (e.g., *Chondrus*, *Palmaria*, *Pyropia*), and green algae (e.g., *Ulva*) support both parasitic and saprobic fungi (Kohlmeyer and Kohlmeyer 1979). Several fungal and oomycete pathogens of *Pyropia* genera impact commercial production, including a recently reported *Alternaria* species causing red rot (Kim et al. 2014; Mo et al. 2016). Fungal pathogens of algae are restricted in host to class but within class may or may not display further host specificity (Kohlmeyer and Kohlmeyer 1979). Saprobic marine fungi occur on algae washed up on shore or floating near shore, although marine yeasts, terrestrial fungi, and protistan thraustochytrids, oomycetes, and bacteria also play a role in decomposition of algae (Kohlmeyer and Kohlmeyer 1979; Seshadri and Sieburth 1971). Fungal species *Lindra thalassiae* and an isolate of the genus *Acremonium*, along with a *Pythium* sp. and two thraustochytrids, all display unique decomposition activities of the brown alga *Sargassum cinereum* (Sharma et al. 1994).

Fungi may also be symbiotic with algae, forming lichens, lichen-like relationships, or mycophycobioses (Kohlmeyer and Kohlmeyer 1979). Marine fungi known from lichens present in the littoral zone of temperate coastal environments include members of *Arthopyrenia*, *Lichina*, *Pyrenocollema*, *Stigmidium*, and *Verrucaria* (Fletcher 1975a, b; Hawksworth 2000; Kohlmeyer and Kohlmeyer 1979). *Wahlenbergiella tavaresiae* was the first reported fungi to form a marine lichen with a brown algae partner and has more recently been physiologically characterized, revealing a distinguishing organization of algal and fungal cells (Moe 1997; Peters and Moe 2001; Sanders et al. 2004). Other fungi that form lichen-like associations (the algal partner may live freely in an independent state) include *Chadefaudia*, *Collemopsidium*, and *Pharcidia* species (Kohlmeyer and Kohlmeyer 1979; Kohlmeyer et al. 2004). Genetic data has resolved the previously confused taxonomy of *Collemopsidium* species into the new order Collemopsidiales, along with the genus *Zwackhiomyces* (Pérez-Ortega et al. 2016). Mycophycobioses, in which the algal partner is dominant, include the fungal species *Stigmidium ascophylli*, which is commonly found living as a symbiont with *Ascophyllum nodosum* and *Pelvetia canaliculata* (Kohlmeyer and Kohlmeyer 1972; Webber 1967). *Mycophycias ascophylii* also forms a symbiotic relationship with *A. nodosum* and *Vertebrata lanosa*, a unique three-way interaction in which the fungus and brown alga *V. lanosa* may prevent tissue damage and provide a materials transfer pathway for their shared host (Garbary et al. 2005).

2.4 Future Perspective

Marine habitats support complex and diverse ecosystems in which fungal species are present and function in dynamic pathogenic, symbiotic, and nutrient cycling roles (Fig. 2.1). The world's oceans are vast, and many areas remain to be explored, providing countless opportunities for marine mycological research: diversity sampling, investigations of fungal ecology, and discovery of novel processes and

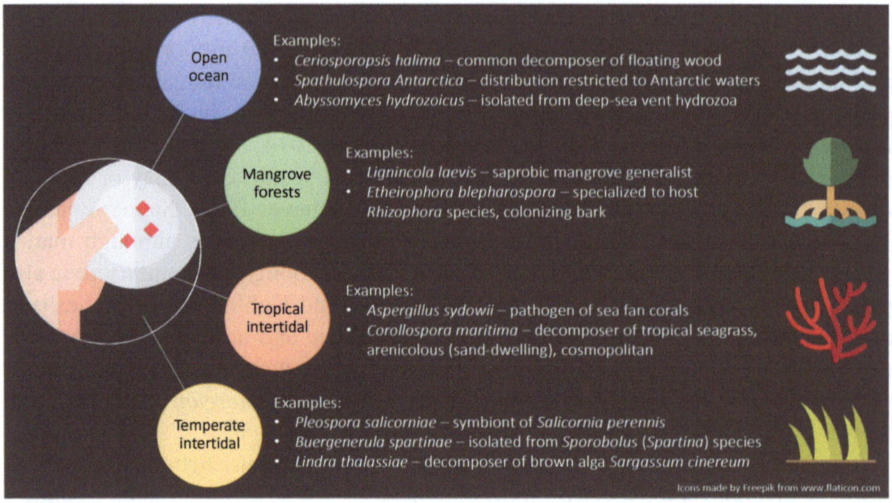

Fig. 2.1 Example of diversity of marine fungal species in four main saline habitats

compounds. Advancements in laboratory techniques occur rapidly and allow for increasingly accurate and precise results (Andreakis et al. 2015; Taylor and Cunliffe 2016). Although the traditional role of fungi in marine environments as decomposers of lignocellulose remains relevant, recent research of marine fungi is uncovering new and exciting avenues, placing greater emphasis on interconnected ecological and biogeochemical effects (Gutiérrez et al. 2016; Hassett and Gradinger 2016; Pointing et al. 1998; Schaumann and Weide 1990; Taylor and Cunliffe 2016; Xu et al. 2018). Future research may provide solutions in a wide variety of fields: conservation, habitat restoration and bioremediation, medicine and health, aquaculture and food production, and manufacturing (see Duarte et al. 2012; Holguin et al. 2001; Meyer 1991; Raghukumar 2000; Rateb and Ebel 2010; Walker and Campbell 2010).

References

Adam P (1990) Saltmarsh ecology. Cambridge University Press, Cambridge

Alker AP, Smith GW, Kim K (2001) Characterization of *Aspergillus sydowii* (Thom et Church), a fungal pathogen of Caribbean Sea fan corals. Hydrobiologia 460:105–111

Andreakis N, Høj L, Kearns P, Hall MR, Ericson G, Cobb RE, Gordon BR, Evans-Illidge E (2015) Diversity of marine-derived fungal cultures exposed by DNA barcodes: the algorithm matters. PLoS One 10:e0136130

Au DWT, Jones EBG, Vrijmoed LLP (1999) Observations on the biology and ultrastructure of the asci and ascospores of *Julella avicenniae* from Malaysia. Mycol Res 103:865–872

Bass D, Howe A, Brown N, Barton H, Demidova M, Michelle H, Li L, Sanders H, Watkinson SC, Willcock S, Richards TA (2007) Yeast forms dominate fungal diversity in the deep oceans. Proc R Soc B 274:3069–3077

Benner R, Newell SY, MacCubbin AE, Hodson RE (1984) Relative contributions of bacteria and fungi to rates of degradation of lignocellulosic detritus in salt-marsh sediments. Appl Environ Microbiol 48:36–40

Buchan A, Newell SY, Butler M, Biers EJ, Hollibaugh JT, Moran MA (2003) Dynamics of bacterial and fungal communities on decaying saltmarsh grass. Appl Environ Microbiol 69:6676–6687

Burgaud G, Arzur D, Durand L, Cambon-Bonavita M, Barbier G (2010) Marine culturable yeasts in deep-sea hydrothermal vents: species richness and association with fauna. FEMS Microbiol Ecol 73:121–133

Burgaud G, Hué NTM, Arzur D, Coton M, Perrier-Cornet JM, Jebbar M, Barbier G (2015) Effects of hydrostatic pressure on yeasts isolated from deep-sea hydrothermal vents. Res Microbiol 166:700–709

Byrne PJ, Jones EBG (1974) Lignicolous marine fungi. Veroff Inst Meeresforsch, Bremerhaven 5:301–320

d'Entremont TW, López-Gutiérrez JC, Walker AK (2018) Examining arbuscular mycorrhizal fungi in saltmarsh hay (*Spartina patens*) and smooth cordgrass (*Spartina alterniflora*) in the Minas Basin, Nova Scotia. Northeast Nat 25:72–86

da Luz Calado M, Carvalho L, Pang KL, Barata M (2015) Diversity and ecological characterization of sporulating higher filamentous marine fungi associated with *Spartina maritima* (Curtis) Fernald in two Portuguese saltmarshes. Microb Ecol 70:612–633

de Hoog GS, Vicente VA, Najafzadeh MJ, Harrak MJ, Badali H, Seyedmousavi S (2011) Water-borne *Exophiala* species causing disease in cold-blooded animals. Persoonia 27:46–72

Duarte K, Rocha-Santos TAP, Freitas AC, Duarte AC (2012) Analytical techniques for discovery of bioactive compounds from marine fungi. Trends Anal Chem 34:97–110

Eppley SM, Mercer CA, Haaning C, Graves CB (2009) Sex-specific variation in the interaction between *Distichlis spicata* (Poaceae) and mycorrhizal fungi. Am J Bot 96:1967–1973

Fazzani K, Jones EBG (1977) Spore release and dispersal in marine and brackish water fungi. Mater Org 12:235–248

Feldmann J (1938) Le *Blodgettia confervoides* Harv. est-il un lichen? Rev Bryol Lichénol 11:155–163

Fell JW, Master IM (1973) Fungi associated with the degradation of mangrove (*Rhizophora mangle* L.) leaves in South Florida. In: Stevenson LH, Colwell RR (eds) Estuarine microbial ecology. USA University of South Carolina Press, Columbia, pp 455–466

Fletcher A (1975a) Key for the identification of British marine and maritime lichens I. Siliceous rocky shore species. Lichenologist 7:1–52

Fletcher A (1975b) Key for the identification of British marine and maritime lichens II. Calcareous and terricolous species. Lichenologist 7:73–115

Gao Z, Li B, Zheng C, Wang G (2008) Molecular detection of fungal communities in the Hawaiian marine sponges *Suberites zeteki* and *Mycale armata*. Appl Environ Microbiol 74:6091–6101

Garbary DJ, Deckett RJ, Hubbard CB (2005) *Ascophyllum* and its symbionts VII. Three-way interactions among *Ascophyllum nodosum* (Phaeophyceae), *Mycophycias ascophylii* (Ascomycetes) and *Vertebrata lanosa* (Rhodophyta). Algae 20:353–361

Gessner RV (1976) In vitro growth and nutrition of *Buergenerula spartinae*, a fungus associated with *Spartina alterniflora*. Mycologia 68:583–599

Gessner RV (1977) Seasonal occurrence and distribution of fungi associated with *Spartina alterniflora* from a Rhode Island estuary. Mycologia 69:477–491

Gessner RV, Kohlmeyer J (1976) Geographical distribution and taxonomy of fungi from salt marsh *Spartina*. Can J Bot 54:2023–2037

Grasso S, La Ferla R, Jones EBG (1985) Lignicolous marine fungi in a harbour environment (Milazzo). Bot Mar 28:259–264

Gunn CR, Dennis JV (1976) World guide to tropical drift in seeds and fruits. A Demeter Press Book. Quadrangle/The New York Times Book, New York

Gutiérrez MH, Jara AM, Pantoja P (2016) Fungal parasites infect marine diatoms in the upwelling ecosystem of the Humboldt current system off central Chile. Environ Microbiol 18(5):1646–1653

Hawksworth DL (2000) Freshwater and marine lichen-forming fungi. Fungal Divers 5:1–7

Hassett BT, Gradinger R (2016) Chytrids dominate arctic marine fungal communities. Environ Microbiol 18:2001–2009

Henssen A, Johns HM (1974) Lichenes: eine einführung in die flechtenkunde. G Thieme

Higgins R (2000) Bacteria and fungi of marine mammals: a review. Can Vet J 41:L105–L116

Hogarth PJ (1999) The biology of mangroves. Oxford University Press, Oxford

Holguin G, Vazquez P, Bashan Y (2001) The role of sediment microorganisms in the productivity, conservation, and rehabilitation of mangrove ecosystems: an overview. Biol Fertil Soils 33:265–278

Hughes GC (1974) Geographical distribution of the higher marine fungi. Veroff Inst Meeresch Bremerhaven 5:419–441

Hyde KD (1989a) Intertidal fungi from the mangrove fern, *Acrostichum speciosum*, including *Massarina acrostichi* sp. nov. Mycol Res 93:435–438

Hyde KD (1989b) Vertical zonation of intertidal mangrove fungi. In: Hattori T, Ishida Y, Maeuyama Y, Morita RY, Uchida A (eds) Recent advances in microbial ecology. Japan Scientific Societies Press, Tokyo, pp 302–306

Hyde KD (1990a) A comparison of the intertidal mycota of five mangrove tree species. Asian Mar Biol 7:93–107

Hyde KD (1990b) A study of the vertical zonation of intertidal fungi on *Rhizophora apiculata* at Kampong Kapok mangrove, Brunei. Aquat Bot 36:255–262

Hyde KD (1992a) Fungi from decaying intertidal fronds of *Nypa fruticans*, including three new genera and four new species. Bot J Linn Soc 110:95–110

Hyde KD (1992b) Intertidal fungi from *Kandelia candel* including *Phomatospora kandelae* sp. nov. Trans Mycol Soc Jpn 33:313–316

Hyde KD, Jones EBG (1998) Marine mangrove fungi. Mar Ecol 9:15–33

Hyde KD, Sarma VV (2006) Biodiversity and ecological observations on filamentous fungi of mangrove palm *Nypa fruticans* wurumb (Liliopsida – Arecales) along the Tutong River, Brunei. Indian J Mar Sci 35:297–307

Hyde KD, Chalermongse A, Boonthavikoon T (1990a) Ecology of intertidal fungi at Ranong mangrove, Thailand. Trans Mycol Soc Jpn 31:17–27

Hyde KD, Chalermongse A, Boonthavikoon T (1990b) The distribution of intertidal fungi on *Rhizophora apiculate*. In: Morton B (ed) The marine biology of the South China Sea. Proceedings of the 1st International Conference on Marine Biology of Hong Kong and South China Sea, pp 643–652

Hyde KD, Jones EBG, Leaño E, Pointing SB, Poonyth AD, Vrijmoed LLP (1998) Role of fungi in marine ecosystems. Biodivers Conserv 7:1147–1161

Jones EBG (1993) Tropical marine fungi. In: Isaac S, Frankland JC, Watling R, Whalley AJS (eds) Aspects of tropical mycology. Cambridge University Press, Cambridge, pp 73–90

Jones EBG (1994) Fungal adhesion. Mycol Res 98:961–981

Jones EBG (1995) Ultrastructure and taxonomy of the aquatic ascomycetous order Halosphaeriales. Can J Bot 73:790–801

Jones EBG (2000) Marine fungi: some factors influencing biodiversity. Fungal Divers 4:53–73

Jones EBG, Tan TK (1987) Observations on manglicolous fungi from Malaysia. Trans Br Mycol Soc 89:390–392

Jones EMG, Kuhne H, Trussell PC, Turner RD (1972) Results of an international cooperative research programme on the biodeterioration of timber submerged in the sea. Mater Org 7:93–118

Khan AG, Belik M (1995) Occurrence and ecological significance of mycorrhizal symbiosis in aquatic plants. In: Varma A, Hock B (eds) Mycorrhiza. Springer, Berlin, pp 628–666

Kim GH, Moon KH, Kim JY, Shim J, Klochkova TA (2014) A revaluation of algal diseases in Korean *Pyropia* (*Porphyra*) sea farms and their economic impact. Algae 29:249–265

Kirk PW (1983) Direct enumeration of marine arenicolous fungi. Mycologia 75:670–682

Kohlmeyer J (1969) Ecological notes on fungi in mangrove forests. Trans Br Mycol Soc 53:237–250

Kohlmeyer J (1972) Marine fungi deteriorating chitin of hydrozoan and keratin-like annelid tubes. Mar Biol 12:277–284
Kohlmeyer J (1983) Geography of marine fungi. Aust J Bot Suppl Ser 13:67–76
Kohlmeyer J, Kohlmeyer E (1971) Marine fungi from tropical America and Africa. Mycologia 63:831–861
Kohlmeyer J, Kohlmeyer E (1972) Is *Ascophyllum nodosum* lichenized? Bot Mar 15:109–112
Kohlmeyer J, Kohlmeyer E (1977) Bermuda marine fungi. Trans Br Mycol Soc 68:207–219
Kohlmeyer J, Kohlmeyer E (1979) Marine mycology: the higher fungi. Academic Press, London
Kohlmeyer J, Volkmann-Kohlmeyer B (1993) Biogeographic observations on Pacific marine fungi. Mycologia 85:337–346
Kohlmeyer J, Volkmann-Kohlmeyer B (1998) Fungi on *Juncus roemerianus*. 10. A new *Orbilia* with ingoldian anamorph. Mycologia 90:303–309
Kohlmeyer J, Volkmann-Kohlmeyer B (1999) Fungi on *Juncus roemerianus*. 13. *Hyphpolynema juncatile* sp. nov. Mycotaxon 70:489–495
Kohlmeyer J, Volkmann-Kohlmeyer B (2001) The biodiversity of fungi on *Juncus roemerianus*. Mycol Res 105:1411–1412
Kohlmeyer J, Volkmann-Kohlmeyer B (2002) Fungi on *Juncus* and *Spartina*: new marine species of *Anthostomella*, with a list of marine fungi known on *Spartina*. Mycol Res 106:365–374
Kohlmeyer J, Volkmann-Kohlmeyer B, Eriksson OE (1995) Fungi on *Juncus roemerianus* 4. New marine ascomycetes. Mycologia 87:532–542
Kohlmeyer J, Hawksworth DL, Volkmann-Kohlmeyer B (2004) Observations on two marine and maritime “borderline” lichens: *Mastodia tessellata* and *Collemopsidium pelvetiae*. Mycol Prog 3:51–56
Le Calvez T, Burgaud G, Mahé S, Barbier G, Vandenkoornhuyse P (2009) Fungal diversity in deep-sea hydrothermal ecosystems. Appl Environ Microbiol 75:6415–6421
Le Campion-Alsumard T, Golubic S, Priess K (1995) Fungi in corals: symbiosis or disease? Interaction between polyps and fungi causes pearl-like skeleton biomineralization. Mar Ecol Prog Ser 117:137–147
Leong WF, Tan TK, Jones EBG (1991) Fungal colonization of submerged *Bruguiera cylindrica* and *Rhizophora apiculata* wood. Bot Mar 34:69–76
Li Q, Wang G (2009) Diversity of fungal isolates from three Hawaiian marine sponges. Microbiol Res 164:233–241
Lightner DV (1988) Black gill syndrome of penaeid shrimp. In: Sindermann CJ, Lightner DV (eds) Disease diagnosis and control in North American marine aquaculture, Developments in aquaculture and fisheries science, vol 17. Elsevier Scientific, New York, pp 86–88
Mcowen CJ, Weatherdon LV, Van Bochove J, Sullivan E, Blyth S, Zockler C, Stanwell-Smith D, Kingston N, Martin CS, Spalding M, Fletcher S (2017) A global map of saltmarshes. Biodivers Data J 5:e11764
Meyer FP (1991) Aquaculture disease and health management. J Anim Sci 69:4201–4208
Meyers SP (1969) Thalassiomycetes XI. Further studies of the genus *Lindra* with a description of *L. marinera*, a new species. Mycologia 61:486–495
Meyers SP, Opurt PA, Simms J, Boral LL (1965) Thalassiomycetes VII. Observations on fungal infestation of turtle grass, *Thalassia testudinum* König. Bull Mar Sci 15:548–564
Miller JD, Whitney NJ (1981) Fungi from the Bay of Fundy 1: lignicolous marine fungi. Can J Bot 59:1128–1133
Mo Z, Li S, Kong F, Tang X, Mao Y (2016) Characterization of a novel fungal disease that infects the gametophyte of *Pyropia yezoensis* (Bangiales, Rhodophyta). J Appl Phycol 28:395–404
Moe RL (1997) *Verrucaria tavaresiae* sp. nov., a marine lichen with a brown algal photobiont. Bull Calif Lichen Soc 4:7–11
Nakagiri A, Tokura R (1994) Taxonomic studies of the genus *Corollospora* (Halosphaeriales, Ascomycotina) with descriptions of seven new species. Trans Mycol Soc Jpn 28:413–436
Nakagiri A, Newell SY, Ito T, Tan TK, Pek CL (1996) Biodiversity and ecology of the oomycetous fungus, *Haltophytophthora*. In: Turner IM (ed) Biodiversity and the dynamics of ecosystems, DIWPA Series, vol 1, pp 273–280

Newell SY (1976) Mangrove fungi: the succession in the mycoflora of red mangrove (*Rhizophora mangle* L.) seedlings. In: Jones EBG (ed) Recent advances in aquatic mycology. Wiley, New York pp, pp 51–91
Newell SY (1992) Autumn distribution of marine *Pythiaceae* across a mangrove-saltmarsh boundary. Can J Bot 70:1912–1926
Newell SY (1996) Established and potential impacts of eukaryotic mycelial decomposers in marine/terrestrial ecotones. J Exp Mar Biol Ecol 200:187–206
Newell SY, Blum LK, Crawford RE, Dai T, Dionne M (2000) Autumnal biomass and potential productivity of salt marsh fungi from 29° to 43° north latitude along the United States Atlantic Coast. Appl Environ Microbiol 66:180–185
Noga EJ (1990) A synopsis of mycotic diseases of marine fishes and invertebrates. In: Perkins FO, Cheng TC (eds) Pathology in marine science. Academic Press, San Diego, pp 143–160
Norton JH, Thomas AD, Barker JR (1994) Fungal infection in the cultured juvenile boring clam *Tridacna crocea*. J Invertebr Pathol 64:273–275
Opurt PA, Meyers SP, Boral LL, Simms J (1964) Thalassiomycetes V. A new species of *Lindra* from turtle grass, *Thalassia testudinum* König. Bull Mar Sci Gulf Caribb 14:405–417
Pang KL, Chow RK, Chan CW, Vrijmoed LP (2011) Diversity and physiology of marine lignicolous fungi in Arctic waters: a preliminary account. Polar Res 30:5859
Pérez-Ortega S, Garrido-Benavent I, Grube M, Olmo R, de los Ríos A (2016) Hidden diversity of marine borderline lichens and a new order of fungi: Collemopsidiales (Dothideomyceta). Fungal Divers 80:285–300
Peters AF, Moe RL (2001) DNA sequences confirm that *Petroderma maculiforme* (Phaeophyceae) is the brown algal phycobiont of the marine lichen *Verrucaria tavaresiae* (Verrucariales, Ascomycota) from central California. Bull Calif Lichen Soc 8:41–43
Petersen KRL, Koch J (1997) Substrate preference and vertical zonation of lignicolous marine fungi on mooring posts of oak (*Querus* sp.) and larch (*Larix* sp.) in Svanemøllen Harbour, Denmark. Bot Mar 40:451–463
Petrini O, Fisher PJ (1986) Fungal endophytes in *Salicornia perennis*. Trans Br Mycol Soc 87:647–651
Picard KT (2017) Coastal marine habitats harbor novel early-diverging fungal diversity. Fungal Ecol 25:1–3
Pointing BS, Vrijmoed LLP, Jones EBG (1998) A quantitative assessment of lignocellulose degrading enzyme activity in marine fungi. Bot Mar 412:293–298
Poon MOK, Hyde KD (1998a) Biodiversity of intertidal estuarine fungi on *Phragmites* at Mai Po marshes, Hong Kong. Bot Mar 41:141–156
Poon MOK, Hyde KD (1998b) Evidence for the vertical distribution of saprophytic fungi on senescent *Phragmites australis* culms at Mai Po marshes, Hong Kong. Bot Mar 41:285–292
Porter D (1986) Mycoses of marine organisms: an overview. In: Moss ST (ed) The biology of marine fungi. Cambridge University Press, Cambridge, pp 141–153
Pugh GJF, Jones EBG (1986) 27 Antarctic marine fungi: a preliminary account. In: Moss ST (ed) The biology of marine fungi. Cambridge University Press, Cambridge, p 323
Raghukumar C (2000) Fungi from marine habitats: an application in bioremediation. Mycol Res 104:1222–1226
Raghukumar C, Lande V (1988) Shell disease of rock oyster *Crassostrea cucullata*. Dis Aquat Org 4:77–81
Raghukumar C, Raghukumar S (1991) Fungal invasion of massive corals. Mar Ecol 12:251–260
Raghukumar C, Raghukumar S (1998) Barotolerance of fungi isolated from deep-sea sediments of the Indian Ocean. Aquat Microb Ecol 15:153–163
Rämä T, Hasset BT, Bubnova E (2017) Arctic marine fungi: from filaments and flagella to operational taxonomic units and beyond. Bot Mar 60:433–452
Ramaiah N (2006) A review on fungal diseases of algae, marine fishes, shrimps and corals. Indian J Mar Sci 35:380–387
Rateb ME, Ebel R (2010) Secondary metabolites of fungi from marine habitats. Nat Prod Rep 28:290–344

Ravikumar DR, Vittal BPR (1996) Fungal diversity on decomposing biomass of mangrove plant *Rhizophora* in Pichavaram estuary, East coast of India. Indian J Mar Sci 25:142–144
Read SJ, Jones EBG, Moss ST, Hyde KD (1994) Ultrastructure of asci and ascospores of two mangrove fungi: *Swampomyces armeniacus* and *Marinosphaera mangrovei*. Mcyol Res 99:1465–1471
Ritchie D (1957) Salinity optima for marine fungi affected by temperature. Am J Bot 44:870–874
Robinson CH (2001) Cold adaptation in Arctic and Antarctic fungi. New Phytol 151:341–353
Rozema J, Aep W, van Diggelen J, van Esbroek M, Broekman R, Punte H (1986) Occurrence and ecological significance of vesicular-arbuscular mycorrhiza in the saltmarsh environment. Acta Bot Neerl 35:457–462
Sadaba RB (1996) An ecological study of fungi associated with the mangrove associate *Acanthus ilicifolius* in Mai Po, Hong Kong. Ph.D. Dissertation, Department of Ecology and Biodiversity, University of Hong Kong, Hong Kong
Sadaba RB, Vrijmoed LLP, Jones EBG, Hodgkiss IJ (1995) Observations on the vertical distribution of fungi associated with standing senescent *Acanthus ilicifolius* stems at Mai Po mangrove, Hong Kong. Hydrobiologia 295:119–125
Sanders WB, Moe RL, Ascaso C (2004) The intertidal marine lichen formed by the pyrenomycete fungus *Verrucaria tavaresiae* (Ascomycotina) and the brown alga *Petroderma maculiforme* (Phaeophyceae): thallus organization and symbiont interaction. Am J Bot 91:511–522
Sarma VV, Hyde KD (2001) A review on frequently occurring fungi in mangroves. Fungal Divers 8:1–34
Sarma VV, Vittal BPR (2001) Biodiversity of mangrove fungi on different substrata of *Rhizophora apiculate* and *Avicennia* spp. from Godavari and Krishna deltas, east coast of India. Fungal Divers 5:23–41
Sarmiento-Ramírez JM, Abella E, Martín MP, Tellería MT, López-Jurado LF, Marco A, Diéguez-Uribeondo J (2010) *Fusarium solani* is responsible for mass mortalities in nests of loggerhead sea turtle, *Caretta caretta*, in Boavista, Cape Verde. FEMS Microbiol Lett 312:192–200
Schaumann K (1975) Olologische untersuchungen uber hohere pilze im meer-und brackwasser der deutschen bucht unter besondere berusksichigung der holzbesiedelen arten. Veroff Inst Meeresch Bremerhaven 15:79–182
Schaumann K, Weide G (1990) Enzymatic degradation of alginate by marine fungi. Hydrobiologia 204:589–596
Schmit JP, Shearer CA (2003) A checklist of mangrove associated fungi, their geographical distribution and known host plants. Mycotaxon 85:423–477
Schneider J (1976) Lignicolous marine fungi (Ascomycetes and Deuteromycetes) from 2 fjords of Western Baltic. Bot Mar 19:295–307
Seshadri R, Sieburth JM (1971) Cultural estimation of yeasts on seaweeds. Appl Microbiol 22:507–512
Sharma S, Raghukumar C, Raghukumar S, Sathe-Pathak V, Chandramohan D (1994) Thraustochytrid and fungal component of marine detritus II. Laboratory studies on decomposition of the brown alga *Sargassum cinereum* J. Ag. J Exp Mar Biol Ecol 175:227–242
Shearer CA, Descals E, Kohlmeyer B, Kohlmeyer J, Marvanová L, Padgett D, Porter D, Raja HA, Schmit JP, Thorton HA, Voglymayr H (2007) Fungal biodiversity in aquatic habitats. Biodivers Conserv 16:49–67
Shields JD (2011) Diseases of spiny lobsters: a review. J Invertebr Pathol 106:79–91
Smolowitz RM, Bullis RA, Abt DA (1992) Mycotic bronchitis in laboratory-maintained hermit crabs (*Pagurus* spp.). J Crustac Biol 12:161–168
Spatafora JW, Volkmann-Kohlmeyer B, Kohlmeyer J (1998) Independent terrestrial origins of the Halosphaeriales (marine Ascomycota). Am J Bot 85:1569–1580
Sridhar KR, Alias SA, Pang K (2012) Mangrove fungi. In: Jones EBG, Pang K (eds) Marine fungi and fungal-like organisms. Walter de Gruyter, Berlin/Boston
Stewart JE (1984) Lobster diseases. Helgoländer Meeresuntersuchungen 37:243

Sundari R (1997) The biology of arenicolous marine fungi. Ph.D. Dissertation, University Malaya, Malaysia

Tan TK, Leong WF, Jones EBG (1989) Succession of fungi on wood of *Avicennia alba* and *A. lanata* in Singapore. Can J Bot 67:2686–2691

Tan TK, Leong WF (1990) Mangrove fungi in Singapore and some possible factors influencing their occurrence. Trans Mycol Soc Jpn 31:35–44

Tan TK, Teng CL, Jones EBG (1995) Substrate type and microbial interactions as factors affecting ascocarp formation by mangrove fungi. Hydrobiologia 295:127–134

Taylor JD, Cunliffe M (2016) Multi-year assessment of coastal planktonic fungi reveals environmental drivers of diversity and abundance. ISME J 10:2118–2128

Tisthammer KH, Cobain GM, Amend AS (2016) Global biogeography of marine fungi is shaped by the environment. Fungal Ecol 19:39–46

Torzilli AP, Sikaroodi M, Chalkey D, Gillevet PM (2006) A comparison of fungal communities from four saltmarsh plants using automated ribosomal intergenic spacer analysis (ARISA). Mycologia 98:690–698

Van Dover CL, Ward MF, Scott JL, Underdown J, Anderson B, Gustafson C, Whalen M, Carnegie RB (2007) A fungal epizootic in mussels at a deep-sea hydrothermal vent. Mar Ecol 28:54–62

Van Ryckegem G, Verbeken A (2005) Fungal diversity and community structure on *Phragmites australis* (Poaceae) along a salinity gradient in the Scheldt estuary (Belgium). Nova Hadwigia 80:173–197

Velez P, González MC, Rosique-Gil E, Cifuentes J, del Rocío R-MM, Capello-García S, Hanlin RT (2013) Community structure and diversity of marine ascomycetes from coastal beaches of the southern Gulf of Mexico. Fungal Ecol 6:513–521

Velez P, González MC, Capello-García S, Rosique-Gil E, Hanlin RT (2015) Diversity of marine ascomycetes from the disturbed sandy beaches of Tabasco, Mexico. J Mar Biol Assoc UK 95:897–903

Vicente VA, Orélis-Ribeiro R, Najafzadeh MJ, Sun J, Guerra RS, Miesch S, Ostrensky A, Meis JF, Klaassen CH, de Hoog GS, Boeger WA (2012) Black yeast-like fungi associated with Lethargic Crab Disease (LCD) in the mangrove-land crab, *Ucides cordatus* (Ocypodidae). Vet Microbiol 158:109–122

Wada S, Nakamura K, Hatai K (1995) First case of *Ochroconis humicola* infection in marine cultured fish in Japan. Fish Pathol 30:125–126

Walker AK, Campbell J (2010) Marine fungal diversity: a comparison of natural and created saltmarshes of the north-central Gulf of Mexico. Mycologia 102:513–521

Walker AK, Vélez P, González MC (2017) Marine fungi. eLS/Wiley, Chichester

Webber FC (1967) Observations on the structure, life history and biology of *Mycosphaerella ascophylii*. Trans Br Mycol Soc 50:583–601

Xu W, Guo S, Gong L, Alias SA, Pang K, Luo Z (2018) Phylogenetic survey and antimicrobial activity of cultivable fungi associated with five scleractinian coral species in the South China Sea. Bot Mar 61:75–84

You YH, Yoon H, Seo Y, Kim M, Shin JH, Lee IJ, Choo YS, Kim JG (2012) Analysis of genomic diversity of endophytic fungal strains isolated from the roots of *Suaeda japonica* and *S. maritima* for the restoration of ecosystems in Buan salt marsh. Microbiol Biotechnol Lett 40:287–295

Yue Y, Yu H, Li R, Xing R, Liu S, Li P (2015) Exploring the antibacterial and antifungal potential of jellyfish-associated marine fungi by cultivation-dependent approaches. PLoS One 10: e0144394

Chapter 3
Microbial Ecology of Saline Ecosystems

Vishnuvardhan Reddy Sultanpuram and Thirumala Mothe

Abstract In studies on microbial diversity and functioning of ecosystems of extreme environments, saline habitats are one of the important model systems. Microbial form of life is found over an extremely wide range of salt concentrations, from that of fresh water to hypersaline environments. Halophilic microorganisms are microbes that live in habitats of high ionic strength, and these organisms cope with hyperosmotic stress by utilizing various strategies. The ecology of these saline ecosystems is studied using various techniques, such as analysis of total community and specific biomarkers. Further, the usual culture-dependent and culture-independent techniques are also helpful in these studies. India has large numbers of biodiversity hotspots including diversified saline ecosystems. There are only limited reports on the culturable and yet to be cultivated halophilic bacteria from the country, which needs great endeavour in the future.

Keywords Saline habitats · Microbial diversity · Characterization studies · India

3.1 Introduction

Living organisms are highly responsive to any serious alterations in their habitats. Increases in salinity, pH, temperature, pressure and drought conditions interrupt the important associations of the biological molecules which make them non-functional, which in turn destroy the cell integrity (Rampelotto 2010). A diverse group of microorganisms are known which can not only tolerate these extreme environmental conditions, but they frequently are required of those conditions for their survival (Pikuta and Hoover 2007). The saline and alkaline lakes are one of the important prototypes for studies on diversity of microorganisms and their functions in the extreme environment ecosystems. Saline systems not only include coastal

V. R. Sultanpuram (✉) · T. Mothe
Microbial Ecology Laboratory, Department of Applied Biosciences, Mahatma Gandhi University, Nalgonda, Telangana, India

B. Giri, A. Varma (eds.), *Microorganisms in Saline Environments: Strategies and Functions*, Soil Biology 56, https://doi.org/10.1007/978-3-030-18975-4_3

environments (viz. marshes, lagoons, estuaries, solar salterns) and inland regions (such as springs, natural lakes and other evaporitic arid habitats) but also the diversity of flora (both micro- and macroflora) and fauna. The coastal as well as inland ecosystems which are good saline environments not only possess a considerable amount of the biodiversity of the world, but they also are an important source of commercial fishing and aquaculture, which are staple food for a sizeable world human populace (Creel 2003). In the present chapter, microbial ecology of the saline ecosystems is highlighted.

3.2 Diversity of Halophiles

Microorganisms may be found over a broad range of salinities, from fresh water (less than 0.5 g/L) through seawater (around 35 g/L) to hypersaline environment (above 350 g/L) salt concentrations (Oren 1999). The multiple properties of these habitats are shown within the microorganism groups which alter their growth to such environmental conditions (Oren 2002b). The microbes that are able to grow in high concentrations of salt (from approx. 10% NaCl to saturation) and some that can remain alive even in salt crystals are called as halophiles (DasSarma 2006). One quarter of the Earth's land mass, which actually represents a big part of the planet's surface, contains salt deposits, saline lakes, seas and oceans; hence, saline ecosystems are of greater environmental significance. Halophiles grow in almost saturated brines, since they are capable of tolerating different types of stresses, like desiccation, temperature, toxic substances, etc. Halophilic microorganisms are reported to even sustain in extraterrestrial conditions. Extreme halophiles that are entrapped within the salt crystal fluid inclusions are highly vigorous and can be isolated from their original source even after the water is completely evaporated (Vauclare et al. 2014).

Hypersaline water contains higher salt concentrations than those of water from the sea (Rodriguez-Valera 1993). The properties of ecosystems containing high salinity mainly depend upon two components, i.e. the overall salt concentrations and the ionic compositions of the respective habitats. A large number of hypersaline ecosystems have emerged by evaporation of seawater, restriction in flow of water, increase in temperatures, limited rainfall, etc. These environments are called as thalassohaline (in Greek *thalasso*, means the sea) (McGenity et al. 2000; Oren 2002a, 2006; Rodriguez-Valera 1993). The salt constitution from these ecosystems is like seawater, in which Na^+ and Cl^- are the predominating ions where the pH is either close to neutral or a bit alkaline (Oren 2002a, 2006). The Great Salt Lake, in Utah, USA, is a case of a thalassohaline ecosystem.

Some of the high-saline ecosystems greatly vary from seawater with respect to their ionic composition. Many of such ecosystems are devoid of the ingredients found in salt from seawater, or sometimes they contain alternate principal ions. Such ecosystems are defined as athalassohaline environments (Rodriguez-Valera 1993; Oren 2002a, 2006). An instance is the Dead Sea brine, which has a pH of approximately 5.8–6.0 and is dominated by divalent cations with Mg^{2+} concentration (approx. 1.9 M) and Ca^{2+} (approx. 0.45 M) which are higher than that of Na^{+} (approx. 1.6 M) and K^{+} (approx. 0.2 M). The high Ca^{2+} concentration results in low solubility of sulphate and domination of monovalent anions such as Cl^{-} and Br^{-} (>99.9% of the total anion) (Oren 2002a, 2006).

Alkaline athalassohaline brine solutions are plentiful and are represented by alkaline soda lakes. These lakes are distributed around the world. The Lake Magadi in Kenya (Eugster 1980), East African Rift Valley lakes (Grant and Tindall 1986; Grant et al. 1990) and the Wadi Natrun lakes in Egypt (Imhoff et al. 1979) are located in Africa. The high-altitude Tibetan soda lakes in China (Wang and Tang 1989; Xu et al. 1999; Zheng et al. 1993) and Lonar, Pangong and Sambhar lakes in India (Jakher et al. 1990; Upasani and Desai 1990) are located in Asia. The Mono Lake, California and Big Soda Lake, Nevada, USA (Oren 2002b), are located in North America. The predominant ions in the brines of the above alkaline soda lakes are Na^{+}, Cl^{-}, HCO_3^{-} and $CO_3^{\ 2-}$, and their pH at times reaches to above 10–11 (Grant and Tindall 1986). Due to this high pH, the divalent cations Mg^{2+} and Ca^{2+} are less soluble, and hence their concentrations may not be detected (Oren 2006).

The Great Salt Lake (Utah, USA) and the Dead Sea (Middle East) are the two largest and modern hypersaline ecosystems, which have been studied in detail. The cold hypersaline ecosystems of Antarctica have high salt content and are supposed to preserve the liquid water at as lower as −20 °C temperatures (Rampelotto 2010). The environments with hypersaline conditions are also found within the subsurface of the Earth (DasSarma 2006).

3.3 Mechanism of Adaptation of Halophiles

Halophilic microorganisms which generally live in habitats of high concentrations of ions cope with hyperosmotic stress by altering their membrane lipid constituents and managing the concentrations of their compatible solutes (Pastor et al. 2010; Vauclare et al. 2014). Halophiles use various strategies to neutralize high salinities in their surroundings, viz. the buildup of inorganic salts (important among which is KCl). The *Halanaerobiales* order members which consist of extremely halophilic obligate anaerobic bacteria and *Salinibacter ruber* use this approach (Oren 1999; Oren et al. 2009; Roberts 2000; DasSarma 2006). Since this adaptation needs to work in habitats of high ionic concentrations, the microorganisms that utilize this strategy have developed large-scale variations in their enzymes and organelles which had made possible their adaption to life under extreme salinities. The enzymes from organisms accumulating KCl are extremely acidic when compared to other orthologous proteins from mesophiles, which is an important adaptive change of these organisms (Dennis and Shimmin 1997; Lanyi 1974).

Similar to the hyperthermophile proteins which remain active near 100 °C, all the halophile proteins are most favourably folded and are functional even at saturated salt conditions. Because of this reason, the sequences, structures and functional characteristics of halophilic proteins in contrast with proteins from thermophiles and mesophiles have been studied, which would also help to understand the approaches adopted by these proteins to stress environments, but the phenomenon is still unclear (Rampelotto 2010). The proteomes from halophilic microorganisms are predominantly acidic containing proteins with a mean pI of 4.2. The study of structures of proteins from halophiles disclosed the cluster of negatively charged residues on their protein surfaces, which is compatible to solvation model suggested by Zaccai and Eisenberg (1990) where hydrated salt ions on surface are responsible for stability and solubility of proteins in the hypersaline environments. The importance of powerful interactions between inter-subunits and intra-ions of the halophilic protein foldings is also accepted (Vauclare et al. 2014). The highly different proteins in the topography of the extremophilic proteome are responsible for its molecular adaptation. The generalization of extremophilic molecular adaptations is impossible because each protein in a cell adopts a very different pathway to reach its stable fold. In contrast, the effect of physicochemical stress on each of the protein will be different, which therefore limits the adaptation of extremophilic microbes to a delicate form of proteins or biological functions. The transcriptomic and proteomic studies under stress conditions are expected to help in identifying these important factors in extremophilic adaptation (Vauclare et al. 2014). Most of the stable and active proteins from halophilic microorganisms exhibit an absolute necessity for increased salt conditions which otherwise would result in inactivation and precipitation of their enzymes due to low-salt conditions (Oren 2011).

Halophiles require maintaining their isosmotic cytoplasm similar to their surroundings which are essential to avert loss of water. To maintain turgor pressure, cytoplasm which is hyperosmotic is essential, and this phenomenon is observed in all the halophiles, except the halophilic archaeal members of the family *Halobacteriaceae* (Noha and Wiegel 2010). Possibly for this reason, it is interpreted that the halophiles always maintain increased salt concentrations in their cytoplasm (Vauclare et al. 2014). However, the above said adaptation is not helpful in moderate or low salinity environments because every time increased intracellular salt concentrations are required for correct overlapping of proteins and hence their activity. Here, an additional flexible strategy of accumulating the so-called osmoprotectors or compatible solutes comes into play. The compatible solutes are organic molecules that are either uncharged or zwitter ionic in nature with high solubility and low molecular weights; these are said to be amassed in the cytoplasm (Pastor et al. 2010). With an aim of protecting themselves from environmental stress, halophiles produce and accumulate these compatible solutes (Pastor et al. 2010). These solutes actually help to maintain osmatic balance, which is done without any interference with the essential normal metabolism and cellular processes of the cells. Compatible solutes have almost no effect on the ionic concentrations of the cytoplasm, and they do not require any special modifications of neither enzymes nor intracellular organelles (Brown 1990; Oren 1999). Compatible solutes are of different configurations like sugars (e.g. trehalose, sucrose), polyols (e.g. glycerol, sorbitol, mannitol, α-glucosylglycerol, mannosyl-glycerol and mannosyl-glyceramide), N-acetylated diamino acids (e.g. N-acetylglutaminylglutamine amide), betaines (e.g. glycine betaine and derivatives), amino acids (e.g. proline, glutamate, glutamine and alanine) and derivatives. The amino acids group includes ectoine and hydroxy-ectoine which widely occur in nature and due to their vast applications are of late getting increased attention (Pastor et al. 2010). Some of the applications include ability to diminish the adverse effects of stress caused due to increase or decrease in temperatures, increase in salinity, presence of oxygen free radicals or urea and different types of radiations and other denaturing agents, which would actually effect the structure of proteins, nucleic acids, membranes and even in some cases the whole cells (da Costa et al. 1998; Lentzen and Schwarz 2006). Further, the ectoines (both ectoine and hydroxyectoine) exhibit some of the most robust stabilizing properties (Lippert and Galinski 1992).

3.4 Ecology

3.4.1 *Community Studies*

The saline systems especially the solar salterns and the alkaline soda lakes in the USA, India, Egypt and Africa are niches for dense communities of halophilic Archaea which often imparts the recognizable pink-red colour of the water in these environments. The prokaryotic counts in these saline systems are generally between 10^7 and 10^8/ml, but higher counts have also been reported (Javor 1983, 1989; Oren 2002b). The enumeration of this overall density of prokaryotic communities in the saline ecosystems was performed using different techniques like simple microscopic quantification (Oren and Gurevich 1995) and enumerating the DAPI (30,6-diamidino-2-phenylindole) stained cells (Pedrós-Alió et al. 2000). Similarly, the phytoplankton cells containing chlorophyll were enumerated using flow cytometry (Estrada et al. 2004). The fluorescence in situ hybridization (FISH) with 16S rRNA-targeted probes was used to estimate the number of organisms belonging to different phylogenetic groups and has been successfully used to enumerate the prokaryotes in salterns in Spain (Antón et al. 2000) and Peru (Maturrano et al. 2006). There is further a report which quantifies the relative numbers of live and dead cells in hypersaline environments using the LIVE/DEAD *Bac*Light kit with certain modifications (Leuko et al. 2004). However, in these types of studies, it is very essential to incorporate proper controls, which not only ensures that the methods used are indeed working at increased salt concentrations but also helps in analysing new cultures in comparison with the properties of known cultures (Oren 2011).

3.4.2 *Biomarker Studies*

Various types of halophilic microorganisms possess particular cellular components that can be used as biomarkers, which actually help in quantification of their abundance. A few of these biomarkers can be identified using simple methods, while many others require advanced analytical techniques (Oren 2011). Halophilic members of the family *Halobacteriaceae* possess diphytanyl diether lipids which are easily identified by thin-layer chromatography. A number of members of this family accumulate the diphytanyl diether derivatives of phosphatidylglycerol and the methyl derivative of phosphatidylglycerol phosphate. Some of them also possess phosphatidylglycerol sulphate which is not present in most of the members of the family. A few members in the family lack glycolipids, while many have different diglycosyl, triglycosyl and/or tetraglycosyl lipids. The glycolipids in few instances possess one or more sulphate groups. Certain polar lipids extracted from total biomass from various ecosystems can also act as biomarkers. These polar lipids from extracts of various environments are especially helpful in estimating the presence and the abundance of halophilic Archaea (Oren 2011).

Another advanced method to identify the polar lipid composition of halophilic communities is using electrospray mass spectrometry (EMS). EMS method was helpful in the identification of a 660 Da sulfonolipid from *Salinibacter ruber* (*Bacteroidetes*), which was also directly estimated in lipid extracts of solar saltern crystallizer biomass (Corcelli et al. 2004). Specific fatty acids from Bacteria and Eucarya can be used as biomarkers for analysing microbial groups in high-saline ecosystems. The approach was used in a study of the fatty acid compositions of halophilic communities within a gypsum crust in a solar saltern evaporation pond. Some monounsaturated and polyunsaturated fatty acids were reported to act as indicators, which were corresponded with the existence of certain forms of cyanobacteria and their in situ way of life (Ionescu et al. 2007).

In particular types of halophilic microorganisms' pigments are also useful biomarkers. Spectrophotometric methods and high-performance liquid chromatography (HPLC) are used to identify and quantify bacterioruberin pigments produced by halophilic Archaea as well as other common and unique pigments produced by halophiles (Oren and Gurevich 1995; Oren and Rodríguez-Valera 2001; Estrada et al. 2004). Spectrophotometric analysis is also used in quantifying different types of common pigments such as chlorophyll, bacteriochlorophyll and carotenoids in benthic microbial mats in salt crystallizer ponds and from the groups of halophilic microorganisms which were lodged within gypsum crusts found in high salinity ponds (Caumette et al. 1994; Oren et al. 2009). The yet to be completely understood anoxygenic phototrophs from high salinity mats which possess chlorosomes were discovered while studying benthic microbial mats in solar salterns using emission spectroscopy and kinetic fluorometry (Prášil et al. 2009). There have been some attempts made based on the absorption spectra or on spectroscopic measurements of pigments to identify the presence of the light-driven proton pump bacteriorhodopsin which is found in *Halobacteriaceae* members. Similar methods were used to assess other retinal proteins in the prokaryotic groups in hypersaline habitats (Oren and Shilo 1981; Javor 1983; Stoeckenius et al. 1985) (Fig. 3.1).

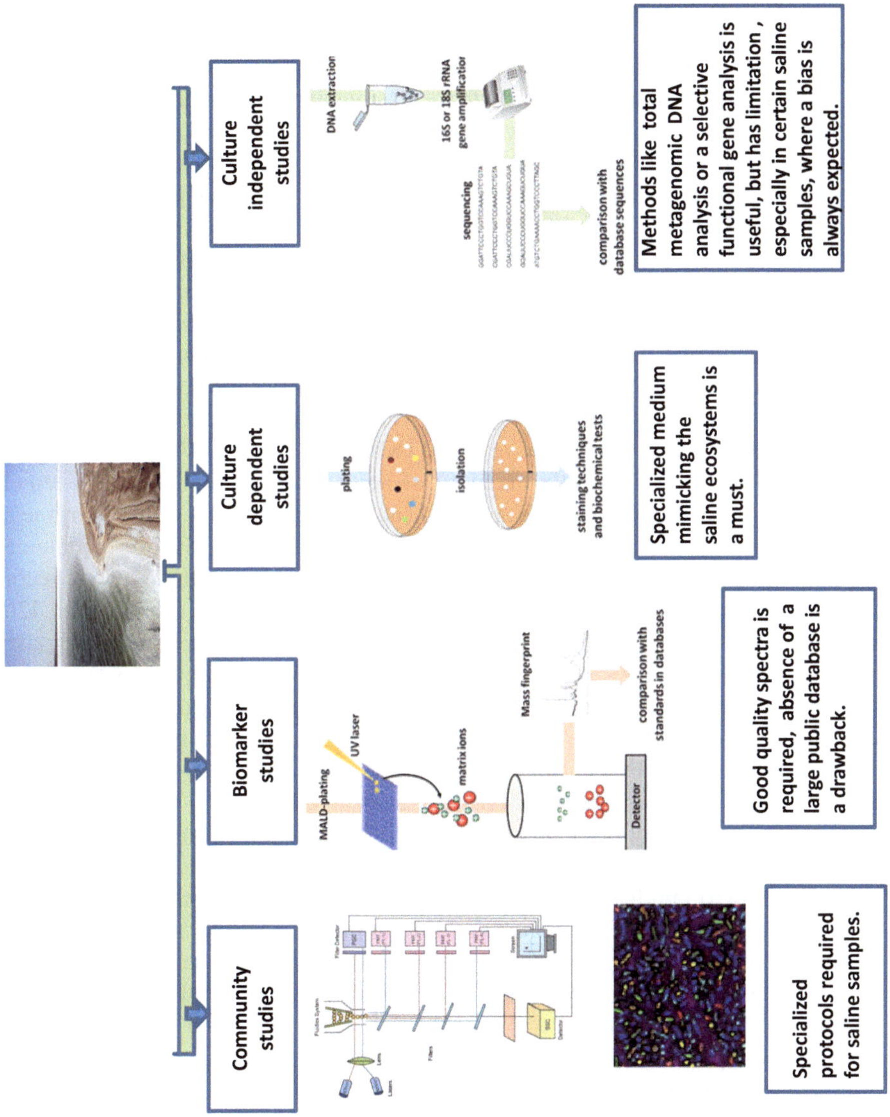

Fig. 3.1 Various techniques used for ecological studies in saline systems

3.4.3 Culture-Dependent Studies

The so-called great plate count anomaly which assumes that only a fraction of the diverse microbes on Earth have been yet cultivated and the number of those microorganisms which form colonies on agar plates is generally of very low magnitude to that of total microscopic or flow cytometric counts seems to no longer exist in case of salt crystallizer ecosystems (Oren 2011). Burns et al. (2004b) demonstrated that using specific growth media containing low nutrients and increasing incubation periods like 8–10 weeks, a majority of prokaryotes estimated in the solar salterns using 16S rRNA gene sequence-based technique can be cultured in the laboratory. A similar sub-cultivation approach with long incubation periods was helpful in the isolation of *Haloquadratum walsbyi*, the square-shaped archaeon that was resisting cultivation on artificial media for nearly two and half decades (Bolhuis et al. 2004; Burns et al. 2004a). Selective media emended with antibiotics, such as anisomycin and bacitracin, were used for isolation of *Salinibacter* and related extremely halophilic bacteria from hypersaline ecosystems (Oren 2011). In a study on microbial groups of alkali-saline soils and halophiles of the Caspian Depression regions, over 40 species and forms of soil microbes, apart from bacteria of the genera *Bacillus* and *Salimicrobium*, were revealed (Khalilova et al. 2017).

In a study on the cultivable diversity of aerobic, heterotrophic and extremely halophilic communities from sediment and salt solution samples of Mediterranean, Canary Islands and Chile, a strategy combining large-scale cultivation, whole cell biomass MALDI-TOF MS and 16S rRNA gene analysis based phylogenetic reconstruction was used. This approach not only allowed identification of more than 4200 strains but also was used to compare microbiota between different sampling sites. Additionally, four novel genera and nine new species belonging to the family *Halobacteriaceae* and a representative of *Salinibacter ruber* phylotype II were identified from the study. Further, the authors based on these results recommended the use of nonredundant environment sampling and large-scale screening for isolation of novel taxa from these habitats (Viver et al. 2015).

In a genetic diversity study of moderate halophilic bacteria from Qrhan Salt Lake, 421 moderate halophilic bacteria were isolated. Eighty-three model strains belonging to 3 phyla, 6 families and 16 genera were isolated. Among them, *Bacillus* sp., *Oceanobacillus* sp. and *Halomonas* sp. were dominant species (Shen 2017). Some 69 halophilic actinobacterial strains were isolated from 19 soil samples collected at two Algerian arid ecosystems (M'zab and Zibans) of septentrional Sahara. Based on 16S rRNA gene sequencing, the strains were identified to be belonging to six clusters corresponding to six genera, *Actinopolyspora*, *Nocardiopsis*, *Prauserella*, *Saccharomonospora*, *Saccharopolyspora* and *Streptomonospora*, some of which were found to represent known species but others were forming distinct phylogenetic arrangements. Additionally, four strains which were forming cluster VII were found to represent an unknown actinobacterial genus (Saker et al. 2015). Using amplified ribosomal RNA restriction analysis (ARDRA) technique, 64 archaeal strains which were isolated from the sediments of a salt crystallizer pond (TS18) and non-crystallizer (M1) pond at Sfax, Tunisia, were characterized. *Halorubrum chaoviator* was the most abundant cultivable species in the non-crystallizer M1 pond which exhibited five distinct patterns in the ARDRA analysis. However, the TS18 crystallizer pond showed ten restriction patterns, which were dominated by haloarchaea EB27K, an yet to be identified genotype. Further, the non-crystallizer pond (M1) exhibited two distinct ARDRA patterns with unique 35 bacterial strains. These strains belonged to the *γ-Proteobacteria* subdivision, with members of *Salicola marasensis* (83%) being the abundant species among the isolated bacterial strains (Boujelben et al. 2014).

It has been now well established that isolation of microorganism from saline systems depends upon the constituents of the media used and the incubation period. Hence, it is a general requirement to use different media so as to discover the parameters that give the highest colonies in the saline environments. Sometimes the isolation media with unusual ionic concentrations give good results as reported in a study of halotolerant and halophilic bacterial groups on ancient monuments (Saiz-Jimenez and Laiz 2000).

3.4.4 Culture-Independent Studies

In microbial ecology, methods based on phylogenetic analysis of the components of microbial communities such as, 16S rRNA (Bacteria, Archaea) and 18S rRNA (Eucarya) gene sequences are considerably used, which are even applied to saline environments. Further, the analysis of functional genes such as *bop* gene which codes for bacteriorhodopsin protein (Pašić et al. 2005), *cbbL*, and *cbbM* coding for form 1 and form 2 of ribulose bisphosphate carboxylase/oxygenase (RuBisCO) (Giri et al. 2004), *drsAB* gene coding for dissimilatory sulphite reductase (Kjeldsen et al. 2006; Scholten et al. 2005; Joye et al. 2009), *apsA* gene coding for adenosine-5′-phosphosulfate reductase (Scholten et al. 2005) and *mcrA* gene coding for methyl coenzyme A reductase (Scholten et al. 2005; Joye et al. 2009) have also been reported. At most times the usual techniques used in such studies for other environments are applicable also to samples from high-saline habitats. However, archaeal members like *Halococcus* (which has a rigid and thick cell wall) require some extraordinary chemical and enzymatic lysis, viz. incubation of 2 h at 63 °C in a potassium ethyl xanthogenate containing buffer (Leuko et al. 2008).

There are a number of studies using 16S rRNA gene sequence analysis which have been described from solar salterns at different geographical locations, viz. Spain (Benlloch et al. 1995, 1996, 2002; Casamayor et al. 2002; Ochsenreiter et al. 2002), Tunisia (Baati et al. 2008), Slovenia and Croatia (Pašić et al. 2005, 2007), Taiwan (Wang et al. 2007) and Peru (Maturrano et al. 2006). Similar methods were also used to study the microbiota of Lake Tebenquiche (Salar de Atacama, Argentina) (Demergasso et al. 2008); the Mono Lake, California (Humayoun et al. 2003); the Nevada Lake, Nevada (Ochsenreiter et al. 2002); high-altitude lakes on the Tibetan plateau (Wu et al. 2006); Baer Lake, Inner Mongolia (Ma et al. 2004);

Lake Magadi and other high salinity soda lakes in Kenya (Rees et al. 2004); alkaline saltern ponds from East Africa (Grant et al. 1999); the Wadi An Natrun, Egypt (Mesbah et al. 2007); solar salterns at Sua Pan, Botswana (Gareeb and Setati 2009); a small pond near a potassium mine in Germany (Ochsenreiter et al. 2002); the salt solutions of Kebrit Deep and Shaban Deep, Red Sea (Eder et al. 1999, 2001, 2002); the northern part of Great Salt Lake (Almeida-Dalmet et al. 2015); the Abijata Shalla Soda ash concentration pond system in Ethopia (Simachew et al. 2016); and the shallow soda lake brines of Kulunda Steppe (Vavourakis et al. 2016). A few similar studies were made on the halophilic microbial groups of sediments from Mediterranean salterns (Mouné et al. 2002), the archaeal groups inhabiting the benthic microbial mats at Guerrero Negro, Baja California Sur, Mexico (Robertson et al. 2009) and the alkaline saline soil of the former Lake Texcoco (Mexico) (Valenzuela-Encinas et al. 2008). DNA fingerprinting methods like denaturing gradient gel electrophoresis (DGGE), ribosomal internal spacer analysis (RISA) and terminal restriction fragments length polymorphism (TRFLP) have also been used to observe changes in the archaeal, bacterial and eukaryal assemblages along a salinity gradient in a multipond solar saltern crystallizer in Spain (Casamayor et al. 2002). The knowledge of the 16S rRNA sequences of the different prokaryotes present in hypersaline habitats helped in their detection using fluorescence in situ hybridization (FISH), which was successfully adapted at high salt concentrations and used in recognizing the presence of Bacteria in saltern ponds which ultimately led to the isolation of *Salinibacter ruber* (Antón et al. 1999).

3.5 The Indian Scenario

3.5.1 Culture Based

Though there are reports from very unusual locations, such as the foot hills of the great Himalayan Mountains, the culture-dependent studies of the halophiles have been very limited in the country. Gupta et al. have reported identification of five halophilic bacterial isolates which were identified as *Halobacillus trueperi* (two isolates), *Shewanella algae*, *Halomonas venusta* and *Marinomonas* sp. The study area was Lunsu in Himachal Pradesh (Gupta et al. 2015). Similar studies were also conducted on solar salterns of Mulund, Maharashtra, which yielded seven carotenoid-producing halophiles, of which an isolate was identified as *Halorubrum* sp., which was further characterized and was found to contain bacterioruberin-type pigment (Pathak and Sardar 2012). A survey of some extreme halophilic environments for some enzyme-producing bacteria of industrial value yielded 108 bacteria. These halophilic bacteria were producing enzymes, such as amylase, lipase and protease. The morphological, biochemical and 16S rRNA gene analysis of 21 isolates which showed salt requirement in the range of 3–20% identified these isolates to be belonging to the genera *Marinobacter*, *Virgibacillus*, *Halobacillus*, *Geomicrobium*, *Chromohalobacter*, *Oceanobacillus*, *Bacillus*, *Halomonas* and *Staphylococcus*.

Initial characterization studies involving crude hydrolases from these isolates identified them to be active and stable under polyextreme conditions like increased salt, pH, temperature and/or presence of organic solvents. The results help in the conclusion that these strains of halophilic microorganisms are diverse in both phylogeny and their enzyme characteristics (Kumar et al. 2012).

A comprehensive culture-dependent study of the brackish water Pulicat Lake was also performed using five different media. The phylogenetic analysis carried out using 16S rRNA sequences disclosed that 19 strains of the 198 total isolates from this study claded to *Firmicutes*, which included low G+C Gram-positive bacteria related to the family *Bacillaceae* that were belonging to *Bacillus*, *Virgibacillus*, *Rummelibacillus*, *Alkalibacillus* and *Halobacillus* genera. Additionally other genus identified in the phylum *Firmicutes* was *Salimicrobium*. The analysis also revealed the presence of ten isolates which were cladding with *γ-Proteobacteria*, all of which belonged to only one genus *Halomonas* and represented by six different species *H. salina*, *H. shengliensis*, *H. salifodinae*, *H. pacifica*, *H. aquamarina* and *H. halophila*. Most of these strains revealed various enzyme activities (Sahay et al. 2011). Similarly, in a study on the inland salt-producing Sambhar Lake, 93 haloalkaliphilic bacteria were isolated. Based on phylogenetic analysis using 16S rRNA sequences, they were further identified as belonging to 3 different phyla which were able to be grouped into 32 different groups. More than half of these isolates showed similarity with *Firmicutes*, followed by *Proteobacteria* and *Actinobacteria*. A group of 32 isolates each of which were representing different phyla were studied for various characteristics, viz. extracellular enzyme production, ability to produce plant growth-promoting factors and substrate utilization by BIOLOG™. Most of the strains isolated had the ability to produce xylanase (68%), followed by protease (56%), cellulase (40%) and amylase (37%). The plant growth-promoting attributes like ammonia production, 1-aminocyclopropane-1-carboxylate (ACC) deaminase activity, production of indole-3-acetic acid (IAA), hydrogen cyanide production, production of iron-chelating siderophores and phosphate-solubilizing capabilities were also identified from the isolates. Further, the strains which exhibited enzymatic and PGP activities were practically used for promotion of plant growth in saline soils (Sahay et al. 2012).

There have also been some studies on the cultivable halophiles confined to certain groups, like Haloarchaea and the *Actinobacteria*. A study of haloarchaea from high-saline solar salterns of Bhandup and Mumbai resulted in isolating and identifying *Haloferax prahovense*, *H. alexandrines*, *H. lucentense*, *Haloarcula tradensis*, *H. marismortui* and *H. argentinensis*. An extreme haloarchaeon *H. marismortui* RR12 identified in this study was utilized to understand adaptation to wide range of salt concentrations (1.5 M and 4.5 M). The results revealed that *H. marismortui* RR12 adapted a 'salt-in' strategy of osmoadaptation. This strategy normally involves increasing the intracellular sequestration of K^+ and Cl^- ions in hyposalinity and hypersalinity, respectively. The increase in these ion concentrations can be identified using energy-dispersive X-ray spectroscopy microanalysis (EDAX) and inductively coupled plasma-atomic emission spectroscopy (ICP-AES). Similarly, this bacterium was identified to produce small heat shock like proteins (sHSP) and increase its red-coloured carotenoid production in response to salinity stress (Thombre et al. 2016). In a study on halotolerant and halophilic *Actinobacteria* producing antibacterial compounds from Ribandar solar saltern, Goa, R2A agar at 35 psu salinity was identified as the optimal medium. Predominantly the identified isolates were from halotolerant *Streptomyces* spp. Apart from members of *Nocardiopsis*, *Micromonospora* and *Kocuria* spp., most of the strains exhibited antimicrobial activity against more than one human pathogens tested. Most antibiotic-producing halotolerant strains were able to grow from 0 to 75 psu. The authors claimed their findings to be a first of its kind, where a halophilic *Kocuria* sp., *Nocardiopsis* sp. and *Micromonospora* sp. producing antibacterial metabolites against *S. aureus*, *S. citreus* and *V. cholerae*, respectively, were identified (Ballav et al. 2015). In another similar study on *Actinobacteria* from Tuticorin solar salterns using critical processing and various types of media, from a total of 69 actinomycete strains, 12 representative isolates were selected using amplified ribosomal DNA restriction (ARDA) analysis. The identification of selective representatives which was carried out using 16S rDNA gene sequence analysis identified many members of the *Actinobacteria* belonging to *Micromonospora*, *Nocardia*, *Nocardiopsis*, *Saccharopolyspora*, *Nonomuraea* and *Streptomyces* genera (Jose and Jebakumar 2012). Further, in a related report, 59 different alkali-halophilic actinomycete morphotypes from water and sediment samples of Chilika brackish water lake were isolated. Of these, 21 isolates were identified as alkali-halophiles growing in media with 10% NaCl and at pH 9.0. Detailed characterization of these identified isolates mostly belonging to the genera *Streptomyces* and *Micromonospora* for plant growth promotion and biocontrol attributes was carried out. The results revealed that a maximum percentage of siderophore producers were harboured in the south sector of the lake, while highest indole 3-acetic acid and extracellular protease producers were found in the central sector of the lake. The sea mouth sector of the lake had a higher portion of actinomycetes which possessed nitrate reductase activity and biocontrol attributes, such as antimicrobial activity against fungal pathogens (*Rhizoctonia solani*, *Fusarium udum* and *Fusarium oxysporum* f. sp. *ciceri*) and chitinase enzyme production capabilities (Malviya et al. 2014). In a different investigation, the aerobic cultivable diversity of heterotrophic *Bacillus* and *Bacillus*-derived genera (BBDG) was studied in various extreme environments of India. Among these environments were halophilic sites like Chilika Lake, Sambhar Lake, Bhitarkanika-Sunderban mangroves and Rann of Kutch sites. All the bacterial isolates from the study were found to be tolerant to 5% NaCl. However, only a limited number of isolates were able to tolerate NaCl concentrations of 10, 15 and 20%. Among these isolates, two strains which were isolated from Rann of Kutch and identified as *Virgibacillus halodenitrificans* ABK-2 and *Bacillus marisflavi* AB-18 were found to grow at high NaCl concentration of 30%. As anticipated, most of the strains isolated from Sambhar Lake and Rann of Kutch were showing resistance to higher salinity levels (20% NaCl or more) which were further grouped into various clusters. The salt tolerance among the representative strains from each cluster varied, all isolates except CS-39 tolerated NaCl concentration of 10%, 25 isolates could tolerate NaCl concentration of 15%, and 8 strains could tolerate NaCl concentration of 20%. The strains identified as *Bacillus halodurans* ABSL-8, *B. methanolicus* ABSL-11 and *Ammoniphilus* sp. ABSL-12 were isolated from Sambhar Salt Lake. Similarly, a *Bacillus mojavensis* AB-10 was isolated from Bhitarkanika mangrove soil. Further, *Halobacillus litoralis* S-91 and *Bacillus hwajinpoensis* S-93 were isolated from Sunderbans mangrove soil, which were found to grow at 20% NaCl concentration and pH 9.0. Various obligate halophilic strains requiring 5% NaCl concentrations in their growth medium were identified as *Bacillus* sp. NSP22.2, *Bacillus* sp. SB47, *Thalassobacillus devorans* strain MSP14, *Bacillus* sp. NSP10, *Bacillus* sp. MSP18.3, *Salinibacillus aidingensis* strain MSP4 and *Bacillus licheniformis* strain NSP4; these strains were isolated from Rann of Kutch (Yadav et al. 2015).

A certain number of studies were dedicated for isolation of biotechnologically potential halophiles like biosurfactant producers, halocin (an antibiotic peptide) producers, etc. Halophilic, biosurfactant-producing bacteria, *Halomonas* sp. BS4, with optimal growth at NaCl concentration of 8% and pH 6.0–8.0, was isolated from Thamaraikulam solar salt works. Further, the GC-MS characterized partially purified biosurfactants from the isolate were found to contain 1, 2-Ethanediamine N, N, N′, N′-tetra, 8-Methyl-6-nonenamide, (*Z*)-9-octadecenamide and a fatty acid derivative. The antimicrobial and anticancer assay tests carried out revealed that this biosurfactant produced by the bacterium could be used in controlling some human pathogenic bacteria, fungi and an aquaculturally important virus, WSSV. The biosurfactant at a concentration of 2.5 μg was also able to suppress the multiplication of mammary epithelial carcinoma cell by 46.77% (Donio et al. 2013). A similar work on Kovalam solar salt works in Kanyakumari yielded eight halophilic bacterial genera members. After initial screening, isolates belonging to *Kocuria*, *Kurthia* and *Halococcus* genera were positively identified for biosurfactant production. The biosurfactant produced by *Kocuria* sp. when compared to the other strains identified in this study could emulsify more than 50% of the crude oil, coconut oil, sunflower oil, olive oil and kerosene. Further, the analysis of the biosurfactant derived from an identified isolate *Kocuria marina* BS-15 revealed that it belonged to the lipopeptide group (Sarafin et al. 2014). In one similar report, a strain BTSH10 identified as *Natrinema* sp. which is a member of extremely halophilic archaea was isolated from a saltern at Kanyakumari, Tamil Nadu. This bacterium could produce halocin (SH10), the peptide antibiotic at 42 °C, pH 8.0 and 104 h of incubation at 200 rpm with 2% (v/v) inoculums. The medium used in the study was Zobell's medium containing 3 M NaCl, galactose, beef extract and calcium chloride as supplement (Karthikeyan et al. 2013). Similarly, a bioemulsifier-producing *Haererehalobacter* sp. JS1, a close relative of *Salinicoccus halophilus* and other *Haererehalobacter* spp., was isolated and identified from a salt pan. This partially purified bioemulsifier was of siloxane type, and it emulsified many different types of oils like coconut, gingelly, olive and palmolein. This strain of bacterium, *Haererehalobacter* sp. JS1, also could degrade various types of dyes that were tested. Response surface methodology (RSM) regression co-efficient and contour plot analysis showed that pH and NaCl concentration could positively affect the production of bioemulsifier (Donio et al. 2018).

Further, there are a number of recent reports which report of successful isolation and characterization of several novel halophilic, halotolerant species and genera from the country (Table 3.1).

3.5.2 Culture Independent

In a study, the prokaryotic diversity from three solar pan crystallizer ponds, which were fed by high salinity water from a spring mixed with water from sea, was examined. Members from the family *Halobacteriaceae* were predominant in the cultivable portion of diversity, and they belonged to various genera like *Haloferax*, *Halorubrum*, *Haloarcula*, *Halobacterium* and *Halogeometricum*. Contrastly *Salinibacter*, *Cytophaga* and *Marinococcus* genera species were isolated from only one sampling site. The culture-independent sampling concluded that genus *Natrinema* members were predominant in the haloarchaeal crystallizer community (Manikandan et al. 2009). The structure and functions of microbial groups from saline desert soil of Rann of Kutch in response to effects of alterations in seasonal climate and physicochemical properties of the soil were analysed. Across the samples, isolates that were affiliated to three phyla, namely, *Proteobacteria*, *Firmicutes* and *Bacteroidetes*, were predominant. The multivariate statistics showed the presence of diverse communities of microorganisms representing a wide diversity of taxa and the archaeal community correlating higher concentration of salt and electrical conductivity. The carbon, nitrogen, methane and sulphur metabolism pathways were displayed in metabolic potentials of all seasons (Narayan et al. 2018). In a similar study, metagenome isolated from a saline desert soil sample of Little Rann of Kutch was analysed. The metagenome showed 633,760 sequences with a total size of 141,307,202 bp and G + C content of 56%. Further analysis of the various microbial groups represented in this particular metagenome revealed the presence of a total of 1802 species which belonged to 43 different phyla. The predominant among them were sequences belonging to *Marinobacter* (48.7%) and *Halobacterium* (4.6%) genus which represented the bacterial and archaeal domains, respectively. The functional analysis of this metagenome revealed 18.2% sequences that were very poorly characterized and 4% gene sequences that encoded for various stress responses (Patel et al. 2015).

A recent study on the microbial groups associated with sediment and water samples from Lonar Lake revealed a difference in their indigenous compositions. The sequences belonging to *Proteobacteria*, *Actinobacteria*, *Firmicutes* and *Cyanobacteria* predominated in the sequencing survey, whereas *Bacteroidetes*, *BD1-5*, *Nitrospirae* and *Verrucomicrobia* were identified in relatively low abundances in the Lonar soda lake habitat. Among the sequences belonging to the phylum *Proteobacteria*, the *γ-Proteobacteria* were predominant within the sediment samples but were poorly detected in the water samples. Abundance of *Proteobacteria* and *Firmicutes* was reported in sediment samples, compared with *Actinobacteria*,

Table 3.1 Recently described novel halophilic, halotolerant species and genera from India (2014–2018)

Name of the bacteria	NaCl tolerance (%)	Optimal NaCl (%)	Source of isolation	References
Allostreptomyces indica	0–10	0	Limestone quarry, Gulbarga	Sahu et al. (2017)
Halobacillus marinus	0–25		Chilaka Lake	Panda et al. (2018)
Bacillus campisalis	0–7	0	Solar saltern, Kanyakumari	Kumar et al. (2015a)
Marinococcus salis			Surajbhari	Vishnuvardhan Reddy et al. (2016)
Tamilnaduibacter salinus	1–20	8–10	Thamaraikulam solar salt pan	Verma et al. (2015)
Salisediminibacterium haloalkalitolerans			Lonar soda lake	Reddy et al. (2015)
Marinobacter nitratireducens	0.5–6	2	Coastal seawater, Visakhapatnam	Vaidya et al. (2015a)
Salinicoccus sediminis	0–25	15	Marine sediment from Sothavilai, Kanyakumari	Kumar et al. (2015b)
Bacillus lacus	0–4		Sambhar lake	Singh et al. (2018)
Salibacterium halotolerans	2–25	12	Khavda	Vishnuvardhan Reddy et al. (2016)
Bacillus pseudoflexus	2–11	5	Compost	Chandna et al. (2016)
Alteribacillus alkaliphilus	0.5–24	3–5	Alkaline soil, Rachalpalli Village, Telangana	Azmatunnisa Begum et al. (2016)
Pontibacillus salipaludis	1–19	2–3	Tuticorin salt pan	Sultanpuram et al. (2016b)
Bacillus maritimus	0–7	5	Kovalam, Kanyakumari	Pal et al. (2017)
Salipaludibacillus aurantiacus	0.5–22	3–5	Narayan Sarovar Lake	Sultanpuram and Mothe (2016)
Bacillus encimensis	0–5		Veraval	Dastager et al. (2015)
Candidatus 'Marispirochaeta associata'	1–8	2–6	Saline habitat, Gujarat	Shivani et al. (2016)
Pelagirhabdus alkalitolerans	0–16	10	Pingaleshwar beach	Sultanpuram et al. (2016a)
Thiorhodococcus fuscus	0.5–7	2–4	Lagoon sediment at Chilika	Lakshmi et al. (2015)
Nesterenkonia cremea	2–11	5–6	Lonar soda lake	Sultanpuram et al. (2017a)

(continued)

Table 3.1 (continued)

Name of the bacteria	NaCl tolerance (%)	Optimal NaCl (%)	Source of isolation	References
Alteromonas pelagimontana		3.5	Southwest Indian Ridge, Indian Ocean	Sinha et al. (2017)
Bacillus lonarensis		4–6	Lonar soda lake	Reddy et al. (2015)
Mangrovibacter phragmitis	0–8	1	Roots of *Phragmites karka*	Behera et al. (2017)
Bacillus caseinilyticus	0–28	6	Lonar soda lake	Vishnuvardhan Reddy et al. (2016)
Desertibacillus haloalkaliphilus		7.5	Little Rann of Kutch	Bhatt et al. (2017)
Zunongwangia mangrovi	0.3–12	2–3	Mangrove (*Avicennia marina*) rhizosphere	RameshKumar et al. (2014)
Bacillus endolithicus	2–10	3	Pebbles from Pingaleshwar beach	Parag et al. (2015)
Alkalispirochaeta cellulosivorans	1–10	1	Gut of a wood-eating cockroach (*Cryptocercus punctulatus*)	Sravanthi et al. (2016)
Rhodovulum aestuarii	0–8	2–4	Brackish water near Somnath	Divyasree et al. (2016)
Bacillus oleivorans	0–8	0–1	Contaminated plate	Azmatunnisa et al. (2015)
Xanthomarina gelatinilytica	0.5–7.5	2–3	Seawater near Kochi	Vaidya et al. (2015b)
Salinimicrobium sediminis	0–9		Deep sea sediment at Vishakhapatnam	Subhash et al. (2014)
Salinicola rhizosphaerae	0.5–30	10–15	Rhizosphere of the mangrove *Avicennia marina* L.	Raju et al. (2016)
Salinicoccus amylolyticus	0–25	8	Saltern of Nari	Srinivas et al. (2016)
Alcanivorax xenomutans	0.5–20	2–5	Shrimp cultivation pond at Ramnad	Rahul et al. (2014)
Zhihengliuella somnathii	0.5–15	6–10	Rhizosphere of *Salicornia brachiata*	Jha et al. (2015)
Rhodovulum mangrovi	0–12	1–9	Mangrove forest sediment	Nupur et al. (2014)
Nitrincola alkalisediminis	0–0.6 M	0.4 M	Lonar soda lake	Joshi et al. (2016)
Halomonas malpeensis			Rhizosphere sand of a costal sand dune plant	Kämpfer et al. (2018)
Bacillus alcaliphilum		2–5	Lonar soda lake	Sultanpuram et al. (2017b)
Streptomyces barkulensis	0–10	8	Estuarine water from Barkul	Ray et al. (2014)

(continued)

Table 3.1 (continued)

Name of the bacteria	NaCl tolerance (%)	Optimal NaCl (%)	Source of isolation	References
Streptomyces chitinovorans	0–18		Estuarine water from Balugaon	Ray et al. (2016)
Photobacterium panuliri	0.5–7	2–4	Eggs of spiny lobster, *Panulirus penicillatus*	Deep et al. (2014)
Bacillus aequororis	0–7		Marine sediment	Singh et al. (2014)
Fictibacillus enclensis	0–12		Marine sediment	Dastager et al. (2014)
Oceanospirillum sanctuarii	2–6	2	Coringa Wildlife Sanctuary	Sidhu et al. (2017)
Thalassorhabdus alkalitolerans	0–11	0.5–2	Marine sediment at Dhuvaran	Sultanpuram and Mothe (2018)

TM7 (a candidate division) and *Cyanobacteria* which were significant in samples from the lake water (Paul et al. 2016).

3.6 Conclusions and Future Directions

While there are many reports on the ecology studies relating to halophiles, still there is a gap of knowledge in understating these versatile microorganisms. There is a large extent of the so-called microbial dark matter (MDW) yet to be discovered in the saline ecosystems. Though some ecological studies using the culture-independent techniques have been successful in predicting the variability and extent of this group, there is still a shortage of studies dealing with the cultivation of some extreme halophiles which are biotechnologically very important.

References

Almeida-Dalmet S, Sikaroodi M, Gillevet PM, Litchfield CD, Baxter BK (2015) Temporal study of the microbial diversity of the north arm of Great Salt Lake, Utah, U.S. Microorganisms 3(3): 310–326

Antón J, Llobet-Brossa E, Rodríguez-Valera F, Amann R (1999) Fluorescence in situ hybridization analysis of the prokaryotic community inhabiting crystallizer ponds. Environ Microbiol 1: 517–523

Antón J, Rosselló-Mora R, Rodríguez-Valera F, Amann R (2000) Extremely halophilic bacteria in crystallizer ponds from solar salterns. Appl Environ Microbiol 66:3052–3057

Azmatunnisa M, Rahul K, Subhash Y, Sasikala C, Ramana C (2015) *Bacillus oleivorans* sp. nov., a diesel oil-degrading and solvent-tolerant bacterium. Int J Syst Evol Microbiol 65(4):1310–1315

Azmatunnisa Begum M, Varshini V, Rahul K, Chandana A, Sasikala C, Ramana C (2016) Description of *Alteribacillus alkaliphilus* sp. nov., reassignment of *Bacillus iranensis* (Bagheri et al. 2012) as *Alteribacillus iranensis* comb. nov. and emended description of the genus *Alteribacillus*. Int J Syst Evol Microbiol 66(11):4772–4778

Baati H, Guermazi S, Amdouni R, Gharsallah N, Sghir A, Ammar E (2008) Prokaryotic diversity of a Tunisian multipond solar saltern. Extremophiles 12:505–518

Ballav S, Kerkar S, Thomas S, Augustine N (2015) Halophilic and halotolerant actinomycetes from a marine saltern of Goa, India producing anti-bacterial metabolites. J Biosci Bioeng 119(3): 323–330

Behera P, Venkata Ramana V, Maharana B, Joseph N, Vaishampayan P, Singh N, Shouche Y, Bhadury P, Mishra S, Raina V, Suar M, Pattnaik A, Rastogi G (2017) *Mangrovibacter phragmitis* sp. nov., an endophyte isolated from the roots of *Phragmites karka*. Int J Syst Evol Microbiol 67(5):1228–1234

Benlloch S, Martínez-Murcia AJ, Rodríguez-Valera F (1995) Sequencing of bacterial and archaeal 16S rRNA genes directly amplified from a hypersaline environment. Syst Appl Microbiol 18: 574–581

Benlloch S, Acinas SG, Martínez-Murcia AJ, Rodríguez-Valera F (1996) Description of prokaryotic biodiversity along the salinity gradient of a multipond saltern by direct PCR amplification of 16S rDNA. Hydrobiologia 329:19–31

Benlloch S, López-López A, Casamayor EO, Øvreås L, Goddard V, Dane FL, Smerdon G, Massana R, Joint I, Thingstad F, Pedrós-Alió C, Rodríguez-Valera F (2002) Prokaryotic genetic diversity throughout the salinity gradient of a coastal solar saltern. Environ Microbiol 4: 349–360

Bhatt H, Azmatunnisa Begum M, Chintalapati S, Chintalapati V, Singh S (2017) *Desertibacillus haloalkaliphilus* gen. nov., sp. nov., isolated from a saline desert. Int J Syst Evol Microbiol 67(11):4435–4442

Bolhuis H, te Poele EM, Rodríguez-Valera F (2004) Isolation and cultivation of Walsby's square archaeon. Environ Microbiol 6:1287–1291

Boujelben I, Martínez-García M, van Pelt J et al (2014) Diversity of cultivable halophilic archaea and bacteria from superficial hypersaline sediments of Tunisian solar salterns. Antonie Van Leeuwenhoek 106:675–692

Brown AD (1990) Microbial water stress physiology. Principles and perspectives. Wiley, Chichester

Burns DG, Camakaris HM, Janssen PH, Dyall-Smith ML (2004a) Cultivation of Walsby's square haloarchaeon. FEMS Microbiol Lett 238:469–473

Burns DG, Camakaris HM, Janssen PH, Dyall-Smith ML (2004b) Combined use of cultivation-dependent and cultivation-independent methods indicates that members of most haloarchaeal groups in an Australian crystallizer pond are cultivable. Appl Environ Microbiol 70:5258–5265

Casamayor EO, Massana R, Benlloch S, Øvreås L, Díez B, Goddard VJ, Gasol JM, Joint I, Rodríguez-Valera F, Pedrós-Alió C (2002) Changes in archaeal, bacterial and eukaryal assemblages along a salinity gradient by comparison of genetic fingerprinting methods in a multipond solar saltern. Environ Microbiol 4:338–348

Caumette P, Matheron R, Raymond N, Relexans J-C (1994) Microbial mats in the hypersaline ponds of Mediterranean salterns (Salins-de-Giraud, France). FEMS Microbiol Ecol 13:273–286

Chandna P, Mayilraj S, Kuhad RC (2016) *Bacillus pseudoflexus* sp. nov., a moderately halophilic bacterium isolated from compost. Ann Microbiol 66:895–905

Corcelli A, Lattanzio VMT, Mascolo G, Babudri F, Oren A, Kates M (2004) Novel sulfonolipid in the extremely halophilic bacterium *Salinibacter ruber*. Appl Environ Microbiol 70:6678–6685

Creel L (2003) Ripple effects: population and coastal regions. Measure Communication, Population Reference Bureau, Washington

da Costa MS, Santos H, Galinski EA, Antranikian G (1998) An overview of the role and diversity of compatible solutes in bacteria and Archaea. Adv Biochem Eng Biotechnol 61:117–153

DasSarma S (2006) Extreme halophiles are models for astrobiology. Microbe 1:120–126

Dastager SG, Mawlankar R, Srinivasan K et al (2014) *Fictibacillus enclensis* sp. nov., isolated from marine sediment. Antonie Van Leeuwenhoek 105:461–469
Dastager S, Mawlankar R, Mual P, Verma A, Krishnamurthi S, Joseph N, Shouche Y (2015) *Bacillus encimensis* sp. nov. isolated from marine sediment. Int J Syst Evol Microbiol 65(5): 1421–1425
Deep K, Poddar A, Das SK (2014) *Photobacterium panuliri* sp. nov., an alkalitolerant marine bacterium isolated from eggs of spiny lobster, *Panulirus penicillatus* from Andaman Sea. Curr Microbiol 69:660–668
Demergasso C, Escudero L, Casamayor EO, Chong G, Balagué V, Pedrós-Alió C (2008) Novelty and spatio-temporal heterogeneity in the bacterial diversity of hypersaline Lake Tebenquiche (Salar de Atacama). Extremophiles 12:491–504
Dennis PP, Shimmin LC (1997) Evolutionary divergence and salinity-mediated selection in halophilic Archaea. Microbiol Mol Biol Rev 61:90–04
Divyasree B, Lakshmi K, Bharti D, Sasikala C, Ramana C (2016) *Rhodovulum aestuarii* sp. nov., isolated from a brackish water body. Int J Syst Evol Microbiol 66(1):165–171
Donio MBS, Ronica FA, Viji VT et al (2013) *Halomonas* sp. BS4, a biosurfactant producing halophilic bacterium isolated from solar salt works in India and their biomedical importance. Springerplus 2:149
Donio MBS, Karthikeyan CS, Michaelbabu M, Uma G, Raja Jeya Sekar R, Citarasu T (2018) *Haererehalobacter* sp. JS1, a bioemulsifier producing halophilic bacterium isolated from Indian solar salt works. J Basic Microbiol 58:597–608
Eder W, Ludwig W, Huber R (1999) Novel 16S rRNA gene sequences retrieved from highly saline brine sediments of Kebrit Deep, Red Sea. Arch Microbiol 172:213–218
Eder W, Jahnke LL, Schmidt M, Huber R (2001) Microbial diversity of the brine-seawater interface of the Kebrit Deep, Red Sea, studied via 16S rRNA gene sequences and cultivation methods. Appl Environ Microbiol 67:3077–3085
Eder W, Schmidt M, Koch M, Garbe-Schönberg D, Huber R (2002) Prokaryotic phylogenetic diversity and corresponding geochemical data of the brine seawater interface of the Shaban Deep, Red Sea. Environ Microbiol 4:758–763
Estrada M, Henriksen P, Gasol JM, Casamayor EO, Pedrós-Alió C (2004) Diversity of planktonic photoautotrophic microorganisms along a salinity gradient as depicted by microscopy, flow cytometry, pigment analysis and DNA-based methods. FEMS Microbiol Ecol 49:281–293
Eugster HP (1980) Lake Magadi, Kenya, and its Pleistocene precursors. In: Nissenbaum A (ed) Hypersaline brines and evaporitic environments. Elsevier, Amsterdam, pp 195–232
Gareeb AP, Setati ME (2009) Assessment of alkaliphilic haloarchaeal diversity in Sua pan evaporator ponds in Botswana. Afr J Biotechnol 8:259–267
Giri BJ, Bano N, Hollibaugh JT (2004) Distribution of RuBisCO genotypes along a redox gradient in Mono Lake, California. Appl Environ Microbiol 70:3443–3448
Grant WD, Tindall BJ (1986) The alkaline saline environment. In: Herbert RA, Codd GA (eds) Microbes in extreme environments. Academic Press, London, pp 25–54
Grant WD, Mwatha AA, Jones BE (1990) Alkaliphiles: ecology, diversity and applications. FEMS Microbiol Rev 75:255–270
Grant S, Grant WD, Jones BE, Kato C, Li L (1999) Novel archaeal phylotypes from an East African alkaline saltern. Extremophiles 3:139–145
Gupta S, Sharma P, Dev K et al (2015) A diverse group of halophilic bacteria exist in Lunsu, a natural salt water body of Himachal Pradesh, India. Springerplus 4:274
Humayoun SB, Bano N, Hollibaugh JT (2003) Depth distribution of microbial diversity in Mono Lake, a meromictic soda lake in California. Appl Environ Microbiol 69:1030–1042
Imhoff JF, Sahl HG, Soliman GSH, Trüper HG (1979) The Wadi Natrun: chemical composition and microbial mass developments in alkaline brines of eutrophic desert lakes. Geomicrobiol J 1: 219–234
Ionescu D, Lipski A, Altendorf K, Oren A (2007) Characterization of the endoevaporitic microbial communities in a hypersaline gypsum crust by fatty acid analysis. Hydrobiologia 576:15–26

Jakher GR, Bhargava SC, Sinha RK (1990) Comparative limnology of Sambhar and Didwana lakes (Rajasthan, NW India). Hydrobiologia 197:245–256

Javor BJ (1983) Planktonic standing crop and nutrients in a saltern ecosystem. Limnol Oceanogr 28: 153–159

Javor B (1989) Hypersaline environments. Microbiology and biogeochemistry. Springer, Berlin

Jha B, Kumar Singh V, Weiss A, Hartmann A, Schmid M (2015) *Zhihengliuella somnathii* sp. nov., a halotolerant actinobacterium from the rhizosphere of a halophyte *Salicornia brachiata*. Int J Syst Evol Microbiol 65:3137–3142

Jose PA, Jebakumar SRD (2012) Phylogenetic diversity of actinomycetes cultured from coastal multipond solar saltern in Tuticorin, India. Aquat Biosyst 8:23

Joshi A, Thite S, Kulkarni G, Dhotre D, Joseph N, Venkata Ramana V, Polkade A, Shouche Y (2016) *Nitrincola alkalisediminis* sp. nov., an alkaliphilic bacterium isolated from an alkaline lake. Int J Syst Evol Microbiol 66(3):1254–1259

Joye SB, Samarkin VA, Orcutt BM, MacDonald IR, Hinrichs K-U, Elvert M, Teske AP, Lloyd KG, Lever MA, Montoya JP, Meile CD (2009) Metabolic variability in seafloor brines revealed by carbon and sulphur dynamics. Nat Geosci 2:349–354

Kämpfer P, Rekha P, Busse H, Arun A, Priyanka P, Glaeser S (2018) *Halomonas malpeensis* sp. nov., isolated from rhizosphere sand of a coastal sand dune plant. Int J Syst Evol Microbiol 68(4):1037–1046

Karthikeyan P, Bhat SG, Chandrasekaran M (2013) Halocin SH10 production by an extreme haloarchaeon *Natrinema* sp. BTSH10 isolated from salt pans of South India. Saudi J Biol Sci 20(2):205–212

Khalilova EA, Kotenko ST, Islammagomedova EA et al (2017) Extremophilic microbial communities of saline soils and their diversity in the regions of the Caspian depression. Arid Ecosyst 7:116–120

Kjeldsen KU, Loy A, Jakobsen TF, Thomsen TR, Wagner M, Ingvorsen K (2006) Diversity of sulfatereducing bacteria from an extreme hypersaline sediment, Great Salt Lake (Utah). FEMS Microbiol Ecol 60:287–298

Kumar S, Karan R, Kapoor S, Singh SP, Khare SK (2012) Screening and isolation of halophilic bacteria producing industrially important enzymes. Braz J Microbiol 43(4):1595–1603

Kumar R, Kaur G, Kumar A, Bala M, Singh N, Kaur N, Kumar N, Mayilraj S (2015a) Taxonomic description and genome sequence of *Bacillus campisalis* sp. nov., a member of the genus *Bacillus* isolated from a solar saltern. Int J Syst Evol Microbiol 65(10):3235–3240

Kumar R, Kaur G, Kumar N, Kumar A, Singh N, Bala M, Kaur N, Mayilraj S (2015b) Taxonomic description and genome sequence of *Salinicoccus sediminis* sp. nov., a halotolerant bacterium isolated from marine sediment. Int J Syst Evol Microbiol 65(11):3794–3799

Lakshmi K, Divyasree B, Sucharita K, Sasikala C, Ramana C (2015) *Thiorhodococcus fuscus* sp. nov., isolated from a lagoon. Int J Syst Evol Microbiol 65(11):3938–3943

Lanyi JK (1974) Salt dependent properties of proteins from extremely halophilic bacteria. Bacteriol Rev 38:272–290

Lentzen G, Schwarz T (2006) Extremolytes: natural compounds from extremophiles for versatile applications. Appl Microbiol Biotechnol 72:623–634

Leuko S, Legat A, Fendrihan S, Stan-Lotter H (2004) Evaluation of the LIVE/DEAD BacLight kit for detection of extremophilic Archaea and visualization of microorganisms in environmental hypersaline samples. Appl Environ Microbiol 70:6884–6886

Leuko S, Goh F, Ibáñez-Peral R, Burns BP, Walker MR, Neilan BA (2008) Lysis efficiency of standard DNA extraction methods for *Halococcus* spp. in an organic rich environment. Extremophiles 12:301–308

Lippert K, Galinski EA (1992) Enzyme stabilization be ectoine-type compatible solutes: protection against heating, freezing and drying. Appl Microbiol Biotechnol 37:61–65

Ma Y, Zhang W, Xue Y, Zhou P, Ventosa A, Grant WD (2004) Bacterial diversity of the inner Mongolian Baer Soda Lake as revealed by 16S rRNA gene sequence analyses. Extremophiles 8:45–51

Malviya N, Yandigeri MS, Yadav AK et al (2014) Isolation and characterization of novel alkali-halophilic actinomycetes from the Chilika brackish water lake, India. Ann Microbiol 64:1829
Manikandan M, Kannan V, Pašić L (2009) Diversity of microorganisms in solar salterns of Tamil Nadu, India. World J Microbiol Biotechnol 25:1007–1017
Maturrano L, Santos F, Rosselló-Mora R, Antón J (2006) Microbial diversity in Maras salterns, a hypersaline environment in the Peruvian Andes. Appl Environ Microbiol 72:3887–3895
McGenity TJ, Gemmell RT, Grant WD, Stan-Lotter H (2000) Origins of halophilic microorganisms in ancient salt deposits. Environ Microbiol 2:243–250
Mesbah NM, Abou-El-Ela SH, Wiegel J (2007) Novel and unexpected prokaryotic diversity in water and sediments of the alkaline, hypersaline lakes of the Wadi An Natrun, Egypt. Microb Ecol 54:598–617
Mouné S, Caumette P, Matheron R, Willison JC (2002) Molecular sequence analysis of prokaryotic diversity in the anoxic sediments underlying cyanobacterial mats of two hypersaline ponds in Mediterranean salterns. FEMS Microbiol Ecol 44:117–130
Narayan A, Patel V, Singh P, Patel A, Jain K, Karthikeyan K, Shah A, Madamwar D (2018) Response of microbial community structure to seasonal fluctuation on soils of Rann of Kachchh, Gujarat, India: representing microbial dynamics and functional potential. Ecol Genet Genomics 6:22–32
Noha MM, Wiegel J (2010) Halophilic thermophiles: a novel group of extremophiles. Saline Syst 9:5
Nupur P, Srinivas T, Takaichi S, Anil Kumar P (2014) *Rhodovulum mangrovi* sp. nov., a phototrophic alphaproteobacterium isolated from a mangrove forest sediment sample. Int J Syst Evol Microbiol 64(9):3168–3173
Ochsenreiter T, Pfeifer F, Schleper C (2002) Diversity of Archaea in hypersaline environments characterized by molecular-phylogenetic and cultivation studies. Extremophiles 6:267–274
Oren A (1999) Bioenergetic aspects of halophilism. Microbiol Mol Biol Rev 63:334–348
Oren A (2002a) Diversity of halophilic microorganisms: environments phylogeny, physiology, and applications. J Ind Microbiol Biotechnol 28:56–63
Oren A (2002b) Halophilic microorganisms and their environments. In: Seckbach J (ed) Cellular origin and life in extreme habitats. Kluwer Academic, Dordrecht
Oren A (2006) Life at high salt concentrations. In: Dworkin M, Falkow S, Rosenberg E, Schleifer K-H, Stackebrandt E (eds) The prokaryotes. A handbook on the biology of bacteria: ecophysiology and biochemistry, vol 2. Springer, New York, pp 263–282
Oren A (2011) Ecology of halophiles. In: Horikoshi K, Antranikian G, Bull AT, Robb FT, Stetter KO (eds) Extremophile handbook, vol 1. Springer, Tokyo, pp 343–361
Oren A, Gurevich P (1995) Dynamics of a bloom of halophilic archaea in the Dead Sea. Hydrobiologia 315:149–158
Oren A, Rodríguez-Valera F (2001) The contribution of *Salinibacter* species to the red coloration of saltern crystallizer ponds. FEMS Microbiol Ecol 36:123–130
Oren A, Shilo M (1981) Bacteriorhodopsin in a bloom of halobacteria in the Dead Sea. Arch Microbiol 130:185–187
Oren A, Sørensen KB, Canfield DE, Teske AP, Ionescu D, Lipski A, Altendorf K (2009) Microbial communities and processes within a hypersaline gypsum crust in a saltern evaporation pond (Eilat, Israel). Hydrobiologia 626:15–26
Pal D, Mathan Kumar R, Kaur N, Kumar N, Kaur G, Singh N, Krishnamurthi S, Mayilraj S (2017) *Bacillus maritimus* sp. nov., a novel member of the genus *Bacillus* isolated from marine sediment. Int J Syst Evol Microbiol 67(1):60–66
Panda AN, Mishra SR, Ray L et al (2018) Taxonomic description and genome sequence of *Halobacillus marinus* sp. nov., a novel strain isolated from Chilika Lake, India. J Microbiol 56: 223–230
Parag B, Sasikala C, Ramana C (2015) *Bacillus endolithicus* sp. nov., isolated from pebbles. Int J Syst Evol Microbiol 65(12):4568–4573

Pašić L, Galán Bartual S, Poklar Ulrih N, Grabnar M, Herzog Velikonja B (2005) Diversity of halophilic archaea in the crystallizers of an Adriatic solar saltern. FEMS Microbiol Ecol 54: 491–498

Pašić L, Poklar Ulrih N, Črnigoj M, Grabnar M, Herzog Velikonja B (2007) Haloarchaeal communities in the crystallizers of two Adriatic solar salterns. Can J Microbiol 53:8–18

Pastor JM, Salvador M, Argandona M et al (2010) Ectoines in cell stress protection: uses and biotechnological production. Biotechnol Adv 28(6):782–801

Patel R, Mevada V, Prajapati D, Dudhagara P, Koringa P, Joshi CG (2015) Metagenomic sequence of saline desert microbiota from wild ass sanctuary, Little Rann of Kutch, Gujarat, India. Genomics Data 3:137–139

Pathak AP, Sardar AG (2012) Isolation and characterization of carotenoid producing Haloarchaea from solar saltern of Mulund, Mumbai, India. Indian J Nat Prod Resour 3(4):483–488

Paul D, Shreyas VK, Mhatre SS, Chowdhury SP, Shetty SA, Marathe NP, Bhute S, Shouche YS (2016) Exploration of microbial diversity and community structure of Lonar Lake: the only hypersaline meteorite crater lake within basalt rock. Front Microbiol 6:1553

Pedrós-Alió C, Calderón-Paz JI, MacLean MH, Medina G, Marassé C, Gasol JM, Guixa-Boixereu N (2000) The microbial food web along salinity gradients. FEMS Microbiol Ecol 32:143–155

Pikuta EV, Hoover RB (2007) Microbial extremophiles at the limits of life. Crit Rev Microbiol 33: 183–209

Prášil O, Bina D, Medová H, Řeháková K, Zapomělová E, Veselá J, Oren A (2009) Emission spectroscopy and kinetic fluorometry studies of phototrophic microbial communities along a salinity gradient in solar saltern evaporation ponds of Eilat, Israel. Aquat Microb Ecol 56: 285–296

Rahul K, Sasikala C, Tushar L, Debadrita R, Ramana C (2014) *Alcanivorax xenomutans* sp. nov., a hydrocarbonoclastic bacterium isolated from a shrimp cultivation pond. Int J Syst Evol Microbiol 64(10):3553–3558

Raju K, Sekar J, Vaiyapuri Ramalingam P (2016) *Salinicola rhizosphaerae* sp. nov., isolated from the rhizosphere of the mangrove *Avicennia marina* L. Int J Syst Evol Microbiol 66(2): 1074–1079

Rameshkumar N, Krishnan R, Lang E, Matsumura Y, Sawabe T, Sawabe T (2014) *Zunongwangia mangrovi* sp. nov., isolated from mangrove (*Avicennia marina*) rhizosphere, and emended description of the genus *Zunongwangia*. Int J Syst Evol Microbiol 64(2):545–550

Rampelotto PH (2010) Resistance of microorganisms to extreme environmental conditions and its contribution to astrobiology. Sustainability 2:1602–1623

Ray L, Mishra S, Panda A, Rastogi G, Pattanaik A, Adhya T, Suar M, Raina V (2014) *Streptomyces barkulensis* sp. nov., isolated from an estuarine lake. Int J Syst Evol Microbiol 64(4):1365–1372

Ray L, Mishra S, Panda A, Das S, Rastogi G, Pattanaik A, Adhya T, Suar M, Raina V (2016) *Streptomyces chitinivorans* sp. nov., a chitinolytic strain isolated from estuarine lake sediment. Int J Syst Evol Microbiol 66(8):3241–3248

Reddy SV, Thirumala M, Farooq M et al (2015) *Bacillus lonarensis* sp. nov., an alkalitolerant bacterium isolated from a soda lake. Arch Microbiol 197:27–34

Rees HC, Grant WD, Jones BE, Heaphy S (2004) Diversity of Kenyan soda lake alkaliphiles assessed by molecular methods. Extremophiles 8:63–71

Roberts MF (2000) Osmoadaptation and osmoregulation in archaea. Front Biosci 5:D796–D812

Robertson CE, Spear JR, Harris JK, Pace NR (2009) Diversity and stratification of archaea in a hypersaline microbial mat. Appl Environ Microbiol 75:1801–1810

Rodriguez-Valera F (1993) Introduction to saline environments. In: Vreeland RH, Hochstein LI (eds) The biology of halophilic bacteria. CRC, Boca Ratom, FL, pp 1–23

Sahay H, Singh S, Kaushik R et al (2011) Characterization of halophilic bacteria from environmental samples from the brackish water of Pulicat Lake, India. Biologia 66:741–747

Sahay H, Mahfooz S, Singh AK et al (2012) Exploration and characterization of agriculturally and industrially important haloalkaliphilic bacteria from environmental samples of hypersaline Sambhar lake, India. World J Microbiol Biotechnol 28:3207–3217

Sahu AK, Quadri SR, Agasar D, Ruwaili JA, Jun-Li W, Dastager SG (2017) *Allostreptomyces* indica sp. nov., isolated from India. J Antibiot 70:1000–1003

Saiz-Jimenez C, Laiz L (2000) Occurrence of halotolerant/halophilic bacterial communities in deteriorated monuments. Int Biodeterior Biodegrad 46:319–326

Saker R, Meklat A, Bouras N et al (2015) Diversity and antagonistic properties of culturable halophilic actinobacteria in soils of two arid regions of septentrional Sahara: M'zab and Zibans. Ann Microbiol 65:2241–2253

Sarafin Y, Donio MBS, Velmurugan S, Michaelbabu M, Citarasu T (2014) *Kocuria marina* BS-15 a biosurfactant producing halophilic bacteria isolated from solar salt works in India. Saudi J Biol Sci 21(6):511–519

Scholten JCM, Joye SB, Hollibaugh JT, Murrell JC (2005) Molecular analysis of the sulfate reducing and archaeal community in a meromictic soda Lake (Mono Lake, California) by targeting 16S rRNA, mcrA, apsA, and dsrAB genes. Microb Ecol 50:29–39

Shen S (2017) Community structure and diversity of culturable moderate halophilic bacteria isolated from Qrhan salt lake on Qinghai-Tibet Plateau. Acta Microbiol Sin 57(4):490–499

Shivani Y, Subhash Y, Sasikala C, Ramana C (2016) Description of '*Candidatus* Marispirochaeta associata' and reclassification of *Spirochaeta bajacaliforniensis*, *Spirochaeta smaragdinae* and *Spirochaeta sinaica* to a new genus *Sediminispirochaeta* gen. nov. as *Sediminispirochaeta bajacaliforniensis* comb. nov., *Sediminispirochaeta smaragdinae* comb. nov. and *Sediminispirochaeta sinaica* comb. nov. Int J Syst Evol Microbiol 66(12):5485–5492

Sidhu C, Thakur S, Sharma G, Tanuku N, Pinnaka A (2017) *Oceanospirillum sanctuarii* sp. nov., isolated from a sediment sample. Int J Syst Evol Microbiol 67(9):3428–3434

Simachew A, Lanzén A, Gessesse A et al (2016) Prokaryotic community diversity along an increasing salt gradient in a soda ash concentration pond. Microb Ecol 71:326–338

Singh NK, Kaur C, Kumar N et al (2014) *Bacillus aequororis* sp. nov., isolated from marine sediment. Curr Microbiol 69:758–762

Singh H, Kaur M, Kaur L, Sharma S, Mishra S, Tanuku N, Pinnaka A (2018) *Bacillus lacus* sp. nov., isolated from a water sample of a salt lake in India. Int J Syst Evol Microbiol 68(3): 801–809

Sinha R, Krishnan K, Singh A, Thomas F, Jain A, John Kurian P (2017) *Alteromonas pelagimontana* sp. nov., a marine exopolysaccharide-producing bacterium isolated from the Southwest Indian ridge. Int J Syst Evol Microbiol 67(10):4032–4038

Sravanthi T, Tushar L, Sasikala C, Ramana C (2016) *Alkalispirochaeta cellulosivorans* gen. nov., sp. nov., a cellulose-hydrolysing, alkaliphilic, halotolerant bacterium isolated from the gut of a wood-eating cockroach (*Cryptocercus punctulatus*), and reclassification of four species of *Spirochaeta* as new combinations within *Alkalispirochaeta* gen. nov. Int J Syst Evol Microbiol 66(4):1612–1619

Srinivas A, Divyasree B, Tushar L, Suresh G, Sasikala C, Ramana C (2016) *Salinicoccus amylolyticus* sp. nov., isolated from a saltern. Int J Syst Evol Microbiol 66(10):3814–3820

Stoeckenius W, Bivin D, McGinnis K (1985) Photoactive pigments in halobacteria from the Gavish sabkha. In: Friedman GM, Krumbein WE (eds) Hypersaline ecosystems. The Gavish sabkha. Springer, Berlin, pp 288–295

Subhash Y, Sasikala C, Ramana C (2014) *Salinimicrobium sediminis* sp. nov., isolated from a deep-sea sediment. Int J Syst Evol Microbiol 64(3):984–988

Sultanpuram V, Mothe T (2016) *Salipaludibacillus aurantiacus* gen. nov., sp. nov. a novel alkali tolerant bacterium, reclassification of *Bacillus agaradhaerens* as *Salipaludibacillus agaradhaerens* comb. nov. and *Bacillus neizhouensis* as *Salipaludibacillus neizhouensis* comb. nov. Int J Syst Evol Microbiol 66(7):2747–2753

Sultanpuram V, Mothe T (2018) *Thalassorhabdus alkalitolerans* gen. nov., sp. nov., a novel *Bacillaceae* member isolated from a marine sediment. Int J Syst Evol Microbiol 68:2969–2979

Sultanpuram V, Mothe T, Chintalapati S, Chintalapati V (2016a) *Pelagirhabdus alkalitolerans* gen. nov., sp. nov., an alkali-tolerant and thermotolerant bacterium isolated from beach sediment, and reclassification of *Amphibacillus fermentum* as *Pelagirhabdus fermentum* comb. nov. Int J Syst Evol Microbiol 66(1):84–90

Sultanpuram V, Mothe T, Mohammed F, Chintalapati S, Chintalapati V (2016b) *Pontibacillus salipaludis* sp. nov., a slightly halophilic bacterium isolated from a salt pan. Int J Syst Evol Microbiol 66(10):3884–3889

Sultanpuram V, Mothe T, Chintalapati S, Chintalapati V (2017a) *Nesterenkonia cremea* sp. nov., a bacterium isolated from a soda lake. Int J Syst Evol Microbiol 67(6):1861–1866

Sultanpuram VR, Mothe T, Chintalapati S et al (2017b) *Bacillus alcaliphilum* sp. nov., a bacterium isolated from a soda lake. Arch Microbiol 199:1303–1309

Thombre RS, Shinde VD, Oke RS, Dhar SK, Shouche YS (2016) Biology and survival of extremely halophilic archaeon *Haloarcula marismortui* RR12 isolated from Mumbai salterns, India in response to salinity stress. Sci Rep 6:25642

Upasani V, Desai S (1990) Sambhar Salt Lake. Chemical composition of the brines and studies on haloalkaliphilic archaebacteria. Arch Microbiol 154:589–593

Vaidya B, Kumar R, Korpole S, Tanuku N, Pinnaka A (2015a) *Marinobacter nitratireducens* sp. nov., a halophilic and lipolytic bacterium isolated from coastal surface sea water. Int J Syst Evol Microbiol 65(7):2056–2063

Vaidya B, Kumar R, Sharma G, Srinivas T, Anil Kumar P (2015b) *Xanthomarina gelatinilytica* gen. nov., sp. nov., isolated from seawater. Int J Syst Evol Microbiol 65(11):3926–3932

Valenzuela-Encinas C, Neria-González I, Alcántara-Hernández RJ, Enríquez-Aragón JA, Estrada-Alvarado I, Hernández-Rodríguez C, Dendooven L, Marsch R (2008) Phylogenetic analysis of the archaeal community in an alkaline-saline soil of the former Lake Texcoco (Mexico). Extremophiles 12:247–254

Vauclare P, Madern D, Girard E, Gabel F, Zaccai G, Franzetti B (2014) New insights into microbial adaptation to extreme saline environments. Bio Web Conf 2:02001

Vavourakis CD, Ghai R, Rodriguez-Valera F, Sorokin DY, Tringe SG, Philip H, Gerard M (2016) Metagenomic insights into the uncultured diversity and physiology of microbes in four Hypersaline Soda Lake brines. Front Microbiol 7:211

Verma A, Mual P, Mayilraj S, Krishnamurthi S (2015) *Tamilnaduibacter salinus* gen. nov., sp. nov., a halotolerant gammaproteobacterium within the family *Alteromonadaceae*, isolated from a salt pan in Tamilnadu, India. Int J Syst Evol Microbiol 65(10):3248–3255

Vishnuvardhan Reddy S, Thirumala M, Farooq M (2015a) *Bacillus caseinilyticus* sp. nov., an alkali- and thermotolerant bacterium isolated from a soda lake. Int J Syst Evol Microbiol 65(8): 2441–2446

Vishnuvardhan Reddy S, Thirumala M, Sasikala C, Venkata Ramana C (2015b) *Salibacterium halotolerans* gen. nov., sp. nov., a bacterium isolated from a salt pan, reclassification of *Bacillus qingdaonensis* as *Salibacterium qingdaonense* comb. nov. and *Bacillus halochares* as *Salibacterium halochares* comb. nov. Int J Syst Evol Microbiol 65(11):4270–4275

Vishnuvardhan Reddy S, Thirumala M, Farooq M, Sasikala C, Venkata Ramana C (2016) *Marinococcus salis* sp., nov., a moderately halophilic bacterium isolated from a salt marsh. Arch Micrbiol 198:1013–1018

Viver T, Cifuentes A, Díaz S, Rodríguez-Valdecantos G, González B, Antón J, Rosselló-Móra R (2015) Diversity of extremely halophilic cultivable prokaryotes in Mediterranean, Atlantic and Pacific solar salterns: evidence that unexplored sites constitute sources of cultivable novelty. Syst Appl Microbiol 38(4):266–275

Wang D, Tang Q (1989) *Natronobacterium* from soda lakes of China. In: Hattori T, Ishida Y, Maruyama Y, Morita RY, Uchida A (eds) Recent advances in microbial ecology. Japan Scientific Societies Press, Tokyo, pp 68–72

Wang C-Y, Ng C-C, Chen T-W, Wu S-J, Shyu Y-T (2007) Microbial diversity analysis of former salterns in southern Taiwan by 16S rRNA-based methods. J Basic Microbiol 7:525–533

Wu QL, Zwart G, Schauer M, Kamst-van Agterveld MP, Hahn MW (2006) Bacterioplankton community composition along a salinity gradient of sixteen high-mountain lakes located on the Tibetan plateau, China. Appl Environ Microbiol 72:5478–5485

Xu Y, Zhou P, Tian X (1999) Characterization of two novel haloalkaliphilic archaea *Natronorubrum bangense* gen. nov., sp. nov. and *Natronorubrum tibetense* gen. nov., sp. nov. Int J Syst Bacteriol 49:261–266

Yadav AN, Verma P, Kumar M et al (2015) Diversity and phylogenetic profiling of niche-specific bacilli from extreme environments of India. Ann Microbiol 65:611–629

Zaccai G, Eisenberg H (1990) Halophilic proteins and the influence of solvent on protein stabilization. Trends Biochem Sci 15:333–337

Zheng MP, Tang JY, Liu JY, Zhang FS (1993) Chinese saline lakes. Hydrobiologia 267:23–36

Chapter 4
Microbial Diversity of Saline Habitats: An Overview of Biotechnological Applications

Shubhrima Ghosh, Sumit Kumar, and Sunil Kumar Khare

Abstract Saline environments across oceans, salt lakes, salt pans, and salterns are home to a myriad variety of microbes, plants, and higher animals. These environments give rise to unusual adaptive properties in microbes, which could be exploited for biotechnological applications. Current advances in technology such as culture-based methods and metagenomics have facilitated the detection and study of distribution of microorganisms in extreme saline habitats. This chapter broadly covers the microbial diversity of saline ecological niches with emphasis on screening and isolation of halophiles, leading to the production of new bioactive molecules such as enzymes, antibiotics, osmolytes, and polymers. Further applications of these microbes in bioremediation, enzyme industry, drug development, biofuels, bioplastics, and compatible solutes have been extensively discussed.

Keywords Microbial diversity · Halophiles · Industrial enzymes · Antibiotics · Bioremediation · Compatible solutes · Bioplastics

4.1 Introduction

The surface of the earth is covered with a number of saline and hypersaline ecosystems. About 70% of the planet's surface is comprised of oceans and seas. Additionally, natural and man-made ecosystems such as salt lakes, soda lakes, salterns, salt marshes, salt mine, solar salt production facilities, and petroleum deposits constitute other saline habitats. The Dead Sea in Israel, the Red Sea, the Great Salt Lake in Utah, Sambhar Lake in India, and lakes of the Atacama Desert in Chile are well-known for their hypersalinity. Owing to their biodiversity and natural resources, these habitats are ecologically, economically, as well as scientifically valuable. The concentration of salt in these environs can vary from minimal

S. Ghosh · S. Kumar · S. K. Khare (✉)
Enzyme and Microbial Biochemistry Laboratory, Department of Chemistry, Indian Institute of Technology Delhi, New Delhi, India
e-mail: skkhare@chemistry.iitd.ac.in

B. Giri, A. Varma (eds.), *Microorganisms in Saline Environments: Strategies and Functions*, Soil Biology 56, https://doi.org/10.1007/978-3-030-18975-4_4

(3.5%) to that of saturation (35%) of total dissolved salts. Depending on their origin, saline and hypersaline aquatic habitats can be broadly divided into athalassohaline and thalassohaline environments. Water in athalassohaline environments is not of marine origin and is mainly formed from salt deposits due to evaporative procedures in inland water bodies. On the other hand, thalassohaline environments are created from the evaporation of seawater due to the inland isolation of seawater bodies and are dominated by sodium chloride and sulfate ions, alike to the seawater. Combined with hypersalinity, often hyper-alkaline waters with high pH such as soda lakes occur due to the presence of carbonates (Le Borgne et al. 2008). Salt is detrimental for microbial growth and survival as it leads to cell death due to osmotic pressure difference between cytoplasm and the surrounding environment. However, such saline habitats are inhabited by organisms which are able to carry out physiological functions in the presence of salts.

Halophiles are a major class of extremophiles, which have the ability to thrive in environments and ecological niches having high salt concentrations. Halophilic microbes occur in all the three domains of archaea, bacteria, and eukarya and are broadly classified into slight, moderate, and extreme, based on their threshold of salt tolerance (Todkar et al. 2012). The most accepted classification of halophiles is proposed by Kushner (1993). He classified them into five groups, namely, non-halophiles (<0.2 M, salt), slight halophiles (0.2–0.5 M salt), moderate halophiles (0.5–2.5 M, salt), borderline extreme halophiles (1.5–4.0 M salt), and extreme halophiles (2.5–5.2 M, salt). Halophiles have been mostly isolated and characterized from habitats like saline water, saline soil, salt lakes, soda lakes, salted foods, and salterns (Ventosa et al. 1998; Oren 2008). They employ two strategies to maintain osmotic balance with the outside salty environment. Some extreme halophiles employ the "salting-in" strategy, where they accumulate KCl equivalent to the concentrations of salt in the surrounding environment. On the other hand, moderate halophiles generally use the "salting-out" strategy where they exclude the salts and accumulate or synthesize compatible solutes such as ectoine, betaine, glycerols, and other polyols (Lanyi 1974; Roberts 2005). Researchers have paid great interest to study adaptive features of these microbes under saline conditions as well as to harness them for various applications.

4.2 Isolation and Screening of Halophiles

Various researchers have carried out isolation and screening of halophilic microorganisms with specific desired properties for industrial as well as biotechnological applications. This interest has led to isolation and further study of these microbes from saline ecosystems across the globe. The efforts have advanced in the direction of isolation of pure colonies in case of culturable microbes and metagenomic analysis for unculturable organisms such as archaea. In most cases, the isolation of halophiles from salty environmental samples is carried out in the presence of high amount of NaCl in the media (100–200 g/L) along with carbon and nitrogen sources.

Mainly halophilic bacteria have been screened for hydrolytic enzymes such as amylase, protease, lipase, cellulase, xylanase, inulinase, pullulanase, and pectinase (Rohban et al. 2009; Kumar et al. 2012). For specific screening of halophilic enzymes, such media are supplemented with their respective substrates. Different serial dilutions of soil or water samples are then added to the media for isolation of single colonies of strains. Many a time, the pH of the soil or water taken for isolation may be different from neutral such as that of alkaline soda lakes. In such cases, care should be taken to adjust the pH of the isolation media to be as close to the native environment of isolation. Similarly, microbes isolated from high-temperature environments such as hydrothermal vents are mostly thermophilic in nature, and the same should be maintained during isolation and screening to avoid loss of the microbes.

However, a major problem encountered during the isolation of pure cultures is the vast diversity of "unculturable" microbes due to the impossibility of replicating the natural living environment or due to the absence of certain necessary signaling molecules. Moreover, many genotypically distinct but phenotypically similar organisms may not be distinguished or identified through conventional biochemical methods (Vartoukian et al. 2010). In fact, only 1% of the bacterial strains on earth can be cultured in vitro (Amann et al. 1995). In case of unculturable bacteria and archaea, metagenomic techniques are employed to study the phylogenetic and metabolic diversity. For this purpose, total DNA is extracted from a soil or water sample and subsequently sequenced using pyrosequencing, shotgun sequencing, or PCR-directed sequencing. Often the 16S rRNA genes are cloned to represent the diversity profile of the sample. After pre-filtration, assembly, gene prediction, data integration of the genomic data, and comparative analysis of the obtained metagenomic datasets with existing databases reveal the unique and shared characteristics of the microbial community present in the environment. A metagenomic study on soil samples from salt marshes found *Halobacterium*, *Balneolaeota*, *Bacteroidetes*, *Rhodothermaeota*, and Nanohaloarchaeota as the dominant prokaryotic groups (Vera-Gargallo and Ventosa 2018). Another report identified salt resistance genes such as glycerol transporters, proton pumps, DNA/RNA helicases, and an endonuclease III, among others, in brines and moderately saline rhizosphere by functional metagenomic approach (Mirete et al. 2015).

However, without the isolation of pure colonies, it is impossible to completely characterize the physiological and pathological properties of microbes. Thus, novel cultivation methods are currently being developed to culture these microbes inherently resistant to artificial culture methods. It is believed that the nutrient-rich media methods of cultivation may favor the growth of fast-growing species at the expense of slow-growing ones. Hence, the use of dilute nutrient media has resulted in the successful cultivation of a number of unculturable strains from marine environments (Connon and Giovannoni 2002). Other methods include extending the incubation time for slow-growing strains (Song et al. 2009), supplementing the media with specific nutrients (Tripp et al. 2008) or addition of signaling molecules like cyclic AMP (cAMP) and acyl homoserine lactones (Bruns et al. 2002). Such combinatorial approaches have led to the discovery and investigation into the microbial diversity of saline environments.

4.3 Applications

Due to their ability to survive in extreme environments, halophilic microbes possess unique properties, which can be exploited in industrial and biotechnological applications requiring extreme conditions such as that of pH, temperature, pressure, and nutritional requirements. Halophilic archaea and bacteria have found wide applications in the production of drugs, antibiotics, enzymes, bioplastics, compatible solutes, biofuels, and fermented foods, some of which are discussed in Table 4.1 and elaborated in the next sections.

4.3.1 Bacteriorhodopsin

Bacteriorhodopsin (BR) is perhaps the most widely known molecule isolated from extremely halophilic archaea, mainly of the genus *Halobacteria*. These are membrane-bound retinal pigments, through which the microorganisms use light energy to generate a proton gradient across the membrane, which is in turn used for ATP production. BRs are integral membrane proteins consisting of hexagonal lattices whose repeating elements are composed of three identical protein chains, each composed of seven transmembrane alpha helices, with one molecule of retinal buried deep within the membrane. They are similar to vertebrate rhodopsins, which are responsible for sensing light in the retina. These proteins are purple in color and absorb green light (wavelength 500–650 nm, with the absorption maxima at 568 nm). Though produced by halophilic microbes, BRs are pretty stable in the absence of salts and are operational within broad pH (1–11) and temperature range (0–45 °C). They form thin layers on polymers, have mutants with enhanced properties, and are robust against environmental changes. All these properties make these proteins find myriad biotechnological and medical applications (Oren 2010; Trivedi et al. 2011). Most research involving BR have been through a remarkable number of patents. Several patents and inventions have exploited the capability of bacteriorhodopsin to convert light energy to chemical energy or electricity. Thus, it can be used to harvest solar energy, which can be consumed directly or stored in batteries (Saeedi et al. 2012). BR has been used in its purple membrane form in photoelectric applications to obtain voltage of around 300 mV through forming covalent cross-links or by coupling to polymers (Hampp et al. 2005). In another patented research, thin layers of BR patches were sandwiched between polymer sheets to be used in image sensing systems such as robotic vehicle navigation, surveillance work, and biosensing (Knopf et al. 2009). Other important optical applications of BR include optical data carriers, data storage in writable and rewritable formats (Harvey and

Table 4.1 Potential biotechnological applications of halophilic microorganisms

Halophilic microorganism(s)	Applications	Reference(s)
	Bioplastics	
Halomonas halophile	PHB	Kucera et al. (2018)
Halomonas smyrnensis	PHB	Tohme et al. (2018)
Natrinema ajinwuensis	PHBHV	Mahansaria et al. (2018)
Haloferax mediterranei ES1	PHBHV	Han et al. (2017)
Halomonas elongata BK AG 18	PHB	Hertadi et al. (2017)
Halomonas boliviensis	PHA	García-Torreiro et al. (2016)
Haloferax sp. MA10	PHB	Tekin et al. (2012)
	Compatible solutes	
Spiribacter salinus	Ectoine	León et al. (2018)
Methanohalophilus portucalensis	Betaine	Lai and Lai (2011)
Halomonas boliviensis	Ectoine	Van-Thuoc et al. (2010)
	Pollutant/Toxicant remediated	
Halorientalis hydrocarbonoclasticus	Hexadecane	Zhao et al. (2017)
Exiguobacterium sp. SH31	Arsenic	Castro-Severyn et al. (2017)
Martelella sp. AD-3	Anthracene	Cui et al. (2014)
Halomonas organivorans, *Arhodomonas aquaeolei*, and *Modicisalibacter tunisiensis*	Phenol	Bonfá et al. (2013)
Ten halophilic Archaea (Haloarchaea) of genus *Haloferax*	Naphthalene, anthracene, phenanthrene, pyrene, and benzoanthracene	Bonfá et al. (2011)
Haloarcula sp., *Haloferax* spp.	Heptadecane	Tapilatu et al. (2010)
Halomonas sp. strain	Vanadate	Ghazvini and Mashkani (2009)
Halophilic bacterial consortium	Phenanthrene	Zhao et al. (2009)
Salinicoccus sp. strain QW6	Toxic oxyanions of tellurium	Amoozegar et al. (2008)
Actinopolyspora sp.	Pentadecane	Al-Mueini et al. (2007)
Nesterenkonia sp. strain MF2	Cr(VI)	Amoozegar et al. (2007)
Marinobacter spp.	Benzene and toluene	Nicholson and Fathepure (2004)

(continued)

Table 4.1 (continued)

Halophilic microorganism(s)	Applications		Reference(s)
Pseudoalteromonas sp.	Tellurium and selenium		Rathgeber et al. (2002)
Haloferax sp. D1227	Aromatic compounds		Emerson et al. (1994)
Halobacterium sp.	n-C10 to n-C30 alkanes		Kulichevskaya (1991)
Halobacterium sp.	Hexadecane, eicosane, and pristane		Bertrand et al. (1990)
	Enzymes	*Potential applications*	
Haloferax volcanii	Alcohol dehydrogenase	Synthesis of enantiopure aromatic alcohols	Alsafadi et al. (2017)
Haloferax sp. HA10	α-Amylases	Detergent stable	Bajpai et al. (2015)
Exiguobacterium sp.		Solvent stable; Bakery industry	Chang et al. (2013)
Marinobacter sp. EMB8		Salt and solvent stable; Maltooligosaccharide (MOS) synthesis	Kumar and Khare (2012)
Saccharopolyspora sp.		Detergent formulations	Chakraborty et al. (2011)
Nesterenkonia sp.		Starch hydrolysis	Shafiei et al. (2010)
Bacillus sp.		Detergent formulations	Kiran and Chandra (2008)
Halomonas meridiana		Alkaline and salt stable	Coronado et al. (2000)
Planococcus rifitoensis strain M2-26	Chitinases	Salt and heat stable	Essghaier et al. (2010)
Thalassobacillus sp. LY18	Cellulases	Salt and solvent stable	Li et al. (2012)
Marinobacter sp. MSI032		Stable at alkaline pH	Shanmughapriya et al. (2010)
Haloarcula marismortui	Esterases	Alkaline and salt stable	Camacho et al. (2009)
Halorubrum lacusprofundi	β-Galactosidase	Salt and solvent stable	Karan et al. (2013)
Haloferax alicantei		Salt stable	Holmes et al. (1997)
Halolactibacillus sp. SK71	Glucoamylase	Solvent stable; bioethanol production	Yu and Li (2014)
Aquisalibacillus elongatus	Laccase	Polyextremotolerant; delignification of sugar beet pulp	Rezaei et al. (2017)

(continued)

Table 4.1 (continued)

Halophilic microorganism(s)	Applications		Reference(s)
Marinobacter sp. EMB5	Lipases	Solvent stable	Hemamalini and Khare (2016)
Marinobacter lipolyticus SM19		Eicosapentaenoic acid (EPA) Production; solvent stable	Pérez et al. (2011)
Salicola strain IC10		Alkaline and salt stable	de Lourdes Moreno et al. (2009)
Bacillus iranensis	Proteases	Alkaline and salt stable	Ghafoori et al. (2016)
Bacillus sp. EMB9		Solvent stable	Sinha and Khare (2014)
Geomicrobium sp. EMB2		Solvent stable; detergent formulations	Karan et al. (2011)
Halobacterium sp.		Alkaline; salt stable; used in fish sauce preparation	Akolkar et al. (2010)
Halobacterium halobium		Solvent stable; used for peptide synthesis	Ryu et al. (1994)
Bacillus sp.	Nucleases	Alkaline, salt, and thermal stable	Onishi et al. (1983)
Micrococcus varians		Alkaline and salt stable	Kamekura and Onishi (1974)
Thermoanaerobacterium saccharolyticum NTOU1	Xylanases	Salt stable	Hung et al. (2011)
Halophilic bacterium CL8		pH, salt, and heat stable	Wejse et al. (2003)

Schmadel 2007), lens, lasers, optical grating, beam reshapers (Kaplan et al. 2010, 2018; Wu et al. 2009), and molecular nanoelectronics (Carmeli et al. 2009).

Bacteriorhodopsin is identical to the rhodopsin found in the retina and thus finds immense applications in the medical field. It has been used for the treatment of blind people suffering from photoreceptor disorders (Chen and Birge 1993). Earlier, silicon retina and retina chip were used as artificial electronic stimulators, which activate retinal ganglion cells, thus helping in the repair of impaired vision. However, due to the possibility of rejection of foreign entities by the body, their applications were limited (Masland 2016). BR films containing an exogenous lipid layer surrounding BR or its mutants situated inside thin polymer layers can be joined to electrodes on glass substrates to obtain different wavelength responses. BR has also been used as a marker for the detection, isolation, and purification of proteins prepared using misaminoacylated tRNAs (Rothschild et al. 2007). These proteins find application in the pharmaceutical industry for the detection and treatment of various diseases. Various applications of BR have been explored through natural homologues or mutants of BR, which possess advanced and diverse properties as compared to the native protein. An example of this is the BR-derived protein light-activated cation channel (LACC) which can be used within cell membranes and subcellular regions and has been patented for application in therapy and drug screening. These variants have been achieved by site-specific mutagenesis using cassette or PCR mutagenesis methods (Boyden and Deisseroth 2007). The intravitreal or subretinal delivery of nucleic acid vectors encoding light-gated cation-selective membrane channels has been used in animal models with degenerated photoreceptors (Pan and Dizhoor 2013). Other patents pertaining to BR involve restoration and modulation of functionality to impaired human nervous system, by sensing and communicating through nerve endings and neural signals (Trivedi et al. 2011).

One potential application of BR could be in the manufacture of microwave-absorbing paints, leading to the development of camouflage military clothing undetectable by infrared detectors and other biomaterials effective in light diffraction and concealment (Armstrong and Warner 2003). Another BR alternative that is being explored is proteorhodopsin, which is easier to express and produce in *Escherichia coli*, making the process efficient as well as economic (Jensen et al. 2010). Thus, there is room for engineering and improvement of BR characteristics for versatile applications.

4.3.2 Bioremediation

A major problem encountered by the saline environments in recent times is that of contamination and pollution, leading to loss of microbial life and biodiversity. With increasing entry of potentially harmful chemicals, plastics, and invasive species, hypersaline environments in oceans, seas, and lagoons are highly impacted by pollution. Effluents from chemical, pesticide, pickling, tanning, pharmaceutical, gas, and petroleum industries release millions of liters of saline to highly saline wastewaters into hypersaline depression and intertidal zones (Lefebvre and Moletta 2006; Lefebvre 2005). Thus, there is an urgent need of microorganisms which are able to thrive and remediate these contaminated hypersaline environments (Le Borgne et al. 2008). About 5% of total effluents in the world are estimated to be saline (Lefebvre 2005). Under such conditions, halotolerant or halophilic microorganisms find immense applications in the degradation of xenobiotic and recalcitrant compounds found in these wastewaters. However, biodegradation capability of microbes decreases with increasing salinity of the soil. For in situ bioremediation, the addition of external microorganisms is not recommended as they might be deleterious to the ecosystem and indigenous species are preferred (Margesin and Schinner 2001). Oil spills have been major contaminants of the oceans since the 1990s and have been successfully mitigated by bioremediation in extreme environments such as the Arctic and the Antarctic (Margesin and Schinner 1999; Delille et al. 1998). A major problem faced by in situ bioremediation is the influence of nutrients. Availability of N and P often act as limiting factors in oil spill bioremediation as reported through laboratory and field studies (Margesin and Schinner 1999). In one study, researchers isolated a non-motile, Gram-positive bacterial strain TM-1 from oil field, capable of growing at 18% NaCl concentration, which was proficient in degrading crude oils and transformed higher n-alkanes (Hao and Lu 2009). The potential of archaeal genus such as *Haloarcula* and *Haloferax* has been explored after their isolation from a shallow crystallizer pond. These strains were capable of 32–95% of heptadecane (0.5 g/l concentration) degradation in 30 days when incubated at 40 °C (Tapilatu et al. 2010). Other classes of microorganisms reported to possess hydrocarbonoclastic activity include *Gammaproteobacteria* such as *Marinobacter hydrocarbonoclasticus* (Gauthier et al. 1992), actinobacteria like *Dietzia maris* (Zvyagintseva et al. 2001), as well as fungi such as *Fusarium lateritium* and *Drechslera* sp. (Obuekwe et al. 2005).

Aerobic treatment of industrial wastewater containing salt concentrations up to 10% have been achieved using aerated percolators or rotating discs. In lower-salt conditions (around 6%), role of bacteria such as *Halobacterium salinarum*, *Halomonas* spp., and *Bacillus* spp. has been implicated (Dincer and Kargi 2001; Hinteregger and Streichsbier 1997; Kubo et al. 2001). In order to remove nitrates from salt water, a membrane anaerobic bioreactor was constructed for denitrification by *Haloferax denitrificans* by Cyplik et al. (2007), while in another instance, *Haloferax mediterranei* was employed for bioremediation of nitrite and nitrate in coastal saline groundwater (Martínez-Espinosa et al. 2007).

Aromatic compounds are another major class of pollutants, generated through natural or anthropogenic activities. This class also includes aromatic acids, such as 4-hydroxybenzoic, cinnamic, p-coumaric, vanillic, and syringic acids. These acids are widely present in plant exudates, natural polymers, agrochemicals, pesticides, pharmaceuticals, and varnishes (Le Borgne et al. 2008). Some of the earliest studies on biodegradation of aromatic acids was accomplished through the bacterium *Halomonas halodurans* (earlier named as *Pseudomonas halodurans*), which was able to cleave aromatic compounds at the *ortho* position at high NaCl concentration of around 2.5 M (Rosenberg 1983; Hebert and Vreeland 1987). This bacterium is also reported to degrade toluene (Emerson and Breznak 1997).

PCBs (polychlorinated biphenyls) are another class of chlorinated hazardous materials which were used for a long time in electronic and electrical parts, thermal insulation, paints, adhesives, and plastics. However, their use was banned in 1979, as they are difficult to degrade and can remain for long periods of time in soil, air, and water. Over the course of time, they accumulate in leaves and other plant parts. When PCBs are taken up by small organisms and fish, their concentrations undergo biomagnification along the higher levels of the food chain. PCBs are known to cause cancer as well as affect the immune system, endocrine system, and nervous system in humans. Thus, there have been numerous efforts to screen microbes which are able to use these recalcitrant compounds as sole carbon source and degrade them. A halophilic bacterium designated as B2.6 was isolated from deep-sea sediments and was found to degrade PCBs to 90% of the initial concentration within 72 h (Zhao and Zhang 2010). Other aromatics include benzene, toluene, ethylbenzene, and xylenes (BTEX), which are released from oil spills or oilfield wastewater and easily contaminate soil and aquifers because of their partial solubility in water. Not many halophilic microbes are reported to degrade BTEX. One study investigated the biodegradation of BTEX by enrichment culture obtained from salt flat soil using benzene as a sole carbon source. Around 33% of benzene was mineralized to CO_2 in 4 weeks, while <1% of naphthalene was converted in the same time in high salt concentrations. Microbial community analysis using denaturing gradient gel electrophoresis (DGGE) revealed the presence of strains closely resembling *Halobacillus salinus*, *Bacillus simplex*, and *Cytophagales* (Nicholson and Fathepure 2005). Some of the halophiles implicated in bioremediation of toxicants are summarized in Table 4.1. Thus, halophilic strains have found promising application in bioremediation of recalcitrant compounds.

4.3.3 Industrial Enzymes

Enzymes from halophilic microorganisms are characterized by the unique property of stability in hypersaline environs, a condition where most proteins may undergo denaturation, aggregation, or precipitation. They have a high resistance to denaturation and have the ability to catalyze in low-water activity (Madern et al. 2000; Karan et al. 2012b). The adaptation of halophilic enzymes can be explained by the presence of acidic over basic amino acid residues, large number of protein charges, and increased hydrophobicity (Bayley et al. 1978; Kumar et al. 2016b). Halophilic proteins generally show low Lys content, a decrease in aliphatic residues, and increase in hydrophobic residues (Madern et al. 1995). At low-salt concentrations, the dissolved ions shield electrostatic interactions while increasing hydrophobicity at high-salt concentrations. Many ion pairs are also thought to stabilize the active site of the halophilic enzymes and help in interactions among subunits. The halophilic proteome is characterized by strong negativity and a mean isoelectric point (pI) of around 5, whereas non-halophilic proteomes generally exhibit an average pI close to neutrality (DasSarma and DasSarma 2015). A key feature maintained by halophilic enzymes is the high solubility and activity in low-water activity conditions, as is evident from their structural and biochemical characterization (Kennedy et al. 2001). Under low-water conditions, hydration shell is maintained by hydrogen bonding between negatively charged side chains of the amino acids and water molecules (Britton et al. 2006). The low water activity also mimics the aqueous-organic solvent mixtures, which makes most halophilic enzymes active in organic media, thus increasing their applicability in industrial biocatalysis (Karan et al. 2012a; de Lourdes Moreno et al. 2013).

In some halophilic enzymes such as malate dehydrogenase from *Haloarcula marismortui* (Mevarech et al. 1977), glucose dehydrogenase from *Haloferax mediterranei* (Rajagopalan and Altekar 1994), 1-phosphofructokinase from *Haloarcula vallismortis* (Rangaswamy and Altekar 1994), and dihydrolipoamide dehydrogenase from *Haloferax volcanii* (Jolley et al. 1997), increase in salt concentrations leads to decrease in activity, whereas in other instances such as β-glucosidase from *Haloferax alicantei*, activity increase was observed with increase in NaCl concentration up to 4 M (Holmes et al. 1997).

Halophilic enzymes have been engaged in immense biotechnological applications owing to their ability to flourish in hypersaline, organic media, which can ideally exclude non-halophilic organisms, thereby increasing the economic benefits of the process. Halophilic cellulolytic enzymes have been used for the biotransformation of plant and animal polymers into useful products such as biofuels. Enzymes from *Haloarcula*, isolated from salt mines, have been found to show considerable cellulolytic activity which was further enhanced by immobilization in sodium alginate (Ogan et al. 2012). Another unique endo-cellulase isolated from the alkaliphilic and halophilic *Bacillus* sp. BG-CS10, exhibiting glucomannanase activity, was found to have enhanced pH and temperature stability in the presence of NaCl (2.5 M). This enzyme acts on β-1,4-linkages and could efficiently hydrolyze β-D-glucan from barley, methyl cellulose, and konjac glucomannan (Zhang et al. 2012). Functional screening of a soil metagenomic library on Luria-Betani (LB) plates containing 0.5% carboxymethyl cellulose revealed the existence of a novel cellulase Cel5R with high halotolerance and thermal stability. In fact, a 24% increase in activity was observed in the presence of 1 M NaCl, along with high stability in LiCl and KCl. The crystal structure of Cel5R obtained at 2.2 Å resolution revealed the presence of a number of acidic amino acid residues with (Asp +Glu = 16.7%) on the protein surface which plausibly contribute to the salt stability of the enzyme (Garg et al. 2016).

Due to their stability in hypersaline, alkaliphilic, solvent-rich, and high-temperature environments, halophilic proteases find wide applications in detergent, beer, food, leather tanning, and pharmaceutical industries. They can be broadly distinguished into acidic, neutral, and basic proteases based on their pH of optimal activity. Many such proteases have been reported from several bacterial genera such as *Bacillus* sp. (Sinha and Khare 2013), *Virgibacillus* sp. (Sinha and Khare 2012), *Halobacillus* sp. (Santos et al. 2013), *Halobacterium* sp. (Akolkar et al. 2008), *Geomicrobium* sp. (Karan and Khare 2011), and *Chromohalobacter* sp. (Vidyasagar et al. 2009). Salt-fermented foods such as fish *budu* served as a source for halophilic bacteria such as *Halobacterium* sp. strain LBU50301 from which protease was produced (Chuprom et al. 2016). Salt-stable proteases have also been identified and purified from archaeal strains like *Halogranum rubrum*, *Natrialba magadii*, and *Natronolimnobius innermongolicus* (Gao et al. 2017; Mellado et al. 2005; Selim et al. 2014). Many of these proteases were found to be activated by metal ions such as Ca^{2+} ions (Selim et al. 2014; Pathak and Sardar 2014).

Another important class of industrial enzymes is α-amylase, which catalyzes the hydrolysis of α-1,4-glycosidic linkages in starch and related polysaccharides. They find wide applications in paper, textile, chemical, pharmaceutical, and beverage industries (Kumar et al. 2016a). The search for halophilic α-amylases comes from the need for stable enzymes and optimal activities in saline and low-water conditions. Several such α-amylases have been reported from *Haloferax mediterranei* (Pérez-Pomares et al. 2003), *Halomonas meridiana* (Coronado et al. 2000), *Natronococcus amylolyticus* (Kanai et al. 1995), *Haloarcula* sp. (Fukushima et al.

2005), and *Marinobacter* sp. (Kumar and Khare 2015). Their molecular weight varies from around 31–100 kDa (Li and Yu 2012; Shafiei et al. 2010). The need for salt for stability and activity of these enzymes makes them difficult to clone and express in non-halophilic hosts such as *E. coli* which may lead to misfolding of the protein due to lack of NaCl/KCl. Thus, strains of *Haloferax volcanii* and *Halomonas elongata* have been used as expression hosts for halophilic α-amylase by many researchers (Kumar et al. 2016a). The major application of halophilic amylases is in the saccharification of starch into maltooligosaccharides, which find application in food and pharmaceuticals. Several halophilic genera, such as *Chromohalobacter* sp. TVSP 101 (Prakash et al. 2009), *Micrococcus halobius* (Onishi and Sonoda 1979), *Marinobacter* sp. EMB8 (Kumar and Khare 2015), and *Amphibacillus* sp. NM-Ra2 (Mesbah and Wiegel 2014), have been explored for their potential to yield maltooligosaccharides by hydrolysis. Polyextremophilic α-amylases such as that from *Aspergillus gracilis* have been found to be suitable for bioremediation of waste and effluents containing starch and organic solvents (Ali et al. 2014).

Lipases from halophilic microorganisms find applications in the production of fatty acids, in baking, in detergents, as biocatalysts in processing diesel (Harding et al. 2008), and for the conversion of vegetable oil into fuel (Gupta et al. 2004). A halophilic lipase LipBl, isolated from *Marinobacter lipolyticus* SM19 and expressed in *E. coli*, was used for the production of eicosapentaenoic acid from fish oil (Pérez et al. 2011). Halophilic strains of *Halomonas* isolated from Spanish and Turkish saltworks were also found to produce salt-tolerant lipolytic enzymes (Gutiérrez-Arnillas et al. 2016). One potential application of halophilic lipases has been in the production of marine-algae-based biofuel systems (Schreck and Grunden 2014). Some studies have also reported the isolation of active aggregates of halophilic lipase from *Marinobacter* sp. EMB5 (Hemamalini and Khare 2016, 2018).

Some representative examples of halophilic enzymes and their potential applications are tabulated in Table 4.1. In consequence, enzymes from halophilic organisms have found versatile industrial and biotechnological applications due to their unique polyextremophilic nature, which have been enhanced by current molecular advances.

4.3.4 Antibiotics and Drugs

With increasing incidences of antibiotic resistance in human pathogens, there is a constant and cyclic need of exploration and discovery of antibiotics from newer ecological niches. In recent times, halophiles, primarily of the actinomycetes class, have been explored as a major source of antibiotics, especially antibacterial compounds. Halophiles mainly of the family Halobacteriaceae and genus *Halomonas* are also reported to be resistant to antibiotics such as kanamycin, neomycin, and tetracycline (Fernandez-Castillo et al. 1992). Many halophilic strains contain a number of plasmids, which are thought to confer antibacterial properties as well as antibiotic resistance. Due to their survival in extreme habitats, these organisms produce a number of secondary metabolites many of which are unique bioactive compounds belonging to a wide variety of chemical classes such as terpenes, polyketides, peptides, and alkaloids with varying structures and pharmacological activities (Velho-Pereira and Furtado 2012). Often, changes in the salty environmental conditions can lead to adaptations in the microbes which can lead to the production of different metabolites than the ones generally reported. Isolation and screening of halophiles from coastal areas of India have led to the discovery of antibiotic compounds, which were found to be effective against several pathogenic strains such as *Staphylococcus aureus*, *Klebsiella pneumoniae*, *E. coli* and *Bacillus subtilis*, and fungi like *Aspergillus niger*, *Penicillium* sp. and *Candida albicans* (Todkar et al. 2012). Culture broths of halophilic bacteria isolated from sponges, corals, and bivalves of Indian origin were tested for their antibiotic potential and were found to be effective against pathogenic bacteria such as *Staphylococcus citreus*, *Serratia marcescens*, *E. coli*, *P. vulgaris*, *Salmonella typhi*, and *Aerobacter aerogenes* (Velho-Pereira and Furtado 2012). A study on *B. subtilis* subsp. *spizizenii* found bioactive components with antibiotic properties against various bacterial and fungal strains such as *Proteus vulgaris*, *B. subtilis*, *A. niger*, *Rhizopus oligosporus*, and *Penicillium chrysogenum*. The bioactive compounds were separated using three-phase partitioning and further analyzed through GC-MS, FTIR, and ^{13}C- NMR (Yadav et al. 2017). Many mixed cultures of halophiles have also been explored for antibiotic production. Instead of isolating individual cultures, a mixed culture comprising of bacteria and archaea was obtained from the solar saltern of Manaure, La Guajira, in the north of Colombia. They were grown in seven different culture media, and 40 different extracts were obtained from them by extraction with ethyl acetate. These extracts were found to be potent in killing multidrug-resistant pathogens such as *K. pneumoniae*, *Enterococcus faecium*, *S. aureus*, and *B. subtilis*. Additionally, the extracts were found to be cytotoxic against human lung cancer cell lines, indicating their high bioactivity (Conde-Martínez et al. 2017). Secondary metabolites with unique scaffold structure have been discovered in marine myxobacteria. These are swarming bacteria and form fruiting bodies in high NaCl concentrations, which are implicated in cell signaling. Myxobacteria are reported to produce a number of antibiotics as well as gephyronic acid, an inhibitor of eukaryotic protein synthesis and a potential anticancer agent (Sasse et al. 1995).

Halophilic and halotolerant actinomycetes have been a valuable resource for the production of bioactive compounds with antimicrobial, antineoplastic, antiviral, antimitotic, and neurotoxic properties (Cragg and Newman 2005). Diverse groups of antibiotics have been reported by several strains of actinomycetes such as chandrananimycin from *Actinomadura* sp. (Maskey et al. 2003), diazepinomicin from *Micromonospora* sp. (Charan et al. 2004), salinosporamides from *Salinispora tropica* (Williams et al. 2005), and saliniquinones produced from *Salinispora arenicola* (Murphy et al. 2010). Strains of the *Actinopolyspora* genus isolated from salty soil and marine sediments are also potent antimicrobial producers (Yoshida et al. 1991; Kokare et al. 2004). The largest number of antibiotics has been detected in the genus *Streptomyces*. One halophilic and alkaliphilic strain, *Streptomyces sannanensis* strain RJT-1, has been reported as a potent antibiotic producer against Gram-positive bacteria. Marine *Streptomyces psammoticus* BT-408 have been found to produce a polyketide antibacterial compound effective against methicillin-resistant *S. aureus*, while antifungal molecules have been reported from *Streptomyces aburaviensis* 1DA-28 (Sujatha et al. 2005; Raytapadar and Paul 2001). New antibiotics such as chinikomycin and lajollamycin have also been isolated from marine *Streptomyces* strains (Li et al. 2005; Manam et al. 2005).

Other classes of drugs such as anticancer molecules have similarly been investigated in case of halophilic microbes. Lomaiviticins isolated from marine *Micromonospora lomaivitiensis* strain LL-37I366 have dimeric diazobenzofluorene glycoside structures and possess DNA-damaging activity against cancer cell lines (He et al. 2001). Researchers have also isolated thiocoraline, which is a thiodepsipeptide antitumor compound, produced by the actinomycetes, *Micromonospora* sp., and then further expressed it heterologously into *Streptomyces albus* and *Streptomyces lividans* (Lombó et al. 2006). Marine microbes and other halophiles are a powerful repository of many bioactive compounds, a miniscule of which have been discovered. Therefore, efforts to uncover this abundant resource would be a boon for the pharmaceutical industry.

4.3.5 Compatible Solutes

In order to counteract external osmotic pressure, and for adapting to moderate to high-salt environments, halophilic microbes use a variety of organic and inorganic solutes. In fact, these compatible solutes are thought to be identifying markers of halophilic archaea, bacteria, and eukarya. Inorganic cations, such as K^+ or Na^+, are involved in cellular osmotic balance and response. Additionally, these microorganisms possess a number of organic solutes (zwitterionic, anionic, and cationic) whose biosynthesis and mechanisms of actions are active areas of current research. They can be synthesized or accumulated from the surrounding medium and act by maintaining proper concentration of electrolytes, turgor pressure, and cell volume. The different types of organic molecules include carbohydrates, amino acids, peptides, polyols, ectoines, betaines, and their derivatives, which are evaluated through gas chromatography (GC) or nuclear magnetic resonance (NMR) (Galinski 1995). They help the microbes withstand salt stress, heat, desiccation, as well as extreme cold. For example, *K. pneumoniae* was found to synthesize betaine de novo in LB medium with salinity range of 0.17–3.5 M NaCl, while ectoine was only found above 2 M NaCl concentration (Bumadian et al. 2014). Moderately halophilic Gram-negative bacterial strains such as *Volcaniella eurihalina*, *Deleya salina*, and *Marinomonas* sp. and Gram-positive strains like *Salinicoccus hispanicus* and *Salinicoccus roseus* have been found to accumulate ectoine, in complex media containing yeast extract. Furthermore, Gram-positive strains produced hydroxyectoine, proline, alanine, glucose, glutamate, glutamine, and trehalose as minor components (Del Moral et al. 1994).

These osmolytes are also reported to increase the stability of proteins, acting as chaperones in cells, the property being exploited for many biotechnological applications (Roberts 2005). Compatible solutes form strong water structure and are probably excluded from the hydration shell of proteins, enabling them to act as functional stabilizers of proteins (Galinski 1993). For example, in one study, use of osmolytes resulted in folded, functional proteins, expressed in the periplasm of *E. coli* (Barth et al. 2000). Similarly, in vitro osmolytes such as betaine, ectoine, citrulline, and trehalose have been shown to inhibit amyloid formation in insulin (Arora et al. 2004). Betaine, ectoine, hydroxyectoine, and homoectoine also find applications as PCR enhancers of GC-rich DNA templates, by lowering the melting point, T_m (Schnoor et al. 2004; Galinski 1995). Other biotechnological applications of osmolytes include use as cryoprotectant for the preservation of bacteria in lieu of serum albumin and trehalose/dextran (Cleland et al. 2004) and as a stabilizer of biomolecules in pharmaceutical and cosmeceutical industries (Buenger and Driller 2004). An interesting application of compatible solutes is the generation of transgenic organisms with the ability to withstand salt stress. Transgenic *Arabidopsis thaliana* transformed with a choline oxidase gene (which is needed to synthesize betaine), and modified tobacco plants with *E. coli* betA and betB genes have been developed which possess increased salt, cold, and heat tolerance (Hayashi et al. 1998; Holmström et al. 2000).

Various researchers have developed processes for effective production of osmolytes from halophilic and halotolerant microbes. A novel process developed by the German biotechnology company bitop exploits the concept of "bacterial milking," where the bacteria are firstly grown to high cell density in high NaCl concentrations, with subsequent transfer to low osmolarity medium, where the osmotic shock induces the cells to rapidly excrete the excess osmolytes. After removal of the solutes, the cells are again subjected to high-salt concentrations to resynthesize the osmolytes. This process could be repeated several times after a defined generation time. Through this technology the researchers were able to produce around 155 mg ectoine/g cell dry wt/cycle, obtained after nine cycles of reuse (Sauer and Galinski 1998; Shivanand and Mugeraya 2011). In fed-batch cultures using the moderate halophile *Halomonas boliviensis*, 6.3 g/l/day of ectoine production was obtained (Van-Thuoc et al. 2010). While achieving recombinant production in non-halophilic hosts such as *E. coli* DH5α, co-expression with feedback-resistant aspartate kinase from *Corynebacterium glutamicum* reduced metabolic bottlenecks (Bestvater et al. 2008). Osmolytes find remarkable biotechnological applications in stabilization of proteins and as PCR enhancers. For proper application, it is pertinent to understand their effects on various systems. Also, genetic engineering pathways to produce solutes in desired microbes are expected to be optimized and commercialized in the future.

4.3.6 Biofuels

The potential of halophilic microorganisms in producing biofuels has been evaluated in terms of oil production or enzymatic biomass hydrolysis activity. The major environmental problem associated with the production of biofuels is the use of large quantities of fresh water especially for enzymatic reactions. This drawback can be easily addressed by the use of halophilic organisms and their enzymes which are able to utilize seawater or residual brine for biofuel production. One strain of halophilic archaea *Haloarcula* 2TK2 isolated from Tuzkoy Salt Mine, Turkey, was found to grow on 2.5 M NaCl and exhibited high cellulolytic activity in free as well as sodium alginate immobilized forms (Ogan et al. 2012). The cellulolytic potential of *Haloarcula* sp. LLSG7, isolated from the saline soil of Yuncheng Salt Lake, China, was explored in the presence of 25% NaCl. The cellulase obtained from this strain was remarkably stable over a broad range of temperature (40–80 °C), pH (7.0–11.0), and nonpolar solvents and was applied to hydrolyze alkali-pretreated rice straw, which was further used for bioethanol production (Li and Yu 2013). In addition to archaea, halophilic bacterial strains such as *Halolactibacillus* sp. SK71, isolated from saline soil, was found to produce glucoamylase which was applied for hydrolysis of raw cornstarch and subsequent bioethanol production using *Saccharomyces cerevisiae* (Yu and Li 2014). Mostly unicellular halophilic algae have been explored for high productivity of lipids. When *Dunaliella* cell material undergoes catalysis at 200–240 °C, the exothermic process produces an oil-like substance soluble in benzene. Up to 75% of the cell material of algal slurry can be converted to extractable oil (Goldman et al. 1981; Ginzburg 1993). Recent investigations into the microalgae, *Aphanothece halophytica*, found the production of lipid content of around 29% by dry weight when grown in seaweed extract and nitrogen-phosphorus-potassium (NPK) medium (Miriam et al. 2017). The use of halophilic archaea, algae, and bacteria for the production of biofuels is an effective solution to the increasing energy crunch while being easy on the environment. Efforts are underway to test these biofuels to run eco-friendly vehicles and as a power source for industries.

4.3.7 Bioplastics

Halophilic microbes have similarly been explored for their potential to produce bioplastics such as polyhydroxyalkanoates (PHA), which are biodegradable in nature. The high-salt content of the growth/production media diminishes the need for sterilization, thus markedly reducing the production costs. Moreover, at lower-salt concentrations, cells are prone to lysis, efficiently releasing the plastics and making downstream processing cost-effective (Don et al. 2006; Fernandez-Castillo et al. 1986). Anaerobic acidogenesis of complex waste by moderately halophilic bacterium *Halomonas boliviensis* resulted in the production of a combination of volatile fatty acids which were then utilized as carbon and energy source for PHA yield of 13 g/L (García-Torreiro et al. 2016). In another study, a co-culture of engineered cyanobacteria *Synechococcus elongates* along with the halophilic bacterium *Halomonas boliviensis* (a natural producer of PHB) resulted in the accumulation of 28.3 mg polyhydroxybutyrate (PHB)/L/d. The use of cyanobacteria helped in the production of sucrose from photosynthetically fixed carbon, thus reducing the use of plant-based carbohydrate feedstocks (Weiss et al. 2017). Other strains of this genus such as *Halomonas* sp. KM-1 have been used to produce PHB utilizing waste glycerol (Kawata and Aiba 2010), while *Halomonas* sp. SK5 was found to produce P (3HB) homopolymer as well as P(3HB-co-3HV) copolymer from carbon sources such fructose, glucose, sucrose, and sodium acetate (Rathi et al. 2013). Hertadi et al. (2017) used *Halomonas elongata* BK AG 18 to produce PHB using modified High Medium containing glucose with obtained yield of 21.36% of the dry bacterial cell weight. In related studies, the potential of archaea isolated from seawater-based salterns, such as *Haloferax* sp. MA10, has been investigated for PHB production. This strain was found to have a high yield of 6.53% of the dry cell weight when grown in PHB medium supplemented with acetate (Tekin et al. 2012). The application of microbes for bioplastic production can help in reducing the use of petrochemicals for plastic production. Additionally, being biodegradable in nature, bioplastics provide an environment-friendly alternative to conventional plastics.

4.3.8 Halophiles in Fermented Foods

A large variety of traditional fermented food products such as "jeotgal," a Korean-fermented seafood; "kimchi," Korean salted-fermented vegetables; the Japanese salted puffer fish ovaries "fugunoko nukazuke"; and Thai fish sauce, "nam pla," are prepared with large amounts of salt, sometimes as high as 25–30% NaCl (Oren 2010). Very little information is available about the kind of microorganisms that populate the food products during their preparation and maturation and the role they play in the characteristics of the product. Halophilic lactic acid bacteria such as *Lactobacillus plantarum* is commonly found in fermented vegetables such as sauerkraut (pickled cabbage) and pickled cucumbers, which can grow in NaCl concentrations of 2.25–2.5% (Margesin and Schinner 2001). The high concentrations of salt found in "nam pla" led to the isolation of the archaeal isolate such as *Halobacterium salinarum*, *Halococcus thailandensis*, and *Natrinema gari* (Thongthai et al. 1992; Namwong et al. 2007; Tapingkae et al. 2008). Another strain, *Halalkalicoccus jeotgali*, was isolated from shrimp jeotgal (Roh et al. 2007). It is predicted that archaea might play a role in the fermentation process as well as the development of aroma, texture, and other characteristics of the food. At relatively lower concentrations of salt in foods, bacterial strains of *Halobacillus*, *Halomonas*, *Tetragenococcus*, and *Lentibacillus* have been discovered, and the involvement of bacterial proteases in the fermentation process is predicted (Lopetcharat et al. 2001; Sinsuwan et al. 2008). *Tetragenococcus halophila* is also reported to be involved in the brining stage during the fermentation of Indonesian soy sauce (kecap) (Röling et al. 1999). Investigations using difference gel electrophoresis (DIGE) into the microbial diversity of kimchi have revealed the presence of diverse communities of archaea, lactic acid bacteria, and yeasts (Chang et al. 2008). The production of many of these indigenous foods has been carried on for centuries with the traditional knowledge of resident microflora. As is evident, these delectable, food preparations would have been impossible without the involvement of halophilic microbes, which give unique flavor, texture, and characteristics to these products.

Other potential biotechnological applications of halophilic microbes include the production of β-carotene, glycerol, and liposomes. Though a number of applications of halophilic organisms have been investigated by researchers, few have been commercially viable, in the current scenario. A new cost-effective technique of production of chemicals from microbes has been explored through non-sterile fermentation. In this technique, a certain stress condition (high-salt concentration in this case) is applied which results in the growth of only those strains which are able to withstand the stress. Thus, the cost of media sterilization is markedly reduced, along with reduced maintenance requirements, simpler reactor designs, simple operations, and reduced environmental impact. Non-sterile fermentation can be achieved by various strategies such as the use of antimicrobial agents; selective nutrition; altering the temperature, salinity, and pH: immobilizing the cells; use of mixed cultures and metabolic engineering (Chen and Wan 2017).

4.4 Conclusions

Saline and hypersaline environments comprise the largest ecosystems on earth with the resident microbes coming under slight, moderate, and extreme halophiles. Both culture-based and metagenomics approaches have been employed to investigate the diversity of these habitats. These habitats are home to diverse species of microorganisms with unique properties which could be exploited for biotechnological applications such as the production of compatible solutes, drugs, enzymes, bioplastics, biofuels, traditional fermented foods and in bioremediation. Genomic- and proteomic-based approaches have opened new possibilities such as the availability of genetic systems susceptible to modulation and desired modifications. It is hoped that the remarkable diversity of halophilic microorganisms would be exploited for more novel, versatile, and viable biotechnological applications in the near future.

Acknowledgements SG is grateful to the Indian Institute of Technology Delhi and MHRD, Government of India for Senior Research Fellowship. Author SK is grateful to the Council of Scientific and Industrial Research (CSIR, Govt. of India) for Senior Research Associateship (Scientists' Pool Scheme).

References

Akolkar AV, Deshpande GM, Raval KN, Durai D, Nerurkar AS, Desai AJ (2008) Organic solvent tolerance of *Halobacterium* sp. SP1 (1) and its extracellular protease. J Basic Microbiol 48 (5):421–425

Akolkar AV, Durai D, Desai AJ (2010) *Halobacterium* sp. SP1 (1) as a starter culture for accelerating fish sauce fermentation. J Appl Microbiol 109(1):44–53

Ali I, Akbar A, Yanwisetpakdee B, Prasongsuk S, Lotrakul P, Punnapayak H (2014) Purification, characterization, and potential of saline waste water remediation of a polyextremophilic α-amylase from an obligate halophilic *Aspergillus gracilis*. Biomed Res Int 2014:1–7

Al-Mueini R, Al-Dalali M, Al-Amri IS, Patzelt H (2007) Hydrocarbon degradation at high salinity by a novel extremely halophilic actinomycete. Environ Chem 4(1):5–7

Alsafadi D, Alsalman S, Paradisi F (2017) Extreme halophilic alcohol dehydrogenase mediated highly efficient syntheses of enantiopure aromatic alcohols. Org Biomol Chem 15 (43):9169–9175

Amann RI, Ludwig W, Schleifer KH (1995) Phylogenetic identification and in situ detection of individual microbial cells without cultivation. Microbiol Rev 59(1):143–169

Amoozegar MA, Ghasemi A, Razavi MR, Naddaf S (2007) Evaluation of hexavalent chromium reduction by chromate-resistant moderately halophile, *Nesterenkonia* sp. strain MF2. Process Biochem 42(10):1475–1479

Amoozegar MA, Ashengroph M, Malekzadeh F, Razavi MR, Naddaf S, Kabiri M (2008) Isolation and initial characterization of the tellurite reducing moderately halophilic bacterium, *Salinicoccus* sp. strain QW6. Microbiol Res 163(4):456–465

Armstrong RE, Warner JB (2003) Biology and the battlefield. Defense Horiz:24–25 S1

Arora A, Ha C, Park CB (2004) Inhibition of insulin amyloid formation by small stress molecules. FEBS Lett 564(1–2):121–125

Bajpai B, Chaudhary M, Saxena J (2015) Production and characterization of α-amylase from an extremely halophilic archaeon, *Haloferax* sp. HA10. Food Technol Biotechnol 53(1):11–17

Barth S, Huhn M, Matthey B, Klimka A, Galinski EA, Engert A (2000) Compatible-solute-supported periplasmic expression of functional recombinant proteins under stress conditions. Appl Environ Microbiol 66(4):1572–1579

Bayley ST, Morton RA, Lanyi JK (1978) Recent developments in the molecular biology of extremely halophilic bacteria. CRC Crit Rev Microbiol 6(2):151–206

Bertrand JC, Almallah M, Acquaviva M, Mille G (1990) Biodegradation of hydrocarbons by an extremely halophilic archaebacterium. Lett Appl Microbiol 11(5):260–263

Bestvater T, Louis P, Galinski EA (2008) Heterologous ectoine production in *Escherichia coli*: by-passing the metabolic bottle-neck. Saline Syst 4:12

Bonfá MRL, Grossman MJ, Mellado E, Durrant LR (2011) Biodegradation of aromatic hydrocarbons by Haloarchaea and their use for the reduction of the chemical oxygen demand of hypersaline petroleum produced water. Chemosphere 84(11):1671–1676

Bonfá MRL, Grossman MJ, Piubeli F, Mellado E, Durrant LR (2013) Phenol degradation by halophilic bacteria isolated from hypersaline environments. Biodegradation 24(5):699–709

Boyden E, Deisseroth K (2007) Light-activated cation channel and uses thereof. U.S. Patent Application 11/459, 637

Britton KL, Baker PJ, Fisher M, Ruzheinikov S, Gilmour DJ, Bonete MJ, Ferrer J, Pire C, Esclapez J, Rice DW (2006) Analysis of protein solvent interactions in glucose dehydrogenase from the extreme halophile *Haloferax mediterranei*. Proc Natl Acad Sci 103(13):4846–4851

Bruns A, Cypionka H, Overmann J (2002) Cyclic AMP and acyl homoserine lactones increase the cultivation efficiency of heterotrophic bacteria from the Central Baltic Sea. Appl Environ Microbiol 68(8):3978–3987

Buenger J, Driller H (2004) Ectoin: an effective natural substance to prevent UVA-induced premature photoaging. Skin Pharmacol Physiol 17(5):232–237

Bumadian MM, Williamson MP, Gilmour DJ (2014) The de novo synthesis and uptake of compatible solutes in *Klebsiella pneumoniae* ATCC 342. Nucleus 12:26

Camacho RM, Mateos JC, González-Reynoso O, Prado LA, Córdova J (2009) Production and characterization of esterase and lipase from *Haloarcula marismortui*. J Ind Microbiol Biotechnol 36(7):901–909

Carmeli C, Carmeli I, Richter S, Frolov L (2009) Optoelectronic device and method of fabricating the same. U.S. Patent 8,212,005

Castro-Severyn J, Remonsellez F, Valenzuela SL, Salinas C, Fortt J, Aguilar P, Pardo-Esté C, Dorador C, Quatrini R, Molina F (2017) Comparative genomics analysis of a new *Exiguobacterium* strain from salar de huasco reveals a repertoire of stress-related genes and arsenic resistance. Front Microbiol 8:456

Chakraborty S, Khopade A, Biao R, Jian W, Liu XY, Mahadik K, Chopade B, Zhang L, Kokare C (2011) Characterization and stability studies on surfactant, detergent and oxidant stable α-amylase from marine haloalkaliphilic *Saccharopolyspora* sp. A9. J Mol Catal B Enzym 68 (1):52–58

Chang HW, Kim KH, Nam YD, Roh SW, Kim MS, Jeon CO, Oh HM, Bae JW (2008) Analysis of yeast and archaeal population dynamics in kimchi using denaturing gradient gel electrophoresis. Int J Food Microbiol 126(1–2):159–166

Chang J, Lee YS, Fang SJ, Park IH, Choi YL (2013) Recombinant expression and characterization of an organic-solvent-tolerant α-amylase from *Exiguobacterium* sp. DAU5. Appl Biochem Biotechnol 169(6):1870–1883

Charan RD, Schlingmann G, Janso J, Bernan V, Feng X, Carter GT (2004) Diazepinomicin, a new antimicrobial alkaloid from a marine *Micromonospora* sp. J Nat Prod 67(8):1431–1433

Chen Z, Birge RR (1993) Protein-based artificial retinas. Trends Biotechnol 11(7):292–300

Chen Z, Wan C (2017) Non-sterile fermentations for the economical biochemical conversion of renewable feedstocks. Biotechnol Lett 39(12):1765–1777

Chuprom J, Bovornreungroj P, Ahmad M, Kantachote D, Dueramae S (2016) Approach toward enhancement of halophilic protease production by *Halobacterium* sp. strain LBU50301 using statistical design response surface methodology. Biotechnol Rep 10:17–28

Cleland D, Krader P, McCree C, Tang J, Emerson D (2004) Glycine betaine as a cryoprotectant for prokaryotes. J Microbiol Methods 58(1):31–38

Conde-Martínez N, Acosta-González A, Díaz LE, Tello E (2017) Use of a mixed culture strategy to isolate halophilic bacteria with antibacterial and cytotoxic activity from the Manaure solar saltern in Colombia. BMC Microbiol 17:230

Connon SA, Giovannoni SJ (2002) High-throughput methods for culturing microorganisms in very-low-nutrient media yield diverse new marine isolates. Appl Environ Microbiol 68 (8):3878–3885

Coronado MJ, Vargas C, Hofemeister J, Ventosa A, Nieto JJ (2000) Production and biochemical characterization of an α-amylase from the moderate halophile *Halomonas meridian*. FEMS Microbiol Lett 183(1):67–71

Cragg GM, Newman DJ (2005) Biodiversity: a continuing source of novel drug leads. Pure Appl Chem 77(1):7–24

Cui C, Ma L, Shi J, Lin K, Luo Q, Liu Y (2014) Metabolic pathway for degradation of anthracene by halophilic *Martelella* sp. AD-3. Int Biodeterior Biodegradation 89:67–73

Cyplik P, Grajek W, Marecik R, Króliczak P, Dembczyński R (2007) Application of a membrane bioreactor to denitrification of brine. Desalination 207(1–3):134–143

DasSarma S, DasSarma P (2015) Halophiles and their enzymes: negativity put to good use. Curr Opin Microbiol 25:120–126

de Lourdes Moreno M, García MT, Ventosa A, Mellado E (2009) Characterization of *Salicola* sp. IC10, a lipase-and protease-producing extreme halophile. FEMS Microbiol Ecol 68 (1):59–71

de Lourdes Moreno M, Pérez D, García MT, Mellado E (2013) Halophilic bacteria as a source of novel hydrolytic enzymes. Life 3(1):38–51

Del Moral A, Severin J, Ramos-Cormenzana A, Trüper HG, Galinski EA (1994) Compatible solutes in new moderately halophilic isolates. FEMS Microbiol Lett 122(1–2):165–172

Delille D, Basseres A, Dessommes A (1998) Effectiveness of bioremediation for oil-polluted Antarctic seawater. Polar Biol 19(4):237–241

Dincer AR, Kargi F (2001) Performance of rotating biological disc system treating saline wastewater. Process Biochem 36(8–9):901–906

Don TM, Chen CW, Chan TH (2006) Preparation and characterization of poly (hydroxyalkanoate) from the fermentation of *Haloferax mediterranei*. J Biomater Sci Polym Ed 17(12):1425–1438

Emerson D, Breznak JA (1997) The response of microbial populations from oil-brine contaminated soil to gradients of NaCl and sodium p-toluate in a diffusion gradient chamber. FEMS Microbiol Ecol 23(4):285–300

Emerson D, Chauhan S, Oriel P, Breznak JA (1994) *Haloferax* sp. D1227, a halophilic Archaeon capable of growth on aromatic compounds. Arch Microbiol 161(6):445–452

Essghaier B, Rouaissi M, Boudabous A, Jijakli H, Sadfi-Zouaoui N (2010) Production and partial characterization of chitinase from a halotolerant *Planococcus rifitoensis* strain M2-26. World J Microbiol Biotechnol 26(6):977–984

Fernandez-Castillo R, Rodriguez-Valera F, Gonzalez-Ramos J, Ruiz-Berraquero F (1986) Accumulation of poly (β-hydroxybutyrate) by halobacteria. Appl Environ Microbiol 51(1):214–216

Fernandez-Castillo R, Vargas C, Nieto JJ, Ventosa A, Ruiz-Berraquero F (1992) Characterization of a plasmid from moderately halophilic eubacteria. Microbiology 138(6):1133–1137

Fukushima T, Mizuki T, Echigo A, Inoue A, Usami R (2005) Organic solvent tolerance of halophilic α-amylase from a Haloarchaeon, *Haloarcula* sp. strain S-1. Extremophiles 9 (1):85–89

Galinski EA (1993) Compatible solutes of halophilic eubacteria: molecular principles, water-solute interaction, stress protection. Experientia 49(6–7):487–496

Galinski EA (1995) Osmoadaptation in bacteria. Adv Microb Physiol 37:273–328

Gao R, Shi T, Liu X, Zhao M, Cui H, Yuan L (2017) Purification and characterisation of a salt-stable protease from the halophilic archaeon *Halogranum rubrum*. J Sci Food Agr 97 (5):1412–1419

García-Torreiro M, Lú-Chau TA, Steinbüchel A, Lema JM (2016) Waste to bioplastic conversion by the moderate halophilic bacterium *Halomonas boliviensis*. Chem Eng Trans 49:163–168

Garg R, Srivastava R, Brahma V, Verma L, Karthikeyan S, Sahni G (2016) Biochemical and structural characterization of a novel halotolerant cellulase from soil metagenome. Sci Rep 6:39634

Gauthier MJ, Lafay B, Christen R, Fernandez L, Acquaviva M, Bonin P, Bertrand JC (1992) *Marinobacter hydrocarbonoclasticus* gen. nov., sp. nov., a new, extremely halotolerant, hydrocarbon-degrading marine bacterium. Int J Syst Evol Microbiol 42(4):568–576

Ghafoori H, Askari M, Sarikhan S (2016) Purification and characterization of an extracellular haloalkaline serine protease from the moderately halophilic bacterium, *Bacillus iranensis* (X5B). Extremophiles 20(2):115–123

Ghazvini PTM, Mashkani SG (2009) Effect of salinity on vanadate biosorption by *Halomonas* sp. GT-83: preliminary investigation on biosorption by micro-PIXE technique. Bioresour Technol 100(8):2361–2368

Ginzburg BZ (1993) Liquid fuel (oil) from halophilic algae: a renewable source of non-polluting energy. Renew Energy 3:249–252

Goldman Y, Garti N, Sasson Y, Ginzburg BZ, Bloch MR (1981) Conversion of halophilic algae into extractable oil. 2. Pyrolysis of proteins. Fuel 60(2):90–92

Gupta R, Gupta N, Rathi P (2004) Bacterial lipases: an overview of production, purification and biochemical properties. Appl Microbiol Biotechnol 64(6):763–781

Gutiérrez-Arnillas E, Rodríguez A, Sanromán MA, Deive FJ (2016) New sources of halophilic lipases: isolation of bacteria from Spanish and Turkish saltworks. Biochem Eng J 109:170–177

Hampp N, Seitz A, Paster-Nack R, Fuchsbauer HL (2005) Linker-free covalent coupling of bacteriorhodopsin in purple membrane form. U.S. Patent 6,927,050

Han J, Wu LP, Liu XB, Hou J, Zhao LL, Chen JY, Zhao DH, Xiang H (2017) Biodegradation and biocompatibility of haloarchaea-produced poly (3-hydroxybutyrate-co-3-hydroxyvalerate) copolymers. Biomaterials 139:172–186

Hao R, Lu A (2009) Biodegradation of heavy oils by halophilic bacterium. Prog Nat Sci 19 (8):997–1001

Harding KG, Dennis JS, Von Blottnitz H, Harrison STL (2008) A life-cycle comparison between inorganic and biological catalysis for the production of biodiesel. J Clean Prod 16 (13):1368–1378

Harvey T, Schmadel D (2007) Holographic storage system with single switch access. U.S. Patent Application 11/750,264

Hayashi H, Sakamoto A, Murata N (1998) Enhancement of the tolerance of *Arabidopsis* to high temperatures by genetic engineering of the synthesis of glycinebetaine. Plant J 16(2):155–161

He H, Ding WD, Bernan VS, Richardson AD, Ireland CM, Greenstein M, Ellestad GA, Carter GT (2001) Lomaiviticins A and B, potent antitumor antibiotics from *Micromonospora lomaivitiensis*. J Am Chem Soc 123(22):5362–5363

Hebert AM, Vreeland RH (1987) Phenotypic comparison of halotolerant bacteria: *Halomonas halodurans* sp. nov., nom. rev., comb. Nov. Int J Syst Evol Microbiol 37(4):347–350

Hemamalini R, Khare SK (2016) Purification and characterization of active aggregates of an organic solvent tolerant lipase from *Marinobacter* sp. EMB5. Insights Enzym Res 1:3

Hemamalini R, Khare SK (2018) Halophilic lipase does forms catalytically active aggregates: evidence from *Marinobacter* sp. EMB5 lipase (LipEMB5). Int J Biol Macromol 119:172–179

Hertadi R, Kurnia K, Falahudin W, Puspasari M (2017) Poly-hydroxybutyrate (PHB) production by *Halomonas elongata* BK AG 18 indigenous from salty mud crater at Central Java Indonesia. Malays J Microbiol 13:26–32

Hinteregger C, Streichsbier F (1997) *Halomonas* sp., a moderately halophilic strain, for biotreatment of saline phenolic waste-water. Biotechnol Lett 19(11):1099–1102

Holmes ML, Scopes RK, Moritz RL, Simpson RJ, Englert C, Pfeifer F, Dyall-Smith ML (1997) Purification and analysis of an extremely halophilic β-galactosidase from *Haloferax alicantei*. Biochim Biophys Acta Protein Struct Mol Enzymol 1337(2):276–286

Holmström KO, Somersalo S, Mandal A, Palva TE, Welin B (2000) Improved tolerance to salinity and low temperature in transgenic tobacco producing glycine betaine. J Exp Bot 51 (343):177–185

Hung KS, Liu SM, Tzou WS, Lin FP, Pan CL, Fang TY, Sun KH, Tang SJ (2011) Characterization of a novel GH10 thermostable, halophilic xylanase from the marine bacterium *Thermoanaerobacterium saccharolyticum* NTOU1. Process Biochem 46(6):1257–1263

Jensen RB, Kelemen BR, McAuliffe JC, Smith WC (2010) Method for preparing solid materials comprising immobilized proteorhodopsin. U.S. Patent 7,745,066

Jolley KA, Russell RJM, Hough DW, Danson MJ (1997) Site-directed mutagenesis and halophilicity of dihydrolipoamide dehydrogenase from the halophilic archaeon, *Haloferax volcanii*. Eur J Biochem 248(2):362–368

Kamekura M, Onishi H (1974) Halophilic nuclease from a moderately halophilic *Micrococcus varians*. J Bacteriol 119(2):339–344

Kanai H, Kobayashi T, Aono R, Kudo T (1995) *Natronococcus amylolyticus* sp. nov., a haloalkaliphilic archaeon. Int J Syst Evol Microbiol 45(4):762–766

Kaplan DL, Omenetto F, Lawrence B, Cronin-Golomb M, Georgakoudi I (2010) Biopolymer optofluidic device and method of manufacturing the same. U.S. Patent Application 12/513,423

Kaplan DL, Omenetto F, Lawrence B, Cronin-Golomb M (2018) Nanopatterned biopolymer optical device and method of manufacturing the same. U.S. Patent 9,969,134

Karan R, Khare SK (2011) Stability of haloalkaliphilic *Geomicrobium* sp. protease modulated by salt. Biochem Mosc 76(6):686–693

Karan R, Singh SP, Kapoor S, Khare SK (2011) A novel organic solvent tolerant protease from a newly isolated *Geomicrobium* sp. EMB2 (MTCC 10310): production optimization by response surface methodology. New Biotechnol 28(2):136–145

Karan R, Capes MD, DasSarma S (2012a) Function and biotechnology of extremophilic enzymes in low water activity. Aquat Biosyst 8:4

Karan R, Kumar S, Sinha R, Khare SK (2012b) Halophilic microorganisms as sources of novel enzymes. In: Satyanarayana T, Johri BN, Prakash A (eds) Microorganisms in sustainable agriculture and biotechnology. Springer, Dordrecht, pp 555–579

Karan R, Capes MD, DasSarma P, DasSarma S (2013) Cloning, overexpression, purification, and characterization of a polyextremophilic β-galactosidase from the Antarctic haloarchaeon *Halorubrum lacusprofundi*. BMC Biotechnol 13:3

Kawata Y, Aiba SI (2010) Poly (3-hydroxybutyrate) production by isolated *Halomonas* sp. KM-1 using waste glycerol. Biosci Biotechnol Biochem 74(1):175–177

Kennedy SP, Ng WV, Salzberg SL, Hood L, DasSarma S (2001) Understanding the adaptation of Halobacterium species NRC-1 to its extreme environment through computational analysis of its genome sequence. Genome Res 11(10):1641–1650

Kiran KK, Chandra TS (2008) Production of surfactant and detergent-stable, halophilic, and alkalitolerant alpha-amylase by a moderately halophilic *Bacillus* sp. strain TSCVKK. Appl Microbiol Biotechnol 77(5):1023–1031

Knopf GK, Wang WW, Bassi AS (2009) Flexible bioelectronic photodetector and imaging arrays based on bacteriorhodopsin (BR) thin films. U.S. Patent 7,573,024

Kokare CR, Mahadik KR, Kadam SS, Chopade BA (2004) Isolation, characterization and antimicrobial activity of marine halophilic *Actinopolyspora* species AH1 from the west coast of India. Curr Sci 86(4):593–597

Kubo M, Hiroe J, Murakami M, Fukami H, Tachiki T (2001) Treatment of hypersaline-containing wastewater with salt-tolerant microorganisms. J Biosci Bioeng 91(2):222–224

Kucera D, Pernicová I, Kovalcik A, Koller M, Mullerova L, Sedlacek P, Mravec F, Nebesarova J, Kalina M, Marova I (2018) Characterization of the promising poly (3-hydroxybutyrate) producing halophilic bacterium *Halomonas halophile*. Bioresour Technol 256:552–556

Kulichevskaya IS (1991) Oxidation of petroleum hydrocarbons by extremely halophilic archaebacteria. Mikrobiologiya 60:860–866

Kumar S, Khare SK (2012) Purification and characterization of maltooligosaccharide-forming α-amylase from moderately halophilic *Marinobacter* sp. EMB8. Bioresour Technol 116:247–251

Kumar S, Khare SK (2015) Chloride activated halophilic α-amylase from *Marinobacter* sp. EMB8: production optimization and nanoimmobilization for efficient starch hydrolysis. Enzym Res 2015:859485

Kumar S, Karan R, Kapoor S, Singh SP, Khare SK (2012) Screening and isolation of halophilic bacteria producing industrially important enzymes. Braz J Microbiol 43(4):1595–1603

Kumar S, Grewal J, Sadaf A, Hemamalini R, Khare SK (2016a) Halophiles as a source of polyextremophilic α-amylase for industrial applications. AIMS Microbiol 2(1):1–26

Kumar S, Khan RH, Khare SK (2016b) Structural elucidation and molecular characterization of *Marinobacter* sp. α-amylase. Prep Biochem Biotechnol 46(3):238–246

Kushner DJ (1993) Growth and nutrition of halophilic bacteria. In: Vreeland RH, Hochstein L (eds) The biology of halophilic bacteria. CRC, Boca Raton, FL, pp 87–103

Lai SJ, Lai MC (2011) Characterization and regulation of the osmolyte betaine synthesizing enzymes GSMT and SDMT from halophilic methanogen *Methanohalophilus portucalensis*. PLoS One 6(9):e25090

Lanyi JK (1974) Salt-dependent properties of proteins from extremely halophilic bacteria. Bacteriol Rev 38(3):272–290

Le Borgne S, Paniagua D, Vazquez-Duhalt R (2008) Biodegradation of organic pollutants by halophilic bacteria and archaea. J Mol Microbiol Biotechnol 15(2–3):74–92

Lefebvre O (2005) Application des micro-organismes halophiles au traitement des effluents industriels hypersalins. Application des micro-organismes halophiles au traitement des effluents industriels hypersalins, Ecole Nationale Supérieure Agronomique de Montpellier 2005

Lefebvre O, Moletta R (2006) Treatment of organic pollution in industrial saline wastewater: a literature review. Water Res 40(20):3671–3682

León MJ, Hoffmann T, Sánchez-Porro C, Heider J, Ventosa A, Bremer E (2018) Compatible solute synthesis and import by the moderate halophile *Spiribacter salinus*: physiology and genomics. Front Microbiol 9:108

Li X, Yu HY (2012) Characterization of an organic solvent-tolerant α-amylase from a halophilic isolate, *Thalassobacillus* sp. LY18. Folia Microbiol 57(5):447–453

Li X, Yu HY (2013) Halostable cellulase with organic solvent tolerance from *Haloarcula* sp. LLSG7 and its application in bioethanol fermentation using agricultural wastes. J Ind Microbiol Biotechnol 40(12):1357–1365

Li F, Maskey RP, Qin S, Sattler I, Fiebig HH, Maier A, Zeeck A, Laatsch H (2005) Chinikomycins A and B: isolation, structure elucidation, and biological activity of novel antibiotics from a marine *Streptomyces* sp. Isolate M045. J Nat Prod 68(3):349–353

Li X, Wang HL, Li T, Yu HY (2012) Purification and characterization of an organic solvent-tolerant alkaline cellulase from a halophilic isolate of *Thalassobacillus*. Biotechnol Lett 34 (8):1531–1536

Lombó F, Velasco A, Castro A, De la Calle F, Braña A, Alfredo F, Sánchez Puelles JM, Méndez C, Salas JA (2006) Deciphering the biosynthesis pathway of the antitumor thiocoraline from a marine actinomycete and its expression in two *Streptomyces* species. Chembiochem 7 (2):366–376

Lopetcharat K, Choi YJ, Park JW, Daeschel MA (2001) Fish sauce products and manufacturing: a review. Food Rev Int 17(1):65–88

Madern D, Pfister C, Zaccai G (1995) Mutation at a single acidic amino acid enhances the halophilic behaviour of malate dehydrogenase from *Haloarcula marismortui* in physiological salts. Eur J Biochem 230(3):1088–1095

Madern D, Ebel C, Zaccai G (2000) Halophilic adaptation of enzymes. Extremophiles 4(2):91–98

Mahansaria R, Dhara A, Saha A, Haldar S, Mukherjee J (2018) Production enhancement and characterization of the polyhydroxyalkanoate produced by *Natrinema ajinwuensis* (as synonym) ≡ *Natrinema altunense* strain RM-G10. Int J Biol Macromol 107:1480–1490

Manam RR, Teisan S, White DJ, Nicholson B, Grodberg J, Neuteboom STC, Lam KS, Mosca DA, Lloyd GK, Potts BCM (2005) Lajollamycin, a nitro-tetraene spiro-β-lactone-γ-lactam antibiotic from the marine actinomycete *Streptomyces nodosus*. J Nat Prod 68(2):240–243

Margesin R, Schinner F (1999) Biological decontamination of oil spills in cold environments. J Chem Technol Biotechnol 74(5):381–389

Margesin R, Schinner F (2001) Potential of halotolerant and halophilic microorganisms for biotechnology. Extremophiles 5(2):73–83

Martínez-Espinosa RM, Zafrilla B, Camacho M, Bonete MJ (2007) Nitrate and nitrite removal from salted water by *Haloferax mediterranei*. Biocatal Biotransformation 25(2–4):295–300

Maskey RP, Li FC, Qin S, Fiebig HH, Laatsch H (2003) Chandrananimycins AC: production of novel anticancer antibiotics from a marine *Actinomadura* sp. isolate M048 by variation of medium composition and growth conditions. J Antibiot (Tokyo) 56(7):622–629

Masland RH (2016) Method for augmenting vision in persons suffering from photoreceptor cell degeneration. U.S. Patent 9,434,781

Mellado E, Sánchez-Porro C, Ventosa A (2005) Proteases produced by halophilic bacteria and archaea. Microbiol Enzym Biotransform 17:181–190

Mesbah NM, Wiegel J (2014) Halophilic alkali-and thermostable amylase from a novel polyextremophilic *Amphibacillus* sp. NM-Ra2. Int J Biol Macromol 70:222–229

Mevarech M, Eisenberg H, Neumann E (1977) Malate dehydrogenase isolated from extremely halophilic bacteria of the Dead Sea. 1. Purification and molecular characterization. Biochemistry 16(17):3781–3785

Mirete S, Mora-Ruiz MR, Lamprecht-Grandío M, de Figueras CG, Rosselló-Móra R, González-Pastor J (2015) Salt resistance genes revealed by functional metagenomics from brines and moderate-salinity rhizosphere within a hypersaline environment. Front Microbiol 6:1121

Miriam LRM, Raj RE, Kings AJ, Visvanathan MA (2017) Identification and characterization of a novel biodiesel producing halophilic *Aphanothece halophytica* and its growth and lipid optimization in various media. Energy Convers Manag 141:93–100
Murphy BT, Narender T, Kauffman CA, Woolery M, Jensen PR, Fenical W (2010) Saliniquinones AF, new members of the highly cytotoxic anthraquinone-γ-pyrones from the marine actinomycete *Salinispora arenicola*. Aust J Chem 63(6):929–934
Namwong S, Tanasupawat S, Visessanguan W, Kudo T, Itoh T (2007) *Halococcus thailandensis* sp. nov., from fish sauce in Thailand. Int J Syst Evol Microbiol 57(10):2199–2203
Nicholson CA, Fathepure BZ (2004) Biodegradation of benzene by halophilic and halotolerant bacteria under aerobic condition. Appl Environ Microbiol 70(2):1222–1225
Nicholson CA, Fathepure BZ (2005) Aerobic biodegradation of benzene and toluene under hypersaline conditions at the Great Salt Plains, Oklahoma. FEMS Microbiol Lett 245 (2):257–262
Obuekwe CO, Badrudeen AM, Al-Saleh E, Mulder JL (2005) Growth and hydrocarbon degradation by three desert fungi under conditions of simultaneous temperature and salt stress. Int Biodeterior Biodegradation 56(4):197–205
Ogan A, Danis O, Gozuacik A, Cakmar E, Birbir M (2012) Production of cellulase by immobilized whole cells of *Haloarcula*. Appl Biochem Microbiol 48(5):440–443
Onishi H, Sonoda K (1979) Purification and some properties of an extracellular amylase from a moderate halophile, *Micrococcus halobius*. Appl Environ Microbiol 38(4):616–620
Onishi H, Mori T, Takeuchi S, Tani K, Kobayashi T, Kamekura M (1983) Halophilic nuclease of a moderately halophilic *Bacillus* sp.: production, purification, and characterization. Appl Environ Microbiol 45(1):24–30
Oren A (2008) Microbial life at high salt concentrations: phylogenetic and metabolic diversity. Saline Syst 4:2
Oren A (2010) Industrial and environmental applications of halophilic microorganisms. Environ Technol 31(8–9):825–834
Pan ZH, Dizhoor AM (2013) Restoration of visual responses by in vivo delivery of rhodopsin nucleic acids. U.S. Patent 8,470,790
Pathak AP, Sardar AG (2014) Isolation and characterization of salt stable protease producing archaea from marine solar saltern of Mulund, Mumbai. Indian J Geo Mar Sci 43(3):412–417
Pérez D, Martín S, Fernández-Lorente G, Filice M, Guisán JM, Ventosa A, García MT, Mellado E (2011) A novel halophilic lipase, LipBL, showing high efficiency in the production of eicosapentaenoic acid (EPA). PLoS One 6(8):e23325
Pérez-Pomares F, Bautista V, Ferrer J, Pire C, Marhuenda-Egea FC, Bonete MJ (2003) α-Amylase activity from the halophilic archaeon *Haloferax mediterranei*. Extremophiles 7(4):299–306
Prakash B, Vidyasagar M, Madhukumar MS, Muralikrishna G, Sreeramulu K (2009) Production, purification, and characterization of two extremely halotolerant, thermostable, and alkali-stable α-amylases from *Chromohalobacter* sp. TVSP 101. Process Biochem 44(2):210–215
Rajagopalan R, Altekar W (1994) Characterisation and purification of ribulose-bisphosphate carboxylase from heterotrophically grown halophilic archaebacterium, *Haloferax mediterranei*. Eur J Biochem 221(2):863–869
Rangaswamy V, Altekar W (1994) Characterization of 1-phosphofructokinase from halophilic archaebacterium *Haloarcula vallismortis*. Biochim Biophys Acta Gen Subj 1201(1):106–112
Rathgeber C, Yurkova N, Stackebrandt E, Beatty JT, Yurkov V (2002) Isolation of tellurite-and selenite-resistant bacteria from hydrothermal vents of the Juan de Fuca Ridge in the Pacific Ocean. Appl Environ Microbiol 68(9):4613–4622
Rathi DN, Amir HG, Abed RMM, Kosugi A, Arai T, Sulaiman O, Hashim R, Sudesh K (2013) Polyhydroxyalkanoate biosynthesis and simplified polymer recovery by a novel moderately halophilic bacterium isolated from hypersaline microbial mats. J Appl Microbiol 114 (2):384–395
Raytapadar S, Paul AK (2001) Production of an antifungal antibiotic by *Streptomyces aburaviensis* IDA-28. Microbiol Res 155(4):315–323

Rezaei S, Shahverdi AR, Faramarzi MA (2017) Isolation, one-step affinity purification, and characterization of a polyextremotolerant laccase from the halophilic bacterium *Aquisalibacillus elongatus* and its application in the delignification of sugar beet pulp. Bioresour Technol 230:67–75

Roberts MF (2005) Organic compatible solutes of halotolerant and halophilic microorganisms. Saline Syst 1:5

Roh SW, Nam YD, Chang HW, Sung Y, Kim KH, Oh HM, Bae JW (2007) *Halalkalicoccus jeotgali* sp. nov., a halophilic archaeon from shrimp jeotgal, a traditional Korean fermented seafood. Int J Syst Evol Microbiol 57(10):2296–2298

Rohban R, Amoozegar MA, Ventosa A (2009) Screening and isolation of halophilic bacteria producing extracellular hydrolyses from Howz Soltan Lake, Iran. J Ind Microbiol Biotechnol 36(3):333–340

Röling WFM, Prasetyo AB, Stouthamer AH, Van Verseveld HW (1999) Physiological aspects of the growth of the lactic acid bacterium *Tetragenococcus halophila* during Indonesian soy sauce (kecap) production. J Appl Microbiol 86(2):348–352

Rosenberg A (1983) *Pseudomonas halodurans* sp. nov., a halotolerant bacterium. Arch Microbiol 136(2):117–123

Rothschild KJ, Olejnik J, Sonar SM (2007) Detection of markers in nascent proteins. U.S. Patent 7,211,394

Ryu K, Kim J, Dordick JS (1994) Catalytic properties and potential of an extracellular protease from an extreme halophile. Enzym Microb Technol 16(4):266–275

Saeedi P, Moosaabadi JM, Sebtahmadi SS, Mehrabadi JF, Behmanesh M, Mekhilef S (2012) Potential applications of bacteriorhodopsin mutants. Bioengineered 3(6):326–328

Santos AF, Valle RS, Pacheco CA, Alvarez VM, Seldin L, Santos A (2013) Extracellular proteases of *Halobacillus blutaparonensis* strain M9, a new moderately halophilic bacterium. Braz J Microbiol 44(4):1299–1304

Sasse F, Steinmetz H, Höfle G, Reichenbach H (1995) Gephyronic acid, a novel inhibitor of eukaryotic protein synthesis from *Archangium gephyra* (myxobacteria). J Antibiot 48(1):21–25

Sauer T, Galinski EA (1998) Bacterial milking: a novel bioprocess for production of compatible solutes. Biotechnol Bioeng 57(3):306–313

Schnoor M, Voss P, Cullen P, Böking T, Galla HJ, Galinski EA, Lorkowski S (2004) Characterization of the synthetic compatible solute homoectoine as a potent PCR enhancer. Biochem Biophys Res Commun 322(3):867–872

Schreck SD, Grunden AM (2014) Biotechnological applications of halophilic lipases and thioesterases. Appl Microbiol Biotechnol 98(3):1011–1021

Selim S, Hagagy N, Aziz MA, El-Meleigy ES, Pessione E (2014) Thermostable alkaline halophilic-protease production by *Natronolimnobius innermongolicus* WN18. Nat Prod Res 28 (18):1476–1479

Shafiei M, Ziaee AA, Amoozegar MA (2010) Purification and biochemical characterization of a novel SDS and surfactant stable, raw starch digesting, and halophilic α-amylase from a moderately halophilic bacterium, *Nesterenkonia* sp. strain F. Process Biochem 45(5):694–699

Shanmughapriya S, Kiran GS, Selvin J, Thomas TA, Rani C (2010) Optimization, purification, and characterization of extracellular mesophilic alkaline cellulase from sponge-associated *Marinobacter* sp. MSI032. Appl Biochem Biotechnol 162(3):625–640

Shivanand P, Mugeraya G (2011) Halophilic bacteria and their compatible solutes-osmoregulation and potential applications. Curr Sci 100(10):1516–1521

Sinha R, Khare SK (2012) Isolation of a halophilic *Virgibacillus* sp. EMB13: characterization of its protease for detergent application. Indian J Biotechnol 11(4):416–426

Sinha R, Khare SK (2013) Characterization of detergent compatible protease of a halophilic *Bacillus* sp. EMB9: differential role of metal ions in stability and activity. Bioresour Technol 145:357–361

Sinha R, Khare SK (2014) Effect of organic solvents on the structure and activity of moderately halophilic *Bacillus* sp. EMB9 protease. Extremophiles 18(6):1057–1066

Sinsuwan S, Rodtong S, Yongsawatdigul J (2008) Production and characterization of NaCl-activated proteinases from *Virgibacillus* sp. SK33 isolated from fish sauce fermentation. Process Biochem 43(2):185–192

Song J, Oh HM, Cho JC (2009) Improved culturability of SAR11 strains in dilution-to-extinction culturing from the East Sea, West Pacific Ocean. FEMS Microbiol Lett 295(2):141–147

Sujatha P, Raju KB, Ramana T (2005) Studies on a new marine streptomycete BT-408 producing polyketide antibiotic SBR-22 effective against methicillin resistant *Staphylococcus aureus*. Microbiol Res 160(2):119–126

Tapilatu YH, Grossi V, Acquaviva M, Militon C, Bertrand JC, Cuny P (2010) Isolation of hydrocarbon-degrading extremely halophilic archaea from an uncontaminated hypersaline pond (Camargue, France). Extremophiles 14(2):225–231

Tapingkae W, Tanasupawat S, Itoh T, Parkin KL, Benjakul S, Visessanguan W, Valyasevi R (2008) *Natrinema gari* sp. nov., a halophilic archaeon isolated from fish sauce in Thailand. Int J Syst Evol Microbiol 58(10):2378–2383

Tekin E, Ateş M, Kahraman Ö (2012) Poly-3-hydroxybutyrate-producing extreme halophilic archaeon: *Haloferax* sp. MA10 isolated from Çamaltı Saltern, İzmir. Turk J Biol 36(3):303–312

Thongthai C, McGenity TJ, Suntinanalert P, Grant WD (1992) Isolation and characterization of an extremely halophilic archaeobacterium from traditionally fermented Thai fish sauce (nam pla). Lett Appl Microbiol 14(3):111–114

Todkar S, Todkar R, Kowale L, Karmarkar K, Kulkarni A (2012) Isolation and screening of antibiotic producing halophiles from Ratnagiri coastal area, state of Maharashtra. Int J Sci Res 2:2250–3153

Tohme S, Hacıosmanoğlu GG, Eroğlu MS, Kasavi C, Genç S, Can ZS, Oner ET (2018) *Halomonas smyrnensis* as a cell factory for co-production of PHB and Levan. Int J Biol Macromol 118:1238–1246

Tripp HJ, Kitner JB, Schwalbach MS, Dacey JWH, Wilhelm LJ, Giovannoni SJ (2008) SAR11 marine bacteria require exogenous reduced sulphur for growth. Nature 452:741–744

Trivedi S, Prakash Choudhary O, Gharu J (2011) Different proposed applications of bacteriorhodopsin. Recent Pat DNA Gene Seq 5(1):35–40

Van-Thuoc D, Guzmán HC, Thi-Hang M, Hatti-Kaul R (2010) Ectoine production by *Halomonas boliviensis*: optimization using response surface methodology. Mar Biotechnol 12(5):586–593

Vartoukian SR, Palmer RM, Wade WG (2010) Strategies for culture of 'unculturable' bacteria. FEMS Microbiol Lett 309(1):1–7

Velho-Pereira S, Furtado I (2012) Antibacterial activity of halophilic bacterial bionts from marine invertebrates of Mandapam-India. Indian J Pharm Sci 74(4):331–338

Ventosa A, Nieto JJ, Oren A (1998) Biology of moderately halophilic aerobic bacteria. Microbiol Mol Biol Rev 62(2):504–544

Vera-Gargallo B, Ventosa A (2018) Metagenomic insights into the phylogenetic and metabolic diversity of the prokaryotic community dwelling in hypersaline soils from the Odiel saltmarshes (SW Spain). Genes 9(3):152

Vidyasagar M, Prakash S, Mahajan V, Shouche YS, Sreeramulu K (2009) Purification and characterization of an extreme halothermophilic protease from a halophilic bacterium *Chromohalobacter* sp. TVSP101. Braz J Microbiol 40(1):12–19

Weiss TL, Young EJ, Ducat DC (2017) A synthetic, light-driven consortium of cyanobacteria and heterotrophic bacteria enables stable polyhydroxybutyrate production. Met Eng 44:236–245

Wejse PL, Ingvorsen K, Mortensen KK (2003) Purification and characterisation of two extremely halotolerant xylanases from a novel halophilic bacterium. Extremophiles 7(5):423–431

Williams PG, Buchanan GO, Feling RH, Kauffman CA, Jensen PR, Fenical W (2005) New cytotoxic Salinosporamides from the marine actinomycete *Salinispora tropica*. J Org Chem 70(16):6196–6203

Wu W, Fattal D, Santori C, Bicknell RN, Wang SY, Williams RS, Quitoriano NJ (2009) Optical device including waveguide grating structure. U.S. Patent 7,548,671

Yadav N, Gupta MN, Khare SK (2017) Three phase partitioning and spectroscopic characterization of bioactive constituent from halophilic *Bacillus subtilis* EMB M15. Bioresour Technol 242:283–286

Yoshida M, Matsubara K, Kudo T, Horikoshi K (1991) *Actinopolyspora mortivallis* sp. nov., a moderately halophilic actinomycete. Int J Syst Evol Microbiol 41(1):15–20

Yu HY, Li X (2014) Characterization of an organic solvent-tolerant thermostable glucoamylase from a halophilic isolate, *Halolactibacillus* sp. SK 71 and its application in raw starch hydrolysis for bioethanol production. Biotechnol Prog 30(6):1262–1268

Zhang G, Li S, Xue Y, Mao L, Ma Y (2012) Effects of salts on activity of halophilic cellulase with glucomannanase activity isolated from alkaliphilic and halophilic *Bacillus* sp. BG-CS10. Extremophiles 16(1):35–43

Zhao Z, Zhang LY (2010) Screening and identification of a PCBs-degrading halophilic bacterium. Adv Mater Res 113–116:2177–2180

Zhao B, Wang H, Mao X, Li R (2009) Biodegradation of phenanthrene by a halophilic bacterial consortium under aerobic conditions. Curr Microbiol 58(3):205–210

Zhao D, Kumar S, Zhou J, Wang R, Li M, Xiang H (2017) Isolation and complete genome sequence of *Halorientalis hydrocarbonoclasticus* sp. nov., a hydrocarbon-degrading haloarchaeon. Extremophiles 21(6):1081–1090

Zvyagintseva IS, Poglazova MN, Gotoeva MT, Belyaev SS (2001) Effect on the medium salinity on oil degradation by nocardioform bacteria. Microbiology 70(6):652–656

Chapter 5
Investigation of the Structural and Functional Microbial Diversity in Indian Mangroves

Shayantan Mukherji, Shyamalina Haldar, and Abhrajyoti Ghosh

Abstract Mangrove forests are excellent examples of extremely productive ecosystems, present on earth. Located along the coastal zones, they provide nutrients that help to nourish the coastal waters, as well as serve as barriers that protect the coastal zones. In the last century, mangrove ecosystems became increasingly threatened due to changes in land usage pattern, different biotic pressure, natural calamities, and finally anthropogenic influences. It is therefore indispensable to monitor and assess mangrove ecosystems at a regular interval. India has a long coastline of which an area of about 4921 km^2 is covered with mangroves, making this country the fourth largest concerning the area covered with mangrove ecotype. Microbes are important components of mangrove ecosystems where they play an integral part in maintaining the biogeochemistry of such ecologically important regions. In Indian mangroves, a considerable effort has been made to understand flora and fauna biodiversity. However, knowledge about the microbial diversity and function within Indian mangrove ecosystem is comparatively less. The present chapter focuses mainly on diversity, distribution, and function of microbes in Indian mangrove ecosystem. Both culture-dependent and culture-independent studies on understanding the diversity and function of microorganisms in Indian mangroves have been discussed.

Keywords Mangrove forests · Microbial diversity · Mangrove ecosystem · Coastal estuarine ecosystems

S. Mukherji · A. Ghosh (✉)
Department of Biochemistry, Bose Institute, Kolkata, West Bengal, India
e-mail: abhrajyoti.ghosh@jcbose.ac.in

S. Haldar
Department of Microbiology, Goa University, Taleigão, India

B. Giri, A. Varma (eds.), *Microorganisms in Saline Environments: Strategies and Functions*, Soil Biology 56, https://doi.org/10.1007/978-3-030-18975-4_5

5.1 Introduction

India is considered to be a nation of megadiversity housing 10 distinct biogeographic zones and 2 of the world's 25 biodiversity hotspots, because of their extraordinarily high levels of species richness, endemicity, and threatened status. At the ecosystem level, India is also well-endowed, with ten distinct biogeographic zones. The harnessing of this biodiversity of India using modern biotechnological tools is important not only to assess the species richness of the country but also to explore their potential applications toward the benefit of mankind.

Mangrove ecosystems of India, cover a total area of 4921 km^2 contributing to 3.3% of the global mangrove cover and is a hotspot for diverse flora and fauna. The mangrove cover of India spreads along east, west, and south coasts alongside the Ganges, Mahanadi, Godavari, and Kaveri rivers with Sundarban at the deltas of the Ganges in East India being the largest mangrove region and the largest single stretch of halophytic mangrove forests, worldwide.

The microfauna which includes the bacteria, archaea, fungi, protozoa, and algae forms an integral and fundamental component of this ecosystem that plays an extremely vital role in maintaining biogeochemical cycling, helping in nutrient transformation and turnover of organic matter, as well as establishing a link between the terrestrial and the marine ecosystems (Bai et al. 2013). In addition, they contribute to the production and consumption of gases that play a role in global climate, degrade and destroy pollutants and waste products, and produce secondary metabolites, viz., antibiotics, pigments, alkaloids, toxins, drugs, plant growth-promoting substances, biodegrading enzymes, enzyme inhibitors, effectors of ecological competition and symbiosis, and pheromones, thereby maintaining the ecological integrity and sustainability (Cabral et al. 2016; Haldar and Nazareth 2018; Kharangate-Lad and Bhosle 2016; Suryanarayanan et al. 2004; Xu et al. 2014). This mangrove microbiome can be exploited for agriculture, industrial, pharmaceutical, and other biotechnological applications.

In spite of the importance of Indian mangrove ecosystems, this ecological domain has not been explored vividly unlike the other ecosystems of the world. Also, man-made activities such as deforestation, coal mining, refineries, oil spills, industries and industrial/agricultural waste disposal, and numerous other anthropogenic activities have resulted in the rapid degradation of mangrove ecosystems of India and other countries (Bhattacharyya et al. 2019). India being a tropical country experiences subtropical monsoon with periodic cyclonic storms. However, the mangroves protect and stabilize the coastal zones and, hence, need proper maintenance and overall restoration.

The following chapter focuses on three aspects to summarize the present scenario of the mangrove ecosystems of India. The first section of the chapter accounts for the (a) proportion of distribution of mangrove ecosystem in three distinct zones of India and (b) diversity and abundance of the true mangrove species and associate mangrove species inhabiting these regions. The second section of the chapter abridges the overall abundance, diversity, and composition of the microorganisms in the

Indian mangrove sediments reported till date. This subdivision emphasizes both the culture-dependent and culture-independent assessments of mangrove microbial population structure. The third section of the chapter deals with the overall functional attributes of the microbial consortia in the mangrove ecosystem elucidated by both the (1) laboratory-based studies of the functional activities of the individual microbial isolates from Indian mangrove sediments and (2) functional metagenomics. This latter approach throws light upon the real scenario of the complex species-species interactions and their diverse activities as a whole in this important ecological niche, thus providing a deep insight into the community dynamics of the resident microbial population of Indian mangrove ecosystems.

5.2 Mangrove Ecosystem

The mangrove ecosystems are an excellent example of a unique ecosystem which differs significantly from all other environmental ecosystems (Pramanik et al. 2019). Located along the coastlines of nations situated in the tropical as well as subtropical regions of the world (Taketani et al. 2009), these ecosystems are considered as some of the most productive ecosystems in the world due to their extremely high rate of organic matter turnover (Guidi-Rontani et al. 2014). These mangrove ecosystems serve as host to an extremely diversified group of microorganisms that contribute to the high productivity of this ecological niche by playing major roles in nutrient recycling (Pramanik et al. 2019). It may be stated that the resident microbial communities form the foundation of the food flora and fauna of mangrove ecosystems (Taketani et al. 2011). In these halophytic forests, the resident microbial communities transform the nutrients available from the dead and decaying mangrove foliage into phosphorus, nitrogen, etc., for use by the resident flora, which on its part secrete root exudates which serve as a source of food for the microbial communities (Pramanik et al. 2019). The resident bacterial and archaeal communities perform a majority of the carbon cycling that takes place in these coastal ecosystems (Taketani et al. 2009). Soil characteristics such as pH, salinity, and redox potential as well as particle size are important factors that contribute to the productivity of these coastal forests. Both natural factors such as the slightly acidic soils and tidal variations as well as the increased levels of human interventions such as barges, accidental oil spillage, transportation using ships, etc. are the reasons behind the diverse microflora of these coastal halophytic forests (Pramanik et al. 2019). Mangrove ecosystems and their resident microbial communities are interdependent.

5.2.1 Mangroves Across the World

Although being widely distributed in the geographical sense, mangrove ecosystems are largely restricted to the tropical and subtropical coastlines (Taketani et al. 2009).

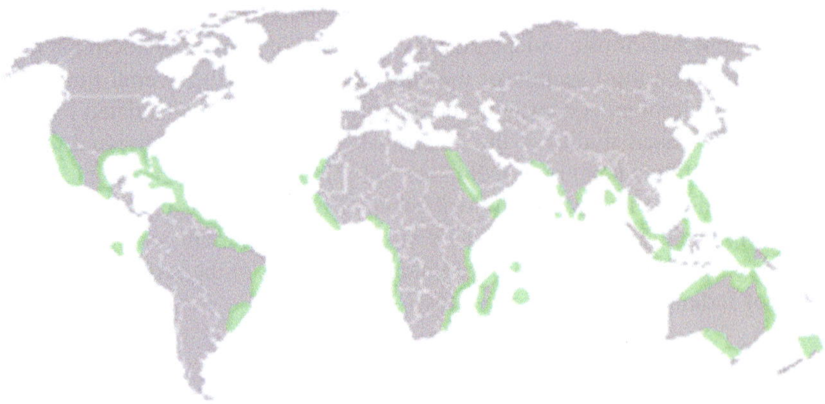

Fig. 5.1 A world map depicting the global distribution pattern of mangroves. Mangroves are the green regions (Huhsunqu, cartographer World map of mangrove distribution: Wikipedia, 2008)

They can be found on all of the continents occupying almost 75% of the tropical coastlines. Mangroves can be found in over 123 different countries with Asia having the largest percentage of the total world's mangroves. At present, the total area under mangrove cover in the world is estimated to be somewhere around 150,000 km^2. The most extensive area under mangrove ecosystems can be found in South and South East Asia that comprise a whopping 41.4% of the global mangroves. This is followed by the mangroves of South America, North Central America, and West and Central Africa (Fig. 5.1).

The countries like Brazil, Indonesia, and Australia have the most abundant mangrove cover (Taketani et al. 2009). Around 54 different mangrove species that belong to around 20 genera covering 11 families have been recorded worldwide. The Brazilian mangroves stretch all the way from the state of Amapá in the North to the state of Santa Catarina in the South. The Brazilian north coast encompasses approximately 8000 km^2 of mangrove ecosystems. Nearly 85% of the total Brazilian mangroves are present as a continuous stretch of about 700,000 ha along the coastline in the states of Pará and Maranhão. Three main species of mangrove trees are the most common in the Brazilian mangroves. They are *Laguncularia racemosa* (white mangrove), *Rhizophora mangle* (red mangrove), and *Avicennia schaueriana* (black mangrove) (Taketani et al. 2009). Mangrove forests of the Siné Saloum estuary cover a total area of 1279 km^2 in Senegal in West Africa. The *Rhizophora* and the *Avicennia* species dominate the mangrove vegetation in this estuary. Other than this, they also host small *Laguncularia racemosa* as well as the mangrove ferns *Acrostichum aureum* and *Conocarpus erectus*. Within these mangrove ecosystems, the composition of the respective mangrove species follows a predictable pattern which is termed as zonation. The *Rhizophora* species are found on the edges of the tidal channels, while the *Avicennia* and *Laguncularia* are more profoundly observed in the more elevated regions of these coastal forests. The mangrove species composition in this Senegalese mangrove estuary follows an interesting pattern which is most possibly influenced by the variations in salinity. *R. mangle* and *R. racemosa* dominate in areas of low variation in salinity while *A. germinans* and *L. racemosa* are more predominant in regions of high salinity variation (Chatellier and Shribman 2018). In the South Asian country of Indonesia, 3 million hectares of mangrove forests spread over the 95,000 km coastline. These Indonesian mangroves constitute 23% of all the mangrove ecosystems in the world (Giri et al. 2011). Within Indonesia, the regionally important mangrove forests are located in Papua, Kalimantan, and Sumatra.

5.2.2 Mangroves in India

The Indian mangroves cover almost 45.8% of the total area under mangroves in South Asia. The total area under mangrove cover in India is around 4921 km^2 which accounts to almost 3.3% of the total area under mangrove cover worldwide. The mangrove cover in India is distributed along all states having a coastline except Lakshadweep. These coastal forests (69°E–89.5°E longitude and 7°N–23°N latitude) are made up of three distinct zones. These zones are the island territories, west coast habitats, and east coast habitats. The east coast, facing the Bay of Bengal, has a coast line of around 2700 km, while the west coast facing the Arabian Sea has a coast line that stretches to about 3000 km. Almost 80% of the total mangroves in India occur along the Bay of Bengal, i.e., the east coast while only 20% along the Arabian Sea, i.e., on the west coast. This differentiation is due to the characteristic terrain and slope of the east coast and due to a large number of river deltas that open up on the east coast leading to the deposition of rich alluvial soil. On the basis of canopy density, the mangrove vegetation in India has been categorized into very dense mangrove cover, moderately dense mangrove cover, and open mangrove cover. Very dense mangrove vegetation is characterized by a canopy density of more than 70% as to that of between 40 and 70% for moderately dense cover, and, finally, a canopy density between 10 and 40% is classified as open mangrove. The overall distribution of mangroves of all three categorized in the different states of India is shown in Table 5.1.

5.2.2.1 Sundarban

The Sundarban mangroves are among the largest single continuous stretches of wetland ecosystem all over the world (Pramanik et al. 2019). These mangrove forests lie in the delta of the Brahmaputra, Ganges, and Meghna Rivers that open up onto the Bay of Bengal. (21°30′ to 22°40′N, 88°05′ to 89°55′E) (Gopal and Chauhan 2018). These mangroves cover an overall area of approximately

Table 5.1 Distribution of mangrove cover in India as of 2017

State/UT	Very dense mangrove (sq km)	Moderately dense mangrove (sq km)	Open mangrove (sq km)	Total (sq km)
Andhra Pradesh	0	213	191	404
Goa	0	20	6	26
Gujarat	0	172	968	1140
Karnataka	0	2	8	10
Kerala	0	5	4	9
Maharashtra	0	88	216	304
Odisha	82	94	67	243
Tamil Nadu	1	25	23	49
West Bengal	999	692	423	2114
A & N Islands	399	169	49	617
Daman and Diu	0	0	3	3
Puducherry	0	0	2	2
Total	**1481**	**1480**	**1960**	**4921**

10,200 km^2 which is spread over the countries of India (approximately 40%) and Bangladesh (approximately 60%), respectively (Ghosh et al. 2010). Overall, these mangroves are restricted to the river Hooghly on the western front and river Baleshwar on the eastern front [in Bangladesh] (Gopal and Chauhan 2018). The international border between India and Bangladesh that passes through these mangrove forests is demarcated by the river Harinbhanga which goes by the name of Raimongal or Ichamati in Bangladesh (Gopal and Chauhan 2018). The extent of the Sundarban mangroves within India is bordered by the river Muriganga (a distributary of river Hooghly) on the west and river Harinbhanga on the east along the Indo-Bangladesh border (Pramanik et al. 2019).

Neotectonic activities over the past millennium along with sediment erosion and accretion have greatly influenced the extent of these mangrove forests. The Sundarban mangrove delta is made up of numerous islands that had formed due to the sediment deposition from the three rivers Brahmaputra, Ganges, and Meghna along with a complex network of smaller rivers, creeks, and channels (Gopal and Chauhan 2018). The Sundarban mangrove delta developed more or less recently, as a result of the sedimentation of the Bengal deltaic basin by the Himalayan rivers. The continuous movement of tectonic plates helped in shaping the hydrology of the Bengal delta, by causing changes in sedimentation pattern and reducing the inflow of freshwater. The maximum elevation of these mangrove forests is only a mere 10 m above the mean sea level, thus tides as high as 5.9 m to 2 m cause periodic flooding of considerable portions of this delta twice a day (Gopal and Chauhan 2018). This high amount of tidal activity leads to the inward flow of a large amount of salt water into these halophytic forests. Several small distributaries flow through these deltaic mangrove forests among which Saptamukhi, Matla, Bidyadhari, and Thakuran are arguably the most noted ones in the Indian portion of the Sundarban mangrove cover. Among these, the two most important rivers of this mangrove estuary are Bidyadhari and Matla which run in the southwest direction. However, due to continued silt deposition, the upstream region of both Bidyadhari and Matla lacks connection with a freshwater source (Manna et al. 2010). The core area of the Sundarbans Tiger Project is divided from the buffer region at the meeting point of these two rivers. During high tides, the whole region gets flooded with brackish water which freely mingles with the freshwater arising from these two rivers. The soil texture of this mangrove estuary is mainly alluvial and highly saline with portions of clayey, silty soil containing both fine and coarse sand particles. These mangrove forests harbor both brackish, saline, and freshwater marshes and submerged macrophyte beds within the creeks (Gopal and Chauhan 2018). Similar to other mangrove forests, the Sundarban mangrove estuary is covered by a layer of trees, while grasses and shrubs make up the lower strata of the forest (Gopal and Chauhan 2018).

Among the total of 69 species of true mangroves (those that occur exclusively in mangroves), about 30 occur in the Sundarban mangrove cover. Among this *Avicennia officinalis*, *Avicennia marina*, and *Avicennia alba* are the dominant ones. Other less dominant species of mangroves that can be seen in the Sundarban forests are *Phoenix paludosa*, *Xylocarpus granatum*, *Sonneratia apetala*, *Heritiera fomes*, and *Aegialitis rotundifolia*, respectively (Ghosh et al. 2015). The distribution pattern of mangrove species in the Sundarban mangroves is greatly dependent on the salinity (Barik et al. 2017). The Sundarban mangrove delta acts host to a wide variety of fauna. These include *Panthera tigris tigris* (Royal Bengal Tiger) which is known to reside in only these saline mudflats and no other environment in the whole world. Other than this, the Sundarban mangroves host almost 600 species of vertebrates, 165 species of fish, 58 species of reptiles, 8 amphibian species, 300 bird species, and almost 40 species of mammals (Gopal and Chauhan 2018). This huge diversity and population of organisms in this mangrove estuary indicate the proper and successful maintenance of a sustainable ecosystem there (Chakraborty et al. 2015).

5.2.2.2 West Coast

Only about 20% of the total area under mangrove cover in India is found along the west coast, due to the presence of a steep gradient. As compared to the deltaic mangroves which are prevalent in the east coast and the insular mangroves of the Andaman and Nicobar Islands, the west coast harbors funnel-shaped estuaries or backwater mangroves (Mulla and Chavan 2017).

5.2.2.2.1 Mangroves in Gujarat

A total area of 1103 km^2 out of the 1600 km coastline in Gujarat is covered by mangrove forests, thus making it the state with the second largest area under mangrove cover. These saline forests cover the regions of South Gujrat that include the Gulf of Khambhat-Dumas Ubharat areas, Gulf of Kachchh, Saurashtra, and Kachchh with Kachchh constituting the largest mangrove share of 71.5%; followed by Gulf of Kachchh (15.6%); Saurashtra (0.3%); and South Gujarat [considering the areas of Gulf of Khambhat and Dumas-Ubharat, 12.6%]. As compared to other states, the high concern about the mangrove conservation and development in Gujarat has resulted in a rise in mangrove cover by 45 km^2.

A large number of mangrove species have been reported from Gujrat such as *Ceriops tagal*, *Avicennia marina*, *Rhizophora apiculata*, *Acanthus ilicifolius*, *Lumnitzera racemosa*, *Avicennia alba*, *Sonneratia apetala*, and *Bruguiera gymnorrhiza*. Other than these, species such as *Avicennia officinalis*, *Kandelia candel*, *Aegiceras corniculatum*, *Ceriops decandra*, *Rhizophora mucronata*, *Bruguiera cylindrica*, and *Excoecaria agallocha* have also been observed to be present there. Among these, *Avicennia marina* is the predominant species, representing as much as 97% of the total repository of mangroves in Gujarat, while *Avicennia ilicifolius, Ceriops tagal*, *Sonneratia apetala*, *Rhizophora mucronata*, and *Avicennia corniculatum* have been observed to be present in localized patches of vegetation. Though Kachchh has the highest abundance of mangroves, heavy freshwater inflow has led to diverse mangrove population in South Gujarat. Only one single type of species from each of *Bruguiera gymnorrhiza*, *Kandelia candel*, and *Lumnitzera racemosa* is present in South Gujarat.

5.2.2.2.2 Mangroves in Goa

The mangrove reserve in Goa covers an area of 2000 ha along the coastline of 120 km aligning the seven estuaries, namely, the Terekhol (30 ha), Chapora (100 ha), Mandovi (700 ha), Zuari (900 ha), Sal (30 ha), Talpona (20 ha), and Galgibag (20 ha). The major species of mangrove plants found in Goa include *Aegiceras majus*, *Bruguiera conjugata*, *Acanthus ilicifolius*, *Excoecaria agallocha*, *Salvadora persica*, and *Avicennia alba*. Other than these, *Sonneratia apetala*, *Avicennia officinalis*, *Clerodendron inermis*, *Derris scandens*, and *Rhizophora mucronata* are also found in pretty high abundance in the coastal mangrove estuaries of Goa. In addition, Goa also houses for a host of associates of mangroves which include *Halophila beccarii*, *Acrostichum aureum*, *Abrus precatorius*, *Derris heterophylla*, and *Thespesia populnea*. In addition, associates of mangrove species such as *Clerodendron inermis*, *Lannea grandis*, *Cyperus* spp., *Caesalpinia crista*, *Porteresia coaretata*, and *Salvadora persica* may also be observed in the coastal mangroves of Goa.

The mangrove species composition of the mangrove forests of Goa vary with the changes in salinity levels of the region, which steadily decreases from the mouth of the rivers to their upstream regions. The gradient is also dependent on the phenomenon of high tides and low tides. Because there exists very distinct longitudinal and vertical strata in the mangroves of Goa, it leads to the formation of a compact canopy from the high tide levels to the low tide levels. The canopy colors also vary according to tidal levels. Pale green and pale yellowish green color is observed in high tide level canopies, while low tide level canopies usually display colors ranging from dark green to blackish green. The varying combination of canopy colors helps to clearly distinguish among the different species from afar. The varying propagule sizes among the different plant species that occupy different tidal levels help to clearly establish a distinction among the high and low tidal levels that are present at variable depths of the silt occurring in the soil substratum.

The most important zones of mangrove cover in Goa are found along the estuaries of Mandovi in the North and Zuari in the South, connected through Kumbharjua canal, the latter harboring a mangrove-rich area of 200 ha. The rest of the estuaries bear sporadic mangrove patches. Of these, the mangroves of Chodon (Chorao) island on Mandovi estuary is a protected mangrove region covering an area of 160 ha harboring a rich flora of mangroves made up of 12 major and 8 associate species of mangrove plants. Interestingly, this is a pristine island with no major anthropogenic threats and thereby forms one of the best mangrove reserves of the country. The mangrove vegetation of this region is interspersed with a water channel, having tidal variations, and has opulent marine fauna and avifauna. The mangroves along Mandovi and Zuari estuaries vary in composition with a higher diversity of species been recorded along Mandovi than Zuari. *Sonneratia alba* followed by *Avicennia alba* and *Avicennia officinalis* is the predominant mangrove species along Mandovi estuary, while the latter two are the dominant species along Zuari estuary, respectively. *Sonneratia caseolaris* is observed in prominence and found in high density in this particular region because of the low salinity conditions that are prevalent there.

In further North along Chapora, which is the third largest mangrove area of the state, *Bruguiera gymnorrhiza* and *Aegiceras corniculatum* form the foremost mangrove species.

5.2.2.2.3 Mangroves in Karnataka

Karnataka harbors fringing type mangrove vegetation in linear formations along its 320 km long coastline covering an area of 300 ha. These are mostly located along the backwaters, estuaries, islands, and other areas under protection, within the intertidal regions, and are dependent on the tidal range, water and soil salinity, substratum, and the expanse of the intertidal region. District-wise distribution of mangroves in Karnataka has shown Kali estuarine complex (605.48 ha), Sharavathi estuarine complex (159.47 ha), Gangavali estuary (94.78 ha), and Aghanashini estuary (707.3 ha) to be the potential reserves of mangrove forests from North Kannada district (Mangrove Mapping of Uttara Kannada District 2016). In total, 14 species of true mangroves, belonging to 9 genera under 7 families have been reported along the estuaries of the Karnataka coast which includes *Acanthus ilicifolius*, *Acrostichum aureum*, *Aegiceras corniculatum*, *Bruguiera cylindrica*, *Rhizophora apiculata*, *Avicennia marina*, *Rhizophora mucronata*, *Avicennia officinalis*, *Sonneratia caseolaris*, *Excoecaria agallocha*, *Avicennia alba*, *Bruguiera gymnorrhiza*, *Sonneratia alba, and Kandelia candel.* North Kannada district harbors all these 14 species (Mangrove Mapping of Uttara Kannada District 2016). The most dominant ones among them include *Acanthus ilicifolius*, *Avicennia marina*, *Rhizophora mucronata*, *Avicennia officinalis*, *Clerodendron inerme*, *and Rhizophora conjugata*. The Karnataka coast is lined by dense mangrove forests comprising of extremely old varieties of *Avicennia officinalis*, *Rhizophora mucronata*, and *Sonneratia caseolaris*, found along the rivers Kali nadi, Belamdar nadi, Gangavali, Agnashini, and Sharavati at Karki, Haldipur, Dhareshvar, and Honavar. A dense mangrove-covered area of 30,700 m^2 is found at the meeting point of the Chakra, Haladi, and Kollur rivers at Coondapur, just before opening onto the Arabian Sea. The major plant species of this region are *Avicennia officinalis*, *Rhizophora mucronata*, *Acanthus ilicifolius*, *and Rhizophora conjugata*.

5.2.2.2.4 Mangroves in Maharashtra

Maharashtra has no dense mangrove cover but has pockets that can be described as open mangroves with sparse coverage along its 720 km long coastline and contributes 45% to the total mangrove growth across India. The mangroves of Maharashtra are distributed over five districts of Thane, Sindhudurg, Ratnagiri, Raigad, and Bombay. Though 40% of mangroves have been destructed due to anthropogenic activities near Thane district over the last 25 years, in regions of Mumbai suburbs and Navi Mumbai, the mangrove cover has increased by 37% during 2015–2017. In Ratnagiri district, about 11 mangrove species are reported, which includes *Aegiceras corniculatum* (L.) Blanco, *Rhizophora mucronata* Lamk., *Ceriops tagal* (Perr.), *Lumnitzera racemosa* Willd, *Avicennia marina* (Forsk.) Vielh, *Avicennia officinalis* L., *Rhizophora apiculata* Blume, *Sonneratia alba* J. Smith, *Avicennia marina* (Dwarf), *Acanthus ilicifolius* L., and *Excoecaria agallocha* L.

5.2.2.3 South Coast

The south coast of India consists of the two states of Kerala and Tamil Nadu that cover an area of 1424.73 km with Tamil Nadu sharing 864.73 km and Kerala covering an area of 560 km, respectively. This long coastline harbors a rich repository of mangrove vegetation, covering an area of 49 km^2 and 9 km^2 in Tamil Nadu and Kerala, respectively.

5.2.2.3.1 Mangroves in Tamil Nadu

The mangroves in Tamil Nadu are divided into two types as moderately dense (16 km^2 area coverage) and open mangrove (23 km^2 area coverage). However, unlike the east coast, no dense mangrove region is found in Tamil Nadu coast. These mangroves are located at the Cauvery deltaic areas of Tamil Nadu. The largest area of mangroves is covered by the district Cuddalore (19 km^2 area covered by mangrove vegetation: 9 km^2 comprising of moderately dense mangrove forests and 10 km^2 comprising of open mangrove cover) followed by Ramanathapuram (8 km^2 area covered by mangrove vegetation: 5 km^2 comprising of moderately dense mangrove cover and 3 km^2 comprising of open mangrove forests) and Chidambaram (7 km^2 area covered entirely by open type mangroves), respectively. Besides this, Chatram, Kodiakarai (Point Calimere), Muthupet, Tuticorin, and Vedaranyam are also rich abodes for mangrove flora.

The Pichavaram mangroves situated at Pichavaram near Chidambaram between the north Vellar estuary and south Coleroon estuary ranks among one of the most exquisite scenic spots of the state and is one of the most well-developed as well as largest mangrove coastal forests of the country. It also abodes for diverse aquatic birds, shell, and finfishes. It consists of some islands interspersing a vast expanse of water covered with mangroves covering a total area of 1100 ha and is distanced from the Bay of Bengal by a sand bar. The dominant mangrove species of this biotope are *Avicennia marina*, *Aegiceras corniculatum*, *Lumnitzera racemosa*, *Ceriops decandra*, *Rhizophora* spp., *Bruguiera cylindrica*, and *Excoecaria agallocha.* About 30% of the mangrove community is occupied by *Avicennia marina*, subsequently followed by 17% of *Bruguiera cylindrica* and 16% of *Avicennia officinalis*. Besides these three, the density of other mangrove species is very poor, nearly on the verge of extinction due to the cyclonic events that resulted in the reduction of the beach areas.

For this same reason, various plant species which were previously reported from the mangrove estuary of Pichavaram have now become extinct. For example, pollen grains analysis from the Pichavaram soil sediments revealed that certain species which were abundantly observed previously, such as *Sonneratia* sp. (Caratini et al. 1973), are presently nearing complete extinction. Furthermore, species such as *Bruguiera gymnorrhiza*, *Xylocarpus granatum*, and *Rhizophora stylosa* once found in this ecosystem are not existing presently (Kannupandi and Kannan 1998). Most of the existing *Rhizophora* sp. are matured varieties with low reproduction rate and are mostly being replaced by *Avicennia marina* which is, in general, a more dynamic species (Kannupandi and Kannan 1998). Besides mangrove plants, a total of 20 dinoflagellates species, 30 types of crabs, 30 specimens of copepods, 30 classes of mollusks, 30 types of prawns, 40 tintinnid species, 100 types of diatoms, and 200 varieties of fish have been documented from Pichavaram mangroves (Anon 1987).

5.2.2.3.2 Mangroves in Kerala

The mangrove forests of Kerala, covering an area of 6.63 km^2, are broadly distributed over the four districts of Kannur, Kozhikode, Ernakulam, and Alleppey. The two most commonly observed species of mangrove plants found in all the mangrove estuaries in Kerala are the *Avicennia* and *Rhizophora* species. Besides, moderately dense mixed mangroves species comprised of *Sonneratia alba*, *Rhizophora mucronata*, *Avicennia officinalis*, *Aegiceras corniculatum*, *Avicennia marina*, *Excoecaria agallocha*, *Sonneratia caseolaris*, and *Rhizophora apiculata* are covering approximately an area of 259.26 ha. From the density of canopy cover, a minor area of 26.90 ha (canopy cover < less than 10%) exclusively inhabited by mixed mangroves community is referred to "degraded" [less than 10%].

The highest mangrove distribution contributing to 80% of the total state's mangrove reserve is found in the Kannur region along four ranges of forests, namely, Kasaragod, Taliparamba, Kottiyur, and Kanhangad. These forest ranges differ concerning the type of mangrove species and the distribution pattern of the mangrove species. A wide spectrum of well-conserved mangrove forests from Kannur district are found in the bridge area of Vallapatnam. The two principal species of this region are *Avicennia officinalis* and *Rhizophora apiculata*, constituting up to 40–70% density.

5.3 Approaches to Assess Microbial Diversity

The study of the complex microbial community from different ecological niche is one of the key goals for ecologists. It is a challenging task to grow and study the functional attributes of environmental microbes under laboratory conditions due to the limitations in the procuring of exact conditions suitable for their growth and functions. Therefore, polyphasic studies combining culture-dependent and culture-independent analyses are applied to assess the community diversity, abundance of species, and their functional contributions in a specific ecosystem. Such analysis allows one to gain insight toward species-species communication, species-environment interactions, and the overall dynamicity of the microbial population in a particular ecosystem.

5.3.1 Culture-Dependent Assessment of Microbial Diversity

The application of culture-based microbiological techniques has been an ancient but highly proficient and robust tool to assess the total composition, diversity, and functional role of microbial consortia in environmental samples for ages. Even in today's modern world, where "omic" technologies are growing in stature as potential scientific tools to study the microbial load in environmental samples, the ancient culture-based assessments are essential to support and validate the findings. Therefore, for unique environments like the mangrove ecosystem, the culture-based analyses of the mangrove-associated and non-associated sediment microbiome help to generate a holistic knowledge about the microbial consortia, its functional activities, and interactions with various environmental components.

5.3.1.1 Bacterial Diversity

A wide variety of studies determining the bacterial diversity in mangrove sediments and rhizosphere from all over the coastlines of India have been performed, and the bacteria with biotechnological potential have been isolated. The culture-dependent study of the resident bacterial population of the mangrove sediments of Bhitarkanika revealed *Desulfotomaculum*, *Desulfomonas*, *Vibrio*, *Bacillus*, *Desulfovibrio*, *Micrococcus*, *Pseudomonas*, *Azotobacter*, *Methylococcus*, and *Klebsiella* as the predominant bacterial genera there (Mishra et al. 2012).

A recent study reported the isolation and characterization of 18 bacterial strains that displayed drug resistance character against multiple antibiotics, namely, vancomycin, tetracycline, kanamycin, and ampicillin from sediments of eight sampling stations of Sundarban (Bhattacharyya et al. 2019). These isolates were identified as *Staphylococcus*, *Nitratireductor*, *Salinivibrio*, *Oceanobacillus*, *Marinobacter*, *Halobacillus*, *Mesorhizobium*, *Thalassocella*, *Bacillus*, *and Jeotgalicoccus.* An interesting study on mosquitocidal properties was performed with the isolates *Bacillus sphaericus*, *Bacillus cereus*, *Bacillus subtilis*, and *Bacillus thuringiensis*, obtained from the mangroves of Vellar estuary (Balakrishnan et al. 2015). Production of stable gold nanoparticles by *Staphylococcus warneri* isolated from Sundarban mangrove regions has been reported (Nag et al. 2018). This was proved to be highly effective for bioremediation of toxic nitro aromatic pollutants. Similarly, the variation in the emulsifying capabilities of *Halobacillus trueperi* isolated from mangroves of Goa was studied in the presence of zerovalent iron nanoparticles (Kharangate-Lad et al. 2016). DNase-free ribonuclease (RNase) was isolated from *Bizionia* sp. belonging to *Cytophaga-Flavobacterium-Bacteroides* phylum from the marine environment of the Sundarban mangroves (Sana et al. 2008). Isolation attempts from the rhizosphere soil of the mangroves of Sundarban resulted in the successful isolation of a *Stenotrophomonas maltophilia* strain that displayed resistance to both antibiotics and heavy metals and showed several plant growth-promoting attributes such as siderophore, protease, indole acetic acid, and ammonia production as well as solubilization of phosphate (Nayak et al. 2016). A notable count of phosphate-solubilizing bacteria and a high phosphatase activity were reported in the saline coastal mangrove forests of Sundarban (Das et al. 2014a).

Plant growth-promoting attributes were studied among a diverse group of diazotrophic bacteria identified as *Bacillus* sp. and *Agrobacterium* sp. isolated from the coastal saline sediments of different forest and agricultural lands in the Sundarbans (Barua et al. 2012). Field trials with these species as bioinoculants showed high productivity for rice and lady's finger. A siderophore-producing

Halobacillus trueperi was isolated from the plant litter of mangrove forests of Goa. The strain displayed considerable potency in chelating iron from the CAS (chrome azurol sulphonate) medium (Kharangate-Lad and Bhosle 2016). The production of siderophore and its characterization was observed for another bacterial isolate from Goan mangrove ecosystem and was identified as *Bacillus amyloliquefaciens* NAR38.1 (Gaonkar and Bhosle 2013). Recently, a novel species, namely, *Microbulbifer mangrovi* strain DD-13T with versatile polysaccharide-degrading abilities, was isolated from mangroves of Goa and sequenced to identify the respective genes (Imran et al. 2017). Similarly, a genome sequencing performed on the *Pseudomonas hussainii* MB3 strain isolated from the rhizosphere soils of mangrove forests of Andaman Islands revealed its versatility for sulfate and nitrate reduction and denitrification (Jaiswal et al. 2017). Nine nitrogen-fixing bacterial isolates from groups displaying a response to O_2 were isolated from a large diazotroph population from the roots of the mangrove trees and other plant species associated with mangroves at the Ganges river estuary (Sengupta and Chaudhuri 1991). Recently, a group reported 35 and 25 bacterial isolates from Mandovi and Zuari mangrove sediments of Goa, respectively, and compared them on the basis of their capabilities to be used as biofertilizers, depending on their abilities to produce antineoplastic enzymes and biodegradatives (Haldar and Nazareth 2018). These strains were identified to be *Paenibacillus* spp., *Lysinibacillus* spp., *Bacillus* spp., and *Staphylococcus* spp. A higher number of isolates from the mangrove sediment of Zuari showed the production of biodegradative and antineoplastic enzymes indicating mangrove sediment of Zuari to be a rich repository for bacterial species of probable biotechnological importance. Screening of 800 rhizobacteria from *Avicennia marina* of South India mangrove ecosystem concerning the production of N-acyl homoserine lactone identified 56 positive isolates belonging to 11 distinct genera from *Gammaproteobacteria*, viz., *Vibrio*, *Serratia*, *Pseudomonas*, *Photobacterium*, *Aeromonas*, *Halomonas* and novel species *Photobacterium halotolerans* MSSRF QS48, *Pseudomonas* sp. MSSRF QS1, and *Vibrio xiamenensis* MSSRF QS47 (Viswanath et al. 2015). Nitrogen-fixing *Azospirillum brasilense*, *Azospirillum irakense*, *Azospirillum halopraeferens*, and *Azospirillum amazonense* were isolated from the rhizospheric soils of *Avicennia officinalis*, *Rhizophora mucronata*, and *Acrostichum aureum* in Manakkudi mangrove ecosystem (Ravikumar et al. 2012). This indicates the supreme necessity for the preservation of resident microbial populations present in the mangroves of India and in-depth research on the isolation of beneficial microbes targeting their application in agriculture, industry, and environment management. From Vellar estuary, *Lysinibacillus* sp. is reported as the first endophyte isolated from the mangrove leaves with the ability to produce adenosine deaminase (Kathiresan et al. 2014). The diazotroph *Vibrio plantipsonsor* sp. nov. containing the key functional nitrogenase gene (*nifH*) was isolated from the Pichavaram mangroves and thereby proposed to have a role in nitrogen fixation (Rameshkumar et al. 2011). L-asparaginase production was studied from *Bacillus cereus* MAB5 isolated from Vellar estuaries (Thenmozhi et al. 2011). A study was performed to understand the effect of growth promotion and pigment production by mangrove seedlings of *Avicennia marina* and *Ceriops decandra* after inoculation with *Azotobacter* sp. isolated from the mangroves of South India. A significant level of increase was noted in the dry biomass of the plant roots (Ravikumar et al. 2007).

A group of pathogenic bacterial species like *Vibrio cholerae* O1 and *Vibrio fluvialis* have recently been reported to be present in the Sundarban mangroves (Sultana et al. 2018; Bhattacharjee et al. 2010). Metal-resistant *Klebsiella pneumoniae* was isolated from the marine shrimp *Penaeus monodon* collected from Sundarban (Choudhury and Kumar 1998). Later, antibiotic-resistant *Vibrio parahaemolyticus* was also isolated from these shrimps (Bhattacharya et al. 2000). Even from the mangrove ecosystems of Goa in the West coast, 14 serotypes and 23 untyped species of *Escherichia coli* were detected (Poharkar et al. 2014). Besides, *Listeria*, *Salmonella* and *Vibrio* spp. were also detected from these mangroves at a higher range of colony-forming units (CFU) above the standards (Poharkar et al. 2016). It is noteworthy to mention that in comparison to India, mangrove-specific *Vibrio* spp. have also been reported from the mangrove waters of Ceará, Brazil (Sousa et al. 2006). Taken together, these studies indicate that mangroves are rich reservoirs of pathogens and are influenced by anthropogenic activities.

5.3.1.2 Cyanobacterial Diversity

In extreme or oligotrophic environment of mangrove ecosystem, where the nutrient availability is poor, the cyanobacteria form an important partner toward the assimilation of carbon, nitrogen fixation, phosphate storage, biosorption of xenobiotics, plant growth promotion, and bioremediation (Alvarenga et al. 2015). Though studies on this group of bacteria in the mangrove ecosystem have been underrated over periods, the reports to date have identified a host of marine cyanobacterial taxa from these unique microhabitats using culture-based methods. Halophilic cyanobacteria with antimicrobial properties were successfully isolated from the intertidal soils of the coastal forests of Sundarban. These isolates morphologically belonged to the *Lyngbya-Phormidium-Plectonema* (LPP) group B and were identified to be *Leptolyngbya* sp., *Oscillatoria* sp., and *Synechocystis* sp. (Pramanik et al. 2011). Hazra and Saha Kesh (2017) reported the production of photosynthetic pigment, phycocyanin, from cyanobacteria isolated from these Sundarban mangrove regions.

5.3.1.3 Actinomycetes Diversity

Mangroves are a hot spot for actinomycetes activities and an important resource for the identification of novel secondary metabolites as well as the investigation into the mechanism of their biosynthesis (Hong 2013). Sixty-six actinomycetes isolates that belonged to *Streptomycetes* genera were recovered using enrichment medium from sediments of Coringa mangrove ecosystem, Andhra Pradesh, India, (Shrivastava et al. 2015). From the mangrove sediments of Chorao Island, Goa, *Agromyces indicus* sp. *nov.* strain NIO-1018 (T) belonging to genus *Agromyces* was isolated (Dastager et al. 2012). Antioxidant-producing *Streptomyces* sp. were also isolated from mangrove soil of Vishakhapatnam region (Rao and Rao 2013). *Brevibacillus brevis* EGS9, a novel bacterial strain exhibiting strong antimicrobial activity against vancomycin and methicillin-resistant actinobacteria, was isolated sediments from the soil sediments of the mangrove forests of Tamil Nadu along South coast of India (Arumugam et al. 2018). A total of 17 actinobacterial strains with amylase inhibiting abilities were successfully isolated from the rhizosphere soils of the mangrove plant, *Rhizophora mucronata* (Poir), in Vellar estuary, on the eastern coast of India (Raja et al. 2010). From Sundarban mangrove forests of both Bangladesh and India, a total of 55 actinomycetes were isolated that showed considerable activity against several multidrug-resistant bacteria, including the latest medicinal hazard methicillin-resistant *Staphylococcus aureus* (Arifuzzaman et al. 2010; Saha et al. 2005). Similar antimicrobial activities were studied with 55 actinobacterial isolates against 15 test organisms including 3 phytopathogens from 5 stations along Matla river in this Sundarban region (Sengupta et al. 2015). Of these, nine morphologically distinct isolates which showed high biological activities were identified as *Streptomyces* sp. Recently, *Streptomyces euryhalinus* sp. *nov*. was isolated on ISP (International Streptomyces Project) medium 2 and characterized (Biswas et al. 2017). A *Streptomyces sundarbansensis* sp. *nov*. strain isolated from Sundarban mangrove forests was observed to be producing the compound 2-allyloxyphenol (Arumugam et al. 2011). The production of antimicrobials through the formation of biofilm by three salt-tolerant strains of actinobacteria that were isolated from Sundarban regions and identified as *Streptomyces parvallus* was studied using rotating disk bioreactor technique (Sarkar et al. 2008). A correlation study of antagonistic behavior of the actinomycetes with the physicochemical characteristics of the Sundarban mangroves revealed soil nitrogen to be the determinant for the antagonistic activity (Mitra et al. 2008). A higher abundance of actinobacteria was reported in the winter season in Bhitarkanika, India, and isolated *Micrococcus* sp. by culture-based methods from these tropical mangrove sediments (Mishra et al. 2012). This was attributed to the seasonal variation of soil pH and salinity (Mishra et al. 2012).

5.3.1.4 Fungal Diversity

Among the different mangrove species, *Rhizophora* sp. provides the most suitable niche for marine fungi. Till date, 201 fungal species have been isolated from *Rhizophora* sp. of which 10 are specific to this host. Endophytic fungi belonging to freshwater hyphomycetes, viz., *Mycocentrospora acerina* and *Triscelophorus acuminatus*, were isolated from the whole root segments of *Rhizophora mucronata*, *Sonneratia caseolaris*, and *Avicennia officinalis* from the mangrove ecosystems of Karnataka (Udyavara) on the western coast of India. This report also showed higher counts of fungi in *Rhizophora mucronata* as compared to the other plant species and also recorded variations in the number of isolates depending on the tidal variations with the highest being from the mid-tide level samples (Ananda and Sridhar 2002). A study showing the seasonal variation of microbes from the sediments of mangroves of Bhitarkanika, Odisha, India, revealed a higher abundance of fungi during rainy season as compared to bacteria which were higher in winter (Mishra et al. 2012).

A novel fungal species classified as *Thraustochytrium gaertnerium* belonging to the class *Labyrinthulomycetes* closely similar to *Schizochytrium aggregatum Goldstein and Belsky* and *Thraustochytrium kinnei Gaertner* has been reported from the mangroves of Goa (Bongiorni et al. 2005). From the mangrove forests of Goa coastline, four fungal isolates belonging to *Ascomycetes* and *Basidiomycetes* were isolated and were reported as powerful decolorizing and detoxifying organisms for the textile effluents, thereby indicating their potential use for bioremediation (Verma et al. 2010). Four morphotypes of halotolerant and heavy metal-resistant *Penicillium* were isolated from the mangroves of Goa coastline. These isolates showed the highest tolerance against lead, followed by copper and cadmium (Nazareth and Marbaniang 2008). A total number of 187 halophilic filamentous fungal isolates were isolated from both the water column and soil sediment samples from the coastal mangroves situated along the Mandovi estuary in Goa and offshoot of the estuary (Nayak et al. 2012; Nazareth and Gonsalves 2014). These were identified as obligate halophile *Aspergillus* species (*A. pennicilloids* and *A. ochraceus*, *A. tamari*, *A. flavipes*, *A. terreus* vs *terreus*, *A. versicolor*, *A. wentii*) and facultative *Eurotium* species (*E. amstelodami*, *E. repens*), *Penicillium* species (*P. asymmetrica* sec *fasciculate*, *P. corylophilum*), and the black yeasts (*Hortaea werneckii*). This study also evaluated the halotolerance of these fungal species which revealed that they could grow in 0 to 20% of salt with optimum growth occurring between 5 and 10% of salt. The production of cellulases was studied among 12 fungal isolates obtained from the debris of mangrove plants and the sediments of Valanthakad Mangroves, Kerala. The positive isolates were identified as *Trichoderma viride* (Nathan et al. 2014). This same genus of fungi isolated from mangroves of Tamil Nadu was found to degrade synthetic dye, malachite green (Saravanakumar and Kathiresan 2014). Mangroves are also a rich hub for pathogenic fungi that occur in the form of foliar endophytes such as *Phyllosticta capitalensis* (teleomorph *Guignardia mangiferae*) and have been studied from the mangroves of Pichavaram in Tamil Nadu (Suryanarayanan et al. 2004). This pathogen was proved to produce melanin as an adaptive mechanism for the stressed environment of saline mangrove ecosystem, exposed to tidal variations.

Paris-type arbuscular mycorrhizal (AM) root association has been reported from the roots of 31 mangrove species and their associates, 23 transported flora species from 25 different families at four levels of succession of the community in mangroves from deltaic regions of the estuary of the Ganges river (Sengupta and Chaudhuri 2002). The intensity of AM association varied from mangrove-type species as well as salinity, with the intensity of association increasing with decreasing salinity. Besides, seven types of AM fungi were isolated from the root-free rhizosphere which was found to take part in phosphorus solubilization and increase in biomass of the plants. Using trypan blue staining method, similar AM association was observed from the mangrove species *Acanthus ilicifolius*, *Ceriops tagal*, and *Excoecaria agallocha* from Sundarban region (Kumar et al. 2008). Forty-four AMF species belonging to six genera, namely, *Glomus*, *Scutellospora*, *Entrophospora*, *Sclerocystis*, *Acaulospora*, and *Gigaspora*, were reported from this mangrove ecosystem (Kumar and Ghose 2008). Antimicrobial activities were evaluated for the endophytic fungi (*Pestalotia* sp.) isolated from the *Heritiera fomes Buch.-Ham.*, which is a species of mangrove plant found in the Sundarban mangroves (Nurunnabi et al. 2018).

5.3.1.5 Protozoan Diversity

Benthic protozoa play an important role in the maintenance of the integrity of mangrove estuarine ecosystems due to their high abundance of occurrence and significant biomass (Chen et al. 2009). In spite of its importance, researches on this community are limited particularly in the Indian scenario. In the phylum Protozoa, ciliates form complex and diverse but popular parasites. Being parasitic, they are mostly found to occur with the invertebrates of marine and freshwater organisms and thereby are reported from marine environments including mangroves. An early report is found on the isolation of *Crassostrea cucullata* belonging to the Ciliates from the island of Sagar, located at the meeting point of the river Hoogly with the Bay of Bengal (Zoological Survey of India 1988). Besides this, *Crassostrea gryphoides* is also reported from Kaikhali in West Bengal, India (Biswas and Bandyopadhyay 2011). Protozoan parasite of the rare and endangered genus *Cristigera*, feeding on edible oyster *Crassostrea gryphoides*, has been recorded from the estuarine areas of Frasergunj, Digha, and Kaikhali in West Bengal during the periods of monsoon and post-monsoon (Biswas and Bandyopadhyay 2014, 2014b, 2016). Due to its temporal and spatial drifts, this species is regarded as a potential bio-indicator of the environment and is used to detect the salinity distribution patterns in the water. A heterogeneous population of microorganisms composed of 12 flagellates, 17 ciliates, and 2 sessile ciliates was isolated from degraded leaf litter of *Avicennia* sp. and *Excoecaria* sp. from Coringa mangrove forest at Kakinada bay facing the Bay of Bengal on India's east coast and were identified using vital staining method (Dorothy et al. 2003). The dominant flagellate species noted were *Chromulina* sp., *Astasia* sp., *Heteronema* sp., *Spumella* sp., *Euglena acus*, and *Paranema* sp., while the ciliates included *Prorodon* sp., *Cyclidium* sp., and *Zoothamnium* sp.

5.3.1.6 Plankton Diversity

The planktons, viz., autotrophic phytoplanktons, heterotrophic zooplanktons, bacterioplanktons, and mycoplanktons, play crucial roles in contributing to the productivity of the mangrove estuarine ecosystem by providing water columns, serving as primary and secondary food sources, and determining the overall health and fertility of the mangroves by maintaining the food web (Saifullah et al. 2016). However, recent literature shows that this group of organisms has been underestimated in research.

In India, Sundarban mangrove forests have been reported as a rich niche for diverse phytoplanktons. However, there is only one report investigating the occurrence of zooplankton associated with the serogroups O1 and O139 of *Vibrio cholera* from this heritage site (Magny et al. 2011). 36 species of planktons were reported in this mangrove ecosystem (Aziz et al. 2012). From this same mangrove estuarine water, a total of 134 species of phytoplankton consisting of 99 different species from 41 different genera of *Bacillariophyta*, 12 different species from 9 different genera of *Chlorophyta*,18 different species from 6 different genera of *Pyrrophyta*, 4 different species from 4 different genera of *Cyanobacteria*, and 1 single species of *Ochrophyta* were identified (Rahman et al. 2013). These planktons had a population density of 3.755×10^3–10^5 cells/L. A total of 15 genera under the class *Chlorophyceae*, *Myxophyceae*, and *Bacillariophyceae* were identified by another group from four main streams of the Sundarban mangrove ecosystems throughout 1 year in 2005. They reported *Cosciodiscus* sp. and *Microcystis* sp. as the most dominant genera in this phytoplankton community (Mamun et al. 2009). The bloom generated by pennate diatom *Thalassiothrix frauenfeldii* and centric diatoms *Chaetoceros lorenzianus* and *Coscinodiscus radiates* due to high nutrient availability due to the effluents from the Hoogly river in addition to aquaculture practices was observed in Sagar Island of western Sundarban mangroves (Biswas et al. 2014). The number of plankton species (15–18) with cell densities (11.4×10^5 to 12.3×10^3) increased during this phase than normal (Biswas et al. 2014). Conversely, a decrease in abundance and diversity of plankton communities (from 4140 to 2997 cells/L; $H' = 2.72$–1.33) from this same island was reported during Ganga Sagar festival from 2012–2014 (Rakshit et al. 2015). Analysis of surface water of northeastern zone of Sundarban resulted in the isolation of 20–200-μm-long mesozooplankton during the periods of pre- and post-monsoon as well as during the proper monsoonal season in 2014. A total of 50 dominant copepods were identified and were grouped under 22 families and 43 genera (Bhattacharya et al. 2014). Recently, similar observations have been reported from 15 different study stations in the Saptamukhi river which is one of the most important estuaries feeding the Sundarban mangrove regions (Nandy et al. 2018). All in all 56 taxa of zooplankton were identified with the predominant group of copepods accounting to 33 different species with 25 different species of calanoids accounting for the bulk of the biomass, followed by 5 different species of cyclopoids and 3 different species of harpacticoids, respectively. The dominant species among copepods were *Bestiolina similis*, *Paracalanus parvus*, and *Oithona similis*. Predominance of diatom species (*Bacillariophyceae*) followed by *Chlorophyceae* and *Pyrrophyceae* (*Dinoflagellates*) among the total 46 taxa from 6 different groups was reported in Sundarban mangrove sediments (Manna et al. 2010). The decrease in abundance and diversity of the species from summer to winter was also observed indicating the effect of temperature on algal population. The presence of species that act as bioindicators such as *Prorocentrum concavum* and *Polykrikos schwartzii* indicate a high level of pollution in this estuary. With the help of light microscopy, a total of 15 different diatom species (4 pennate and 11 centric forms) were documented, from the Indian portion of Sundarban Biosphere Reserve (Samanta and Bhadury 2018).

A high abundance of phytoplanktons (750–321,000 cells/L) has been recorded from Pichavaram mangrove waters of Tamil Nadu, with the highest abundance being in summer and the lowest during monsoons (Rajkumar et al. 2009). This study also evaluated a high gross net primary production of 16.5–826.8 mg Cm^{-3} h^{-1} by the planktons in this estuary. Investigation along Kaduviyar estuary at Nagapattinam identified 85 different species of phytoplanktons in total composed of 58 species of diatom, 16 species of dinoflagellates, and only 7 species of *Chlorophyceae* species. As compared to Pichavaram mangrove waters, this estuary holds a higher density of planktons [14,135–74,697 cells/L] but a low gross net productivity of 55 $mgCm^{-3}$ h^{-1} as compared to Pichavaram estuary (Perumal et al. 2009). A study from Muthupettai mangrove regions of Tamil Nadu over a 1-year period (2010–2011) identified 95 plankton species belonging to the families of *Coscinodisceae*, *Ceratiaceae*, *Chaetoceraceae*, *Biddulphoidae*, *Naviculaceae*, *Triceratiinae*, *Solenoidae*, *Fragilariaceae*, *Dinophyceae*, *Cyanophyceae*, *Euodicidae*, *Eucambiinae*, *Prorocentraceae*, *Triadiniaceae*, *Isthiminae*, and *Gonyaulacaceae* (Varadharajan and Soundarapandian 2015). A total of 69 zooplankton species with *Copepods* as the predominant species were identified from the creek waters of three sites along the western mangroves of Kachchh, Gujarat from 1999–2000 with a population density of 3×10^4 to 21×10^4 per m^3 (Saravanakumar et al. 2007). Recently, marine microalgae-like protist, viz., *Thraustochytrium* sp. with the potential of producing docosahexaenoic acid (DHA), has been reported from the mangroves of South India (Chandrasekaran et al. 2018).

5.3.1.7 Archaeal Diversity

Most research on mangrove environments targets the evaluation of functional and structural diversity of bacteria and fungi. However, archaea also form a significant component in this halophilic environment playing a very important role in contributing to the ongoing biogeochemical cycles, nutrient trapping, and turnover of organic matter (Bhattacharyya et al. 2015). Few reports are available describing the occurrence of haloarchaea from the solar salterns/salt pans and estuaries of Goa and Mumbai and their metal-tolerance capabilities (Singh and Singh 2018; Das et al. 2014b; Mani et al. 2012; Salgaonkar et al. 2013; Thombre et al. 2016). *Halobacterium* spp. were isolated from water samples of evaporation salt pans from the four regions, Bheemunipatnam, Parawada, Kakinada, and Iskapalli, spread over a belt of 650 km on the south eastern coast of India covering from Visakhapatnam to Nellore districts in Andhra Pradesh (Keerthi et al. 2018). Siderophore production assays were performed using nine halophilic archaea, viz., *Halococcus saccharolyticus*, *Halogeometricum* sp., *Halobacterium* sp.1, *Halobacterium* sp.2, *Halobacterium salinarum*, *Halobaculum* sp., *Haloterrigena turkmenica*, *Natrialba* sp., and *Halorubrum saccharovorum* isolated from the marine salterns found around the Bhavnagar coast region of Gujarat. Of these five isolates, viz., *Halogeometricum* sp., *Halococcus saccharolyticus*, *Natrialba* sp., *Haloterrigena turkmenica*, and *Halorubrum saccharovorum* were observed to possess the attribute of being capable of producing siderophores (Dave et al. 2006). There is one report of isolation and characterization of methanogenic archaea from eight mangrove sediment samples of Ennore, Chennai, and this study evaluated the production of methane from these isolates (Ramkumar 2014).

Nevertheless, there is no other report on the archaeal diversity from mangrove sediments in India using culture-based methods. This indicates the lacuna present in the research concerning the study of the diversity of archaea and elucidation of their ecological role in mangrove ecosystems of India and therefore points toward an immediate concern to be taken in this field.

5.3.2 Culture-Independent Assessment of Microbial Diversity

The culture-based methods are unreliable when it comes to the proper representation of the resident microbial population of a given environment. This shortcoming arises from the fact that it is extremely difficult to provide an organism with the exact natural conditions that it encounters in its host environment. This may be because of the laboratory conditions provided the interdependency of microbes on one another for growth or for that matter even the overgrowth of dominant species on isolation plates thus masking the rarer species and preventing their successful isolation (Taketani et al. 2009). Only about 1 to 5% of the microbes present in an environment are successfully isolated under laboratory conditions. To get over these problems, newer methods started developing that ultimately introduced the possibility of studying the resident microbial diversity of a given environment without the need to depend on pure culture studies, i.e., the culture-independent techniques.

Although different culture-independent techniques are now used worldwide, all of them fundamentally depend on the same basic principle of successful detection followed up with analysis of a given environmental sample. Past studies would reveal the use of both DNA and RNA for this purpose. While using DNA as the nucleic acid of choice for the assessment of the microbial diversity of a given environment provides one with an overall picture of the resident microbial community, choosing RNA for the same purpose provides one with the knowledge of the active population of resident microbes in the given environment. All of these techniques rely heavily on the successful isolation of DNA and/or RNA from environment samples followed by the amplification of a target gene. Following this the steps diverge into clone library-based analysis, fingerprinting techniques, or even different high-throughput sequencing techniques (next-gen sequencing). Till today, several environmental microbiology groups have employed one or more of these culture-independent approaches to understand the community and structure of microbial groups in different environmental samples both abroad and in India. A summary of the various culture-independent methods employed to address the microbial community diversity of mangroves in India is summarized in Table 5.2.

Table 5.2 Microbial community studies of Indian mangroves using culture-independent methods

Study by	Location	Year	Method	Gene
Bharathkumar et al.	Pichavaram mangroves, Tamil Nadu	2008	Single-strand conformation polymorphism (SSCP)	16S rRNA
Ghosh et al.	Sundarban mangroves, West Bengal	2010	Clone library	16S rRNA
Singh et al.	Mangrove system in Goa	2010	Clone library	16S rRNA
Fernandes et al.	Mangrove system in Goa	2014	Pyrosequencing	16S rRNA
Basak et al.	Sundarban mangroves, West Bengal	2015a, b	Pyrosequencing	16S rRNA
Chakraborty et al.	Sundarban mangroves, West Bengal	2015	Clone library	16S rRNA
Basak et al.	Sundarban mangroves, West Bengal	2015a, b	Pyrosequencing	16S rRNA
Bhattacharyya et al.	Sundarban mangroves, West Bengal	2015	Pyrosequencing	16S rRNA
Basak et al.	Sundarban mangroves, West Bengal	2015a, b	Pyrosequencing	16S rRNA
Loganathachetti et al.	Mangrove system in Tamil Nadu	2016	Pyrosequencing	16S rRNA
Imchen et al.	Mangrove system in Kerala	2017	Illumina sequencing	Whole-genome metagenome and comparative metagenomics
Ghosh and Bhadury	Sundarban mangroves, West Bengal	2017	Clone library and Illumina Sequencing	16S rRNA
Imchen et al.	Mangrove system in Kerala	2018	Illumina sequencing	Whole-genome metagenome and comparative metagenomics
Haldar and Nazareth	Mangrove system in Goa	2018	Illumina sequencing	16S rRNA
Vipindas et al.	Mangrove system in Kerala	2018	Clone library and DGGE	16S rRNA and amoA

5.3.2.1 Fingerprinting and Clone Library-Based Methods

Fingerprinting methods provide an overall idea of the microbial diversity as well as the structure of given environmental samples. It also allows for the successful comparison of the microbial community structure between different environmental samples based on the comparison of patterns that appear corresponding to the most abundant microbial groups in environmental samples. A variety of fingerprinting techniques are now available, all of which provide information on microbial structure based on a set of bands or peaks (Taketani et al. 2009). Examples of some such methods used worldwide are single-strand conformation polymorphism (SSCP), terminal restriction fragment length polymorphism (T-RFLP), temperature gradient gel electrophoresis (TGGE), automated rRNA intergenic spacer analysis (ARISA), and denaturing gradient gel electrophoresis (DGGE). Clone library-based studies allow for the successful comparison of sequences from different environmental samples at different taxonomic levels (Taketani et al. 2009). Clone library construction and analysis allows for both the comparison of distinct samples as well as the accurate phylogenetic affiliation of the members belonging to these environmental communities. These clone libraries may be even used for statistical comparisons such as rarefaction analysis and ecological index estimation. Although in the past decade, different fingerprinting and clone library-based studies have been undertaken worldwide, to reveal the resident microbial populations of different mangroves ecosystems (Brito et al. 2006; Castine et al. 2009; Flores-Mireles et al. 2007; Gomes et al. 2007, 2008; Ikenaga et al. 2009; Lyimo et al. 2009; Liao et al. 2007; Liang et al. 2007; Luo et al. 2009; Sjöling et al. 2005; Sousa et al. 2006; Taketani et al. 2009; 2010a, b; Tian et al. 2008; Yan et al. 2006; Zhang et al. 2009) such studies in India are still few and far between.

The first attempt to understand the microbial diversity of Indian mangroves was performed by characterizing the predominant fraction of the resident bacterial candidates present in different mangrove rhizosphere soils of the Pichavaram mangrove forest of Tamil Nadu with the help of the single-strand conformation polymorphism (SSCP) technique based on the 16S rRNA gene (Bharathkumar et al. 2008). Five different mangrove rhizospheres, namely, *Sonneratia apetala*, *Excoecaria agallocha*, *Avicennia marina*, *Rhizophora mucronata*, and *Porteresia coarctata*, were chosen from four different human-influenced sites within the mangrove forests for analysis. SSCP banding patterns revealed greater spatial variation between the different sites even among the same rhizosphere samples. The excision and sequencing of the bands with a greater intensity that represented the dominant bacterial population revealed sequences with maximum relatedness to soil and marine bacteria as well as unidentified bacterial clones. *Planococcus* and *Bacillus* sequences were observed in all the rhizospheres analyzed. Overall, *Firmicutes* dominated the 16S rRNA gene sequences, while several sequences also revealed the presence of *Gammaproteobacteria*. In 2010, a diversity study of the resident archaeal population of the saline sediments of the mangrove estuaries of Goa, at sites Mandovi and Zuari, was performed using the clone library technique (Singh et al. 2010). The first culture-independent study into the resident bacterial population of the Sundarban mangroves of West Bengal was performed by Ghosh et al. in 2010, using the clone library method. Here, the authors generated two 16S rRNA gene-based clone libraries: one using universal 16S rRNA bacterial primers and the other using proteobacterial 16S rRNA specific primers to selectively enrich the proteobacterial sequences for further insights into the division of sequences among the different classes of *Proteobacteria*. Sequencing of selected clones followed by phylogenetic analysis of the sequences revealed that the Sundarban mangrove sediments house an extremely rich diversity of soil bacteria. The 8 major bacterial phyla to be detected were, namely, *Proteobacteria*, Flexibacteria, *Gemmatimonadetes*, *Chloroflexi*, *Planctomycetes*, *Acidobacteria*, *Firmicutes*, and *Actinobacteria*. The gamma class of *Proteobacteria* was identified to be the most abundant bacterial group present in these saline sediments. This was also the first study that hinted at a probable hydrocarbon and oil contamination of these saline mudflats. This issue was further investigated by Chakraborty et al. in 2015, where they took a clone library-based approach to unravel the resident bacterial diversity from three geographically different locations that had prior reports of being exposed to anthropogenic pressures. The authors constructed 16S rRNA gene-based clone libraries and followed it up with sequencing of the clones in an attempt to identify the microbial players present. The results revealed the presence of bacterial strains related to biodegradation/biotransformation processes. Strains documented to be linked to degradation of hydrocarbons and resistance to heavy metals were identified from these saline mangroves. The microbial diversity patterns observed when correlated to environmental parameters suggested that anthropogenic interventions had a pivotal contribution in shaping of the resident microbial diversity present in the Sundarban mangroves.

Very recently, the bacterioplankton diversity and structure were assessed in the Sundarban mangroves with the help of both clone library construction technique and high-throughput sequencing technique (Ghosh and Bhadury 2017). Out of the 544 clones that were sequenced, *Proteobacteria* and specially *Gammaproteobacteria* and *Alphaproteobacteria* dominated the study across both the post and pre monsoon samples in both of the sampling stations. Other phyla that were observed included *Bacteroidetes*, *Alteromonadales*, *Sphingomonadales*, *Oceanospirillales*, and *Chromatiales*, all of which are reported to take part in the cycling of carbon in coastal regions.

5.3.2.2 High-Throughput Sequencing Techniques

The field of microbial ecology underwent a landmark change with the introduction of the high-throughput techniques which allowed for the generation of large quantities of data in a single analysis, thus providing the opportunity for better comprehension of microbial communities in a given environmental niche (Taketani et al. 2009). These next-generation sequencing technologies thus introduced aimed to reduce the price of DNA sequencing beyond normally achievable limits which are reached while using the standard dye terminator methods (Schuster 2008). Another advantage of these methods is that it allows for the DNA sequencing to take place without the cloning step, thus making the life of the researcher much easier and reducing the bias in the analysis step (Taketani et al. 2009). For the past decade, several environmental ecology groups have utilized these high-throughput techniques to understand the microbial diversity of different environmental systems.

Studies into the mangrove diversity using these techniques are prevalent both abroad (Cleary et al. 2012; Dos Santos et al. 2011; Gomes et al. 2010; Jiang et al. 2013; Pires et al. 2012) and in India.

The first study on the microbial diversity of Indian mangroves using these high-throughput sequencing methods was performed by Fernandes et al. in 2014. The authors applied pyrosequencing to decipher the microbial diversity of pristine and anthropogenically affected mangrove ecosystems in Goa. Consistent with past studies using clone library methods, Proteobacteria were observed to be the dominant phyla in both the pristine and polluted sediments of the mangroves in Goa. However, two striking observations came out of this particular study. One was that contrary to beliefs that anthropogenically affected sediments were pressure environments and would harbor lesser microbial diversity compared to more pristine environments (Hu et al. 2007; Greer 2010), here the total bacterial diversity was observed to be much more in the human-affected sediments in comparison to the naturally preserved sediments. The second observation was the predominance of *Deltaproteobacteria* in the pristine sediments and *Gammaproteobacteria* in the polluted mangrove sediments, respectively. Fernandes et al. (2014) explained that the availability of a wide variety of substrates in the anthropogenically influenced sediments could be responsible for this higher taxonomic diversity there. Also, the higher presence of *Gammaproteobacteria* in these human-influenced sediments could be because of their role in modulating the excess nitrate available in these anthropogenically influenced sediments (Fernandes et al. 2014). Round about the same period, Basak et al. (2015a, b) reported the investigation of the resident bacterial diversity of Sundarban mangrove sediments in West Bengal using the 454-pyrosequencing high-throughput platform to sequence the 16S rRNA gene amplicon. *Proteobacteria* was the most predominant phyla, within which the classes of *Gammaproteobacteria*, *Alphaproteobacteria*, and *Deltaproteobacteria* displayed a higher occurrence. Some of the other predominant phyla that were observed were *Cyanobacteria*, *Bacteroidetes*, *Acidobacteria*, *Planctomycetes*, *Firmicutes*, *Chloroflexi*, *Nitrospira*, and *Actinobacteria*.

Upon performing further taxonomic analysis, the authors observed the abundance of anaerobic and microaerophilic bacterial populations, within the layers of soil close to the surface in the Sundarban mangroves, which ultimately led them to conclude the anaerobic nature of the Sundarban mangrove sediments. This particular study provided in-depth information of the microbial communities that reside in the Sundarban mangrove sediments as well as preliminary insights into the possible parts played by the resident bacterial communities in successfully transforming nutrients in these saline coastal forests. With the goal of understanding the microbial diversity of some of the least explored regions of Sundarban mangroves, Basak et al. (2016) performed metagenomic analysis of soil samples obtained from some of the most pristine regions of these mangroves via special permission from government authorities (Basak et al. 2016). This endeavor revealed the presence of a unique bacterial community structure in these pristine mudflats having a high percentage of unassigned sequences, thus throwing forth the relishing prospect of isolation of novel bacterial species of potential biotechnological utility in the future.

With the progress of scientific methods to understand microbial diversity in different environments, it became increasingly evident that archaea were much more ubiquitous and diverse than earlier believed (Chaban et al. 2006; Leininger et al. 2006). Hence it was commonly believed that archaea might be present in mangrove environments as well. This was confirmed when Pires et al. (2012) reported the archaeal diversity prevalent in mangrove sediments and rhizospheres in Brazil (Pires et al. 2012).

In Indian mangroves, the first report on the archaeal diversity and distribution pattern was put forth in Sundarban mangrove sediment (Bhattacharyya et al. 2015). In this study, the authors described the diversity and distribution pattern of the resident archaeal population, present in the halophytic sediments of Sundarban mangroves, using the amplicon sequencing technique applied to the 16S rRNA gene amplicon. Taxonomic analysis revealed the dominance of *Euryarchaeota* and *Thaumarchaeota* (Marine Group I) in these saline sediments. Several statistical analyses also put forth the picture of a clear shift in the community structure of the resident archaeal populations along the different sampling stations (Bhattacharyya et al. 2015). The authors observed that sampling stations that had a history of anthropogenic influence such as oil/hydrocarbon pollution displayed an archaeal community pattern that was dominated by *Haloarchaea* in comparison to stations that were more pristine where methanogens dominated the archaeal communities (Bhattacharyya et al. 2015). This led the authors to conclude that at least in the Sundarban mangroves, the resident archaeal community structure and diversity were influenced by anthropogenic factors. Loganathachetti et al. (2016) employed the pyrosequencing technique to study the seasonal changes in the microbial diversity of pneumatophore-associated soils of *Avicennia marina* (Loganathachetti et al. 2016). Samples were collected in both monsoon and summer seasons from the soils associated to the pneumatophores of *A. marina* which is native to the coastal mangrove forests between the Vellar River and the Bay of Bengal. *Proteobacteria* was observed to be the most dominant phyla in both the seasons followed by *Acidobacteria* in summer and *Cyanobacteria* in monsoon. The authors claim that *Proteobacteria* play an active role in detoxification as part of the sulfur cycle ongoing in the mangroves which explains their dominance. Among the archaeal population, abundance of phylum *Thaumarchaeota* was maximum followed by the phyla of *Crenarchaeota* and *Euryarchaeota*. The authors explain that *Thaumarchaeota* that includes the ammonia oxidizers are abundant in marine environments where they play active roles in the nitrogen cycle, and hence their high abundance in pneumatophore-associated soils of mangroves is justified due to the coastal location of these forests. Imchen et al. (2017) studied the microbial diversity of mangrove sediment samples taken from the mangrove ecosystem in Kerala using next-gen sequencing on an Illumina HiSeq platform 2500 (Imchen et al. 2017). The authors compared this data to other available mangrove microbial diversity data as well as to the data obtained from rainforest sediment and ocean sediment. Within the four samples from the Kerala mangroves, a strong similarity was observed at the higher taxonomic levels. All the four samples were dominated by the phylum *Proteobacteria*. A similar pattern was reported by groups working on the microbial diversity of both Indian mangroves (Ghosh et al. 2010; Fernandes et al. 2014; Loganathachetti et al. 2016) as well as Brazilian mangroves (Dos Santos et al. 2011). Other dominant bacterial phyla observed were the *Bacteroides*, *Firmicutes*, and *Actinobacteria*. In case of the archaeal diversity, the majority of the reads mapped to the phylum *Euryarchaeota*, followed by the *Crenarchaeota* and *Thaumarchaeota* phyla. Upon comparison of this data with that of a single sample from tropical rainforest and ocean sediment, *Proteobacteria* was observed to dominate in all the forest sediment samples, while the archaeal phylum of *Euryarchaeota* dominated the ocean sediment samples. Various statistical analyses revealed the similarity in microbial diversity pattern among the mangrove forests and rainforest samples from that of ocean sediment samples. Imchen et al. (2018) also performed another study where they investigated the resident microbial diversity of the Kerala mangrove forests using whole metagenome next-generation sequencing techniques followed by a comparison of the taxonomic and functional diversity to that of mangrove metagenomic samples obtained from Saudi Arabia and Brazil. Although certain unique patterns of diversity were observed in the Saudi Arabian and Brazilian mangroves, an overall similarity was observed among all the mangrove samples regarding shared microbial phyla. *Proteobacteria* and *Euryarchaeota* were observed to be the two most dominant phyla common to all the mangrove samples among bacteria and archaea, respectively. Very recently Ghosh and Bhadury (Ghosh and Bhadury 2017) studied the variability of bacterioplankton distribution over the monsoon and post-monsoon periods in the Sundarban mangrove estuary applying both a combination of the clone library and high-throughput Illumina sequencing techniques. The authors observed a prominent variation in the bacterioplankton community structure over a season scale which they linked to the variations in the precipitation level. The monsoons harbored a drastic increase in the numbers of soil sediment-associated bacteria such as the *Desulfovibrio* and *Firmicutes*. The authors claim that this hints at the possible movement of the sediment-residing bacteria into the water column in a suspended state during this wet period. Statistical analysis revealed that variations in the community structure of bacterioplankton are under the influence of the levels of dissolved ammonium and nitrate in these saline coastal forests. Haldar and Nazareth (2018) studied the taxonomic diversity of the resident bacterial community from the Zuari and Mandovi mangrove sediments of Goa with the help of paired-end 16S rRNA amplicon sequencing on an Illumina MiSeq system (Haldar and Nazareth 2018). Taxonomic analysis of the obtained sequences revealed that *Proteobacteria* were the abundant phylum in both the sediment samples. Such results were also observed by Fernandes et al. (2014) in their study of the resident microbial diversity of the mangrove sediments of Goa. Haldar and Nazareth observed that *Actinobacteria* and *Firmicutes* were the other phyla present in predominance in both the sediment samples. However, an eye-catching difference in distribution was observed among the two sediment samples in case of the minor phyla as well as lower bacterial taxa under each of the phyla. The authors observed the presence of *Vibrio* sp. (a pathogenic strain), in the soil sediment of Mandovi, while ecologically significant bacterial classes such as the *Gammaproteobacteria*, *Alphaproteobacteria*, and *Deltaproteobacteria* were observed in the Zuari sediment. These observations led the authors to conclude that the Zuari sediment which was a much more pristine environment harbored a greater bacterial taxonomic richness compared to the Mandovi sediment which was highly influenced by anthropogenic pressures.

5.4 Functional Diversity

Mangrove microbial community is the key determinant of efficient productivity, preservation, and restoration of mangrove ecosystems. The wide spectrum of microflora supported by the tidal variations, typical acidic pH and salinity of the mangrove sediments, and the natural root exudates of the diverse mangrove species plays an important part in the biogeochemical cycles, thereby assisting the nutrient turnover between the soil, plants, and water (Liu et al. 2013; Bai et al. 2013). This intercommunication between mangroves and the microbial community results in the development of a highly dynamic and productive ecosystem at the transition between the terrestrial and marine environments (Huergo et al. 2018). Therefore, investigation of this vast microflora is not only helpful in developing a better understanding of the ecological role of microbes in coastal mangrove ecosystems but is also highly significant in evaluating the importance of these saline coastal forests toward biofertilization, bioremediation, and industrial applications. Besides, the biodiversity of the microbes present in soil serves as a very useful indicator of changes in a given ecosystem (Castaneda and Barbosa 2017).

Till date, most of the studies to assess the functions of microbes in mangrove ecosystem in India are dependent on the isolation of bacterial or fungal species in specific nutrient or enrichment medium and study the antibiotic resistance, metal tolerance, antioxidant properties, plant growth-promoting traits, the production of biodegrading/antimicrobial enzymes, and secondary metabolites with immunostimulatory and antitumor activities (Arumugam et al. 2018; Deshmukh et al. 2018; Haldar and Nazareth 2018; Kharangate-Lad et al. 2016; Saravanakumar et al. 2016; Shrivastava et al. 2015; Nathan et al. 2014; Gaonkar and Bhosle 2013; Kathiresan et al. 2014; Rao and Rao et al. 2013; Dastager et al. 2012; Pramanik et al. 2011; Rameshkumar et al. 2011; Thenmozhi et al. 2011; Verma et al. 2010; Sithranga Boopathy and Kathiresan 2010; Raja et al. 2010; Nazareth and Marbaniang 2008; Suryanarayanan et al. 2004).

The present advancement in molecular analyses of environmental genomes is successful in elucidating the functional genes from the mangrove ecosystem. This helps to evaluate the ongoing activities of the sediment microbes and the interactions between them. An initial study by Imchen et al. (2017) identified functional genes associated with respiration and membrane transport, metabolism of carbohydrate, protein, RNA, DNA amino acids, and their derivatives, synthesizing genes for vitamins, cofactors, prosthetic groups and pigments, genes related to production of cell wall, and capsules and genes related to virulence, defense, and disease using metagenomic analyses from the mangroves of Kerala (Imchen et al. 2017). This same group has recently published comparative functional analyses of the biotic communities of mangroves from four districts of Kerala using the Illumina HiSeq platform (Imchen et al. 2018). This study annotated a total of 7410 protein-coding genes from the mangrove metagenome. Of these, the most diversified gene found in this sediment belonged to protein metabolism with 11 subfunctions including ATP-dependent protease and tRNA synthetase such as valyl, glycyl, leucyl, and class II lysyl (class II). The frequent occurrence of DNA primase and DNA-directed RNA polymerase beta subunit under replication and transcription categories while carbamoyl-phosphate synthase large chain under cell cycle and division category were identified.

Interestingly, long-chain-fatty-acid-CoA ligase associated with the metabolism of aromatic compounds and three genes related to resistance to antibiotics and efflux of toxic materials like the heavy metals like copper and arsenic, viz., cation efflux system protein CusA, acriflavine resistance protein, and topoisomerase IV subunit A, were identified. Similar patterns of the occurrence of resistome with a highly enriched acriflavine resistance protein have also been observed from other mangrove ecosystems (Andreote et al. 2012; Alzubaidy et al. 2016). There is a very recent study evaluating the pattern of distribution and diversity of ammonia-oxidizing archaea (AOA) over a gradient of salinity in the water column of the mangrove estuary in Cochin, a tropical monsoonal estuary situated along the south-east Arabian Sea (Vipindas et al. 2018). Haldar and Nazareth also studied the functional metagenome from Mandovi and Zuari mangrove sediment (unpublished data). This study identified 814 and 646 gene clusters from these two mangrove sediments. Genes belonging to metabolism category (4200 genes) followed by environmental and information (1000 genes) and genetic information processing (1200 genes) were abundant in Mandovi sediment, while cellular processes (500 genes), organismal systems (300 genes), and human diseases (200 genes) formed less abundant categories. In comparison, 1700 genes belonged to metabolism category, 1100 genes belonged to genetic information processing, and 400 genes were of information processing from Zuari sediment.

5.5 Future Scope of Mangrove Microbial Resources in India

Located at the meeting point of coastlines and the sea, mangrove estuaries provide various ecosystem services. They prevent soil erosion and contribute to the proper development of the coastal deltaic regions. Besides, they serve as a shelter for numerous flora and fauna as well as provide direct support to the coastal human habitations in the form of both protections from cyclonic storms as well as providing direct means of livelihood such as fishery, rice cultivation, etc. (Gopal and Chauhan

2018). Although sparse studies have been performed from mangrove stretches in Goa, Sundarban, Tamil Nadu, Kerala, etc., no proper study addressing the complete mangrove microbial load of India has been performed. Being extremely complex ecosystems that harbor different interactions between the different resident microbial species, a thorough understanding of the microbes that reside in the different mangrove systems of this country and their possible interactive roles with each other would help to provide a clear picture of the ecology of these saline estuarine ecosystems. As most of the organisms contribute to one or more stages of the different biogeochemical cycles that are ongoing in these mangrove covers, getting proper information regarding these microbial players would help researchers to judge the productive health of these ecosystems, which is extremely necessary considering the ecosystem services they provide. The population boom in India is affecting the health of the mangrove forests in the country. To accommodate the increasing number of people, mangrove areas are getting continuously reduced as well as being subjected to various human interventions. These anthropogenic pressures disturb the natural functioning of these coastal regions. Hence studies into the ecology of the resident microbial population and the effects that these human interventions have on them would help in understanding how best certain microbial players adapt to the stress. This, in turn, would help to provide researchers with better know-how to build bioremediation strategies for preservation of these coastal forests. Finally, the natural complexity of these mangrove forests along with the anthropogenic stresses that they are being subjected to would lead one to believe that the resident microbes would have an arsenal of novel compounds and pathways to survive in these habitats. Hence, these mangroves act as gold mines containing still uncultured organisms of vast biotechnological potential.

5.6 Conclusion

Mangrove ecosystems have been extensively under the scanner of the scientific community over the past decade, but still, the complexity of these coastal estuarine ecosystems elicits the need for further in-depth exploration of their resident microbial populations. With emerging high-throughput sequencing techniques, in-depth analysis of the microbial loads of the different mangrove stretches in coastal regions of India should not be too difficult a task. Also, the benefits of such endeavors would be the chances of developing better strategies for the protection of these endangered forests especially from the ever-expanding population of the country.

Acknowledgments Authors duly acknowledge the support provided by an extramural grant (no. 38(1391)/14/EMR-II) from the Council of Scientific and Industrial Research (CSIR), Government of India, and research grant of Ramanujan Fellowship to AG from the Science and Engineering Research Board (SERB), India (SR/S2/RJN-106/2012). SM was supported by Senior Research Fellowship from University Grants Commission, Govt. of India.

References

Alvarenga DO, Rigonato J, Branco LHZ, Fiore MF (2015) Cyanobacteria in mangrove ecosystems. Biodivers Conserv 24:799–817

Alzubaidy H, Essack M, Malas TB, Bokhari A, Motwalli O, Kamanu FK, Jamhor SA, Mokhtar NA, Antunes A, Simões MF, Alam I, Bougouffa S, Lafi FF, Bajic VB, Archer JA (2016) Rhizosphere microbiome metagenomics of gray mangroves (*Avicennia marina*) in the Red Sea. Gene 576(2 Pt 1):626–636

Ananda K, Sridhar KR (2002) Diversity of endophytic fungi in the roots of mangrove species on the west coast of India. Can J Microbiol 48:871–878

Andreote FD, Jiménez DJ, Chaves D, Dias AC, Luvizotto DM, Dini-Andreote F, Fasanella CC, Lopez MV, Baena S, Taketani RG, de Melo IS (2012) The microbiome of Brazilian mangrove sediments as revealed by metagenomics. PLoS One 7:e38600

Anon (1987) National aquaculture association: U.S. aquaculture and environmental stewardship. Accessed on 21 Dec 2018

Arifuzzaman M, Khatun MR, Rahman H (2010) Isolation and screening of actinomycetes from Sundarbans soil for antibacterial activity. Afr J Biotechnol 9:4615–4619

Arumugam M, Mitra A, Pramanik A, Saha M, Gachhui R, Mukherjee J (2011) *Streptomyces sundarbansensis* sp. nov., an actinomycete that produces 2-allyloxyphenol. Int J Syst Evol Microbiol 61:2664–2669

Arumugam T, Senthil Kumar P, Hemavathy RV, Swetha V, Karishma Sri R (2018) Isolation, structure elucidation and anticancer activity from *Brevibacillus brevis* EGS 9 that combats multi drug resistant actinobacteria. Microb Pathog 115:146–153

Aziz A, Rahman M, Ahmad A (2012) Diversity, distribution and density of estuarine phytoplankton in the Sundarban mangrove forests in Bangladesh. Bang J Bot 41:87–95

Bai S, Li J, He Z, Van Nostrand JD, Tian Y, Lin G, Zhou J, Zheng T (2013) GeoChip-based analysis of the functional gene diversity and metabolic potential of soil microbial communities of mangroves. Appl Microbiol Biotechnol 97:7035–7048

Balakrishnan S, Indira K, Srinivasan M (2015) Mosquitocidal properties of Bacillus species isolated from mangroves of Vellar estuary, Southeast coast of India. J Parasit Dis 39:385–392

Barik J, Mukhopadhyay A, Ghosh T, Mukhopadhyay SK, Chowdhury SM, Hazra S (2017) Mangrove species distribution and water salinity: an indicator species approach to Sundarban. J Coastal Conserv 22:361–368

Barua S, Tripathi S, Chakraborty A, Ghosh S, Chakrabarti K (2012) Characterization and crop production efficiency of diazotrophic bacterial isolates from coastal saline soils. Microbiol Res 167:95–102

Basak P, Majumder NS, Nag S, Bhattacharyya A, Roy D, Chakraborty A, SenGupta S, Roy A, Mukherjee A, Pattanayak R, Ghosh A, Chattopadhyay D, Bhattacharyya M (2015a) Spatiotemporal analysis of bacterial diversity in sediments of Sundarbans using parallel 16S rRNA gene tag sequencing. Microb Ecol 69:500–511

Basak P, Pramanik A, Roy R, Chattopadhyay D, Bhattacharyya M (2015b) Cataloguing the bacterial diversity of the Sundarbans mangrove, India in the light of metagenomics. Genomics Data 4:90–92

Basak P, Pramanik A, Sengupta S, Nag S, Bhattacharyya A, Roy D, Pattanayak R, Ghosh A, Chattopadhyay D, Bhattacharyya M (2016) Bacterial diversity assessment of pristine mangrove microbial community from Dhulibhashani, Sundarbans using 16S rRNA gene tag sequencing. Genomics Data 7:76–78

Bharathkumar S, RameshKumar N, Paul D, Prabavathy VR, Nair S (2008) Characterization of the predominant bacterial population of different mangrove rhizosphere soils using 16S rRNA gene based single-strand conformation polymorphism (SSCP). World J Microbiol Biotechnol 24: 387–394

Bhattacharjee S, Bhattacharjee S, Bal B, Pal R, Niyogi SK, Sarkar K (2010) Is Vibrio fluvialis emerging as a pathogen with epidemic potential in coastal region of eastern India following cyclone Aila? J Health Popul Nutr 28:311–317

Bhattacharya M, Choudhury P, Kumar R (2000) Antibiotic- and metal-resistant strains of *Vibrio parahaemolyticus* isolated from shrimps. Microb Drug Resist 6:171–172

Bhattacharya BD, Hwang JS, Tseng LC, Sarkar SK, Rakshit D, Mitra S (2014) Bioaccumulation of trace elements in dominant mesozooplankton group inhabiting in the coastal regions of Indian Sundarban mangrove wetland. Mar Pollut Bull 87:345–351

Bhattacharyya A, Majumder NS, Basak P, Mukherji S, Roy D, Nag S, Haldar A, Chattopadhyay D, Mitra S, Bhattacharyya M, Ghosh A (2015) Diversity and distribution of archaea in the mangrove sediment of Sundarbans. Archaea 2015:1–14. https://doi.org/10.1155/2015/968582

Bhattacharyya A, Haldar A, Bhattacharyya M, Ghosh A (2019) Anthropogenic influence shapes the distribution of antibiotic resistant bacteria (ARB) in the sediment of Sundarban estuary in India. Sci Total Environ 647:1626–1639

Biswas T, Bandyopadhyay PK (2011) On the occurrence of a protozoan parasite from edible oysters of Sunderbans region of West Bengal. Proc 22nd Nat Cong Parasitol:315–318

Biswas T, Bandyopadhyay PK (2014) Prevalence of a protozoan parasite Cristigera sp. (Ciliophora: Ciliatea) from edible oysters (Mollusca:Bivalvia) of Sundarbans, West Bengal, India. J Parasit Dis 38:297–301

Biswas T, Bandyopadhyay PK (2016) First record of protozoan parasites, *Tetrahymena rostrata* and *Callimastix equi* from the edible oyster in Sundarbans region of West Bengal, India. J Parasit Dis 40:971–975

Biswas SN, Rakshit D, Sarkar SK, Sarangi RK, Satpathy KK (2014) Impact of multispecies diatom bloom on plankton community structure in Sundarban mangrove wetland, India. Mar Pollut Bull 85:306–311

Biswas K, Choudhury JD, Mahansaria R, Saha M, Mukherjee J (2017) *Streptomyces euryhalinus sp. nov.*, a new actinomycete isolated from a mangrove forest. J Antibiot (Tokyo) 70:747–753

Bongiorni L, Jain R, Raghukumar S, Aggarwal RK (2005) *Thraustochytrium gaertnerium* sp. nov.: a new thraustochytrid stramenopilan protist from mangroves of Goa, India. Protist 156:303–315

Brito EM, Guyoneaud R, Goni-Urriza M, Ranchou-Peyruse A, Verbaere A, Crapez MA, Wasserman JC, Duran R (2006) Characterization of hydrocarbonoclastic bacterial communities from mangrove sediments in Guanabara Bay, Brazil. Res Microbiol 157:752–762

Cabral L, Júnior GV, Pereira de Sousa ST, Dias AC, Lira Cadete L, Andreote FD, Hess M, de Oliveira VM (2016) Anthropogenic impact on mangrove sediments triggers differential responses in the heavy metals and antibiotic resistomes of microbial communities. Environ Pollut 216:460–469

Caratini C, Blasco F, Thanikaimoni G (1973) Relation between the pollen spectra and the vegetation of a south Indian mangrove. Pollen Spores 15(2):281–292

Castaneda LE, Barbosa O (2017) Metagenomic analysis exploring taxonomic and functional diversity of soil microbial communities in Chilean vineyards and surrounding native forests. Peer J 30:e3098

Castine SA, Bourne DG, Trott LA, McKinnon DA (2009) Sediment microbial community analysis: establishing impacts of aquaculture on a tropical mangrove ecosystem. Aquaculture 297:91–98

Chaban B, Ng SYM, Jarrell KF (2006) Archaeal habitats—from the extreme to the ordinary. Can J Microbiol 52:73–116

Chakraborty A, Bera A, Mukherjee A, Basak P, Khan I, Mondal A, Roy A, Bhattacharyya A, SenGupta S, Roy D, Nag S, Ghosh A, Chattopadhyay D, Bhattacharyya M (2015) Changing bacterial profile of Sundarbans, the world heritage mangrove: impact of anthropogenic interventions. World J Microbiol Biotechnol 31:593–610

Chandrasekaran K, Dhanraj M, Chadha A (2018) Effects of organic and inorganic salts on docosahexaenoic acid (DHA) production by a locally isolated strain of Thraustochytrium sp. T01. Prep Biochem Biotechnol 5:1–6

Chatellier J, Shribman Z (2018) Siné-Saloum delta: an illustrated guide to Senegal's remarkable mangrove estuary. Streetlib

Chen QH, Tam NF, Shin PK, Cheung SG, Xu RL (2009) Ciliate communities in a constructed mangrove wetland for wastewater treatment. Mar Pollut Bull 58:711–719

Choudhury P, Kumar R (1998) Multidrug- and metal-resistant strains of *Klebsiella pneumoniae* isolated from *Penaeus monodon* of the coastal waters of deltaic Sundarban. Can J Microbiol 44:186–189

Cleary DFR, Smalla K, Mendonca-Hagler LCS, Gomes NCM (2012) Assessment of variation in bacterial composition among microhabitats in a mangrove environment using DGGE fingerprints and barcoded pyrosequencing. PLoS One 7:e29380

Das S, Jana TK, De TK (2014a) Vertical profile of phosphatase activity in the Sundarban mangrove forest, North East Coast of Bay of Bengal, India. Geomicrobiol J 31:716–725

Das D, Salgaonkar BB, Mani K, Braganca JM (2014b) Cadmium resistance in extremely halophilic archaeon Haloferax strain BBK2. Chemosphere 112:385–392

Dastager SG, Qiang ZL, Damare S, Tang SK, Li WJ (2012) *Agromyces indicus* sp. nov., isolated from mangroves sediment in Chorao Island, Goa, India. Antonie Van Leeuwenhoek 102:345–352

Dave BP, Anshuman K, Hajela P (2006) Siderophores of halophilic archaea and their chemical characterization. Indian J Exp Biol 44:340–344

Deshmukh SK, Gupta MK, Prakash V, Reddy MS (2018) Mangrove-associated fungi: a novel source of potential anticancer compounds. J Fungi (Basel) 4(3):E101. https://doi.org/10.3390/jof4030101

Dorothy KP, Satyanarayana B, Kalavati C, Raman AV (2003) Protozoa associated with leaf litter degradation in Coringa mangrove forest, Kakinada bay, east coast of India. Ind J Mar Sci 32:45–51

Dos SHF, Cury JC, doCarmo FL, dos Santos AL, Tiedje J, van Elsas JD, Rosado AS, Peixoto RS (2011) Mangrove bacterial diversity and the impact of oil contamination revealed by pyrosequencing: bacterial proxies for oil pollution. PLoS One 6:e16943

Fernandes SO, Kirchman DL, Michotey VD, Bonin PC, LokaBharathi PA (2014) Bacterial diversity in relatively pristine and anthropogenically-influenced mangrove ecosystems (Goa, India). Braz J Microbiol 45:1161–1171

Flores-Mireles AL, Winans SC, Holguin G (2007) Molecular characterization of diazotrophic and denitrifying bacteria associated with mangrove roots. Appl Environ Microbiol 73:7308 21

Forest Survey of India Report (2012). http://fsi.nic.in/isfr2017/isfr-forest-cover-2017.pdf. Accessed 21 Dec 2018

Forest Survey of India Report (2013). http://fsi.nic.in/cover_2013/sfr_forest_cover.pdf. Accessed 21 Dec 2018

Forest Survey of India Report (2017). http://fsi.nic.in/isfr2017/isfr-forest-cover-2017.pdf. Accessed 21 Dec 2018

Gaonkar T, Bhosle S (2013) Effect of metals on a siderophore producing bacterial isolate and its implications on microbial assisted bioremediation of metal contaminated soils. Chemosphere 93:1835–1843

Ghosh A, Bhadury P (2017) Investigating monsoon and post-monsoon variabilities of bacterioplankton communities in a mangrove ecosystem. Environ Sci Pollut Res Int 25:5722–5739

Ghosh A, Dey N, Bera A, Tiwari A, Sathyaniranjan KB, Chakrabarti K, Chattopadhyay D (2010) Culture independent molecular analysis of bacterial communities in the mangrove sediment of Sundarban, India. Saline Syst 6:1

Ghosh A, Schmidt S, Fickert T, Nüsser M (2015) The Indian Sundarban mangrove forests: history, utilization, conservation strategies and local perception. Diversity 7:149–169

Giri C, Ochieng E, Tieszen LL, Zhu Z, Singh A, Loveland T, Masek J, Duke N (2011) Status and distribution of mangrove forests of the world using earth observation satellite data. Glob Ecol Biogeogr 20:154–159

Gomes NCM, Borges LR, Paranhos R, Pinto FN, Krogerrecklenfort E, Mendonca-Hagler LCS, Smalla K (2007) Diversity of ndo genes in mangrove sediments exposed to different sources of polycyclic aromatic hydrocarbon pollution. Appl Environ Microbiol 73:7392–7399

Gomes NC, Borges LR, Paranhos R, Pinto FN, Mendonca-Hagler LCS, Smalla K (2008) Exploring the diversity of bacterial communities in sediments of urban mangrove forests. FEMS Microbiol Ecol 66:96–109

Gomes NCM, Cleary DFR, Pinto FN, Egas C, Almeida A, Cunha A, Mendonca-Hagler LC, Smalla K (2010) Taking root: enduring effect of rhizosphere bacterial colonization in mangroves. PLoS One 5(11):e14065

Gopal B, Chauhan M (2018) The transboundary Sundarbans mangroves (India and Bangladesh). In: Finlayson C, Milton G, Prentice R, Davidson N (eds) The wetland book. Springer, Dordrecht

Greer CW (2010) Bacterial diversity in hydrocarbon-polluted rivers, estuaries and sediments. In: Timmis KN (ed) Handbook of hydrocarbon and lipid microbiology, vol 2329. Springer, Berlin, p 2338

Guidi-Rontani C, Jean MR, Gonzalez-Rizzo S, Bolte-Kluge S, Gros O (2014) Description of new filamentous toxic Cyanobacteria (Oscillatoriales) colonizing the sulfidic periphyton mat in marine mangroves. FEMS Microbiol Lett 359:173–181

Haldar S, Nazareth SW (2018) Taxonomic diversity of bacteria from mangrove sediments of Goa: metagenomic and functional analysis. 3 Biotech 8(10):436

Hazra P, Saha Kesh G (2017) Isolation and purification of phycocyanin from cyanobacteria of a mangrove forest. Appl Biol Chem 60:631–636

Hong K (2013) Actinomycetes from mangrove and their secondary metabolites – A review. Acta Microbiol Sin 53:1131–1141

Hu Q, Qi H, Zeng J, Zhang H (2007) Bacterial diversity in soils around a lead and zinc mine. J Environ Sci 19:74–79

Huergo LF, Rissi DV, Elias AS, Gonçalves MV, Gernet MV, Barreto F, Dahmer GW, Reis RA, Pedrosa FO, Souza EM, Monteiro RA, Baura VA, Balsanelli E, Cruz LM (2018) Influence of ancient anthropogenic activities on the mangrove soil microbiome. Sci Total Environ 645:1–9

Ikenaga M, Guevara R, Dean AL, Pisani C, Boyer JN (2009) Changes in community structure of sediment bacteria along the Florida coastal everglades marsh-mangrove-seagrass salinity gradient. Microb Ecol 59:284–295

Imchen M, Kumavath R, Barh D, Avezedo V, Ghosh P, Viana M, Wattam AR (2017) Searching for signatures across microbial communities: metagenomic analysis of soil samples from mangrove and other ecosystems. Sci Rep 7:8859

Imchen M, Kumavath R, Barh D, Vaz A, Góes-Neto A, Tiwari S, Ghosh P, Wattam AR, Azevedo V (2018) Comparative mangrove metagenome reveals global prevalence of heavy metals and antibiotic resistome across different ecosystems. Sci Rep 8:11187

Imran M, Pant P, Shanbhag YP, Sawant SV, Ghadi SC (2017) Genome sequence of microbulbifer mangrovi DD-13T reveals its versatility to degrade multiple polysaccharides. Mar Biotechnol (NY) 19:116–124

Jaiswal SK, Saxena R, Mittal P, Gupta A, Sharma VK (2017) Draft genome sequence of *Pseudomonas hussainii* strain MB3, a denitrifying aerobic bacterium isolated from the rhizospheric region of mangrove trees in the Andaman Islands, India. Genome Announc 5:e01527-16

Jiang X-T, Peng X, Deng G-H et al (2013) Illumina sequencing of 16S rRNA tag revealed spatial variations of bacterial communities in a mangrove wetland. Microb Ecol 66:96–104

Kannupandi T Kannan R (1998) Hundred years of Pichavaram mangrove forest. In: an anthropology of Indian mangroves. ENVIS Special Publication. Annamalai University, India, p 66

Kathiresan K, Saravanakumar K, Sahu SK, Sivasankaran M (2014) Adenosine deaminase production by an endophytic bacterium (Lysinibacillus sp.) from *Avicennia marina*. 3 Biotech 4: 235–239

Keerthi S, Koduru UD, Nittala SS, Parine NR (2018) The heterotrophic eubacterial and archaeal co-inhabitants of the halophilic *Dunaliella salina* in solar salterns fed by Bay of Bengal along south eastern coast of India. Saudi J Biol Sci 25:1411–1419

Kharangate-Lad A, Bhosle S (2016) Studies on siderophore and pigment produced by an adhered bacterial strain *Halobacillus trueperi* MXM-16 from the mangrove ecosystem of Goa, India. Indian J Microbiol 56:461–466

Kharangate-Lad A, Pereira F, Fernandes J, Bhosle S (2016) Studies on the effects of zerovalent iron nanoparticles on bacteria from the mangrove ecosystem. Environ Sci Pollut Res Int 23:927–931

Kumar T, Ghose M (2008) Status of arbuscular mycorrhizal fungi (AMF) in the Sundarbans of India in relation to tidal inundation and chemical properties of soil. Wetl Ecol Manag 16: 471–483

Kumar T, Majumdar A, Das P, Sarafis V, Ghose M (2008) Trypan blue as a fluorochrome for confocal laser scanning microscopy of arbuscular mycorrhizae in three mangroves. Biotech Histochem 83:153–159

Leininger S, Urich T, Schloter M, Schwark L, Qi J, Nicol GW, Prosser JI, Schuster SC, Schleper C (2006) Archaea predominate among ammonia oxidizing prokaryotes in soils. Nature 442: 806–809

Liang JB, Chen YQ, Lan CY, Tam NFY, Zan QJ, Huang LN (2007) Recovery of novel bacterial diversity from mangrove sediment. Mar Biol 150:739–747

Liao PC, Huang BH, Huang S (2007) Microbial community composition of the Danshui river estuary of Northern Taiwan and the practicality of the phylogenetic method in microbial barcoding. Microb Ecol 54:497–507

Liu Y, Li F, Huang Q (2013) Allelopathic effects of gallic acid from *Aegiceras corniculatum* on *Cyclotella caspia*. J Environ Sci (China) 25:776–784

Loganathachetti DS, Balamurugan Sadaiappan B, Anbu Poosakkannu A, Muthuraman S (2016) Pyrosequencing-based seasonal observation of prokaryotic diversity in pneumatophore-associated soil of *Avicennia marina*. Curr Microbiol 72:68–74

Luo YR, Tian Y, Huang X, Yan CL, Hong HS, Lin GH, Zheng TL (2009) Analysis of community structure of a microbial consortium capable of degrading benzo(a)pyrene by DGGE. Mar Pollut Bull 58:1159–1163

Lyimo TJ, Pol A, Jetten MS, Den Camp HJ (2009) Diversity of methanogenic archaea in a mangrove sediment and isolation of a new Methanococcoides strain. FEMS Microbiol Lett 291:247–253

Magny GC, Mozumder PK, Grim CJ, Hasan NA, Naser MN, Alam M, Sack RB, Huq A, Colwell RR (2011) Role of zooplankton diversity in *Vibrio cholerae* population dynamics and in the Incidence of cholera in the Bangladesh Sundarbans. Appl Environ Microbiol 77:6125–6132

Mamun MM, Sarower MG Ali MA, Rahman SMB, Haq KA (2009) Abundance and distribution of plankton in the Sunderbans mangrove forest. J Innov Dev Strategy 3:43–54

Mani K, Salgaonkar BB, Braganca JM (2012) Culturable halophilic archaea at the initial and crystallization stages of salt production in a natural solar saltern of Goa, India. Aquat Biosyst 8:15

Manna S, Chaudhuri K, Bhattacharyya S, Bhattacharyya M (2010) Dynamics of Sundarban estuarine ecosystem: eutrophication induced threat to mangroves. Saline Syst 6:8

Mishra RR, Swain MR, Dangar TK, Thatoi H (2012) Diversity and seasonal fluctuation of predominant microbial communities in Bhitarkanika, a tropical mangrove ecosystem in India. Rev Biol Trop 60:909–924

Mitra A, Santra SC, Mukherjee J (2008) Distribution of actinomycetes, their antagonistic behaviour and the physico-chemical characteristics of the world's largest tidal mangrove forest. Appl Microbiol Biotechnol 80:685–695

Mulla TM, Chavan NS (2017) Mangrove diversity along the coast of Ratnagiri, Maharashtra. Curr Bot 8:123–126

Nag S, Pramanik A, Chattopadhyay D, Bhattacharyya M (2018) Green-fabrication of gold nanomaterials using *Staphylococcus warneri* from Sundarbans estuary: an effective recyclable nanocatalyst for degrading nitro aromatic pollutants. Environ Sci Pollut Res Int 25:2331–2349

Nandy T, Mandal S, Chatterjee M (2018) Intra-monsoonal variation of zooplankton population in the Sundarbans Estuarine System, India. Environ Monit Assess 190:603

Nathan VK, Esther Rani M, Rathinasamy G, Dhiraviam KN, Jayavel S (2014) Process optimization and production kinetics for cellulase production by *Trichoderma viride* VKF3. Springerplus 3:92. https://doi.org/10.1186/2193-1801-3-92

Nayak S, Gonsalves V, Nazareth SW (2012) Isolation and salt tolerance of halophilic fungi from mangroves and solar saltern in Goa-India. Ind J Geo Mar Sc 41:164–172

Nayak B, Roy S, Mitra A, Roy M (2016) Isolation of multiple drug resistant and heavy metal resistant Stenotrophomonas maltophilia strain BN1, a plant growth promoting rhizobacteria, from mangrove associate *ipomoea pes-caprae* of Indian Sundarbans. J Pure Appl Microbiol 10:3131–3139

Nazareth SW, Gonsalves V (2014) Halophillic *Aspergillus penicillioides* from athalassohaline, thalassohaline, and polyhaline environments. Front Microbiol 5:412

Nazareth SW, Marbaniang T (2008) Effect of heavy metals on cultural and morphological growth characteristics of halotolerant Penicillium morphotypes. J Basic Microbiol 48:363–369

Nurunnabi TR, Nahar L, Al-Majmaie S, Rahman SMM, Sohrab MH, Billah MM, Ismail FMD, Rahman MM, Sharples GP, Sarker SD (2018) Anti-MRSA activity of oxysporone and xylitol from the endophytic fungus *Pestalotia* sp. growing on the Sundarbans mangrove plant *Heritiera fomes*. Phytother Res 32:348–354

Perumal VN, Rajkumar M, Perumal PK, Rajasekar KT (2009) Seasonal variations of plankton diversity in the Kaduviyar estuary, Nagapattinam, south east coast of India. J Environ Biol 30: 1035–1046

Pires AC, Cleary DF, Almeida A, Cunha A, Dealtry S, Mendonca-Hagler LC, Smalla K, Gomes NC (2012) Denaturing gradient gel electrophoresis and barcoded pyrosequencing reveal unprecedented archaeal diversity in mangrove sediment and rhizosphere samples. Appl Environ Microbiol 78:5520–5528

Poharkar KV, Kerkar S, Doijad SP, Barbuddhe SB (2014) Prevalence and genetic profiles of *Escherichia coli* from mangroves and mangrove associated foods off Goa, India. Mar Pollut Bull 85:86–91

Poharkar KV, Kerkar S, D'Costa D, Doijad S, Barbuddhe SB (2016) Mangrove ecosystems: an adopted habitat for pathogenic *Salmonella* spp. Water Environ Res 88:264–271

Pramanik A, Sundararaman M, Das S, Ghosh U, Mukherjee J (2011) Isolation and characterization of cyanobacteria possessing antimicrobial activity from the Sundarbans, the world's largest tidal mangrove forest. J Phycol 47:731–743

Pramanik A, Sengupta S, Bhattacharyya M (2019) Microbial diversity and community analysis of the Sundarbans mangrove, a world heritage site. In: Das S, Dash HR (eds) Dash microbial diversity in the genomic era. Academic, London, pp 65–76

Raghava Rao KV, Raghava Rao T (2013) Molecular characterization and its antioxidant activity of a newly isolated *Streptomyces coelicoflavus* BC 01 from mangrove soil. J Young Pharm 5: 121–126

Rahman SMB, Golder J, Rahman MS, Hasanuzzaman AFM, Huq KA, Begum S, Islam SS, Bir J (2013) Spatial and temporal variations in phytoplankton abundance and species diversity in the Sundarbans mangrove forest of Bangladesh. J Mar Sci Res Dev 3:126

Raja S, Ganesan S, Sivakumar K, Thangaradjou T (2010) Screening of marine actinobacteria for amylase enzymes inhibitors. Indian J Microbiol 50:233–237

Rajkumar M, Perumal P, Prabu VA, Perumal NV, Rajasekar T (2009) Phytoplankton diversity in Pichavaram mangrove waters from the south-east coast of India. J Env Bio 30:489–498

Rakshit D, Sarkar SK, Bhattacharya BD, Jonathan MP, Biswas JK, Mondal P, Mitra S (2015) Human-induced ecological changes in western part of Indian Sundarban megadelta: a threat to ecosystem stability. Mar Pollut Bull 99:186–194

Rameshkumar N, Gomez-Gil B, Spröer C, Lang E, Dinesh Kumar N, Krishnamurthi S, Nair S, Roque A (2011) *Vibrio plantisponsor* sp. nov., a diazotrophic bacterium isolated from a mangrove associated wild rice (*Porteresia coarctata Tateoka*). Syst Appl Microbiol 34: 487–493

Ramkumar VS (2014) Studies on methanogenic consortia associated with mangrove sediments of Ennore. J Environ Biol 35:649–654

Ravikumar S, Kathiresan K, Alikhan SL, Williams GP, Gracelin NA (2007) Growth of *Avicennia marina* and *Ceriops decandra* seedlings inoculated with halophilic azotobacters. J Environ Biol 28:601–603

Ravikumar S, Gnanadesigan M, Ignatiammal ST, Sumaya S (2012) Population dynamics of free living, nitrogen fixing bacteria Azospirillum in Manakkudi mangrove ecosystem, India. J Environ Biol 33:597–602

Saha M, Ghosh D, Ghosh D, Garai D, Jaisankar P, Sarkar KK, Dutta PK, Das S, Jha T, Mukherjee J (2005) Studies on the production and purification of an antimicrobial compound and taxonomy of the producer isolated from the marine environment of the Sundarbans. Appl Microbiol Biotechnol 66:497–505

Saifullah ASM, Kamal AHM, Idris MH, Rajaee AH, Bhuiyan MKA (2016) Phytoplankton in tropical mangrove estuaries: role and interdependency. For Sci Technol 12(2):104–113. https://doi.org/10.1080/21580103.2015.1077479

Salgaonkar BB, Mani K, Bragança JM (2013) Accumulation of polyhydroxyalkanoates by halophilic archaea isolated from traditional solar salterns of India. Extremophiles 17:787–795

Samanta B, Bhadury P (2018) Study of diatom assemblages in Sundarbans mangrove water based on light microscopy and rbcL gene sequencing. Heliyon 27(4):e00663

Sana B, Ghosh D, Saha M, Mukherjee J (2008) Purification and characterization of an extracellular, uracil specific ribonuclease from a Bizionia species isolated from the marine environment of the Sundarbans. Microbiol Res 163:31–38

Saravanakumar K, Kathiresan K (2014) Bioremoval of the synthetic dye malachite green by marine *Trichoderma sp*. Springerplus 3:631

Saravanakumar A, Rajkumar M, Sesh Serebiah J, Thivakaran GA (2007) Abundance and seasonal variations of zooplankton in the arid zone mangroves of Gulf of Kachchh-Gujarat, westcoast of India. Pak J Biol Sci 10:3525–3532

Saravanakumar K, Rajendran N, Kathiresan K, Chen J (2016) Bioprospects of microbial enzymes from mangrove-associated fungi and bacteria. Adv Food Nutr Res 79:99–115. https://doi.org/10.1016/bs.afnr.2016.08.003

Sarkar S, Saha M, Roy D, Jaisankar P, Das S, Gauri Roy L, Gachhui R, Sen T, Mukherjee J (2008) Enhanced production of antimicrobial compounds by three salt-tolerant actinobacterial strains isolated from the Sundarbans in a niche-mimic bioreactor. Mar Biotechnol (NY) 10:518–526

Schuster SC (2008) Next-generation sequencing transforms today's biology. Nat Methods 5:16–18

Sengupta A, Chaudhuri S (1991) Ecology of heterotrophic dinitrogen fixation in the rhizosphere of mangrove plant community at the Ganges river estuary in India. Oecologia 87:560–564

Sengupta A, Chaudhuri S (2002) Arbuscular mycorrhizal relations of mangrove plant community at the Ganges river estuary in India. Mycorrhiza 12:169–174

Sengupta S, Pramanik A, Ghosh A, Bhattacharyya M (2015) Antimicrobial activities of actinomycetes isolated from unexplored regions of Sundarbans mangrove ecosystem. BMC Microbiol 15:170

Shrivastava P, Kumar R, Yandigeri MS, Malviya N, Arora DK (2015) Isolation and Characterization of Streptomycetes with-plant growth promoting potential from Mangrove ecosystem. Pol J Microbiol 64:339–349

Singh A, Singh AK (2018) Isolation, characterization and exploring biotechnological potential of halophilic archaea from salterns of western India. 3 Biotech 8:45

Singh SK, Verma P, Ramaiah N, Chandrashekar AA, Shouche YS (2010) Phylogenetic diversity of archaeal 16S rRNA and ammonia monooxygenase genes from tropical estuarine sediments on the central west coast of India. Res Microbiol 161:177–186

Sithranga Boopathy N, Kathiresan K (2010) Anticancer drugs from marine flora: an overview. J Oncol 2010:214186

Sjöling S, Mohammed SM, Lyimo TJ, Kyaruzi JJ (2005) Benthic bacterial diversity and nutrient processesin mangroves: impact of deforestation. Estuar Coast Shelf Sci 63:397–406

Sousa OV, Macrae A, Menezes FG, Gomes NC, Vieira RH, Mendonça-Hagler LC (2006) The impact of shrimp farming effluent on bacterial communities in mangrove waters, Ceará, Brazil. Mar Pollut Bull 52:1725–1734

Sultana M, Nusrin S, Hasan NA, Sadique A, Ahmed KU, Islam A, Hossain A, Longini I, Nizam A, Huq A, Siddique AK, Sack DA, Sack RB, Colwell RR, Alam M (2018) Biofilms comprise a component of the annual cycle of *Vibrio cholerae* in the Bay of Bengal Estuary. MBio 9: e00483-18

Suryanarayanan TS, Ravishankar JP, Venkatesan G, Murali TS (2004) Characterization of the melanin pigment of a cosmopolitan fungal endophyte. Mycol Res 108:974–978

Taketani RG, Dos Santos HF, Van Elsas JD, Rosado AS (2009) Characterisation of the effect of a simulated hydrocarbon spill on diazotrophs in mangrove sediment mesocosm. Anton Leeuw Int J Gen Mol Microbiol 96:343–354

Taketani RG, Franco NO, Rosado AS, Van Elsas JD (2010a) Microbial community response to a simulated hydrocarbon spill in mangrove sediments. J Microbiol 48:7–15

Taketani RG, Yoshiura CA, Dias ACF, Andreote FD, Tsai SM (2010b) Diversity and identification of methanogenic archaea and sulphate-reducing bacteria in sediments from a pristine tropical mangrove. Antonie Van Leeuwenhoek 97:401–411

Taketani RG, Dias ACF, Andreote FD (2011) Microbial diversity from mangroves sediments: insights from culture independent approaches. In: Metras JN (ed) Mangroves ecology, biology and taxonomy. Nova Science, New York

Thenmozhi C, Sankar R, Karuppiah V, Sampathkumar P (2011) L-asparaginase production by mangrove derived *Bacillus cereus* MAB5: optimization by response surface methodology. Asian Pac J Trop Med 4:486–491

Thombre RS, Shinde VD, Oke RS, Dhar SK, Shouche YS (2016) Biology and survival of extremely halophilic archaeon *Haloarcula marismortui* RR12 isolated from Mumbai salterns, India in response to salinity stress. Sci Rep 6:25642

Tian Y, Liu HJ, Zheng TL, Kwon KK, Kim SJ, Yan CL (2008) PAHs contamination and bacterial communities in mangrove surface sediments of the Jiulong River Estuary, China. Mar Pollut Bull 57:707–715

Varadharajan D, Soundarapandian P (2015) Biodiversity and abundance of phytoplankton from Muthupettai mangrove Region, South East coast of India. J Aquac Res Devel 6:383

Verma AK, Raghukumar C, Verma P, Shouche YS, Naik CG (2010) Four marine-derived fungi for bioremediation of raw textile mill effluents. Biodegradation 21:217–233

Vipindas PV, Jabir T, Jasmin C, Balu T, Rehitha TV, Adarsh BM, Nair S, Abdulla MH, Abdulaziz A (2018) Diversity and seasonal distribution of ammonia-oxidizing archaea in the water column of a tropical estuary along the southeast Arabian Sea. World J Microbiol Biotechnol 34:188

Viswanath G, Jegan S, Baskaran V, Kathiravan R, Prabavathy VR (2015) Diversity and N-acyl-homoserine lactone production by Gammaproteobacteria associated with *Avicennia marina* rhizosphere of South Indian mangroves. Syst Appl Microbiol 38:340–345

World Atlas of Mangroves (2010). https://www.cbd.int/cooperation/pavilion/.../2010-10-21-session1-mangroves-en.pdf. Accessed 21 Dec 2018

Xu DB, Ye WW, Han Y, Deng ZX, Hong K (2014) Natural products from mangrove actinomycetes. Mar Drugs 12:2590–2613. Review

Yan B, Hong K, Yu ZN (2006) Archaeal communities in mangrove soil characterized by 16S rRNA gene clones. J Microbiol 44:566–571

Zhang Y, Dong J, Yang B, Ling J, Wang Y, Zhang S (2009) Bacterial community structure of mangrove sediments in relation to environmental variables accessed by 16S rRNA gene-denaturing gradient gel electrophoresis fingerprinting. Sci Mar 73:487–498

Part II
Impact of Salinity on Plants and Microbial Life

Chapter 6
Exopolysaccharides and Biofilms in Mitigating Salinity Stress: The Biotechnological Potential of Halophilic and Soil-Inhabiting PGPR Microorganisms

Aparna Banerjee, Shrabana Sarkar, Sara Cuadros-Orellana, and Rajib Bandopadhyay

Abstract Soil salinity is a key environmental factor restraining the productivity of soil and crop plants. In different parts of the world, agricultural productivity is decreasing mostly because of drought and salinity increase. The situation may become worse for global warming in the future. A wide range of adaptation and mitigation strategies have been adopted to cope up with these circumstances. Among the measures to control soil salinity, those that involve water management and changes in micro relief can be time-consuming and cost-intensive. In this scenario, biological methods of salinity stress management appear as a good alternative, and halotolerant bacteria may play an important role in mitigating salinity stress. The natural ability of these microorganisms to cope with harsh environmental conditions has been attributed to their high genetic diversity and physiological adaptations, such as the synthesis of compatible solutes, growth in biofilm consortia and production of exopolysaccharides (EPSs) and halotolerant enzymes. Furthermore, they are able to produce plant growth-promoting hormones. Salt-tolerant plant growth-promoting rhizobacteria (PGPR) can play an important role in alleviating soil salinity stress during plant growth, and bacterial EPS can help to mitigate salinity stress by reducing the content of Na^+ available for plant uptake. To compensate the stress imposed by salinity, biofilm formation and EPS production are significant strategies of salt-tolerant bacteria to assist metabolism. Halophilic microorganisms can form

A. Banerjee (✉)
Vicerrectoría de Investigación y Postgrado, Centro de Biotecnología de los Recursos Naturales, Universidad Católica del Maule, Talca, Chile
e-mail: abanerjee@ucm.cl

S. Sarkar · R. Bandopadhyay
UGC-Center of Advanced Study, Department of Botany, The University of Burdwan, Burdwan, West Bengal, India

S. Cuadros-Orellana
Facultad de Ciencias Agrarias y Forestales, Centro de Biotecnología de los Recursos Naturales, Universidad Católica del Maule, Talca, Chile

B. Giri, A. Varma (eds.), *Microorganisms in Saline Environments: Strategies and Functions*, Soil Biology 56, https://doi.org/10.1007/978-3-030-18975-4_6

biofilm and accumulate EPS at increasing salt stress. EPSs promote bacterial colonization of plant roots and soil particles, and they can eventually be added to soil to improve its structure and, consequently, to improve plant growth. In a long-term vision, alleviation of derelict soil, improved crop yield and restoration of mangrove diversity are possible by mechanistic use of EPS-producing halotolerant microorganisms.

Keywords Biofilm · Exopolysaccharide · Salinity stress · Halophiles · PGPR

6.1 Introduction

Soil salinity is defined as the total salt content of the soil, and the process of increasing the salt content in soil is known as soil salinization. Salts do occur naturally in soils and water, while salinization may be triggered by natural courses of mineral weathering, evaporation or the gradual withdrawal of ocean. The early twenty-first century is regarded as an era of global scarcity of water, exacerbated environmental pollution and increased soil and water salinization. These factors, together with population growth and the reduction of available land area, pose new challenges to crop production worldwide. Other than these, several environmental factors, viz. extreme temperature, drought, soil salinity and high wind pressure, also negatively affect plant growth, and global warming may contribute to their amplification in years to come. However, among the aforementioned abiotic factors, soil salinity is one of the most devastating factors to crop productivity and quality and probably the most difficult to overcome.

Saline soil is the one in which electrical conductivity (EC) of the saturation extract (EC_e) in the root zone surpasses 4 dS/m (~40 mM NaCl) at 25 °C and consists of an exchangeable sodium of 15%. Yield of most crop plants is reduced at this saturation point, and many plants even display reduced yield at lower EC_e values too (Munns 2005; Jamil et al. 2011). It has been predicted that around 20% of total cultivated and 33% of irrigated agricultural lands worldwide will be affected by high salinity in the near future. Moreover, this negative impact will increase continuously as the saline areas grow at a rate of 10% annually for several reasons, including low precipitation, high surface evaporation, weathering of native rocks, irrigation of saline water, etc. By the year 2050, it is estimated that more than 50% of the arable land will be saline (Jamil et al. 2011).

A wide range of adaptation and mitigation strategies to cope up with these circumstances have been adopted. Among the measures to control soil salinity, those that involve water management and changes in micro relief can be time-consuming and cost-intensive. A possible way to manage soil salinity is the rational use of saline soil, by adopting agroforestry practices in the marginal landscapes and introducing cost-effective innovative technologies. Biosaline agriculture, for instance, is an option for using salt-affected agricultural lands, as they focus on the growth of slightly halotolerant plants, such as quinoa, sorghum, African millet, triticale, sesame, amaranth, pea, mung beans, salt bush etc. Halophytes are sometimes introduced to marginal land to increase soil productivity. They are good desalinating and phyto-ameliorator agents as they are capable of uptaking salts from soil and irrigation water and concentrating them over the ground. On the other hand, the process of phyto-amelioration is highly time-consuming, and it usually requires ~3–5 years for moderately saline soil and up to ~7 years for highly saline soils (Vargas et al. 2018). This raises concerns about the need to control marginal saline soils in order to fulfil the increasing food demands of exponentially growing human population earth.

Several groups of microorganisms have the ability to thrive in different extreme environmental conditions, viz. extreme temperature, radiation, pressure, pH, salinity, oxygen concentration and water activity. Some of them often produce a biofilm or network consisting of exopolysaccharides (EPSs), which act as a protecting sheath to the microorganisms for survival and growth in such environments. Bacterial EPSs have several important biotechnological applications, which include medical use (Moscovici 2015), probiotics (Şengül et al. 2011), food production (Zannini et al. 2016) and commercial bioremediation (Banerjee and Bandopadhyay 2016a, b; Gupta and Diwan 2017). Their use in mitigating saline stress in plants has recently been reviewed (Meena et al. 2017). Halotolerant and halophilic microorganisms found in thalasohaline and athalasohaline environments, such as salt marshes and saline soils, may play an important role in mitigating salinity stress (Makhdoumi-Kakhki et al. 2012). Their genetic makeup can be highly diverse, and their adaptive strategies include the synthesis of compatible solutes, production of plant growth promoting hormones, growth in biofilm consortium or production of extrapolymeric substances. In this chapter, we discuss soil salinity, the management of saline soils and the use of standard and innovative methods based on microorganisms to mitigate salinity stress. Special focus is given to biofilms and exopolysaccharides produced by halophilic archaea/bacteria and important soil PGPR strains. The biocontrol of derelict soil and the restoration of the potential for crop production are also discussed in the context of microorganism-mediated processes, as possible long-term, cheap, fast and eco-friendly options for mitigating salinity stress.

6.2 Soil Salinity and Crop Production

Salinity is one of the most common abiotic stressors in an era of evident climate change. Also, it is among the foremost reasons behind stunted plant growth and reduced crop yield in many regions of the earth. The term 'soil salinity' is a simple terminology to understand the salt concentration naturally present in soil. Optimum salt concentration is required for plant growth as salt helps to balance the osmotic potential of the cell via continuous water influx. However, when the salt concentration in soil exceeds the maximum tolerable values, plants fail in absorbing water, and wilting takes place (Gupta and Huang 2014).

Saline soils have a high osmotic pressure, which leads to an imbalance in the plants nutrient uptake (Nouri et al. 2017). It occurs primarily under excessive Na^+ and Cl^- rhizospheric concentration, as these ions interact with other nutrient ions (K^+, NO_3^- and $H_2PO_4^-$) for binding sites and transport proteins in root cells and thereafter for translocation, deposition and partitioning within the plant (Tester and Davenport 2003). When the water availability to plants growing in salt-supplemented soil under elevated temperature reaches critical levels of 100 mM, it delays plant developmental stages and decreases plant dry biomass and seed yield, stomatal size, internal CO_2 concentration and dark respiration rates in quinoa (Becker et al. 2017).

The term 'salinity' refers to the concentration of soluble salts in soil or water. Classifying according to the cause, salinization is of three types: (1) primary salinization, which is natural salinity due to rainfall of thousand years and accumulation of salt or weathering of rocks; (2) secondary salinization, which is dryland salinity due to rise of groundwater level and accumulation of salt to the surface; and (3) tertiary salinization, which is irrigated salinity due to reapplying water for several cycles, resulting in evaporation of water and accumulation of concentrated salts in the surface (Zhou et al. 2013). In the coastal regions of many countries, salinity is a major issue inhibiting the production of standing crops (Mahmuduzzaman et al. 2014). Thus, the long-term continuous irrigation with saline water may result in significant loss of soil parameters, which is a type of secondary soil salinization. Saline irrigation water initially gets absorbed by soil and is deposited in ground level. At high temperature, water evaporates, and salt molecules remain trapped at the soil surface; hence the field becomes highly saline, rough and dry (Fig. 6.1). In the presence of high concentration of NaCl in soil, K^+ uptake gets replaced by Na^+ uptake, causing intoxication (Tavakkoli et al. 2010). As many enzymes in plant cells typically use K^+ ion as cofactors, some intracellular biochemical reactions are also impaired by the competitive action of Na^+. Additionally, an increase in Na^+ ions and a decrease in K^+ ions cause turgor pressure to drop, thus affecting stomatal activity (Shrivastava and Kumar 2015). When this happens, the overall water availability is

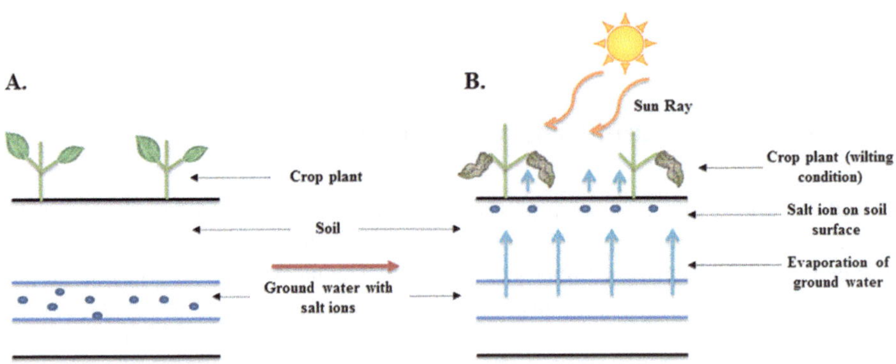

Fig. 6.1 A schematic representation on effect of temperature, salinity and evaporation of groundwater on plant growth

low, and the production of biomass substantially falls off. Cl^- ion, released by the weathering of NaCl, disturbs photosynthesis by reducing the efficiency of photosystem (PS) II. Salinity also affects the rate of chlorophyll production in plants, disturbing their key metabolic pathways (Netondo et al. 2004). Salinity stress affects the cell at the molecular level by reducing cyclin-dependent kinase activity, thereby arresting the cell cycle (Shrivastava and Kumar 2015). Accumulation of salt in plant cells may also contribute to the formation of reactive oxygen species (ROS) and lead to cellular oxidative stress. This abrupt change in oxidative status hinders critical metabolic processes in plant body, viz., photosynthesis, carbohydrate metabolism, protein production and many more. It restricts overall biomass production and diminishes the ultimate crop yield (Duarte et al. 2015; Becker et al. 2017).

Salts are mainly present in the soil as ions that are released from the breakdown of complex mineral forms (Shrivastava and Kumar 2015). Plants absorb salt as per their requirements and have a certain threshold salt level that varies from plant to plant. The excessive accumulation of salt prevents plant growth. Even in low salt concentrations, some plants grow poorly in the soil (Feng et al. 2017). Salinity stress is known to affect gene expression in plants (Kreps et al. 2002), and this may be a possible reason for low crop yield (Duarte et al. 2015). High salinity affects crop plants in every aspect, starting from germination to reproduction by increasing the level of cellular toxicity (Shrivastava and Kumar 2015).

Shrivastava and Kumar (2015) reported that almost 1 billion ha of land, i.e. nearly 7% of the world's total land area, are affected by salt stress, and about 20% of total farming land worldwide is reported to have soil salinity-related issues. Most of the staple foods, viz., sugarcane, peas, beans, rice, wheat, potato and maize, are either highly or moderately sensitive to salt stress (Feng et al. 2017; Becker et al. 2017).

Irrigated agriculture is a major human activity that often leads to salt accumulation by means of irrigation water, fertilizers or sometimes migration of salts from shallow groundwater to the soil surface. When precipitation is insufficient to leach ions from the soil profile, salts accumulate in the soil, resulting in increased soil salinity (Blaylock 1994). This affects human health greatly (Shrivastava and Kumar 2015). Increased salinity exposure has happened due to anthropological and industrial activities, such as drinking, cooking, bathing and farming (e.g. shrimp). Increased human exposure to salinity may result in high blood pressure, hypertension, tendency for miscarriage, skin disease, acute respiratory infection and diarrhoea (Vineis et al. 2011). As the worldwide population grows, the reduction of staple food production due to high soil salinity is an issue that needs to be addressed. Therefore, salt stress is becoming the fiercest environmental factor affecting plant growth as well as human health. Considering the current scenario on problems associated with soil salinization, concerted future research is needed in this area, particularly on evaluation and application of potential organisms in stressed soil.

6.3 Soil Salinity Management: Standard and Innovative Methods

In order to reduce soil salinity, research focus is on the management of saline soils. The main aims of these processes are to reduce salt concentration of soil to the threshold value and help the plant to grow healthy. Additionally, for betterment of agricultural field, the key objective in soil salinity management is to increase biomass of the crops and meet the worldwide food demand. There are few traditional methods of removing high salt concentration for better plant growth. For soil salinity management, firstly, it is required to measure the present salt concentration in the soil. Leaching is one of the most useful standard procedures for mitigating soil salinity.

Excess irrigation in arid or semiarid soil may cause increase in salt level over threshold value. Leaching requirement (LR) is the amount of water required to bring the soil salinity down to a level of acceptable value so that the salt-tolerant crop can be grown (Corwin et al. 2007). Spectral behaviour of salt helps to find radiological mineralogy of salt ions helping in measurement of salinity of the field (Metternicht and Zinck 2003).

Application of different irrigation methods in addition to water circulation is a traditional method of management of land affected with salinity stress. Rethinking the use of nonconventional water resources, like wastewater treatment, may help reduce salt deposition in groundwater (Malash et al. 2008). Cyclic water management is another traditional process of decreasing salt concentration in the field. In arid condition, water circulation and irrigation is used from a long ago as the most common method in order to manage soil salinity. In many Asian countries, furrow irrigation methods are applied mainly in cotton field and other crop field to leach accumulated salt from soil. Every furrow irrigation (EFI) is a conventional method for saline land management. Permanent skip furrow irrigation (PSFI) is another effective irrigation method used to reduce salt stress (Devkota et al. 2015). According to Devkota et al. (2015) results, EFI is much more effective than other irrigation processes. Under traditional farming, salt leaching is an effective method which already has been used in saline crop field of Uzbekistan (Devkota et al. 2015). The fields which show permanent rising in salt concentrations use PSFI management strategy for increasing the agricultural quality. This method has high potential to leach out accumulated salts even at the time of initial farming. Using these conventional methods, salinity is reduced, and roots can extend properly deep in the soil. As reported by Devkota et al. (2015), more than 30% of the saline lands use different furrow irrigation management for reducing the effects of salt stress.

One unique method was proposed by some scientists to cultivate sunflower on salinized soil (Zhao et al. 2016). They used plastic film on the upper surface of crop field and straw layer buried in soil below 40 cm. The straw layer was nearly 5 cm in width, and the plastic layer was shown to reduce the evaporation as well as salt accumulation. The diversity and scale of positive outcomes from afforestation demonstrated within the framework of the ZEF/UNESCO projects may also help to increase the awareness of financial and ecological advantages of agroforestry practices on degraded arable lands. From the overall studies, it may be observed that innovative methods have been proven better than standard strategies of saline soil management.

6.4 Microorganisms in Mitigating Salinity Stress

In order to address the aforementioned problem and considering that halotolerant bacteria may play an important role in mitigating salinity stress, biological methods for salinity stress management appear as a good alternative. They are physiologically adapted to the harsh environment by means of forming biofilms and synthesizing compatible solutes, extrapolymeric substances and halophilic enzymes. Furthermore, some halophilic bacteria are able to produce plant growth-promoting hormones, viz. auxin, gibberellin, cytokinin and more.

Salt-tolerant plant growth-promoting rhizobacteria (PGPR) play an important role in alleviating soil salinity stress during plant growth (Sen and Chandrasekhar, 2014), while bacterial exopolysaccharide (EPS) can also help to mitigate salinity stress by reducing the content of Na^+ ion available for uptake by plants (Upadhyay et al. 2011). To counteract the negative effects of the stress imposed by salinity, biofilm formation and hydrophilic EPS production are used by salt-tolerant bacteria to retain water and assist their metabolism. Halophilic microorganisms can use both strategies when exposed to increasing salt stress. In addition, EPSs contribute to bacterial colonization of plant roots and soil particles, and they can eventually be added to soil to promote plant growth and improve soil structure. The structure of EPSs produced by halophiles and PGPR is different from those from other sources, and their biosynthesis mechanism is also different. In a long-term vision, alleviation of derelict soil, improved crop yield and restoration of mangrove diversity are possible by mechanistic use of EPS-producing halotolerant microorganisms (Khan et al. 2009).

The use of soil using water with high content of soluble salts is a major limiting factor to crop productivity, particularly in the semiarid areas of the world. Thus, the identification and exploitation of halotolerant microorganisms that may alleviate salinity stress will open a new alternative for stress tolerance in plants. The first bacterial halotolerance gene was described in the mesophilic prokaryotic model *Escherichia coli*. This gene, namely, *proB-74*, determines the accumulation of osmolytes such as proline in cells (Serrano and Gaxiola 1994). *E. coli* also has some salt-sensor proteins which act like molecular springs, i.e. under low salt conditions, they hold a ‘stretched’ condition and under high salt stress become ‘compressed’ (Wang et al. 2012). A functional metagenomics study of saline soil microbiome identified the homologs of osmotolerant genes *BCAA_ABCtp*, *GSDH*, *STK_Pknb* and *duf344* (Ahmed et al. 2018). These possibly enhance host osmotolerance by means of the cytosolic accumulation of amino acids and other osmolytes. In the case of the photosynthetic bacteria *Rhodopseudomonas*, an

Table 6.1 Microbial taxa that were reported to produce exopolysaccharides and form biofilm under salt stress during the past decade

Name of bacteria	Mechanism to adapt in salt stress	References
Pseudomonas putida GAP-P45	Produces exopolysaccharide and alleviates salt stress in sunflower seedlings	Sandhya et al. (2009)
Achromobacter, *Azospirillum* and *Microbacterium*	Help plants grow under different types of abiotic stress like high salt concentration and drought	Grover et al. (2011)
Halomonas variabilis (HT1), *Planococcus rifietoensis* (RT4)	Forms biofilm and produces exopolysaccharides under high salt stress	Qurashi and Sabri (2012)
Pseudomonas stutzeri and *Pseudomonas putida*	PGPR strains, improve the growth of tomato plants	Nadeem et al. (2014)
Acinetobacter sp.	PGPR strain, improves the growth of oat plant by forming mycorrhizal associations	Xun et al. (2015)
Pseudomonas alcaligenes PsA15, *Bacillus polymyxa* BcP26 and *Mycobacterium phlei* MbP18	PGPR strains able to tolerate high temperatures and salt concentrations in arid soil	Shrivastava and Kumar (2015)
Rhodopseudomonas palustris TN114 and PP803	Produces exopolysaccharides that absorb Na^+ ions under salt stress	Nunkaew et al. (2015)
Pseudomonas sp. and *Bacillus* spp.	Potassium-solubilizing bacteria that alleviate salt stress by triggering production of antioxidant enzymes	Jha and Subramanian (2016)
Paenibacillus fujiensis and *Paenibacillus durus*	Nitrogen-fixing bacteria that help improve production of crop biomass	Navarro-Noya et al. (2016)
Pseudomonas fluorescens 002 (P.f.002.)	Grow in the rhizosphere of date plant and helps the plant to grow in saline soil	Zerrouk et al. (2016)
Pseudomonas extremorientalis TSAU20 and *Pseudomonas chlororaphis* TSAU13	Survive highly in saline conditions in the rhizosphere of tomato plants and reduce oxidative stress	Egamberdieva et al. (2017)
Smithella, *Halomonas*, *Pseudomonas* and *Comamonas*	Show good growth in high salt conditions	Zhao et al. (2018)

increase in the content of carotenoid and bacteriochlorophyll is detected when cells are exposed to salinity stress, and this has been regarded as a self-protection mechanism. Furthermore, the biotechnological potential of this halotolerant strain was demonstrated in the removal of organic compounds from saline wastewater (Wang et al. 2017a).

Endophytic bacteria can also contribute to salt tolerance in plants. In a study on the endophytic bacterial diversity of *Avicennia marina*, *Bacillus pumilus* and *Exiguobacterium* were the predominant endophytes that survive in high-saline media (Ali et al. 2017). A selection of studies reporting the mitigation of salinity stress by microorganisms is shown in Table 6.1. Decoding the genetic elements present within these halotolerant microorganisms can contribute to the development of new technologies for ameliorating saline soils or modifying the crops microbiome to improve resistance and survival in saline/hypersaline environments. These organisms likely play an important role in alleviating salinity stress in plants growing in saline environments, viz. mangrove soil, while halophilic PGPR bacteria represent an important support to plant growth in arid/semiarid soils with high salt content.

6.5 Exopolysaccharide- and Biofilm-Producing Plant Growth-Promoting Rhizobacteria and Rhizosphere Microorganisms

Some additional mechanisms leading to salinity resistance in halotolerant or halophilic microorganisms are biofilm formation and exopolysaccharide production (Steele et al. 2014). The use of plant growth-promoting rhizobacteria (PGPR) is a promising biological approach to mitigate salinity stress in plants. These bacteria can elicit tolerance to a number of abiotic stress factors other than salinity, through a mechanism termed as induced systemic tolerance (IST). PGPRs induce broad physical and chemical modifications that result in enhanced plant tolerance to a group of abiotic stresses. Indirectly, they help plants to fight against pathogens through the production of siderophores and facilitate nutrient uptake through the production of auxin, gibberellin and cytokinin-like phytohormones (Goswami et al. 2016), and they also contribute to the regulation of ethylene levels in the plants through the action of ACC deaminase (Glick 2005). Under stress conditions, plant hormone ethylene regulates the cellular homeostasis and thereby reduces the shoot and root growth. Exopolysaccharides (EPSs) produced by PGPR strains bind to soil particles and form both micro- and macroaggregates. Plant roots perfectly fit in between the microaggregates, which in turn stabilize the macroaggregates. Plants treated with EPSs display increased resistance to both water and salinity stress and improve the soil structure (Sandhya et al. 2009). Under saline condition, acidic EPSs bind to the Na^+ ions and hence make the salts unavailable to plants, thus recovering plant growth (Shrivastava and Kumar 2015). Rhizospheric bacteria and PGPRs are mostly reported to produce succinoglycan-type acidic EPSs. The commercial use of succinoglycan could improve nodule establishment in leguminous plants to be cultivated in saline soils, in agreement with sustainable agricultural practices (Halder et al. 2017). A study on rhizospheric bacteria *Bacillus insolitus*, *Aeromonas hydrophila* and *Bacillus* sp. isolated from saline soils demonstrated that they substantially increase the dry matter of shoots and roots of wheat seedling growing under salt stress. Actually, the EPSs produced by these microorganisms induced the uptake of Na^+, Ca^{2+} and K^+ ions from salts, leading to an increase in yields, as measured using plant dry matter (Ashraf et al. 2004). Glucose-, rhamnose- and mannose-containing EPSs produced by *Pseudomonas putida* are reported to induce tolerance against several abiotic stresses including salt stress, and they also seem to help in soil aggregation (Sandhya and Ali 2015). *Azospirillum* (Bashan et al. 2004) and *Microbacterium arborescens* (Sudhamani et al. 2004) also exhibit good EPS-mediated soil aggregation. Vaishnav et al. (2016) investigated the effect of the biofilm-producing PGPR *Pseudomonas simiae* strain AU on the expression of salt-tolerant genes in soybean plants growing under salinity stress. The authors observed that all genes except nitrate reductase were upregulated in the presence of salt, when compared to the nonsaline control. However, a more pronounced increase in the relative expression of antioxidant enzymes (peroxidase, catalase) and defence protein (vegetative storage protein) occurred in seedlings growing in saline soil treated with *P. simiae*, especially when the nitric oxide donor, sodium nitroprusside, was also added. The high-affinity K^+ transporter was downregulated both in the nonsaline and in the saline soils in the presence of bacteria. Interestingly, lipoxygenase and polyphenol oxidase, though upregulated by salt, showed a less marked increase in their relative expression in saline soil treated with bacteria, when compared to the non-inoculated saline control, showing that different mechanisms may take place to modulate the expression of defence- and antioxidant-related genes under high salt conditions. The authors report that an overall improvement in growth of soybean plants occurs under salinity stress in the presence of *P. simiae* and the nitric oxide donor. A similar effect was observed by Hungria et al. (2013) in soybeans and common beans that were co-inoculated with rhizobia and azospirilla.

A different set of microorganisms are present in soil endosphere, rhizoplane and rhizosphere. Root-microbe symbiosis facilitates a positive growth to both plant and microorganisms. Biofilms formed and EPSs produced by soil microflora in soil affected with abiotic factors like salinity stress both help in soil aggregation, leading to better nutrient uptake in plants and also making salt ions unavailable to plants by bioaccumulating them in the polysaccharide network and thus alleviating salinity stress to plants. Inoculation of crop plants with *Thalassobacillus denorans* and *Oceanobacillus kapialis* PGPR strains in saline conditions facilitates seed germination and improves nutrient uptake, as well as plant vigour, chlorophyll content and biomass yield (Shah et al. 2017). Thus, cocultivation of plant growth-promoting bacteria and plants in saline soils may reduce the severity of saline stress together with enhancing crop yield, which could be both a cheap and eco-friendly approach for salinity stress management.

6.6 EPS and Biofilm-Producing Halophiles in Mitigating Salinity Stress

Bacterial biofilm has a co-dependent lifestyle that coordinates between bacterial cells and EPS matrix. As mentioned above, EPS production and biofilm formation by salt-tolerant halophiles can potentiate plant growth improvement in saline soil. Different salinity levels result into different types of biofilm formation. In a study with *Halomonas meridiana*, *Kushneria indalinina* and *Halomonas aquamarina*, the authors report that the best EPS production and biofilm formation occurred at 1 M salinity. They also report that these halophiles colonize the plant rhizosphere more efficiently in saline condition, helping the plants grow under such stress (Qurashi and Sabri 2011). The production of biofilm and EPS by *Halomonas variabilis* and *Planococcus rifietoensis* increased proportionally to the increase in salinity. It resulted into higher rate of soil aggregation and growth for the plant *Cicer arietinum* var. CM-98 (Qurashi and Sabri 2012).

Halophilic bacteria were used as a tool for bioremediation of salt-affected soils in several earlier studies. One of such reports discusses the effectiveness of the halophilic bacteria *Thalassobacillus devorans* and *Oceanobacillus kapialis* in improving plant growth in terms of physio-biochemical characteristics. These strains produce EPS molecules that get bound to the roots, thereby decreasing the uptake of Na^+ ion and alleviating salt stress (Shah et al. 2017). In a similar study on the co-inoculation of *Halomonas maura* and *Ensifer meliloti*, the yield of alfalfa plants was improved in saline soil. *H. maura* is well known for mauran type of EPS production which favours soil aggregation under both water and salinity stress. As the main composition of mauran is glucuronic acid, mannose, galactose and glucose (Arias et al. 2003) and they all are metabolic products, they could act as plant nutrients. A similar type of increase in crop productivity has been studied in other leguminous plants too. Sivaramaiah et al. (2007) observed an increase in the effectiveness of symbiosis in chickpea by co-inoculating six different halotolerant *Bacillus* strains. Moreover, it has been described by Ahmad et al. (2011) that the co-inoculation actually enhances plant growth, thus reducing the inhibitory effect of salinity.

The overall scenario of halophilic microorganisms and their EPS/biofilm production hints the need of their probable commercial applications in mitigation of salinity stress and improved plant growth. Halotolerant and halophilic microorganisms provide multiple potential applications in the field of biotechnology. They produce bacteriorhodopsin, biosurfactants, exopolysaccharides and many more, which have great commercial interest. Keeping in mind the adsorption of salt ions to the biofilm matrix and the effect of making the salts unavailable to plants, thus alleviating cytotoxicity, opens the door for halophilic bacterial community commercial use as inoculants in salinity stress management to achieve increased crop yield. This will also solve the questions of meeting the food demand of the ever-growing human population and the utilization of leftover saline soils for farming.

6.7 Biocontrolling Derelict Soil and Restoration of Crop Yield

The use of biological organisms like bacteria and fungi in procurement of the environment is known as 'biocontrolling'. Without depending on surface water, continuous irrigation by groundwater is one of the most critical reasons behind soil salinization. As groundwater contains various types of salts in different concentrations, after water evaporation, salt molecules get deposited in soil and increase salinity, making the soil derelict. This condition is responsible for changing the microflora of soil. Plant root can tolerate salt concentration up to 1.9 g/kg of soil. Exceeding this concentration, rhizosphere bacterial community changes abruptly (Yang et al. 2016). Bacteria, which are growing in that derelict soil, normally change their enzymes and metabolism to live in that adverse condition. Usually plants or other organisms cannot grow in highly saline soil, but the microorganisms which are able to cope with it have some special mechanisms or resistant structures to protect themselves and thrive. In order to control derelict soil biologically, researchers are studying microorganism-mediated mechanisms, based on the fact that bacteria play an important role in decreasing high salt concentration in soil.

It is reported that salt-tolerant halophilic bacteria mainly form biofilms which confer salt-stress resistance. Actually, biofilms contain exopolysaccharides (EPSs) which have ability to protect bacterial cells against high salinity. The bacterial strains *Halomonas variabilis* HT1 and *Planococcus rifietoensis* RT4 are reported to have the capability to improve soil structure (Qurashi and Sabri 2011). Application of these two bacteria in soil was found to reduce salt concentrations in the saline soil. It is also reported that after application of these two bacterial strains in saline soil, *Cicer arietinum* var. CM-98 has shown adequate growth (Qurashi and Sabri 2012). Microbial management thus helps to increase fertility of derelict soil. By the formation of exopolysaccharide-containing biofilm, normally halophilic bacteria aggregate in the root zone. *Pseudomonas* sp. and *Bacillus* spp. are most dominating strains used as plant growth-promoting bacteria (PGPB). They are mainly potassium-solubilizing microorganisms/bacteria (KSM). Under high salt stress, KSM strains reduce lipid peroxidation and increase cell membrane stability (Jha and Subramanian 2016). These bacterial associations with root also increase productivity of the plants (Mayak et al. 2004). *Paenibacillus durus* and *Paenibacillus fujiensis* are two significant isolates from saline soil that have great potential for nitrogen fixation. Even these two microorganisms have shown satisfactory nitrogenase activity due to having *nifD* and *nifH* gene (Navarro-Noya et al. 2016). In order to increase the growth of crop plants, these two bacteria can be inoculated in the farming lands. Presence of *Pseudomonas extremorientalis* TSAU20 and *Pseudomonas chlororaphis* TSAU13 in soil of agricultural field shows beneficial effect by increasing antioxidant enzymes produced by plants (Egamberdieva et al. 2017). This helps to reduce oxidative damage by decreasing the formation of reactive oxygen species (ROS) under saline soil. As reported by Egamberdieva et al. (2017), these two *Pseudomonas* species normally aggregate in the root region of tomato plants and protect them from salt stress by increasing the concentration of the amino acid proline in the cell. Accumulation of proline helps to adjust osmotic balance which otherwise becomes disturbed in high salt concentration. In addition, it also acts as scavenger of reactive oxygen species (Singh et al. 2014). Therefore, biocontrolling of saline soil using *Pseudomonas* spp. is a very effective way to manage derelict farming land. *Smithella*, *Halomonas*, *Pseudomonas* and *Comamonas* are some other important bacterial strains which have also shown significant growth in high saline stress (Zhao et al. 2018). Thus growth and farming of these bacteria in agricultural field may show a momentous positive effect in biomass production of crop plants.

Bacteria which grow in high salt stress normally produce EPS around their cell. It is reported that EPS binds to soil and increase macro-pores in the soil, thus increasing the fertility of farming land (Vimal et al. 2017). Therefore, they may be called as PGPR strains as they have a positive role in plant growth promotion and agricultural production. Aggregation of soil and EPS produced by microorganisms has potentiality to increase fertility in rhizospheric region which amplifies the availability of fertilizers to the crops (Vimal et al. 2017). Bacterial EPS also has promising water retention capacity. Therefore, EPS-treated plant shows significant tolerance to abiotic stress like drought and salinity (Sandhya et al. 2009). Accumulation of EPS-producing bacteria in the root zone of plant reduces the accessibility of Na^+ ion in rhizospheric region and alleviates salt stress (Nunkaew et al. 2015). Hence, it is well understood that EPS alleviates NaCl concentration in the soil and helps crop plants to grow properly. Therefore, the use of EPS-producing bacteria in the management of salt stress may be an innovative biocontrolling strategy of derelict soil.

6.8 Patenting Trends on Use of Microbial Technology in Saline Soil Management

The use of salt-tolerant microbes in soil salinity management has only recently received scientific attention. One early, if not the first, patent on using microorganism in saline soil has been recorded almost 10 years ago, in 2009. After that, the patenting trend started on 2013 with a peak on 2017 (Table 6.2). This hints that the use of salt-tolerant microorganisms in saline soil management is a modern and promising scientific approach that deserves more scientific attention. A better industry-academia interaction could enable the commercialization of microbial technology for soil salinity management.

Table 6.2 Patenting trends on the use of salt-tolerant microbes on soil salinity management

Patent number	Year	Patent title	Mechanism of action	References
CN108410470A	2018	Species for soil conditioner and a preparation method of microorganism saline	By this invention soil conditioner can be made by halophilic microbes which improve the soil fertility. As being a biological product, this soil management system does not harm overall soil environment	Zhang et al. (2018)
CN108251126A	2018	Composite soil amendments and microbial preparation containing licorice extract waste	Natural growth of microorganisms improves soil micro-environment, quality and physical activity of soil. In addition with this, bacterial enzyme activity also decreases the chance of plant disease	Wang (2018c)
CN108323262A	2018	The method of planting wild Mao sandy saline soil and fertilizer and improved conditions	Microbial powder used to increase fertility of sandy saline soil. This is a good pollution-free mechanism which has positive economic impact on crop productivity by increasing biomass production	Wang (2018b)
CN108235903A	2018	The method of planting clover beans sandy saline soil and fertilizer and improved conditions	Improvement of sandy saline soil by implication of naturally grown microorganisms and large-scale growth of crop plant	Wang (2018a)
CN108330141A	2018	Species encoding a protein identification method and application Pm_MF	Genetic engineering of crop plant by using salinity tolerant gene of naturally growing microorganisms under high salt stress	Jiang et al. (2018)
CN106866300A	2017	Desert area saline-alkali soil amelioration-modifying agent and preparation method	Soil amelioration of saline desert soil by different types of halophilic bacteria. This process improves the organic content of soil and increases productivity	Ni et al. (2017)
CN107324927A	2017	Method for planting *Sophora alopecuroides* by improving saline-alkali soil	A compound fertilizer, consists of natural microbial organisms, used to improve the quality of saline soil which also prevents environmental pollution	Wang (2017d)
CN108085278A	2017	High salt-resistant phosphate-solubilizing bacterial strain and application thereof	*Bacillus licheniformis* and *Bacillus megaterium* are used as compound inoculum which is applied as a microbial fertilizer to improve plant growth	Wang et al. (2017b)

CN108207503A	2017	Coastal sandy saline method of planting grass and sand cited improved soil and fertilizer conditions	Dried powder of microorganism used with other chemicals in order to improve salt fertility which helps in increment of biomass production	Wang (2017a)
CN108029493A	2017	Method for planting peanuts and improving soil fertility condition on sandy saline-alkali soil	Mixture of microbial powder and hydroxyethyl helps to increase fertility of soil very prominently	Wang (2017c)
CN108260373A	2017	Methods of sandy saline soil and fertilizer and improved soybean planting conditions	Invention of pollution-free mechanisms by using bacterial dust and different other natural products with inorganic fertilizers to improve saline sandy soil for better farming	Wang (2017e)
CN107324926A	2017	Improved fertilizer suitable for saline-alkali soil and production method of improved fertilizer	Fertilizers made up of mycorrhiza bacteria, rhododendron mycorrhiza bacteria and orchid mycorrhiza bacteria help to improve agricultural condition of saline soil	Wang (2017b)
CN107162649A	2017	Irrigation water for planting blueberry and preparation method thereof	Irrigation of saline salt by innovative mixtures of different products where halophilic bacteria are present nearly 5% of total mixture	Xie and Yan (2017)
CN108157106A	2017	Methods of sandy saline soil and fertilizer to grow cotton and improved conditions	Microbial powder has been used for better productivity and high growth rate of cotton plant under high salt stress	Wang (2017f)
CN106008821A	2016	Method for carrying out raw soil greening on slight saline and alkaline land by using halophilic microorganism-containing water retaining agent	Water-retaining agent has been made by using halophilic bacteria and many other reagents. This mechanism helps to reduce soil salinity by retaining saline water from field	Wang (2017g)
CN105018096A	2015	Halophilic bacteria-containing soil modifier for saline land and applications of halophilic bacteria-containing soil modifier in *Cichorium intybus* L. planting	Different halophilic bacterial strains have been mixed to modify productivity of agricultural field	Wang (2015)

(continued)

Table 6.2 (continued)

Patent number	Year	Patent title	Mechanism of action	References
CN104017750A	2014	Salt-resistant *Bacillus methylotrophicus* strain LJ and application thereof	Salt-resistant *Bacillus methylotrophicus* strain has a property of salt resistance, and it also acts as plant growth-promoting bacteria, viz. PGPR strain	Shan et al. (2014)
CN103436450A	2013	Alkaline-tolerant and halophilic aspergillus strain and application thereof in environmental management	Bioengineering of halophilic *Aspergillus* strain helps to improve farming land with high salt stress and also reduces the environmental pollution	Wei et al. (2013)
CN103232960B	2013	Oil-degrading composite bacterium and microbial inoculum applied to high salt environment	*Gordonia alkanivorans* has a capacity to grow in high oil-containing saline condition. In high salt condition (8%), this bacterium can grow easily	Wang et al. (2013)

6.9 Conclusions and Future Perspective

Soil salinity is a serious environmental threat to human health. Because of high salt stress, plants cannot grow properly, and the amount of their biomass gets significantly reduced. Due to the effect of saline soil in agricultural system, increasingly growing human population may face severe scarcity of food in the near future. There are some conventional irrigation methods which are applicable in the reduction of soil salinity, but those practices are both time-consuming and costly. To compensate the stress imposed by salinity, biofilm formation and EPS production are significant strategies of salt-tolerant bacteria that can assist plant metabolism. Halophilic microorganisms can form biofilm and accumulate EPS at increasing salt stress. EPS actually contributes to the colonization of plant roots and soil particles by bacteria. Bacterial colonization eventually results in improved plant growth and better soil structure. The structure of EPS from halophile and PGPR bacterial origin is different compared to EPS from other origin, and their biosynthesis mechanism is also different.

PGPR strains and halophilic microorganisms show significant resistance to salt stress and increase soil fertility by decreasing the available NaCl concentration in soil in terms of accumulation of salt ions in biofilm/EPS matrix and also soil aggregation. Bacteria like *Azospirillum*, *Microbacterium* and *Rhizobium* grow normally in the rhizospheric zone and produce EPS. But sometimes halophilic bacteria like the previously mentioned *Halomonas*, *Kushneria*, *Planococcus*, *Thalassobacillus* and *Oceanobacillus* alleviate salinity stress when co-cultivated with plants under salt stress. These EPSs have unique property of retaining water along with salt, which in turn help to decrease salinity of soil, so that the plants can grow better and give better yields. In the future, better farming in saline soil could be achieved by adding EPS-producing bacteria that are able to associate with the plants and reduce salt stress. The developments on understanding the role of PGPR and halophilic bacteria in mitigating salinity stress may open a new avenue for capitalizing on the cultivable microbiome to strengthen plants' salt tolerance and thus to refine agricultural practices and biomass production under saline conditions. In a long-term vision, the alleviation of derelict soil, improved crop yield and restoration of mangrove diversity are possible by the mechanistic use of EPS-producing halotolerant microorganisms. More research effort is needed in the field of biological management of saline soils using soil-inhabiting PGPR and halophilic microorganisms to commercialize the cocultivation of plant and halotolerant microorganisms in order to improve the quality and yield of crops cultivated in such conditions.

Acknowledgements Authors are thankful to Facultad de Ciencias Agrarias y Forestales, Universidad Católica del Maule and Department of Botany, The University of Burdwan, for pursuing research activities. AB is also thankful to Vicerrectoría de Investigación y Posgrado (VRIP), Universidad Católica del Maule, for postdoctoral fellowship.

References

Ahmad M, Zahir ZA, Asghar HN, Asghar M (2011) Inducing salt tolerance in mung bean through coinoculation with rhizobia and plant-growth-promoting rhizobacteria containing 1-aminocyclopropane-1-carboxylate deaminase. Can J Microbiol 57:578–589

Ahmed V, Verma MK, Gupta S, Mandha V, Chauhan NS (2018) Metagenomic profiling of soil microbes to mine salt stress tolerance genes. Front Microbiol 9:159

Ali A, Shahzad R, Khan AL, Halo BA, Al-Yahyai R, Al-Harrasi A, Al-Rawahi A, Lee IJ (2017) Endophytic bacterial diversity of *Avicennia marina* helps to confer resistance against salinity stress in *Solanum lycopersicum*. J Plant Interact 12:312–322

Arias S, Del Moral A, Ferrer MR, Tallon R, Quesada E, Bejar V (2003) Mauran, an exopolysaccharide produced by the halophilic bacterium *Halomonas maura*, with a novel composition and interesting properties for biotechnology. Extremophiles 7:319–326

Ashraf M, Hasnain S, Berge O, Mahmood T (2004) Inoculating wheat seedlings with exopolysaccharide-producing bacteria restricts sodium uptake and stimulates plant growth under salt stress. Biol Fertil Soils 40:157–162

Banerjee A, Bandopadhyay R (2016a) Use of dextran nanoparticle: a paradigm shift in bacterial exopolysaccharide based biomedical applications. Int J Biol Macromol 87:295–301

Banerjee A, Bandopadhyay R (2016b) Bacterial exopolysaccharides: major types and future prospects. In: Shukla P (ed) Microbial biotechnology: an interdisciplinary approach. CRC Press, Boca Raton, FL, pp 1–20

Bashan Y, Holguin G, De-Bashan LE (2004) Azospirillum-plant relationships: physiological, molecular, agricultural, and environmental advances (1997–2003). Can J Microbiol 50:521–577

Becker VI, Goessling JW, Duarte B, Caçador I, Liu F, Rosenqvist E, Jacobsen SE (2017) Combined effects of soil salinity and high temperature on photosynthesis and growth of quinoa plants (*Chenopodium quinoa*). Funct Plant Biol 44:665–678

Blaylock AD (1994) Soil salinity, salt tolerance, and growth potential of horticultural and landscape plants. University of Wyoming, Cooperative Extension Service, Department of Plant, Soil, and Insect Sciences, College of Agriculture

Corwin DL, Rhoades JD, Šimůnek J (2007) Leaching requirement for soil salinity control: steady-state versus transient models. Agric Water Manag 90:165–180

Devkota M, Gupta RK, Martius C, Lamers JPA, Devkota KP, Sayre KD, Vlek PLG (2015) Soil salinity management on raised beds with different furrow irrigation modes in salt-affected lands. Agric Water Manag 152:243–250

Duarte B, Goessling JW, Marques JC, Caçador I (2015) Ecophysiological constraints of *Aster tripolium* under extreme thermal events impacts: merging biophysical, biochemical and genetic insights. Plant Physiol Biochem 97:217–228

Egamberdieva D, Davranov K, Wirth S, Hashem A, Abd_Allah EF (2017) Impact of soil salinity on the plant-growth–promoting and biological control abilities of root associated bacteria. Saudi J Biol Sci 24:1601–1608

Feng G, Zhang Z, Wan C, Lu P, Bakour A (2017) Effects of saline water irrigation on soil salinity and yield of summer maize (*Zea mays* L.) in subsurface drainage system. Agric Water Manag 193:205–213

Glick BR (2005) Modulation of plant ethylene levels by the bacterial enzyme ACC deaminase. FEMS Microbiol Lett 251:1–7

Goswami D, Thakker JN, Dhandhukia PC (2016) Portraying mechanics of plant growth promoting rhizobacteria (PGPR): a review. Cogent Food Agric 2(1):1127500

Grover M, Ali SZ, Sandhya V, Rasul A, Venkateswarlu B (2011) Role of microorganisms in adaptation of agriculture crops to abiotic stresses. World J Microbiol Biotechnol 27:1231–1240

Gupta P, Diwan B (2017) Bacterial exopolysaccharide mediated heavy metal removal: a review on biosynthesis, mechanism and remediation strategies. Biotechnol Rep 13:58–71

Gupta B, Huang B (2014) Mechanism of salinity tolerance in plants: physiological, biochemical, and molecular characterization. Int J Genomics 2014:701596

Halder U, Banerjee A, Bandopadhyay R (2017) Structural and functional properties, biosynthesis, and patenting trends of Bacterial succinoglycan: a review. Indian J Microbiol 57:278–284
Hungria M, Nogueira MA, Araujo RS (2013) Co-inoculation of soybeans and common beans with rhizobia and azospirilla: strategies to improve sustainability. Biol Fertil Soils 49:791–801
Jamil A, Riaz S, Ashraf M, Foolad MR (2011) Gene expression profiling of plants under salt stress. Crit Rev Plant Sci 30:435–458
Jha Y, Subramanian RB (2016) Regulation of plant physiology and antioxidant enzymes for alleviating salinity stress by potassium-mobilizing bacteria. In: Potassium solubilizing microorganisms for sustainable agriculture. Springer, New Delhi, pp 149–162
Jiang J, Bo H, Rui Z, Zhang Z, Xu T, Meng L, Zou Q, Zheng XT (2018) Species encoding a protein identification method and application Pm_MF. CN Patent CN108330141A, 30 Jan 2018
Khan MS, Zaidi A, Musarrat J (2009) Microbial strategies for crop improvement. Springer, Berlin
Kreps JA, Wu Y, Chang HS, Zhu T, Wang X, Harper JF (2002) Transcriptome changes for Arabidopsis in response to salt, osmotic, and cold stress. Plant Physiol 130:2129–2141
Mahmuduzzaman M, Ahmed ZU, Nuruzzaman AKM, Ahmed FRS (2014) Causes of salinity intrusion in coastal belt of Bangladesh. Int J Plant Res 4:8–13
Makhdoumi-Kakhki A, Amoozegar MA, Kazemi B, Pasic L, Ventosa A (2012) Prokaryotic diversity in Aran-Bidgol Salt Lake, the largest hypersaline playa in Iran. Microbes Environ 27:87–93
Malash NM, Flowers TJ, Ragab R (2008) Effect of irrigation methods, management and salinity of irrigation water on tomato yield, soil moisture and salinity distribution. Irrigation Sci 26:313–323
Mayak S, Tirosh T, Glick BR (2004) Plant growth-promoting bacteria confer resistance in tomato plants to salt stress. Plant Physiol Biochem 42:565–572
Meena KK, Sorty AM, Bitla UM, Choudhary K, Gupta P, Pareek A, Singh DP, Prabha R, Sahu PK, Gupta VK, Singh HB, Krishanani KK, Minhas PS (2017) Abiotic stress responses and microbe-mediated mitigation in plants: the omics strategies. Front Plant Sci 8:172
Metternicht GI, Zinck JA (2003) Remote sensing of soil salinity: potentials and constraints. Remote Sens Environ 85:1–20
Moscovici M (2015) Present and future medical applications of microbial exopolysaccharides. Front Microbiol 6:1012
Munns R (2005) Genes and salt tolerance: bringing them together. New Phytol 167:645–663
Nadeem SM, Ahmad M, Zahir ZA, Javaid A, Ashraf M (2014) The role of mycorrhizae and plant growth promoting rhizobacteria (PGPR) in improving crop productivity under stressful environments. Biotechnol Adv 32:429–448
Navarro-Noya YE, Luna-Guido M, Dendooven L (2016) Cultivable nitrogen fixing bacteria from extremely alkaline-saline soils. Adv Microbiol 6:412
Netondo GW, Onyango JC, Beck E (2004) Sorghum and salinity: II. Gas exchange and chlorophyll fluorescence of sorghum under salt stress. Crop Sci 44:806–811
Ni Z, Peng Y, Xia J, Wang Y, Yao H (2017) Desert area saline-alkali soil amelioration modifying agent and preparation method. CN Patent CN106866300A, 23 Jan 2017
Nouri H, Borujeni SC, Nirola R, Hassanli A, Beecham S, Alaghmand S, Mulcahy D (2017) Application of green remediation on soil salinity treatment: a review on halophytoremediation. Process Saf Environ 107:94–107
Nunkaew T, Kantachote D, Nitoda T, Kanzaki H, Ritchie RJ (2015) Characterization of exopolymeric substances from selected *Rhodopseudomonas palustris* strains and their ability to adsorb sodium ions. Carbohydr Polym 115:334–341
Qurashi AW, Sabri AN (2011) Biofilm formation in moderately halophilic bacteria is influenced by varying salinity levels. J Basic Microbiol 52:566–572
Qurashi AW, Sabri AN (2012) Bacterial exopolysaccharide and biofilm formation stimulate chickpea growth and soil aggregation under salt stress. Braz J Microbiol 43:1183–1191
Sandhya V, Ali SZ (2015) The production of exopolysaccharide by *Pseudomonas putida* GAP-P45 under various abiotic stress conditions and its role in soil aggregation. Microbiology 84:512–519

Sandhya VZAS, Grover M, Reddy G, Venkateswarlu B (2009) Alleviation of drought stress effects in sunflower seedlings by the exopolysaccharides producing *Pseudomonas putida* strain GAP-P45. Biol Fertil Soils 46:17–26

Sen S, Chandrasekhar CN (2014) Effect of PGPR on growth promotion of rice (*Oryza sativa* L.) under salt stress. Asian J Plant Sci Res 4:62–67

Şengül N, Işık S, Aslım B, Uçar G, Demirbağ AE (2011) The effect of exopolysaccharide-producing probiotic strains on gut oxidative damage in experimental colitis. Dig Dis Sci 56:707–714

Serrano R, Gaxiola R (1994) Microbial models and salt stress tolerance in plants. CRC Crit Rev Plant Sci 13:121–138

Shah G, Jan M, Afreen M, Anees M, Rehman S, Daud MK, Malook I, Jamil M (2017) Halophilic bacteria mediated phytoremediation of salt-affected soils cultivated with rice. J Geochem Explor 174:59–65

Shan J, Yan X, Wang H (2014) Salt-resistance *Bacillus methylotrophicus* strain LJ and application thereof. CN Patent CN104017750A, 14 Apr 2014

Shrivastava P, Kumar R (2015) Soil salinity: a serious environmental issue and plant growth promoting bacteria as one of the tools for its alleviation. Saudi J Biol Sci 22:123–131

Singh M, Kumar J, Singh VP, Prasad SM (2014) Proline and salinity tolerance in plants. Biochem Pharmacol 3:1000170

Sivaramaiah N, Malik DK, Sindhu SS (2007) Improvement in symbiotic efficiency of chickpea (Cicer arietinum) by coinoculation of Bacillus strains with *Mesorhizobium* sp. Cicer. Indian J Microbiol 47:51–56

Steele DJ, Franklin DJ, Underwood GJ (2014) Protection of cells from salinity stress by extracellular polymeric substances in diatom biofilms. Biofouling 30:987–998

Sudhamani SR, Tharanathan RN, Prasad MS (2004) Isolation and characterization of an extracellular polysaccharide from *Pseudomonas caryophylli* CFR 1705. Carbohydr Polym 56:423–427

Tavakkoli E, Rengasamy P, McDonald GK (2010) High concentrations of Na+ and Cl− ions in soil solution have simultaneous detrimental effects on growth of faba bean under salinity stress. J Exp Bot 61:4449–4459

Tester M, Davenport R (2003) Na+ tolerance and Na+ transport in higher plants. Ann Bot 91:503–527

Upadhyay SK, Singh JS, Singh DP (2011) Exopolysaccharide-producing plant growth-promoting rhizobacteria under salinity condition. Pedosphere 21:214–222

Vaishnav A, Kumari S, Jain S, Varma A, Tuteja N, Choudhary DK (2016) PGPR-mediated expression of salt tolerance gene in soybean through volatiles under sodium nitroprusside. J Basic Microbiol 56:1274–1288

Vargas R, Pankova EI, Balyuk SA, Krasilnikov PV, Khasankhanova GM (2018) Handbook for saline soil management. Food and Agriculture Organization of the United Nations, Lomonosov Moscow State University. http://www.fao.org/3/i7318en/I7318EN.pdf. Accessed 28 Dec 2018

Vimal SR, Singh JS, Arora NK, Singh S (2017) Soil-plant-microbe interactions in stressed agriculture management: a review. Pedosphere 27:177–192

Vineis P, Chan Q, Khan A (2011) Climate change impacts on water salinity and health. J Epidemiol Glob Health 1:5–10

Wang S (2015) Halophilic bacteria-containing soil modifier for saline land, and applications of halophilic bacteria-containing soil modifier in cichorium intybus l planting. CN Patent CN105018096A, 23 Jun 2015

Wang S (2017a) Coastal sandy saline method of planting grass and sand cited improved soil and fertilizer conditions. CN Patent CN108207503A, 30 Nov 2017

Wang S (2017b) Improved fertilizer suitable for saline-alkali soil and production method of improved fertilizer. CN Patent CN107324926A, 31 Aug 2017

Wang S (2017c) Method for planting peanuts and improving soil fertility condition on sandy saline-alkali soil. CN Patent CN108029493A, 30 Nov 2017

Wang S (2017d) Method for planting sophora alopecuroides by improving saline-alkali soil. CN Patent CN107324927A, 31 Aug 2017

Wang S (2017e) Methods of sandy saline soil and fertilizer and improved soybean planting conditions. CN Patent CN108260373A, 30 Nov 2017
Wang S (2017f) Methods of sandy saline soil and fertilizer to grow cotton and improved conditions. CN Patent CN108157106A, 30 Nov 2017
Wang S (2017g) Method for carrying out raw soil greening on slight saline and alkaline land by using halophilic microorganism-containing water retaining agent. CN Patent CN106008821A, 31 May 2016
Wang S (2018a) The method of planting clover beans sandy saline soil and fertilizer and improved conditions. CN Patent CN108235903A, 31 Jan 2018
Wang S (2018b) The method of planting wild Mao sandy saline soil and fertilizer and improved conditions. CN Patent CN108323262A, 31 Jan 2018
Wang Y (2018c) Composite soil amendments and microbial preparation containing licorice extract waste. CN Patent CN108251126A, 19 Mar 2018
Wang LC, Morgan LK, Godakumbura P, Kenney LJ, Anand GS (2012) The inner membrane histidine kinase EnvZ senses osmolality via helix-coil transitions in the cytoplasm. The EMBO journal 31:2648–2659
Wang X, Wu L, Yang Y, An W, Niu Z, Ding D (2013) Oil-degrading composite bacterium and microbial inoculum applied to high salt environment. CN Patent CN103232960B, 23 Apr 2013
Wang H, Yang A, Zhang G, Ma B, Meng F, Peng M, Wang H (2017a) Enhancement of carotenoid and bacteriochlorophyll by high salinity stress in photosynthetic bacteria. Int Biodeterior Biodegradation 121:91–96
Wang H, Cao J, Zheng Y, Zhang W, Yang Z, Wang M, Jiang S, Wang B (2017b) High salt-resistant phosphate solubilizing bacterial strain and application thereof. CN Patent CN108085278A, 29 Dec 2017
Wei Y, Liu X, Jia B, Zhang S, Liu J, Gao W (2013) Alkaline-tolerant and halophilic aspergillus strain and application thereof in environmental management. CN Patent CN103436450, 5 Jun 2013
Xie Z, Yan Y (2017) Irrigation water for planting blueberry and preparation method thereof. CN patent CN107162649, 27 May 2017
Xun F, Xie B, Liu S, Guo C (2015) Effect of plant growth-promoting bacteria (PGPR) and arbuscular mycorrhizal fungi (AMF) inoculation on oats in saline-alkali soil contaminated by petroleum to enhance phytoremediation. Environ Sci Pollut Res Int 22:598–608
Yang H, Hu J, Long X, Liu Z, Rengel Z (2016) Salinity altered root distribution and increased diversity of bacterial communities in the rhizosphere soil of Jerusalem artichoke. Sci Rep 6:20687
Zannini E, Waters DM, Coffey A, Arendt EK (2016) Production, properties, and industrial food application of lactic acid bacteria-derived exopolysaccharides. Appl Microbiol Biotechnol 100:1121–1135
Zerrouk IZ, Benchabane M, Khelifi L, Yokawa K, Ludwig-Müller J, Baluska F (2016) A *Pseudomonas* strain isolated from date-palm rhizospheres improves root growth and promotes root formation in maize exposed to salt and aluminum stress. Plant Physiol 191:111–119
Zhang Q, Zhang Z, Yin L, Pan Y, Gao P, Li Z (2018) Species for soil conditioner and a preparation method of microorganism saline. CN Patent CN108410470A, 5 Mar 2018
Zhao Y, Li Y, Wang J, Pang H, Li Y (2016) Buried straw layer plus plastic mulching reduces soil salinity and increases sunflower yield in saline soils. Soil Tillage Res 155:363–370
Zhao S, Liu JJ, Banerjee S, Zhou N, Zhao ZY, Zhang K, Tian CY (2018) Soil pH is equally important as salinity in shaping bacterial communities in saline soils under halophytic vegetation. Sci Rep 8:4550
Zhou D, Lin Z, Liu L, Zimmermann D (2013) Assessing secondary soil salinization risk based on the PSR sustainability framework. J Environ Manag 128:642–654

Chapter 7
Impact of Salinity Stress on Growth and Development of Aquatic Fungi

Patricia Velez

Abstract In most ecosystems salinity shapes biotic assemblages, representing a key environmental factor. This variable is regarded as a major threat to microbial communities in terrestrial systems, modifying considerably a number of important ecosystem processes, including turnover of organic matter and nutrients acquisition. However, complex effects of salinity remain poorly understood, especially for non-model aquatic microbial assemblages, which account for most of the biodiversity in natural systems. Aquatic fungi are a widespread and phylogenetically heterogeneous group of microorganisms, occurring in marine, estuarine, and freshwater systems. These osmotrophs are completely adapted to rapidly colonize, grow, and reproduce in aquatic systems, where salinity represents a frequently fluctuating environmental variable. Some investigations have approached aquatic fungal response to salinity, suggesting that despite these microorganisms are able to survive under osmotic stress conditions, this variable may select for distinctive community compositions. At large, fungal responses to salinity stress are determined by taxon-specific underlying physiological traits, leading to distinctive tolerance thresholds. Herein, we review the impact of salinity on growth and development of aquatic fungi, integrating literature reports on marine and freshwater species, and recent advances introducing molecular techniques to provide better understanding of the phenomenon of aquatic fungal salinity tolerance.

Keywords Environmental stress · Marine fungi · Osmotolerance · Salt adaptation mechanisms

7.1 Introduction

Salinity, which may be defined as the amount of dissolved inorganic substances in 1 kg of water (Stumm and Morgan 1996), represents a key environmental variable affecting most natural communities. On the one hand, the accumulation of

---

P. Velez (✉)
Department of Botany, Universidad Nacional Autónoma de México, Mexico City, Mexico
e-mail: pvelez@ib.unam.mx

B. Giri, A. Varma (eds.), *Microorganisms in Saline Environments: Strategies and Functions*, Soil Biology 56, https://doi.org/10.1007/978-3-030-18975-4_7

water-soluble salts in the soil, regarded as salinization, causes negative impacts on agricultural production, environmental health, and economic welfare. Currently, salt-affected soils occur in more than 100 countries of the world with a variety of extents, nature, and properties (Rengasamy 2006), posing a threat for food security (Cheeseman 2016). On the other hand, in aquatic systems salinity is considered as a driver of ecosystem structure (McEvoy and Goonan 2003), being regarded as a master element determining the distribution of marine and estuarine species and constraining the occurrence of freshwater organisms. Accordingly, the terms stenohaline (organisms with narrow tolerance to changes in salinity, including most marine species) and euryhaline (species with the ability to tolerate large and rapid changes in salinity, such as estuarine taxa) have been implemented to attempt a classification of aquatic biodiversity based on its response to this variable (Smyth and Elliott 2016).

Presently, no consensus has been reached regarding the meanings of the terms fresh and saline water. For inland aquatic systems, salinity is measured as chloride (Cl^-) content or total dissolved solids (Herbert et al. 2015), whereas in marine systems salinity is measured on the practical salinity scale (PSS) as the ratio of the conductivity of a sample to that of a reference (Lewis 1980). Moreover, in marine ecosystems dissolved salts include two major elements: sodium (Na^+) and Cl^-, accounting for 85.7%. Nevertheless, further seawater components also include magnesium (Mg^{2+}), calcium (Ca^{2+}), potassium (K^+), and sulfate (SO_4^{2-}), which together with Cl^- and Na^+ make up 99.4% of the total. Furthermore, the concentration and availability of these elements are constantly changing, being influenced by a number of factors such as rainfall, wind, evaporation, inflow of river water, and melting of glaciers (Stewart 2008). Accordingly, differences across continents have been recognized. For instance, Na^+ and Cl^- seem to dominate in high-salinity surface waters of arid regions of the subtropics, where evaporation rates are high, as well as in low-salinity rivers and surface waters of tropical regions with high rainfall rates (e.g., South America, Africa, and Australia). While in regions with low evaporation rates, dominated by sedimentary rocks and limestones (e.g., Europe and North America), Ca^{2+} and HCO_3^- tend to dominate (Harris 2009).

The evaluation of biotic responses to temporal and geographical fluctuations in salinity, as well as tolerance mechanisms, represents a fundamental task to prevent loss of agricultural productivity and safeguard ecosystems health and services. This is especially true for neglected, invisible members of the community, who sustain ecosystem processes and represent the vast majority of biodiversity. In this chapter, we review information on the impact of salinity on growth and development of aquatic fungi, integrating literature reports on marine and freshwater species, introducing molecular techniques as an attempt to unravel fungal adaptation mechanisms to salinity stress.

7.2 Fungi in Aquatic Habitats

Aquatic fungi are an ecological rather than a taxonomic grouping of heterotrophs with representatives in all classes of kingdom *Fungi*. Over 3000 species have been reported from freshwater, brackish, and marine habitats, being widely distributed from Arctic to temperate and tropical areas (Shearer et al. 2007). These microorganisms represent a key element of aquatic trophic chains, acting principally as saprobes, yet fewer species have been also recognized as parasites and symbionts (Fisher and Likens 1973; Bärlocher and Kendrick 1974; Kohlmeyer and Kohlmeyer 1979; Ittner et al. 2018). Based on their physiological capacities to tolerate salinity, two general ecological groups have been distinguished: freshwater and marine taxa.

The definition of a freshwater fungus has been long discussed, since physiological and developmental studies are essential to support the theory that all of the so-called freshwater fungi depend on water to complete their life cycle (Shearer 1993). Anyhow, this ecological assemblage has been divided into two clusters: the first one referring to species that have been exclusively reported from the aquatic environment (residents or indwellers) and the second one comprising fungi that occur in water fortuitously by being washed or blown in (transients or immigrants; Shearer et al. 2007).

Likewise former definitions of a marine fungus were merely based on physiological capacities such as the ability to grow (Johnson and Sparrow 1961), sporulate (Kohlmeyer and Kohlmeyer 1979), or be metabolically active in the marine environment or under marine/estuarine salinity conditions (Mahé et al. 2014). However, the recently revised definition by Pang et al. (2016) also considers distinct genetic and evolutionary traits, defining a marine fungi those that are recovered repeatedly from marine habitats because of their ability to grow and/or sporulate (on substrata) in marine environments; establish symbiotic relationships with other marine organisms; and adapt, evolve, and be metabolically active in the marine milieu.

Independently from the definition, this ecological assemblage of microorganisms plays an irreplaceable role in energy flow in aquatic systems (Shearer et al. 2007; Danger et al. 2016) and represents valuable genetic resources as bioindicators (González and Hanlin 2010) and prospects for bioremediation (Kirk et al. 1991) and bioprospecting (Overy et al. 2014). Increasing anthropogenic impacts such as salinization of aquifers and habitat modifications have highlighted the importance of understanding fungal diversity in these ecosystems. However, in relation to other aquatic species, there is still little information about aquatic fungal physiological capacities in response to osmotic stress, due to limitations and biases of collecting data in situ. Nevertheless, these limitations may be overpassed by integrating (in addition to classical in vitro procedures) novel sequencing technologies that assess microbial diversity and gene expression profiles from the environment, such as meta-transcriptomics and metagenomics (e.g., Frias-Lopez et al. 2008; Gilbert et al. 2008; Shi et al. 2009). This approach has been successfully implemented to explore non-fungal microbial gene abundance and expression patterns across salinity gradients, allowing the identification of distinct microbial assemblages and gene expression profiles from river to ocean communities and the upregulation of genes encoding for catabolic pathways, osmoregulators, and metal transporters in the ocean (Fortunato and Crump 2015).

7.3 Aquatic Fungal Response to Salinity Variations

Understanding species ecological tolerances to environmental extremes represents an important first step in conservation and management. In general, salinization has been recognized as a threat for freshwater organisms, which in most cases lack adaptive traits toward osmotic stress (Hintz and Relyea 2017). However, some species have developed strategies to cope with elevated salt concentrations, providing freshwater communities with resilience to salinity disturbance (James et al. 2003). Aquatic fungal tolerance to a wide range of salinities has been recognized in a number of ecosystems and confirmed in vitro through the comparison of growth differences between marine and freshwater taxa under different salinity regimes (Barghoorn and Linder 1944; Ritchie 1959; Jones and Jennings 1964).

Salinity, as other stressors, often leads to the evolution of adapted ecotypes (e.g., Dewey 1960; Tanimoto 1969). Congruently, marine fungal optimal growth is reached under marine salinity concentrations, though it seems that these organisms are capable of growing under freshwater conditions as well (Jones and Jennings 1964; Masuma et al. 2001; De la Cruz et al. 2006; Burgaud et al. 2013; Huang et al. 2011; Velez et al. 2015a). This faculty may confer adaptive and competitive ecological advantages over their terrestrial counterparts in ecotones, which constantly receive freshwater inputs (Calado et al. 2015). For instance, based on their occurrence on different portions of *Spartina maritima*, fungal isolates were in vitro tested on media lacking and containing diluted seawater. Remarkably, the majority of strains showed significantly higher growth under saline conditions (Calado et al. 2015).

The niche overlap hypothesis states that maximal tolerable overlap should vary inversely with the intensity of competition. In addition, ecological theory expands this view by incorporating environmental stability as a key element and suggests that maximal tolerable niche overlap should be relatively insensitive to environmental variability (Pianka 1974). In aquatic environments distinctive fungal assemblages have been recognized in relation to salinity, such as those occurring chiefly in fully saline waters, brackish waters, and freshwater (reviewed by Jones 2000). Usually, these fungal ecotypes do not overlap (Shearer et al. 2007), yet some freshwater species have been sporadically recorded from marine or brackish waters and vice versa, implying that aquatic fungi are strong competitors, which in addition are resilient to environmental heterogeneity. An example is the typically freshwater species *Ascosacculus heteroguttulatus* obtained from a marine sandy beach (Velez et al. 2015b). It is important to mention that information on the viability and ecological role of freshwater fungi in marine ecosystems and vice versa remains unclear. So, detailed work is encouraged in this area to fully understand aquatic fungal ecology and niche overlap.

Coastal ecosystems such as sandy beaches, mangroves, and estuaries epitomize natural salinity gradients, where biota is commonly shaped by environmental heterogeneity (Elliott and Quintino 2007). Hence, these ecosystems represent an ideal model to investigate biological responses to salinity fluctuations. Since pioneer experimental fieldwork on the vertical distribution of marine fungi (e.g., Schaumann 1968, 1969), salinity-based niche differentiation among marine, freshwater, and terrestrial species has been of interest to mycologists. In this regard, fungal communities in salt marsh grasses and perennial herbaceous plants have been well documented (reviewed by Jones 2000). Still further evidence has been accumulating for other systems including oak and larch (Petersen and Koch 1997), mangrove trees (Hyde 1989, 1990a, b, 1993), and *Spartina maritima* (Calado et al. 2015). All of these reports agree, recognizing salinity as a key element enhancing aquatic fungal diversity.

Numerous efforts have been made in order to understand fungal response and adaptation to salinity stress in marine ecosystems. In this regard mangrove forests have been commonly assessed, as they represent marine fungal hotspots. Remarkably, fungal occurrence in this ecosystem is strongly linked to salinity (Hyde and Jones 1988; Tan et al. 1989; Hyde 1988, 1990a, b), where seasonal patterns and daily tidal-related changes in sea level may lead to niche differentiation. Moreover, changes in community composition accounting for differences between the rainy (lower salinity) and dry season (higher salinity) have been recognized as well, with marine fungi as dominant component in dry season while in rainy season typical terrestrial fungi occur (Ananda and Sridhar 2004).

Similarly, sandy beaches represent a dynamic ecosystem where variations in salinity from pluvial and fluvial discharges affect intertidal diversity (Lercari and Defeo 2006). Marine arenicolous fungi constitute a key element of the indigenous biota in sandy beaches. This ecological grouping of saprobes inhabits between or on the surface of sand grains, being perfectly adapted to complete their life cycle under saline conditions (Fig. 7.1a–c; Kohlmeyer and Kohlmeyer 1979). Early works using the species *C. maritima* as a model have revealed that this fungus may be affected by changes in salinity but can grow under freshwater conditions (Fig. 7.1d–g; Johnson and Sparrow 1961; Jones and Jennings 1964) and that these wide tolerance ranges have slight impacts on morphology (vegetative growth). This may represent an adaptive trait enabling *C. maritima* to cope with extreme environmental conditions, including exposure to high temperatures, desiccation, and variation in salinity (Sundari et al. 1996a, b).

Initial efforts approaching marine fungal physiology in response to osmotic stress were solely concentrated on salt requirements of individual species. These investigations were based on the hypothesis that marine fungi depend on sodium chloride concentrations contained in seawater to develop (Jones and Jennings 1964; Meyers 1968; Jones et al. 1971; Byrne and Jones 1975a, b; Jones and Harrison 1976; Jennings 1983, 1986). Subsequently, further information accumulated incorporating a number of variables such as carbohydrate metabolism (Holligan and Jennings 1972), growth on cellulose substrata (Pointing et al. 1998), the relationship between

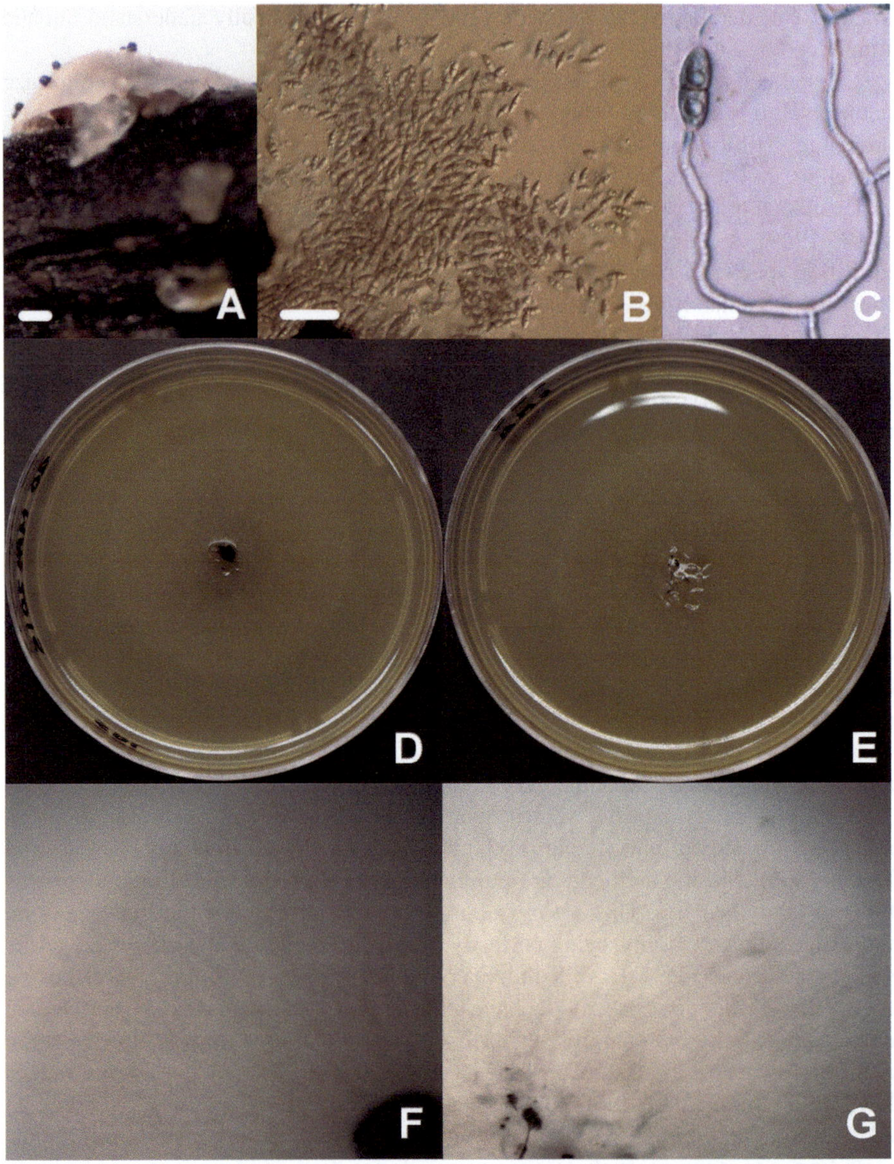

Fig. 7.1 Morphological traits of the salt-tolerant cosmopolitan marine arenicolous species *C. maritima*. Dark, carbonaceous ascomata attached to a shell as substrate from an environmental sample (**a**). Ascomatal squash releasing a mass of hyaline ascospores with conspicuous appendages, oil guttules and volutin bodies (**b**). One-septated germinating ascospore on culture media supplemented with sea salts (**c**). Isolates exposed to in vitro salt fluctuations showing minor differences in mycelial growth (increase in volume) on agar potato-dextrose supplemented with artificial seawater (**d** and **f**), in contrast to freshwater conditions (**e** and **g**); top views (**d** and **e**) and back views (**f** and **g**) of the Petri dishes. Scale bars, (**a**) = 900 μm, (**b**) = 90 μm, and (**c**) = 30 μm

salinity tolerance and temperature (Lorenz and Molitoris 1992; Edwards et al. 1998), ion concentration in mycelia (Wethered et al. 1985; Gibb et al. 1986), hyphal tips development (Galpin and Jennings 1975), and the role of sugar alcohols (polyols) in hyphal osmoregulation (Wethered and Jennings 1985; Wethered et al. 1985).

The physiological and molecular mechanisms of tolerance to osmotic and ionic components of salinity stress have been extensively investigated in plants, leading to the identification of key enzymes regulating reactive oxygen species and Na^{+}-sensitive enzymes in the cytoplasm, among others (Munns and Tester 2008; Türkan and Demiral 2009). Even though not much is known about aquatic fungal mechanisms of control to osmotic stress, significant breakthroughs have been emerged to figure out enzymatic response toward salinity. These include the examination of cellulolytic and ligninolytic activity in 15 marine fungal taxa isolated from lignocellulose substrates in mangroves and pelagic environments, demonstrating that enzyme activity (e.g., cellulases) may be affected by salinity (Pointing et al. 1998, 1999).

Likewise, differential effects of salinity have been addressed on the reproduction of terrestrial, freshwater, and marine fungi. Byrne and Jones (1975a) demonstrated that some terrestrial ascomycetes (e.g., *Chaetomium globosum*) and freshwater hyphomycetes (*Neonectria lugdunensis*) possess a limited tolerance to high salinity, concluding that the percentage spore germination in terrestrial fungi decreases with increasing salinities. Moreover, ranges of morphological characteristics in response to salinity have been reported in *Varicosporina ramulosa*, a marine fungal species (Meyers and Hoyo 1966). The above-described results suggest that salinity represents a key environmental variable modeling fungal fitness (*sensu* Pringle and Taylor 2002) and explain why so few terrestrial fungi are found in the sea (Hughes 1969; Kohlmeyer and Kohlmeyer 1971; Jones 1972).

High-throughput RNA sequencing allows the examination of differentially expressed genes across experimental conditions with unprecedented possibilities and advantages for non-model organisms (Rodríguez-García et al. 2017). This approach has been extensively used to assess osmotic stress response in a number of systems (reviewed in plants by Deyholos 2010). Recently, the underling mechanisms associated with physiological plasticity toward salinity have been evaluated in the endopsamous species *C. maritima* using comparative transcriptomics (Velez et al. 2015a). In this work, the authors identified candidate genes conferring tolerance to osmotic stress such as osmoprotectant molecules (e.g., aspartic-type endopeptidases and choline sulfatase family proteins from the endoplasmic reticulum), providing insights into gene expression profiles under contrasting salinity conditions.

In addition to physiological adaptations, morphological modifications have been reported in response to distinct salinity regimes. Examples include alterations in the stability of ascospore appendages in some species of *Aniptodera* and *Halosarpheia* (Nakagiri and Ito 1994; Jones 2006) and growth and formation of fruiting bodies in *Nia vibrissa* (Doguet 1968, 1969). Some investigations have even documented specific salinity requirements for sporulation, concluding that in spite of experimental bias, reproduction and spore germination of higher marine fungi are very much less affected by salinity than are the same processes of terrestrial counterparts (Jennings 1983).

7.4 Terrestrial Fungi: A Glimpse into Salinity Stress Response

Osmotolerance has been considered as a widespread trait of terrestrial filamentous fungi and yeasts, especially among the *Eurotiomycetes* (Tresner and Hayes 1971; Clipson and Jennings 1992; Kis-Papo et al. 2003; Kutty and Philip 2008), particularly members of penicillia and aspergilli (Tresner and Hayes 1971). In contrast, *Basidiomycota* representatives, seem to be more sensitive to salinity stress being unable to withstand more than 2% NaCl (Tresner and Hayes 1971; Jennings 1983). Nonetheless, few exceptions have been observed at low salinities in estuaries (Shearer 1972).

An amalgamation of factors has been associated with the incapability of some osmotically tolerant terrestrial fungi (which exhibit a broad tolerance to increasing salinities during vegetative growth) to complete their life cycles in the sea. These restraints may be associated with reproduction and spore germination, suggesting that terrestrial fungi have low tolerance toward salinity during propagative stages. Examples include the ubiquitous species *Aspergillus nidulans*, which can grow in culture media supplemented with 4 M NaCl, yet the elongation of germ tube in spores is severely inhibited by 0.5 M NaCl (Cox et al. 1995). Likewise, salinity-associated restrictions in perithecia development have been reported in the terrestrial ascomycete *Neurospora crassa* (Byrne and Jones 1975a).

Further reported effects of salinity in terrestrial fungi include alterations in pathogenicity, as demonstrated for Egyptian and German isolates of soil-borne root-rotting fungi (El-Abyad et al. 1988). Interesting cases are arbuscular mycorrhizal fungi (AMF). These plant symbionts occur naturally in saline environments (*Glomus* spp.). Even though negative effects of salinity have been reported for AMF (such as alteration of root colonization, suppression during arbuscular mycorrhizal formation, and spore germination delay), they confer host with tolerance toward salinity stress. Several surveys have evidenced that AMF provide their plant hosts with protection against salt stress by principally increasing nutrient uptake and improving photosynthetic rate and water-use efficiency, as well as accumulating osmoregulators (reviewed by Evelin et al. 2009; Porcell et al. 2012).

7.5 Final Remarks

Generally, little is known about the effect of salinity on complex aquatic ecosystem processes involving microorganisms (James et al. 2003). We still have limited knowledge of the impacts of salinity stress on fungal species interactions, food web structures, and the integrity of communities. Understanding fungal physiology and biochemistry, in response to salinity, has been appointed as the greatest challenge for marine mycology (Jones 2011). Nonetheless, a collection of factors regulates fungal biodiversity in aquatic ecosystems. These include physical variables such as temperature, pH, nutrient availability, and salinity (Jones 2000). All of these elements shape species distributions simultaneously and act at different spatial scales. Therefore, untangling the effects of an individual variable, such as salinity, represents a challenging task.

The exploration of hypersaline systems represents an area of opportunity in order to understand fungal adaptation to osmotic stress. These environments harbor unique microbial communities with halotolerant and halophilic taxa. Gostinčar et al. (2011) reported the occurrence of *Hortaea werneckii* from hypersaline water of solar salterns, nominating it as a model of cellular responses to high salt concentrations. This black yeast has an outstanding ability to overcome the turgor loss and sodium toxicity that are typical for hypersaline environments (including an increase in energy supply to drive the energy-demanding export of ions and synthesis of compatible solutes and changes in membrane lipid composition), which facilitates its growth even in solutions that are almost saturated with NaCl.

Despite an interrelationship between temperature and salinity as has been demonstrated (Molina and Hughes 1982; Lorenz and Molitoris 1992; Torzili 1997), most studies on the effect of salinity on fungal growth have been conducted at a fixed temperature (25 °C; Jones 2000). The long-term sustainability of aquatic fungal communities in marine ecotones depends on their ability to adapt to shifts in salinity regimes, particularly in light of climate change scenarios. This makes the evaluation of marine fungal response to salinity stress under several temperature scenarios a relevant topic for future research.

There is still little quantitative information on aquatic fungal upper and lower salinity tolerance thresholds. One of the limitations in defining specific upper salinity tolerances in coastal marine environments is the difficulty of conducting controlled field experiments, particularly in subtropical and tropical environments where rainfall is spatially variable and salinities are highly dynamic. In these cases, a mesocosm experimental approach may be efficiently implemented (e.g., Koch et al. 2007). This bounded and partially enclosed outdoor experimentation allows bridging the gap between the laboratory and the real world and can be replicated. Moreover, it allows investigating simultaneously both populations and ecosystems (Odum 1984; Crossland and La Point 1992). In addition, culture-independent research has an enormous potential shedding light into the fungal community assemblages and response to salinity gradients. So, a mesocosm experimental approach coupled with molecular genetics and functional genomics may provide a new opportunity to generate molecular and physiological knowledge on aquatic fungal response to salinity stress.

References

Ananda K, Sridhar KR (2004) Diversity of filamentous fungi on decomposing leaf and woody litter of mangrove forests in the southwest coast of India. Curr Sci 87:1431–1437

Barghoorn ES, Linder DH (1944) Marine fungi: their taxonomy and biology. Farlowia 1:395–467

Bärlocher F, Kendrick B (1974) Dynamics of the fungal populations on leaves in a stream. J Ecol 62:761–791

Burgaud G, Woehlke S, Rédou V, Orsi W, Beaudoin D, Barbier G, Biddle JF, Edgcomb VP (2013) Deciphering the presence and activity of fungal communities in marine sediments using a model estuarine system. Aquat Microb Ecol 70:45–62

Byrne PJ, Jones EBG (1975a) Effect of salinity on the reproduction of terrestrial and marine fungi. Trans Br Mycol Soc 65:185–200

Byrne PJ, Jones EBG (1975b) Effect of salinity on spore germination of terrestrial and marine fungi. Trans Br Mycol Soc 64:497–503

Calado M, Carvalho L, Pang KL, Barata M (2015) Diversity and ecological characterization of sporulating higher filamentous marine fungi associated with *Spartina maritima* (Curtis) Fernald in two Portuguese salt marshes. Microb Ecol 70:612–633

Cheeseman J (2016) Food security in the face of salinity, drought, climate change, and population growth. In: Uham M (ed) Halophytes for food security in dry lands. Academic Press, San Diego, pp 111–123

Clipson NJW, Jennings DH (1992) *Dendryphiella salina* and *Debaryomyces hansenii*: models for ecophysiological adaptation to salinity by fungi which grow in the sea. Can J Bot 70:2097–2105

Cox PW, Clipson NJW, Hooley P, Thomas CR (1995) Effects of growth medium solute concentration upon the spore germination and colony growth characteristics of *Aspergillus nidulans*. Proceedings of the Institution of Chemical Engineers Research Event, Edinburgh 2:986–988

Crossland NO, La Point TW (1992) The design of mesocosm experiments. Environ Toxicol Chem 11:1–4

Danger M, Gessner MO, Bärlocher F (2016) Ecological stoichiometry of aquatic fungi: current knowledge and perspectives. Fungal Ecol 19:100–111

De la Cruz TE, Wagner S, Schulz B (2006) Physiological responses of marine *Dendryphiella* species from different geographical locations. Mycol Prog 5:108–119

Dewey DR (1960) Salt tolerance of twenty-five strains of Agropyron 1. Agron J 52:631–635

Deyholos MK (2010) Making the most of drought and salinity transcriptomics. Plant Cell Environ 33:648–654

Doguet G (1968) *Nia vibrissa* Moore et Meyers, Gasteromycete marin. I. Conditions generates de formation des carpophores en culture. Bull Soc Mycol Fr 84:343–351

Doguet G (1969) *Nia vibrissa* Moore et Meyers, Gasteromycete marin. II. Developpement des carpophores et des basides. Bull Soc Mycol Fr 85:93–104

Edwards J, Chamberlain D, Brosnan G, West D, Stanley MS, Clipson NJW, Hooley P (1998) A comparative physiological and morphological study of *Dendryphiella salina* and *D. arenaria* in relation to adaptation to life in the sea. Mycol Res 102:1198–1202

El-Abyad MS, Hindorf H, Rizk MA (1988) Impact of salinity stress on soil-borne fungi of sugarbeet. Plant Soil 110:27–32

Elliott M, Quintino V (2007) The estuarine quality paradox, environmental homeostasis and the difficulty of detecting anthropogenic stress in naturally stressed areas. Mar Pollut Bull 54: 640–645

Evelin H, Kapoor R, Giri B (2009) Arbuscular mycorrhizal fungi in alleviation of salt stress: a review. Ann Bot 104:1263–1280
Fisher S, Likens G (1973) Energy flow in Bear Brook, New Hampshire: an integrative approach to stream ecosystem metabolism. Ecol Monogr 43:421–439
Fortunato CS, Crump BC (2015) Microbial gene abundance and expression patterns across a river to ocean salinity gradient. PLoS One 10:e0140578
Frias-Lopez J, Shi Y, Tyson GW, Coleman ML, Schuster SC, Chisholm SW, DeLong EF (2008) Microbial community gene expression in ocean surface waters. PNAS 105:3805–3810
Galpin MF, Jennings DH (1975) Histochemical study of the hyphae and the distribution of adenosine triphosphatase in *Dendryphiella salina*. Trans Br Mycol Soc 65:477–483
Gibb FM, Wethered JM, Jennings DH (1986) The effect of monovalent ions on enzyme activity in *Dendryphiella salina*. In: Moss ST (ed) The biology of marine fungi. Cambridge University Press, Cambridge, pp 27–33
Gilbert JA, Field D, Huang Y, Edwards R, Li W, Gilna P, Joint I (2008) Detection of large numbers of novel sequences in the metatranscriptomes of complex marine microbial communities. PLoS One 3:e3042
González MC, Hanlin RT (2010) Potential use of marine arenicolous ascomycetes as bioindicators of ecosystem disturbance on sandy Cancun beaches: *Corollospora maritima* as a candidate species. Bot Mar 53:577–580
Gostinčar C, Lenassi M, Gunde-Cimerman N, Plemenitaš A (2011) Fungal adaptation to extremely high salt concentrations. Adv Appl Microbiol 77:71–96
Harris G (2009) Salinity. In: Tochner K, Likens GE (eds) Encyclopedia of inland waters, vol 2. Oxford University Press/Elsevier, Oxford, pp 79–84
Herbert ER, Boon P, Burgin AJ, Neubauer SC, Franklin RB, Ardón M, Hopfensperger KN, Lamers LPM, Gell P (2015) A global perspective on wetland salinization: ecological consequences of a growing threat to freshwater wetlands. Ecosphere 6:1–43
Hintz WD, Relyea RA (2017) A salty landscape of fear: responses of fish and zooplankton to freshwater salinization and predatory stress. Oecologia 185:147–156
Holligan PM, Jennings DH (1972) Carbohydrate metabolism in the fungus *Dendryphiella salina*. II. The influence of different carbon and nitrogen sources on the accumulation of mannitol and arabitol. New Phytol 71:583–594
Huang J, Lu C, Qian X, Huang Y, Zheng Z, Shen Y (2011) Effect of salinity on the growth, biological activity and secondary metabolites of some marine fungi. Acta Oceanol Sin 30: 118–123
Hughes GC (1969) Marine fungi from British Columbia: occurrence and distribution of lignicolous species. Syesis 2:121–140
Hyde KD (1988) Studies on the tropical marine fungi of Brunei. Bot J Linn Soc 98:135–151
Hyde KD (1989) Vertical zonation of intertidal mangrove fungi. In: Hattori T, Ishida Y, Maeuyama Y, Morita RY, Uchida A (eds) Recent advances in microbial ecology. Japan Scientific Societies Press, Tokyo, pp 302–306
Hyde KD (1990a) A study of the vertical zonation of intertidal fungi on *Rhizophora apiculata* at Kampong Kapok mangrove, Brunei. Aquat Bot 36:255–262
Hyde KD (1990b) A comparison of the intertidal mycota of five mangrove tree species. Asian Mar Biol 7:93–107
Hyde KD (1993) The distribution of intertidal fungi on *Rhizophora apiculata*. In: Morton B (ed) The marine biology of the South China Sea. Proceedings of the First International Conference on the Marine Biology of Hong Kong and the South China Sea. Hong Kong University Press, Hong Kong, pp 643–652
Hyde KD, Jones EBG (1988) Marine mangrove fungi. Mar Ecol 9:15–33
Ittner LD, Junghans M, Werner I (2018) Aquatic fungi: a disregarded trophic level in ecological risk assessment of organic fungicides. Front Environ Sci 6:105
James KR, Cant B, Ryan T (2003) Responses of freshwater biota to rising salinity levels and implications for saline water management: a review. Austral J Bot 51:703–713

Jennings DH (1983) Some aspects of the physiology and biochemistry of marine fungi. Biol Rev 58:423–459
Jennings DH (1986) Some aspects of the physiology and biochemistry of marine fungi. Biol Rev 58:423–459
Johnson TW, Sparrow FK (1961) Fungi in oceans and estuaries. Cramer, Weinheim
Jones EBG (1972) The decay of timber in aquatic environments. Br Wood Presery Assoc Annu Cony:1–18
Jones EBG (2000) Marine fungi: some factors influencing biodiversity. Fungal Divers 4:53–73
Jones EBG (2006) Form and function of fungal spore appendages. Mycoscience 47:167–183
Jones EBG (2011) Fifty years of marine mycology. Fungal Divers 50:73–112
Jones EBG, Harrison JL (1976) Physiology of marine Phycomycetes. In: EBG J (ed) Recent advances in aquatic mycology. Wiley, New York, pp 261–278
Jones EBG, Jennings DH (1964) The effect of salinity on the growth of marine fungi in comparison with non-marine species. Trans Br Mycol Soc 47:619–625
Jones EBG, Byrne P, Alderman DJ (1971) The response of fungi to salinity. Vie Milieu Suppl 22:265–280
Kirk PW, Dyer BJ, Noé J (1991) Hydrocarbon utilization by higher marine fungi from diverse habitats and localities. Mycologia 83:227–230
Kis-Papo T, Oren A, Wasser SP, Nevo E (2003) Survival of filamentous fungi in hypersaline Dead Sea water. Microb Ecol 45:183–190
Koch MS, Schopmeyer SA, Kyhn-Hansen C, Madden CJ, Peters JS (2007) Tropical seagrass species tolerance to hypersalinity stress. Aquat Bot 86:14–24
Kohlmeyer J, Kohlmeyer E (1971) Synoptic plates of higher marine fungi. Lehre Verlag von J Cramer, 87 pp
Kohlmeyer J, Kohlmeyer E (1979) Marine mycology. The higher fungi. Academic Press, New York
Kutty SN, Philip R (2008) Marine yeasts—a review. Yeast 25:465–483
Lercari D, Defeo O (2006) Large-scale diversity and abundance trends in sandy beach macrofauna along full gradients of salinity and morphodynamics. Estuar Coast Shelf Sci 68:27–35
Lewis E (1980) The practical salinity scale 1978 and its antecedents. J Ocean Eng 5:3–8
Lorenz R, Molitoris HP (1992) Combined influence of salinity and temperature (*Phoma*-pattern) on growth of marine fungi. Can J Bot 70:2111–2115
Mahé S, Rédou V, Le Calvez T, Vandenkoornhuyse P, Burgaud G (2014) Fungi in deep-sea environments and metagenomics. In: Martin F (ed) The ecological genomics of fungi. Wiley-Blackwell, Ames, pp 325–354
Masuma R, Yamaguchi Y, Noumi M, Omura S, Namikoshi M (2001) Effect of seawater concentration on hyphal growth and antimicrobial metabolite production in marine fungi. Mycoscience 42:455–459
McEvoy P, Goonan P (2003) Salinity is not necessarily bad for biodiversity: case studies of invertebrates from South Australian streams and River Murray wetlands. Rec South Aus Mus 7:131–134
Meyers SP (1968) Degradative activities of filamentous marine fungi. In: Walters AH, Elpjhick JJ (eds) Biodeterioration of materials. Elsevier, Amsterdam, pp 594–609
Meyers SP, Hoyo L (1966) Observations on the growth of the marine hyphomycete *Varicosporina ramulosa*. Can J Bot 44:1133–1140
Molina FI, Hughes GC (1982) The growth of *Zalerion maritimum* (Linder) Anastasiou in response to variation in salinity and temperature. J Exp Mar Biol Ecol 61:147–156
Munns R, Tester M (2008) Mechanisms of salinity tolerance. Annu Rev Plant Biol 59:651–681
Nakagiri A, Ito T (1994) *Aniptodera salsuginosa*, a new mangrove-inhabiting ascomycete, with observations on the effect of salinity on ascospore appendage morphology. Mycol Res 98: 931–936
Odum EP (1984) The mesocosm. BioScience 34:558–562

Overy DP, Bayman P, Kerr RG, Bills GF (2014) An assessment of natural product discovery from marine (*sensu strictu*) and marine-derived fungi. Mycology 5:145e167

Pang KL, Overy DP, Jones EG, da Luz CM, Burgaud G, Walker AK, Johnson JA, Russell GK, Cha H, Bills GF (2016) 'Marine fungi' and 'marine-derived fungi' in natural product chemistry research: toward a new consensual definition. Fungal Biol Rev 30:163–175

Petersen KRL, Koch J (1997) Substrate preference and vertical zonation of lignicolous marine fungi on mooring posts of oak (*Querens* sp.) and larch (*Larix* sp.) in Svanemøllen Harbour, Denmark. Bot Mar 40:451–464

Pianka ER (1974) Niche overlap and diffuse competition. PNAS 71:2141–2145

Pointing BS, Vrijmoed LLP, Jones EBG (1998) A quantitative assessment of lignocellulose degrading enzyme activity in marine fungi. Bot Mar 412:293–298

Pointing BS, Buswell JA, Jones EBG, Vrijmoed LLP (1999) Extracellular cellulolytic enzyme profiles of five lignicolous mangrove fungi. Mycol Res 103:690–700

Porcel R, Aroca R, Ruiz-Lozano JM (2012) Salinity stress alleviation using arbuscular mycorrhizal fungi. A review. Agron Sustain Dev 32:181–200

Pringle A, Taylor JW (2002) The fitness of filamentous fungi. Trends Microbiol 10:474–481

Rengasamy P (2006) World salinization with emphasis on Australia. J Exp Bot 57:1017–1023

Ritchie D (1959) The effect of salinity and temperature on marine and other fungi from various climates. Bull Torrey Bot Cl 86:367–373

Rodríguez-García A, Sola-Landa A, Barreiro C (2017) RNA-Seq-based comparative transcriptomics: RNA preparation and bioinformatics. Methods Mol Biol 1645:59–72

Schaumann K (1968) Marine hohere Pilze (Ascomycetes und Fungi imperfecti) aus Dem Weser-Astuar. Veroeff Institue Meeresfrosch Bremerh 11:93–117

Schaumann K (1969) Ober marine hogere Pilze von Holzsubstraten der Nordsee-Insel Helgoland. Ber Deutsch Bot Ges 82:307–327

Shearer CA (1972) Fungi of the Chesapeake Bay and its tributaries. III. The distribution of wood-inhabiting ascomycetes and fungi imperfecti of the Patuxent River. Am J Bot 59:961–969

Shearer CA (1993) The freshwater ascomycetes. Nova Hedwigia 56:1–33

Shearer CA, Descals E, Kohlmeyer B, Kohlmeyer J, Marmanová L, Padgett D, Porter D, Raja HA, Schmit PJ, Thorton HA, Voglymayr H (2007) Fungal biodiversity in aquatic habitats. Biodivers Conserv 16:49–67

Shi Y, Tyson GW, DeLong EF (2009) Metatranscriptomics reveals unique microbial small RNAs in the ocean's water column. Nature 459:266

Smyth K, Elliott M (2016) Effects of changing salinity on the ecology of the marine environment. In: Solan M, Whiteley NM (eds) Stressors in the marine environment. Oxford University Press, Oxford, pp 161–174

Stewart RH (2008) Introduction to physical oceanography. Texas A & M University, College Station

Stumm W, Morgan JJ (1996) Aquatic chemistry: chemical equilibria and rates in natural waters. Wiley, New York

Sundari R, Vickineswary S, Yusoff M, Jones EBG (1996a) *Corollospora besarispora* (Halosphaeriales, Ascomycotina), a new marine fungus from Malaysia. Mycol Res 100: 1259–1262

Sundari R, Vickineswary S, Yusoff M, Jones EBG (1996b) Observations on tropical arenicolous marine fungi on driftwood from Malaysia and Singapore. Bot Mar 39:361–375

Tan TK, Leong WF, Jones EBG (1989) Succession of fungi on wood of *Avicennia alba* and *A. lanata* in Singapore. Can J Bot 67:2686–2691

Tanimoto TT (1969) Differential physiological response of sugarcane varieties to osmotic pressures of saline media. Crop Sci 9:683–688

Torzili AP (1997) Tolerance to high temperature and salt stress by a salt marsh isolate of *Aureobasidium pullulans*. Mycologia 89:786–792

Tresner HD, Hayes JA (1971) Sodium chloride tolerance of terrestrial fungi. Appl Environ Microbiol 22:210–213

Türkan I, Demiral T (2009) Recent developments in understanding salinity tolerance. Environ Exp Bot 67:2–9

Velez P, Alejandri-Ramírez ND, González MC, Estrada KJ, Sanchez-Flores A, Dinkova TD (2015a) Comparative transcriptome analysis of the cosmopolitan marine fungus *Corollospora maritima* under two physiological conditions. G3-Genes Genom Genet 5:1805–1814

Velez P, Gonzalez MC, Capello-García S, Rosique-Gil E, Hanlin RT (2015b) Diversity of marine ascomycetes from the disturbed sandy beaches of Tabasco, Mexico. J Mar Biol Assoc UK 95: 897–903

Wethered JM, Jennings DH (1985) Major solutes contributing to solute potential of *Thraustochytrium aureum* and *T. roseum* after growth in media of different salinities. Trans Br Mycol Soc 85:439–446

Wethered JM, Metcalf EC, Jennings DH (1985) Carbohydrate metabolism in the fungus *Dendryphiella salina*: VIII. The contribution of polyols and ions to the mycelial solute potential in relation to the external osmoticum. New Phytol 101:631–649

Chapter 8
Cellular Mechanisms of Plant Salt Tolerance

Mohamed Magdy F. Mansour and Karima Hamid A. Salama

Abstract Salinity is a severe threat to agricultural production of over 80 million ha arable land area globally, especially in irrigated arid and semiarid regions. Understanding how plants cope with salt stress is therefore imperative. As the glycophytes and halophytes have salt tolerance mechanisms occur at the cellular level, understanding cellular mechanisms of salt tolerance becomes indispensable for crop improvement programs. Recent challenges such as climate change, increased world population, and demand for foods all over the globe aggravate the salinity impacts on plant production which urge scientists to understand deeply and fully the effects of salt stress on plants and how plants combat these effects. This chapter hence discusses the cellular mechanisms that may contribute to plant salt adaptation, namely, osmotic adjustment and osmolyte production, ion homeostasis and involvement of membrane transport systems, antioxidant defense systems and reactive oxygen species (ROS) detoxification, membrane lipids and proteins modeling to maintain membrane integrity, transport system activities, and signaling and cell wall modifications to regulate cell wall integrity, cell expansion, and stress sensing and signaling transduction. Disagreements among published data concerning these cellular mechanisms are also addressed. We believe that adopting one cellular mechanism for crop improvement programs or to understand plant adaptation to high salinity is not sufficient and premature. Alternatively, several genes must be pyramided to ensure the efficiency of any cellular salt tolerance trait, and testing hypotheses instead of just collecting repetitive data must be considered as promising feasible strategies to improve productivity in crop plants and unmask the puzzle of salt tolerance mechanism. Moreover, testing the results in realistic growth conditions such as the natural field saline conditions must be performed and focused in our future research in order to verify the underlying mechanism and its proposed functions in salt adaptation.

M. M. F. Mansour (✉) · K. H. A. Salama
Faculty of Science, Department of Botany, Ain Shams University, Cairo, Egypt

B. Giri, A. Varma (eds.), *Microorganisms in Saline Environments: Strategies and Functions*, Soil Biology 56, https://doi.org/10.1007/978-3-030-18975-4_8

Keywords Antioxidant · Compatible solutes · Cell wall components · Ion homeostasis · Membrane lipids and proteins · Membrane transport systems · Osmotic adjustment · Reactive oxygen species

8.1 Introduction

Salinity is one of the most serious environmental threats that have adverse effects on germination, plant growth, and productivity. About 80 million ha of arable lands have been damaged by soil salinity worldwide, and 1.5 million ha become nonproductive each year because of high salinity levels in the soil (Munns and Tester 2008). In addition, a severe shortage of good-quality water is also forcing farmers to use poor-quality groundwater for irrigation, which further aggravates the salinity problem. Salt stress is first perceived by the root system and affects plants by osmotic stress or water deficit, ion toxicity, nutritional imbalance and deficiency, metabolic and membrane disruptions, oxidative stress as well as cell division and cell elongation inhibition, or a combination of these factors (Hasegawa et al. 2000; Flowers 2004; Mansour and Ali 2017a, b). Short-term exposure to salinity primarily initiates water stress which reduces cell division and expansion as well as stomatal closure resulting in leaf expansion and growth reduction (Flowers 2004). Although salinity exposure-induced stomatal closure declines growth in several plant species, stomatal closure can contribute to a decline in the flow of toxic ions in the transpiration stream (Acosta-Motos et al. 2017), and plants therefore should compromise between both situations in order to withstand high salts and continue their life cycle. One response resulted from salt-induced water stress is that as the conformation and the correct functioning of proteins and cellular membranes depend on an intact hydration shell, severe water stress can lead to an impairment of amino acid synthesis, protein metabolism, the dark reaction of photosynthesis or respiration, and membrane functions (Koyro et al. 2012). Ionic stress results as a consequence of long-term salinity exposure leading to accumulation of toxic ions, in particular Na^+ and Cl^-, which causes severe ion toxicity when these ions are not readily sequestered in vacuoles or pumped out of the cells into the apoplast (Flowers 2004). Buildup of high concentrations of toxic ions (i.e., ion toxicity) under saline environment reduces beneficial ones leading also to nutritional disorder (Munns and Tester 2008). Ion toxicity causes premature senescence of adult leaves reducing the photosynthetic area needed to support continued growth, inhibition of plant enzyme activity, disruption of protein synthesis, and other physiological alterations (Hasegawa et al. 2000). Salt stress frequently induces formation of singlet oxygen, superoxide anion, hydrogen peroxide, and hydroxyl radical and hence causes oxidative stress in many plants (Das et al. 2016). These ROS attack proteins, nucleic acids, and membrane lipids leading to their oxidation, damage, and in turn impaired functions (Mansour and Ali 2017a, b). Moreover, several evidences indicate that salinity also induces alterations in membrane fluidity, permeability, and stability and thus changes the degree of salt tolerance of the plants (Mansour 2013; Mansour et al. 2015). It is also known that the previous responses to salinity vary depending on the stage of plant growth, the growth conditions, the agronomic and irrigation management practices, and the degree of tolerance of the species (glycophytes vs. halophytes), which can even differ among the different cultivars of the same species (Acosta-Motos et al. 2017). Salt-resistant species/cultivars cope with these injurious responses through different mechanisms, which are most likely lacking in the sensitive ones. Unfortunately, crop plants belong to glycophytes which means that most of them are salt-sensitive species. Halophytes, however, are plants adapted to grow and complete their life cycle in habitats with soil salinity equivalent to, at least, 200 mM NaCl (Flowers and Colmer 2008).

The significance of understanding the cellular mechanisms of salt tolerance comes from the fact that glycophytes and halophytes have salt tolerance mechanisms occur at the cellular level (Hasegawa et al. 2000; Mansour and Salama 2004; Mansour 2013; Mansour and Ali 2017a, b). It is therefore pivotal to identify the cellular master traits for salt tolerance, which urges plant physiologists to carry out intensive and comprehensive studies in this promising area in order to unmask the puzzle of salt tolerance mechanism. This will also enable us to recognize how to deal and control the impacts of high salinity on plants. Furthermore, understanding different cellular mechanisms associated with salt stress will contribute to developing new tolerant crops through simultaneous engineering of these tolerance traits by transgenic technologies and/or breeding programs (Mansour and Ali 2017b; Gilliham et al. 2017). One important cellular issue that has to be dealt with in the ionic relations is that achieving tolerance to high salinity needs the management of Na^+ which must be balanced against the protection and continued uptake of K^+ under saline environments (Hasegawa et al. 2000; Gilliham et al. 2017). The importance of that comes from the fact that a high Na^+/K^+ ratio in the cytosol is toxic to plants because it inhibits various processes such as K^+ absorption, vital enzyme reactions, protein synthesis, and photosynthesis and also ROS detoxification is critically dependent on K^+ availability (Shabala et al. 2016; Ali and Yun 2017). The restoration of ion balance or ion homeostasis in the cell is therefore one of the crucial cellular mechanism works against salinity imposition (Gilliham et al. 2017). Additionally, accumulation of enzymatic and nonenzymatic defense systems, synthesis of compatible osmoprotectants, osmotic adjustment, stress proteins and metabolites, cell wall modifications, and alterations to gene expression are other key cellular salt tolerance mechanisms in plants (Fig. 8.1; Hasegawa et al. 2000; Flowers 2004; Munns and Tester 2008). Increasing the activity and biosynthesis of antioxidant enzymes as well as accumulating nonenzymatic antioxidants ensures lower level of ROS, which protect cells against oxidation of their macromolecules. It is important to note that ROS have also beneficial impacts as protective or signaling factors under stress conditions when they present in low concentrations (Sah et al. 2016). Compatible solutes are able to protect membrane lipids, proteins, and other cellular structures; maintain cell turgor by osmotic adjustment, chaperones, storage form of nitrogen, and redox metabolism to remove excess levels of ROS; and

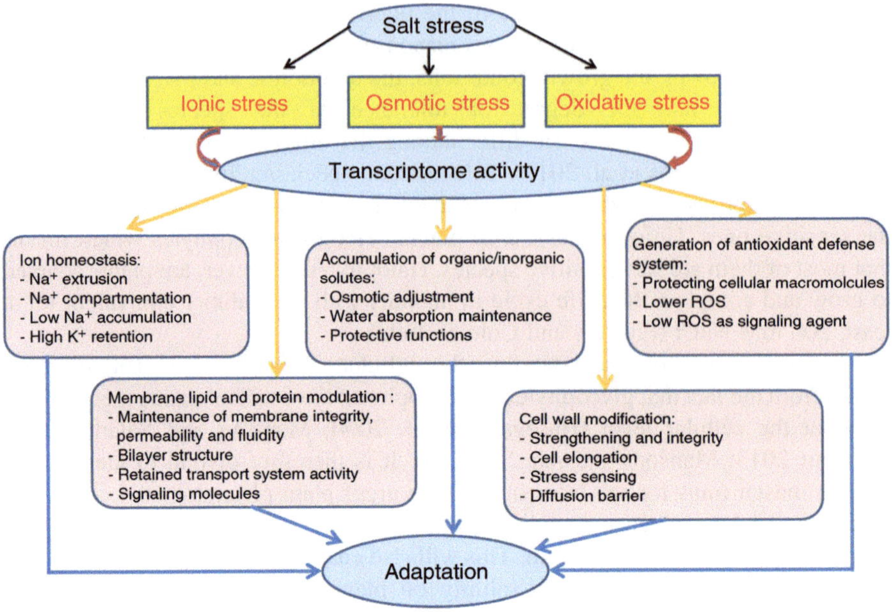

Fig. 8.1 Effects and cellular tolerance mechanisms in plants in response to salinity stress

re-establish the cellular redox balance under salinity (Mansour 1998; Hasegawa et al. 2000; Ashraf and Foolad 2007; Munns and Tester 2008; Mansour et al. 2015).

Cellular responses to high salinity include also adjustments of the membrane transport systems (Mansour 2014) and membrane lipids (Mansour et al. 2015). Membrane lipids are responsible for determining biological properties of the membrane, and their composition was remarkably different and underlying variations in stress tolerance among crop species (Mansour 2013; Mansour et al. 2015). Moreover, a prominent role for the membrane lipid composition in regulating the activity of membrane-associated transport proteins is documented, which greatly affects ion homeostasis and in turn tolerance to high salinity (Mansour 2013, 2014; Mansour et al. 2015). Membrane lipids also serve as substrates for the generation of numerous signaling lipids such as phosphatidic acid, phosphoinositides, sphingolipids, lysophospholipids, oxylipins, *N*-acylethanolamines, free fatty acids, and others, which control the expression of gene clusters and activate plant adaptation processes (Hou et al. 2016). It is reported that all cellular responses rely heavily on the regulation of transport activity across cellular membranes: specifically those for Na^+ and K^+ ions (Mansour 2014; Shabala et al. 2016). Evidence also indicates that changes in the cell wall composition and elasticity have occurred in response to high salinity and are considered as a cellular mechanism for tolerance to salinity (Acosta-Motos et al. 2017 and references therein). It is of great interest to mention that like most stresses, salinity tolerance is a polygenic and physiologically multifaceted trait, and high salinity therefore has more than one effect on the plants. This makes it unlikely there is a one-gene solution, and thus no one trait for salt stress tolerance can perform efficiently (Gilliham et al. 2017). It is hence expected that one cellular mechanism cannot work alone and separately to induce salt adaptation; rather all mechanisms collectively can work together in order to contribute to tolerance to saline environments. That is, considering only one mechanism, as a salt tolerance trait, may not result in a real tolerance in terms of continuing plant whole life cycle and yield because other mechanisms (increased ROS damage, mineral deficiencies (e.g., K^+, Ca^{2+}, Mg^{2+}), hormonal imbalances, and the energetic requirements of either ion transport or the synthesis of compatible solutes) can severely reduce growth and eventually death. In this connection, Hanin et al. (2016) provide strong evidence that higher salt tolerance in plants is conferred by at least three complementary physiological mechanisms. Different cellular traits that underpin salinity tolerance are diagramed in Fig. 8.1. An update on recent findings related to the cellular responses of plants to salt stress is therefore provided in this chapter. How these different stress responses are combined to shed light on the cellular mechanisms and also how they affect the growth and development of plants in saline environments have been addressed.

8.2 Osmotic Adjustment, Compatible Solutes, and Salt Tolerance

Owing to lowering of the soil water potential induced by high concentrations of salts in the soil solution, plants have difficulty to absorb water and in turn experience water deficit. In response to that, plants have to reduce their water potential to maintain water gradient and subsequently water absorption going on. Plants counteract water deficit or osmotic action of salinity by accumulation of compatible solutes and other organic compounds, which are not injurious to the cytoplasm and do not interfere with cellular metabolism even at high concentrations (Rhodes et al. 2002). Plants thus lower their solute potential via accumulating organic and/or inorganic solutes under salt stress, the process termed osmotic adjustment. In halophytes, the major contributors for osmotic adjustment and maintaining the cellular osmotic potential are Na^+ and Cl^-, whereas compatible solutes most probably accumulated in the cytoplasmic compartments (Flowers et al. 2015). For instance, efficient osmotic adjustment, achieved mainly through ion uptake, is associated with salt tolerance of halophyte *Centaurea ragusina* (Radic et al. 2013). On the other hand, glycophytes achieve osmotic adjustment predominantly by increased de novo synthesis of compatible solutes (Shabala 2013). The three major inorganic ions, Na^+, K^+, and Cl^-, constitute 80–95% of the osmotic pressure of the cell sap in halophytes, while in glycophytes the contribution is between 50% and 70% (Shabala and Mackay 2011; Shabala 2013). High accumulation of compatible solutes and subsequently osmotic adjustment has been reported to be important cellular mechanisms for plant salt adaptation under saline environments (Turkan and Demiral 2009; Acosta-Motos et al. 2017; D'Amelia et al. 2018; Annunziata et al. 2017; Nikalje et al. 2018). For example, accumulation of glycine betaine, sucrose, asparagine, and proline was responsible for the osmotic adjustment, the assimilation of the excess of ammonium, and the scavenging of ROS in the cytosol under salinity, and correlated with salt adaptation in durum wheat (Annunziata et al. 2017). Moreover, higher osmotic adjustment (i.e., greater decline in solute potential) has been reported in salt-tolerant wheat genotype relative to sensitive one under salt stress, which was attributed to higher levels of organic osmolytes, namely, proline and soluble sugars (Maswada et al. 2018). In agreement, osmotic adjustment is associated with salt tolerance in rice lines contrasting in salt tolerance, and proline content was greater in tolerant lines compared with sensitive ones (Nounjan et al. 2018). The importance of osmotic adjustment for salt tolerance relies on the fact that osmotic adjustment reduces cellular water potential and hence retains water potential gradient between plant and soil and thus maintains water absorption under salt stress. This situation maintains turgescent cells and ensures balanced water relations, turgor, and thus cell expansion under salt stress. As shown in Fig. 8.2, the high concentration of compatible solutes mainly exists in the cytosol and other metabolically active cellular compartments (e.g., chloroplasts, mitochondria) to balance the lower water potentials of the apoplast and vacuole, as mentioned above, due to Na^+ and Cl^- accumulation (Turkan and Demiral 2009; Shabala 2013; Flowers et al. 2015). Toxic ions are concentrated in the vacuoles relative to the cytosol and other cellular compartments because these ions have the potentiality to disrupt cellular

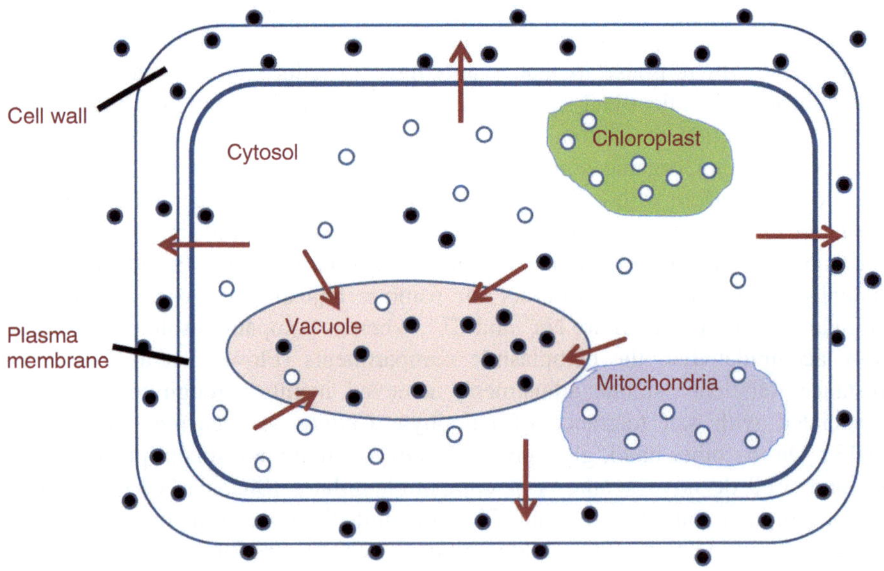

Fig. 8.2 Cells accumulate compatible solutes in metabolically active components (cytosol, chloroplast, mitochondria) and export toxic ions into the vacuole or out of the cells via active transport systems. Open circles, compatible solutes; closed circles, toxic ions; arrows, active transport

metabolism if present at high concentrations in the cellular compartments (i.e., ion toxicity) (Acosta-Motos et al. 2017).

Salt stress affects metabolic responses and associated genomic regulations in plants, with several metabolite classes. These metabolites could be either organic ions or other low molecular weight organic solutes. The compatible solutes are highly soluble in water, electrically neutral in the physiological pH range, and noninhibitory to enzymes even at high concentrations, so that they do not interfere with essential metabolic or enzymatic reactions. The compatible solutes may include polyamines (putrescine, spermidine, spermine), sugar alcohols (mannitol, glycerol, and methylated inositols), quaternary amino acid derivatives (proline, glycinebetaine, β-alanine betaine, and proline betaine), tertiary amines (1,4,5,6-tetrahydro-2-mehyl-4-carboxyl pyrimidine) and sulfonium compounds (choline-*O*-sulfate, dimethylsulfoniopropionate), and soluble sugars (glucose, fructose, sucrose, trehalose, raffinose, fructans), which demonstrate to have positive correlation between their accumulation and salt tolerance (Mansour 2000; Nahar et al. 2016; D'Amelia et al. 2018; Nikalje et al. 2018). The chemical structure of abundant representative compatible solutes is given in Fig. 8.3. The compatible solutes have two main functions: First, they lower the cellular solute potential and hence contribute to osmotic adjustment under the stress conditions. Second, they function as chaperons by attaching to the cellular macromolecules and thus preventing their denaturation, i.e., protective function (Mansour 2000; Koyro et al. 2012; Nahar et al. 2016). This protective function of compatible solutes can also alleviate ion-specific effects of salt stress caused by ion toxicity and ion imbalance such as the precipitation of proteins due to changes in charge or the destruction of membranes caused by

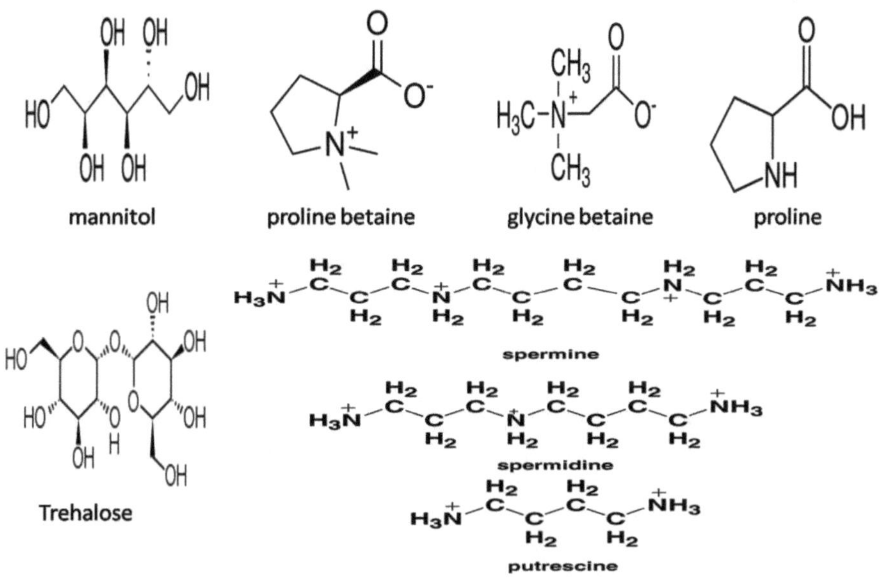

Fig. 8.3 Chemical structure of representative compatible solutes accumulating under salt stress

alterations of the membrane potential (Koyro et al. 2012). Moreover, compatible solutes mitigate the oxidative stress damage to membrane transporters under salt imposition (Cuin and Shabala 2008). This ameliorative effect may be due to ROS scavenging (Ashraf and Foolad 2007; Nahar et al. 2016) and/or direct protection of transporters themselves (Mansour and Ali 2017a, b; D'Amelia et al. 2018). Many studies demonstrated that the protective effect and osmotic adjustment involvement of the compatible solutes (e.g., proline, glycinebetaine, polyamines, trehalose) were also achieved by exogenous application and transgenic approaches in several plant species under saline conditions (Mansour 1998; Turkan and Demiral 2009; Koyro et al. 2012; Lai et al. 2014; Salama et al. 2011, 2015; Nahar et al. 2016; Mansour and Ali 2017a, b). For example, polyamines and trehalose displayed a protective role and improved salt tolerance in various species (Garg et al. 2002; Nahar et al. 2016; D'Amelia et al. 2018). Further, overproduction of trehalose enhances salt tolerance in the rice transgene, and more favorable mineral balance under salt stress was one of the cellular mechanisms that contributed to salt tolerance in transgenic rice seedlings (Garg et al. 2002). The study illustrated also that trehalose level was low and hence did not contribute to osmotic adjustment. Also, polyamine treatment enhanced the response of mung bean seedlings to salt stress (Nahar et al. 2016). Shabala and Mackay (2011) and Shabala et al. (2016) indicate that physiologically relevant concentrations of compatible solutes by their regulating K^+ transport across the plasma membrane enabling maintenance of an optimal K^+/Na^+ ratio, and thus contribute to plant adaptive responses to salinity. Consistently, exogenous application of polyamines enhances salt tolerance of pistachio, and the enhancement effect was through increasing the activities of antioxidant enzymes, decreasing ROS, Na^+: K^+ ratio, and Cl^- in leaves under high salinity (Kamiab et al. 2014). Similarly, foliar application of spermidine had a protective effect on photosynthetic apparatus of tomato seedlings after salt stress (Hu et al. 2016). In addition, compatible solutes have been involved in a complex signaling system as well as they are very efficient in ion homeostasis and hence have an important role in the regulation of salt stress tolerance (Cuin et al. 2009; Turkan and Demiral 2009; Sah et al. 2016).

Among the organic solutes, proline and glycinebetaine are the most abundant and efficient compatible ones (Mansour and Ali 2017a, b; Acosta-Motos et al. 2017; D'Amelia et al. 2018; Nikalje et al. 2018). Proline synthesis is mainly catalyzed by Δ-1-pyrroline-5-carboxylate synthetase (P5CS), which converts glutamate into Δ-1-pyrroline-5-carboxylate (P5C) in the cytoplasm or chloroplast. Then, P5C reductase (P5CR) reduces P5C to proline (Acosta-Motos et al. 2017; Mansour and Ali 2017b). In addition, proline may be also synthesized in mitochondria from ornithine by ornithine δ-aminotransferase (OAT), catalyzing the formation of P5C, which is converted to proline by P5CR. Proline degradation in the mitochondria is catalyzed by proline dehydrogenase (ProDH), which produces P5C from proline, and then P5C dehydrogenase converts P5C to glutamate. Glycinebetaine is synthesized in the chloroplast stroma from choline by the action of choline monooxygenase producing betaine aldehyde. Betaine aldehyde is oxidized to glycinebetaine by NAD^+-dependent betaine aldehyde dehydrogenase (Mansour and Ali 2017a). Although several evidences have been presented by Mansour and Ali (2017a, b and references therein) that accumulation of proline and glycinebetaine is not associated with salt tolerance, other works illustrated their positive involvement in plant responses to saline conditions (Mansour 1998; Kumar et al. 2010; Salama et al. 2011, 2015; Khan et al. 2015a; Nahar et al. 2016). For example, co-expression of two enzymes involved in glycinebetaine biosynthesis resulted in increased glycinebetaine level and enhanced salt tolerance of transgenic rice seedlings under high salinity (Niu et al. 2014). Similarly, glycinebetaine foliar application in onion plants confers tolerance to salt stress (Rady et al. 2018). In the same trend, exogenous proline significantly improved salinity tolerance of *Eurya emarginata*; the improvement effect was through protective impacts on water relations, ionic and osmotic adjustment, as well as antioxidant defense (Zheng et al. 2015). On contrast, proline was not effective in alleviating the adverse effects of salt stress on eggplant cultivars under saline and nonsaline regime (Shabaz et al. 2013). Also, although glycinebetaine enhanced salt tolerance in onion via increased glycinebetaine level, enzymatic and nonenzymatic antioxidants, and maintained membrane stability, proline content was not affected under high salt stress (Rady et al. 2018). The above discussion demonstrates that compatible solutes could regulate antioxidant enzyme activities, modulate intracellular ion concentrations, lower cell solute potential, protect cellular macromolecules, and act as signaling molecules, thereby protecting plants from salt stress damages and hence having a positive correlation with salt tolerance. However, despite the evident roles of compatible solutes as a significant cellular mechanism of salt adaptation, some inconsistencies in the results concerning compatible solute responses to saline conditions warrant further future intensive investigations.

It is important to note that despite the majority are organic solutes, potassium can also serve as compatible solute in plants (Cuin et al. 2009). Besides its contribution to osmotic adjustment, potassium is known to act as a cofactor for several metabolic enzymes and plays a major role in regulating stomatal mechanics and ROS detoxification as well as phloem transport (Zorb et al. 2014; Shabala et al. 2016). Although plants can overcome water deficit in response to salt stress as described earlier via solute accumulation and osmotic adjustment, one more hazard of water stress is the restriction of CO_2 uptake, and hence plants should compromise the dilemma of maintaining a low transpiration and a high net photosynthetic rate. Actually, various plant species owing to stomatal closure exhibit reduced assimilation rate, but their complete life cycle is maintained under salt conditions (Acosta-Motos et al. 2017). It is worth mentioning that although osmotic adjustment is a common cellular mechanism reported in various plant species/cultivars under high salinity, its association with salt tolerance is argued (Mansour et al. 1993; Mansour and Stadelmann 1994; Mansour 2000; Mansour and Ali 2017a, b). In the previous studies, there is no significant difference in solute potential reduction of wheat or barley cells of either salt-sensitive or salt-tolerant cultivars under salt stress (Mansour et al. 1993; Mansour and Stadelmann 1994). Similarly, Mansour and Ali (2017a, b) and references therein present different lines of evidence that accumulation of two common important compatible solutes (i.e., proline and glycinebetaine) in response to salt stress shows no correlation with salt adaptation and may not be even involved in osmotic adjustment. Further, osmotic stress or water deficit caused by salinity was reported not to be the limiting factor for plant growth under saline conditions; rather salt toxicity caused the damage (Termaat et al. 1985; Munns and Termaat 1986). Also, Lee et al. (2008) do not recommend the accumulation of the most abundant proline and glycinebetaine for accurate use for physiological/biochemical marker-assisted salt screening. The study showed that glycinebetaine did not relate to osmotic adjustment differences among nine genotypes of seashore paspalum. In addition, the contribution of compatible solutes to osmotic adjustment was negligible in cabbage varieties, contrasting in salt tolerance, grown under salt stress, rather was ion accumulation (Sanoubar et al. 2016). Consistently, overexpression of trehalose improved rice salt tolerance, but trehalose was not high enough to participate to osmotic adjustment (Garg et al. 2002). It seems that high concentrations of compatible solutes, and possibly osmotic adjustment, might not guarantee association with salt tolerance and even may induce salt sensitivity in certain species/cultivars under salt stress (Mansour 2000; Koyro et al. 2012; Mansour and Ali 2017a, b). Consequently, there is still much to be learned about the exact role of compatible solutes in osmotic adjustment and adaptation of plants to salinity. Also, confirming the importance of long-term studies on the response of plants to salinity is necessary to uncover the master cellular traits of salt tolerance.

8.3 Ion Homeostasis, Membrane Transport Systems, and Salt Tolerance

To minimize the deleterious effects of toxic ions, plants have evolved various adaptive mechanisms to regulate ion concentrations. Under salt stress, excess accumulation of Na^+ ions is considered to be toxic for cell metabolism and can disrupt plant nutrient balance by affecting the uptake of other essential elements, e.g., K^+ (Flowers and Colmer 2015; Shabala et al. 2016). Further, a high Na^+/K^+ ratio in the cytosol inhibits various processes such as vital enzyme reactions, protein synthesis, and photosynthesis (Shabala and Mackay 2011; Wang et al. 2017). In a broad range of salt-tolerant plant species, membrane transport systems regulate such conditions via maintaining important elements and minimizing toxic ones (Mansour 2014; Gilliham et al. 2017; Han et al. 2018). Reducing Na^+ concentrations in the cytosol, K^+ retention, and cell ability to maintain cytosolic K^+ homeostasis therefore correlate with salt adaptation in various plant species (Munns and Tester 2008; Shabala et al. 2016; Han et al. 2018). These transport systems mostly include the ion channels and transporters; both work either at the plasma membrane or at the vacuolar membrane, i.e., tonoplast (Mansour 2014; Fan et al. 2015), which ensure selective passage of specific ions and subsequently modulate their concentrations. Ion homeostasis is hence stringently aiming to accumulate essential ions while the overall concentrations of toxic ions remain low under salt conditions (Roy and Chakraborty 2015). Reports ascribed therefore ion homeostasis as an essential factor or cellular mechanism of salt stress adaptation in salt-tolerant glycophytes and halophytes (Turkan and Demiral 2009; Shabala and Mackay 2011; Roy and Chakraborty 2015; Ali and Yun 2017). For example, ion homeostasis is associated with superior salinity tolerance in *Cynodon dactylon*, relative to the glycophyte *Oryza sativa* (Roy and Chakraborty 2015). Supporting to that, overall, tolerant glycophytes use highly developed, complex systems to tolerate salinity by maintaining a low cytosolic Na^+/K^+ ratio, sequestration of Na^+ into vacuoles, which lowers the cell water potential and thus sustaining continuous water influx under saline conditions (Araujo et al. 2015; Shabala et al. 2016). Moreover, the importance of K^+ retention, rather than net K^+ loss, for cytosolic ion homeostasis, and the regulation of Na^+ and Cl^- uptake while maintaining cytoplasmic K^+ and Mg^{2+} concentrations at levels required for activation of essential enzyme activities are emphasized (Munns and Tester 2008; Flowers and Colmer 2015). Ion homeostasis can be impaired by generation of ROS, as chloroplast-generated ROS dominate NaCl-induced K^+ efflux under salinity (Wu et al. 2015a, b). It is of great interest to mention that as chloroplastic Na^+ accumulation is noxious to the chloroplast, and therefore exclusion of Na^+ from chloroplast is important for salt stress tolerance (Assaha et al. 2017a, b). The authors present in their review thylakoid transporters that fulfill this task to export Na^+ out of the chloroplast. They also illustrate that inability to export Na^+ out of the chloroplast can impair the salt tolerance, photosynthesis, and growth of the plants. As we have already noted, a key aspect of salt tolerance under salt stress is ion homeostasis, i.e., the ability of salt-tolerant species/cultivars to remove toxic ions (either export ions out of the cells into the apoplast or compartmentalize them in the vacuoles) and to retain essential elements (e.g., K^+, Ca^{2+}, Mg^{2+}). These resistant features are summarized in Fig. 8.4 and will be reviewed below.

To cope with the ionic action of salinity, ion exclusion or ion inclusion (i.e., intracellular compartmentalization) mechanisms are activated. While salt exclusion is a whole plant level mechanism, it is discussed here because of its importance and association with ion homeostasis and salt tolerance. Salt exclusion mechanism means that salt-tolerant species/cultivars respond to salt stress by accumulating Na^+ in the roots and thus preventing Na^+ transport to the leaves (Munns and Tester 2008). For example, Na^+ retention in the root has been found to be a key adaptive mechanism to low and high salinity in the glycophyte, *Talinum paniculatum* (Assaha et al. 2017a). However, under high salinity, the study shows that Na^+ exclusion, osmotic adjustment, and antioxidant defense act in concert to minimize salt stress damage, which highlights the complexity of salt tolerance mechanism. Salt exclusion as a crucial mechanism of salt adaptation is also confirmed by and Flowers and Colmer (2015): "all plants are able, up to a point, to regulate the ions entering the xylem stream and this ability may be a determining factor in salt tolerance." Accordingly, at high external salt concentrations, halophytes and tolerant glycophytes are able to remove Na^+ from the xylem sap and thus exclude it from entering leaves, which can contribute to genotypic differences in apparent exclusion of Na^+ (Flowers and Colmer 2015). Also, the formation of salt glands in the epidermis of halophytes can facilitate the excretion of salts, preventing salt accumulation in the leaves (Flowers and Colmer 2015; Flowers et al. 2015), a form of salt

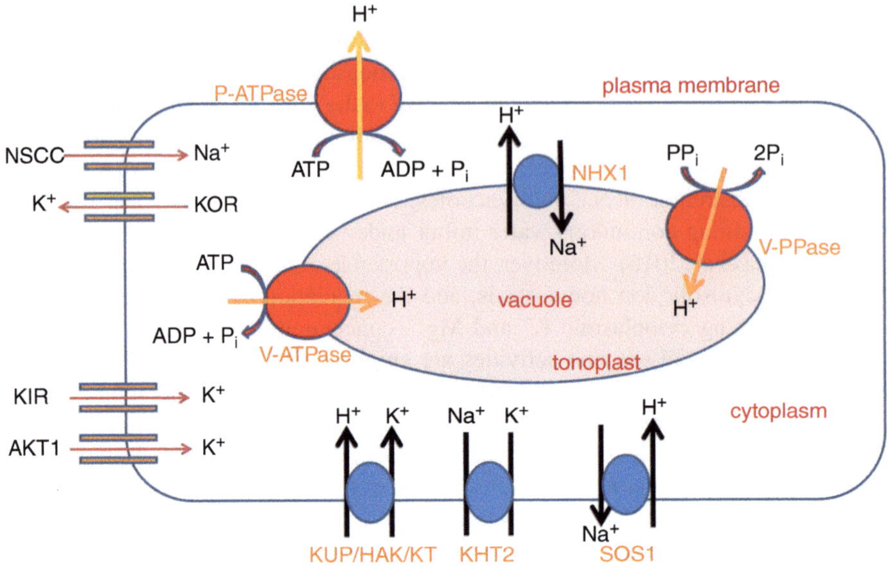

Fig. 8.4 A root cell summarizes the plasma and vacuolar membrane transport proteins contributing to ion homeostasis and Na^+ compartmentation in the vacuole as a cellular mechanism of salt tolerance

exclusion. Another feature of salt exclusion mechanism is the accumulation of salts in the old leaves causing their death and fall (Acosta-Motos et al. 2017). In the mechanism of salt exclusion, cellular membranes and their transporters play a major role in determining the ions that pass into the xylem (Acosta-Motos et al. 2017). In most species of glycophytes, the K^+/Na^+ and Ca^{2+}/Na^+ ratios as well as uptake selectivity essentially discriminate salt-tolerant from salt-sensitive genotypes (Araujo et al. 2015). On the other hand, in ion inclusion mechanism, halophytes and few tolerant glycophytes accumulate more ions of both Na^+ and Cl^- in shoots and roots under saline conditions, and their salt tolerance is partially explained by the compartmentalization of Na^+ into cell vacuoles (Flowers et al. 2015). For instance, a gradual increase was observed in the Na^+ content of leaf with increasing salt concentration up to 750 mM NaCl reported in the halophyte *Salvadora persica* (Rangani et al. 2016). In this study, it seems that sequestration of Na^+ into the vacuole for osmotic adjustment reduced Na^+ concentration in the cytosol, and organic solutes are most probably accumulated in the cytoplasmic compartments to balance the reduced solute potential in the vacuole (Flowers et al. 2015). Although salt inclusion mechanism is frequently adopted by halophytes, Horei et al. (2012) argue that salt accumulation in the leaf apoplast could be responsible for the death of leaves in glycophytes (in particular salt-sensitive species) under salt exposure. To conclude, while mechanisms of minimizing the entry of salt into the plant (i.e., salt exclusion) or decreasing the concentrations of salt in the cytoplasm of the cells (i.e., salt compartmentalization in the vacuoles) are adopted by the halophytes, most glycophytes have a low ability to exclude salt and thus accumulate toxic ions in the leaves through the transpiration flow contributing to their salt sensitivity (Acosta-Motos et al. 2017).

As the transport systems of the plasma membrane and tonoplast play key role in ion homeostasis mechanism, their contribution to this mechanism is discussed in the following sections. Plant cells in order to prevent the harmful effects of Na^+ ions in the cytosol remove excess Na^+ into the apoplast by the activity of the plasma membrane H^+-ATPase and Na^+/H^+ antiporter or sequester Na^+ within the vacuole via H^+-ATPase, H^+-pyrophosphatase (H^+-PPase), and Na^+/H^+ antiporter (Silva and Gerós 2009; Silva et al. 2010; Mansour 2014; Assaha et al. 2017a, b). The first members of the important membrane transporters are plasma membrane H^+-ATPase, vacuolar H^+-ATPase, and H^+-pyrophosphatase which have reported to play decisive functions in plant salt tolerance (Silva and Gerós 2009; Mansour 2014; Assaha et al. 2017a, b). In addition to their roles in different physiological processes, these proton pumps generate an electrochemical potential gradient across the membrane, which provides membrane potential and proton gradient necessary for a large set of secondary transporters (Shabala et al. 2016). The two sources of energy are used by ion channels, carriers, and Na^+/H^+ antiporters, respectively, for Na^+ and K^+ transport, thus modulating cytosolic ion homeostasis under salt stress (Janicki-Russak 2011; Assaha et al. 2017b). The crucial role of plasma membrane H^+-ATPase in salt adaptation is supported by the findings that an increase in its activity and expression has been indicated in various halophytes and tolerant glycophytes under saline conditions (Janicki-Russak 2011; Bose et al. 2013, 2015). The importance of the proton gradient generated by plasma membrane H^+-ATPase comes from the fact that it supplies the energy for transport of Na^+ from the cytosol out into the free space by the plasma membrane Na^+/H^+ antiporters, which is encoded by *SOS1* gene in many plant species (Janicki-Russak 2011; Mansour 2014). For example, transcriptional upregulation of *SOS1* and *NHX1* genes under high salinity in roots and leaves has been shown to be involved in controlling Na^+ levels in *Sesuvium portulacastrum* and hence contribute to ion homeostasis and salt tolerance (Nikalje et al. 2018). Other plasma membrane channels and transporters that have key functions in ion homeostasis include AKT1 (K^+_{in} channels) and KUP1/HAK/KT (high affinity K^+-H^+ symporters) which contribute to K^+ influx and homeostasis under high salinity (Shabala and Mackay 2011; Mansour 2014). In addition, nonselective cation channels (NSCCs) for passive Na^+ influx, outward-rectifying K^+ channels (KOR) for K^+ efflux, and inward-rectifying K^+ channels (KIR) for K^+ influx also mediate Na^+ and K^+ homeostasis in saline conditions (Fig. 8.4) (Mansour 2014). The response of one of these channels (i.e., K^+ inward rectifier) to saline conditions, however, was not different between two rice cultivars contrasting in salt tolerance (Kavitha et al. 2012), referring to high complexity of ion homeostasis under salt imposition (Mottaleb et al. 2013).

Besides the plasma membrane transport systems, the vacuolar H^+-ATPase and H^+-PPase also provide the proton-motive force for exclusion of Na^+ from the cytosol in exchange for H^+ by a vacuolar Na^+/H^+ antiporter which is encoded by *NHX1* gene (Silva et al. 2010; Almeida et al. 2017) and hence contribute to cellular adaptation to salt stress. In the study of Silva et al. (2010), salt stress-induced vacuolar H^+-ATPase activity and Na^+/H^+ exchange activity in the tonoplast of *Populus euphratica* cultured cells demonstrate that both vacuolar transport systems are essential to Na^+ sequestration in the vacuole and hence salt tolerance. Na^+ accumulation in the vacuole under salt stress is also demonstrated by the study. In addition, under salt stress, an increase in vacuolar H^+-ATPase activity has been reported in many plant species, which in turn enhances their salt tolerance (Silva and Gerós 2009; Almeida et al. 2017). A recent work similarly demonstrated that higher activity of both V-H^+-ATPase and V-H^+-PPase and greater Na^+/H^+ exchange activity are important strategies explaining the differences in salt tolerance of potato cultivars (Jaarsma and de Boer 2018). Wu et al. (2015a, b) also report that salinity tolerance correlated positively with vacuolar Na^+ sequestration ability in wheat roots. In support to the important role of vacuolar proton pump and vacuolar Na^+/H^+ antiporter in salt adaptation mechanism are the findings that co-expression of both tonoplast NXH1 and H^+-PPase into alfalfa lines (Bao et al. 2016) and tomato (Gouiaa and Khoudi 2015) improves their salt tolerance in response to saline conditions. The improved performance of transgenic alfalfa plants in saline environment was associated with more K^+ and Ca^{2+} accumulation in leaves and roots, which contributed to maintaining intracellular ion homeostasis, osmotic adjustment, enhanced photosynthesis capacity, stomatal conductance, and water-use efficiency (Bao et al. 2016). Moreover, overexpression of a vacuolar H^+-pyrophosphatase (*SbVPPase*) in transgenic finger millet facilitates efficient sequestration of excess Na^+ ions into the vacuoles and enhances salt tolerance (Anjaneyulu et al. 2014). Similarly, transgenic tomato plants showed increased biomass production, chlorophyll content, relative water content, catalase activity, and K^+ in leaves and fruits while decreased Na^+ in leaves and fruits as well as decreased electrolyte leakage from transgenic plants relative to wild type under salt stress (Gouiaa and Khoudi 2015). The previous studies clearly report the decisive role of vacuolar transport systems in ion homeostasis as a cellular mechanism of salt tolerance.

Preventing Na^+ accumulation in shoots is also performed by HKT (high-affinity potassium transporter family), which is another member of the membrane transport systems contributing to salt adaptation (Mansour 2014; Han et al. 2018). HKT-type transporters play crucial role in Na^+, K^+, and Ca^{2+} homeostasis (Han et al. 2018). HKT transporters include two subclasses: subclass 1 that are sodium-selective transporters (typically HKT1 family members) and subclass 2 that are K^+/Na^+ symporters (typically HKT2 family members) (Mansour 2014; Gilliham et al. 2017). Under salt stress, members of subclass 1 retrieve Na^+ from the xylem stream to reduce its transport to the shoots, whereas members of subclass 2 transporters maintain a balanced Na^+/K^+ ratio in the cytoplasm (Sunarpi et al. 2005; Mansour 2014; Han et al. 2018). In agreement with the crucial role of HKT1 transporters in salt tolerance mechanism is the finding that knocking out or down regulation of HKT1 leads to salt sensitivity in some glycophytes and halophytes (Ali et al. 2013; Han et al. 2018). For example, knockdown of *HvHKT1;1* in barley led to higher Na^+ accumulation in both roots and leaves, while overexpression of *HvHKT1;1* in salt-sensitive *Arabidopsis* resulted in significantly less shoot and root Na^+ accumulation (Han et al. 2018). In addition, homologs of HKT1 are isolated from different plant species which play key role in regulating Na^+ movement throughout plants (Ali and Yun 2017). Moreover, overexpression of these transporters (i.e., HKT1) contributes to Na^+ uptake which leads to hypersensitivity to salinity in salt-sensitive plants, while their downregulation restricts Na^+ influx in tolerant plants and hence salt tolerance (Plett and Moller 2010; Mansour 2014). Other studies on HKT provided important background information supporting the key role of transport systems in salt adaptation and subsequently led to the development of durum wheat with improved yield in the field under saline conditions (Munns et al. 2012; Gilliham et al. 2017).

Another transport protein that has role in salt tolerance is the plasma membrane aquaporins (PIPs) and vacuolar aquaporins (TIPs) (Martinez-Ballesta et al. 2006; Mansour 2014). Aquaporins or water channels increase the permeability of membranes to water and other small molecules such as urea, CO_2, glycerol, silicon, boron, and hydrogen peroxide (Martinez-Ballesta et al. 2006; Mansour 2014). Consistently, a significant reduction in hydraulic conductance was coupled with a 60–75% decrease in *PIP* and *TIP* aquaporin transcripts abundance in *Arabidopsis* after exposure to salt stress (Afzal et al. 2016). Afzal et al. (2016) also showed that in addition to TIPs role in increased water transport across the tonoplast, they are also involved in the accumulation of ions in vacuoles in response to salt stress. It is therefore expected that aquaporins may play a role in maintaining osmosis and turgor of halophyte and glycophyte cells under salt stress (Diray-Arce et al. 2016; Afzal et al. 2016). It seems that the response of plants to salt stress is most likely related to aquaporin expression and activity. Consistently, protein level of plasma and vacuolar membrane aquaporins of radish roots was increased by NaCl treatment, which may be a salt stress responsive mechanism (Suga et al. 2002). More evidence confirming the role of aquaporins in acclimation to saline conditions is that: First, Si-induced salinity tolerance was via enhancement of root hydraulic conductance and water uptake, suggestive of aquaporin involvement as a key point in the amelioration of the adverse effects of salinity stress (Rios et al. 2017). Second, Lopez-Perez et al. (2009) indicated salt-induced plasma membrane lipid modulation significantly affected aquaporin expression and activity which allows maintaining the membrane permeability to water as a mechanism for acclimation to salinity stress. Further, salinity-increased expression of aquaporins PIP1 and PIP2 subfamilies in roots of different salt-sensitive species was associated with their improved adaptation to salt stress (Hu et al. 2012; Bhaedwaj et al. 2013). Moreover, exposure of plant roots to salt stress induced changes in the plasma membrane aquaporin expression, which has been shown to be correlated with salt tolerance (Xu et al. 2014). Additionally, overexpressing plasma membrane intrinsic protein *PIP2;7* and tonoplast intrinsic protein *TIP1;3* from *Jatropha* in *Arabidopsis* improved salt tolerance and seed yield in transgenic plants (Khan et al. 2015a, b). Reviewing the literature, however, revealed controversial results among publications. For instance, cases of failure of improving plant tolerance to saline environment due to overexpressing aquaporins are provided by Mansour (2014) and Afzal et al. (2016). Also, a salinity-tolerant ice plant (*Mesembryanthemum crystallinum*) also showed downregulation of many PIP genes in roots (Yamada et al. 1995) and a TIP gene (MIPF) in leaves (Kirch et al. 2000) under salt stress. These discrepancies are interpreted by the possibility that salt-induced modifications in the lipid microenvironment surrounding the transport proteins may affect aquaporins performance rather than a direct effect of salinity on the aquaporin themselves or even their expression (Lopez-Perez et al. 2009; Mansour 2014). In conclusion, as the water deficit stress is the primary response of plants to salt stress, inhibition of root-water permeability and uptake is manifested, and hence aquaporin role in plant salt tolerance is to enhance the water uptake capacity of plant roots which contribute to osmotic homeostasis and turgor maintenance, and thus the role of aquaporins in adaptation of plant species/cultivars to high salinity cannot be ruled out.

8.4 Reactive Oxygen Species, Antioxidants, and Salt Tolerance

Salinity induces oxidative stress, as a secondary stress, in plants by accumulation of superoxide radicals ($O_2\cdot^-$), hydrogen peroxide (H_2O_2), or hydroxyl radical ($OH\cdot$) in chloroplasts, mitochondria, and apoplast, which results in lipid peroxidation and protein oxidation (Mittova et al. 2003; López-Gómez et al. 2007; Rubio et al. 2009). Noteworthy, it is evident that chloroplasts and mitochondria are main sources of ROS in plants subjected to salt stress: as salt stress induces impairment of electron transport chains and in absence of final electron acceptor, electron transfer from ferredoxin to oxygen is favored to form ROS (Wu et al. 2015a, b). Besides these sites of ROS production, they can also be formed at peroxisomes, plasma membrane, and apoplast (Mansour et al. 2019). Although their role as a major signaling agent in plants is now acknowledged (Foyer and Noctor 2005), elevated ROS levels can result in an impairment of cellular macromolecules integrity (Kibinza et al. 2011; Mittler et al. 2011) and disturbed ion homeostasis (Wu et al. 2015a, b). That is, depending on the cellular circumstances and the balance between ROS generation and detoxification, ROS may either exert toxic effects on plant cells or act as signaling molecules in a broad range of adaptive and developmental responses as well as regulators of numerous stress-responsive genetic networks (Kibinza et al. 2011; Mittler et al. 2011). Accumulation of various ROS is documented under saline conditions. For example, under saline conditions, the chloroplast H_2O_2 and mitochondrial superoxide radical contents increased in two pea cultivars, more so in salt-sensitive cultivar (Acosta-Motos et al. 2017). Exposing a mangrove plant, *Suaeda maritima*, to 255 mM NaCl resulted in H_2O_2 accumulation (Mallik et al. 2011). Also, H_2O_2 and $O_2\cdot^-$ contents have been reported to be increased under salinity stress (Hernández et al. 2000; Rahman et al. 2016). When triticale genotypes contrasting in salt tolerance were treated with long-term salt stress, the salt-tolerant genotypes exhibited lower H_2O_2 and lipid peroxidation relative to salt-sensitive genotypes, which was associated with greater increased activities of the enzymes participating in ROS scavenging in tolerant genotypes (Kiani-Pouya 2015).

Furthermore, salt sensitivity of wild-type wheat was shown to be related to higher superoxide radicles, H_2O_2, lipid peroxidation, and lower antioxidant enzyme activities comparing with salt-tolerant wheat stay-green mutant *tasg1* (Wang et al. 2018). Conversely, salt-tolerant Jerusalem artichoke exhibited significantly more lipid peroxidation than sensitive one, and antioxidant enzymes were inadequate to correlate with salt tolerance mechanism (Chen et al. 2011), suggestive of lower efficiency of scavenging ROS in tolerant cultivar. While there are few exceptions, salt stress usually increased the level of ROS in various plant species, and as a cellular mechanism for salt adaption, antioxidant defense systems frequently enhanced in tolerant species/cultivars in order to scavenge ROS. The following section reviews the latter issue.

Several reports indicate that one of the cellular mechanisms involved in plant salt tolerance is to scavenge the salt-induced ROS via upregulation of antioxidant enzymes and/or nonenzymatic antioxidants (Rady et al. 2018). Enzymatic antioxidants include the following: superoxide dismutase converts superoxide radicals ($O_2\cdot^-$) into hydrogen peroxide (H_2O_2), peroxidase reduces H_2O_2 to water using various substrates as electron donors, ascorbate peroxidase uses ascorbate as electron donor to reduce H_2O_2 to water, and catalase dismutates H_2O_2 into water and oxygen (Mansour et al. 2019). On the other hand, nonenzymatic antioxidants may include ascorbic acid, glutathione, proline, carotenoids, total phenolics, flavonoids, and tocopherols that are able to eliminate or reduce ROS (Gill and Tuteja 2010). In support to the crucial role of antioxidants as a cellular mechanism to plant salt tolerance is the finding that upregulation of the antioxidative machinery and/or higher constitutive antioxidant enzyme levels is usually greater in the salt-tolerant species/cultivars, whereas the antioxidant defense in salt-sensitive ones always exhibits insignificant response or lower constitutive levels of antioxidant enzymes (Mittova et al. 2003; López-Gómez et al. 2007; Rubio et al. 2009; Bazrafshan and Ehsanzadeh 2016). Other different works indicated that antioxidant defenses were increased in salt-tolerant species/cultivars (Roy and Chakraborty 2015; Yousuf et al. 2017; Nikalje et al. 2018), whereas a decline was observed in these antioxidants in salt-sensitive ones (Mittova et al. 2002; Yousuf et al. 2017; Wang et al. 2018). In the previous studies, antioxidant enzyme activity or expression in salt-tolerant plant species/cultivars has been illustrated in response to salt stress. In the same trend, numerous reports demonstrate that enhanced antioxidant enzyme activity is associated with tolerance to stress conditions (Gill and Tuteja 2010; Tanou et al. 2012; Hasanuzzaman et al. 2017; Roy and Chakraborty 2015). Also, lower ROS accumulation, lipid peroxidation, and higher membrane stability have been reported in salt-tolerant wheat genotype comparing with sensitive one under high salinity (Rasool et al. 2013; Maswada et al. 2018). In rice seedlings, decreased ROS and lipid peroxidation through increased activities of antioxidant enzymes was associated only with the supplementation of calcium (Rahman et al. 2016), suggestive of Ca^{2+} role in improving salt tolerance through enhancing the antioxidant defense. Also, exogenous application of polyamines, as compatible solutes, to mung bean seedlings grown under salt stress showed increased salt tolerance by increasing the levels of nonenzymatic antioxidants (i.e., glutathione and ascorbate) and promoting the activity of different antioxidant enzymes (Nahar et al. 2016). In addition, elevation of antioxidant enzyme activities as well as nonenzymatic antioxidants (glutathione, ascorbic acid) was one of the strategies contributed to improve salt tolerance in response to foliar glycinebetaine application to onion plants (Rady et al. 2018). In the same trend, exogenous NO alleviated the damage effects of salt stress by increasing antioxidant capacity in cucumber (Lin et al. 2012). Salt tolerance of halophyte *Centaurea ragusina* is also associated with antioxidative defense (Rodic et al. 2013). In these works, enzymatic and nonenzymatic defense systems have been found to scavenge ROS resulting in maintained cell ultrastructures. Consistently, nonenzymatic antioxidants have been upregulated under high salinity in salt-tolerant species/cultivars, but not in sensitive ones (Hernández et al. 2000; Rasool et al. 2013). For example, a highly significant correlation was recorded between total phenol content and antioxidant activity in leaves and roots of olive cultivars under NaCl stress (Petridis et al. 2012), suggestive of nonenzymatic antioxidants crucial role as defense system under stress conditions. The importance of nonenzymatic antioxidants as the only effective means to prevent detrimental effects of hydroxyl radicals on cellular structures has been also illustrated by Bose et al. (2014). Additionally, the importance of the scavenging ROS by antioxidant defense systems comes from the finding that ROS dominate NaCl-induced K^+ efflux and thus disturb ion homeostasis under salinity (Wu et al. 2015a, b). Further, a rice mutant defective in antioxidant-defense system exhibited increased sensitivity to salt stress (Lin et al. 2016). Tabulated review of the responses of enzymatic and nonenzymatic antioxidants in plant species/cultivars contrasting in salt tolerance under saline conditions is given in Table 8.1. Although there are contrasting reports on antioxidant production in response to salinity (Table 8.1), the antioxidative protection system as an adaptation strategy to high salinity is obvious.

8.5 Membrane Lipids and Protein Modeling and Salt Tolerance

Membranes are highly ordered structures and involve mosaic lipids and proteins. High salinity modulates the properties of these membranes, i.e., fluidity and permeability, via alterations in the lipid composition and/or interactions between lipids and specific membrane proteins (Mansour et al. 2015). As salt tolerance is complex mechanism and involves changes in primary metabolism required for cellular synthesis of different metabolites, it also needs to change membrane proteins and lipids as well as their metabolism in order to adjust to changing cellular water and ionic status (Barkla et al. 2018). Membrane composition remodeling therefore participates not only in cell perceiving salt changes in the surrounding milieu but also in intracellular responses and thus determines the cell fate. Although plasma, vacuolar, and organelle membranes are possibly involved in these responses, most studies concern the PM, and hence our focus will be on the plasma membrane components.

Table 8.1 Effect of high salinity on enzymatic and nonenzymatic antioxidants in salt-sensitive and salt-tolerant species/cultivars and their contribution to salt tolerance

Species	Rating	Salt treatment	Responses	References
Medicago sativa	Moderate	200 mM NaCl	Tolerant cultivar showed enhanced germination and growth, low H_2O_2, low lipid peroxidation, higher activities of antioxidant enzymes	Wang et al. (2009)
Prunus dulcis	Sensitive, moderate, and tolerant species	40, 80, 120 mM NaCl	Increased antioxidative enzyme activity and δ- and ɣ-tocopherols (nonenzymatic antioxidants)	Sorkheh et al. (2012)
Mesembryanthemum edule	Halophyte	300, 600 mM NaCl	Increased activities of antioxidative enzymes and phenolic compounds (nonenzymatic antioxidants) were greater in salt-tolerant provenance	Falleh et al. (2012)
Oryza sativa	Sensitive and tolerant cultivars	12 dSm^{-1}	Tolerant cultivar exhibited greater enzymatic and nonenzymatic defense systems	Sarkar et al. (2013)
Hordeum vulgare	Sensitive and tolerant varieties	80 mM NaCl	Sensitive variety showed higher antioxidative enzymes activity. Enzymatic antioxidant activity therefore does not correlate with salinity tolerance	Maksimovic et al. (2013)
Hordeum vulgare	Sensitive and tolerant varieties	240 mM NaCl	Enzymatic antioxidants were similar in sensitive and tolerant varieties. Antioxidative enzymes are not biochemical indicators and not correlated with salt tolerance	Fan et al. (2014)
Oryza sativa	Sensitive and tolerant lines	150 mM NaCl	Enhanced antioxidative enzyme activity was more in tolerant line	Pushpalatha et al. (2013)

(continued)

Table 8.1 (continued)

Species	Rating	Salt treatment	Responses	References
Triticum aestivum	Sensitive and tolerant genotypes	8, 160 mM NaCl	Tolerant genotype had more enzymatic antioxidant activities	Bhutta (2011)
Zea mays	Sensitive	75, 150 mM NaCl	Increased enzymatic and nonenzymatic (ascorbate, glutathione, tocopherol, polyphenols) antioxidants under salt stress	AbdElgawad et al. (2016)
Solanum tuberosum	Sensitive and tolerant cultivars	30, 60, 90, 120 mM NaCl	Greater activation of antioxidant enzymes in tolerant cultivar than sensitive one	Aghaei et al. (2009)
Ocimum basilicum	Sensitive	25 mM Na_2SO_4, 50 mM NaCl	Higher sensitivity to Na_2SO_4 was associated with inhibition of antioxidative enzymes and lower production of ascorbate and glutathione. Antioxidative defense systems (enzymatic and nonenzymatic) were more efficient under NaCl	Tarchoune et al. (2010)
Suaeda maritima	Halophyte	255 mM NaCl	Increased antioxidative enzyme activities in response to salinity	Mallik et al. (2011)
Brassica rapa	Sensitive and tolerant cultivars	80, 160 mM NaCl	Decreased antioxidant enzyme activities. Salt tolerance does not relate to antioxidant capacity	Noreen et al. (2010)
Chenopodium album	Tolerant	50, 300 mM NaCl	Salt tolerance was due to increased activities of antioxidant enzymes and nonenzymatic antioxidants (carotenes, ascorbate) and increased expression of antioxidant genes	Yao et al. (2010)
Oryza sativa	Sensitive and tolerant cultivars	200 mM NaCl	Activity and expression of antioxidative enzymes were increased, more so in tolerant cultivar. Ascorbate was also	Turan et al. (2012)

(continued)

Table 8.1 (continued)

Species	Rating	Salt treatment	Responses	References
			increased in tolerant cultivar, but ascorbate and tocopherol were decreased in sensitive one	
Brassica juncea	Sensitive and tolerant genotypes	50, 100, 150, 200 mM NaCl	Tolerant genotypes had less lipid peroxidation and increased antioxidant enzyme activities and nonenzymatic antioxidants (ascorbate and glutathione) relative to sensitive ones	Yousuf et al. (2017)
Carthamus tinctorius	Sensitive	25, 50, 75 mM NaCl	Enhanced enzymatic and nonenzymatic ascorbate (for short-term treatment) and glutathione (for long-term salt stress) antioxidants contribute to tolerate NaCl stress	Zaoui et al. (2016)
Triticum aestivum	Sensitive	200 mM NaCl	Less ROS, lipid peroxidation, and higher activities of antioxidant enzymes contribute to salt tolerance of *Tasg1*comparing with wild type	Wang et al. (2018)
Solanum tuberosum	Salt adapted and non-adapted cell lines	150 mM NaCl	Antioxidative enzyme activity decreased in both lines and does not correlate to salt tolerance. Tolerance to salinity associated with increased levels of ascorbate, glutathione, and α-tocopherol	Queiros et al. (2011)
Chenopodium quinoa	Tolerant	300 mM NaCl	Enhanced antioxidant enzyme activities associated with salt tolerance	Amjad et al. (2015)
Ipomoea batatas	Sensitive and tolerant genotypes	150, 300, 450 mM NaCl	Activity and expression of ascorbate peroxidase were higher in salt-tolerant genotypes than sensitive ones and correlate with salt adaptation	Lin and Pu (2010)

(continued)

Table 8.1 (continued)

Species	Rating	Salt treatment	Responses	References
Oryza sativa	Sensitive	200 mM NaCl + 2 mM $CaCl_2$	Ca-induced salt tolerance was due to enhanced antioxidant defense systems, which decreases ROS and lipid peroxidation	Rahman et al. (2016)
Limonium delicatulum	Halophyte	50–500 mM NaCl	Increased enzymatic and nonenzymatic (polyphenols) antioxidants was correlated with salt adaptation, in particular enzymatic ones	Souid et al. (2016)
Sesamum indicum	Sensitive and tolerant genotypes	30, 60 mM NaCl	Lower lipid peroxidation and enhanced enzymatic antioxidants were greater in salt-tolerant genotypes and associated with salt adaptation	Bazrafshan and Ehsanzadeh (2016)
Sesuvium portulacastrum	Halophyte	150, 500 mM NaCl	Enhanced enzymatic and nonenzymatic (flavonoids) antioxidants correlated with salt tolerance	Nikalje et al. (2018)
Solanum stoloniferum	Sensitive and tolerant species	40, 80, 120 mM NaCl	Nonenzymatic antioxidant capacity (proline, flavonoids, phenolic compounds) was more crucial than enzymatic antioxidants capacity for salt tolerance	Daneshmand et al. (2010)
Helianthus tuberosum	Sensitive and tolerant cultivars	250 mM NaCl	Salt-sensitive cultivar exhibited lower lipid peroxidation and higher antioxidant enzyme activities than tolerant one, indicative of absence of antioxidative enzyme role in salt tolerance	Chen et al. (2011)

Overall, signaling pathways reach the nucleus and change gene expression related to the PM lipid synthesis resulting in its lipid alterations, and thus PM bilayer structure, fluidity, permeability, transport system activity, and signaling molecules are retained

Table 8.2 Representative examples of plasma membrane lipid classes modified in response to high salinity and their role in salt tolerance

Species	Rating	Tissue	Lipid class	Response and remark	Reference
Spartina patens	Halophyte	Callus	Phosphatidylcholine (PC)	No effect, maintained PC to form membrane lamellar structure	Wu et al. (2005)
			Campesterol	Increase, one of planar sterols, which have tighter packing in the phospholipid bilayer, and reduce permeability to ions	
			Sitosterol	Decrease, one of non-planar forming lipid, its low abundance correlates with bilayer maintenance	
Zea mays	Tolerant cultivar	Root	Phosphatidylglycerol (PG)	Increase, one of bilayer forming phospholipids, correlates with salt tolerance	Salama et al. (2007)
			Phosphatidylethanolamine (PE)	Decrease, one of non-lamellar forming phospholipids, its increase disrupts membranes	
	Sensitive cultivar	Root	Linoleic acid (18:2)	Decrease, sensitive cultivars showed low linoleic acid experienced a reduction in membrane fluidity	
Solanum lycopersicum	Tolerant	Callus	Campesterol	Increase, one of planar sterols and correlates with adaptation to salinity	Kerkeb et al. (2001)
Brassica napus	Tolerant cultivar	Root	Phosphatidylglycerol (PG)	Increase, one of bilayer forming phospholipids, correlates with salinity tolerance	Zamani et al. (2010)
	Sensitive cultivar	Root	Phosphatidylinositol (PI)	Increase, correlates with salt sensitivity, and plays a role as signaling molecule in response to stresses	
Hordeum vulgare	Tolerant	Root	Linolenic acid (18:3)	Increase, maintained membrane stability of salt-tolerant cultivar was associated with increase in polyunsaturated fatty acid (e.g., 18:3)	Yu et al. (1999)
Mesembryanthemum crystallinum	Halophyte	Epidermal bladder cells	Phosphatidylcholine (PC) Phosphatidylglycerol (PG)	Increase, both phospholipids maintain the proper structure and function of the membranes, correlate with salt tolerance	Barkla et al. (2018)

(continued)

Table 8.2 (continued)

Species	Rating	Tissue	Lipid class	Response and remark	Reference
Triticum aestivum	Sensitive	Root	Phosphatidylcholine (PC)	Decrease, PC decrease disrupts the membrane integrity of the sensitive cultivars	Mansour et al. (2015)
			Phosphatidylinositol (PI)	Increase, PI increase correlates with salt sensitivity	
			Stigmasterol	Increase, increased more planar sterol (e.g., stigmasterol) reduces the ATPase activity, and thus no membrane energization for Na^+ extrusion by Na^+/H^+ antiporter	
			Linolenic acid (18:3)	Decrease, sensitive plants under salt stress show decreased levels of C18:3 in their membranes and thus experience membrane fluidity reduction	

under salinity. Lipidomic studies have therefore shed light on the importance of lipids in the response of plants to salt treatment (Table 8.2), providing evidence that plant lipid metabolism is responsive to salinity at the gene, protein, and metabolite levels (Kosová et al. 2013; Mansour et al. 2015; Barkla et al. 2018). Implication of the PM lipids and proteins as a cellular mechanism of salt adaptation relies on the contention that both PM components are the most important agents regulating ion homeostasis under salinity stress (Mansour 2014; Mansour et al. 2015), and it is known that ion homeostasis is a pivotal strategy for plant tolerance to saline environments. In support to that, PM qualitative and quantitative adjustments of its components have been shown to be different in salt-sensitive and salt-tolerant plant species/cultivars under salt stress, as they are mostly in favorable and sustainable directions for the tolerant species/cultivars (Mansour et al. 2015). Also, evidences indicate that salinity stress is perceived at the PM, and the PM components hence play crucial role in cellular signal transduction under salinity (Okazaki and Saito 2014; Mansour et al. 2015). The previous issues will be reviewed in the following sections.

The prominent role of the PM lipid composition in regulating the PM properties and the activity of PM-associated transport systems is documented (Table 8.2), as it greatly affects ion homeostasis and in turn tolerance to high salinity (Russell 1989; Mansour 2014; Mansour et al. 2015). In addition, the PM components have a pivotal role in cellular signal transduction under stress (Ruelland et al. 2014). The key role of PM lipids as a strategy for salt adaptation comes from the fact that the lipid requirements of transport systems have been documented, and membrane lipid microenvironment is more effective than the membrane bulk lipids in this respect (Cook et al. 1993; Gutla et al. 2012). For instance, PM lipids correlate with salt tolerance through lipid effect on the PM H^+-ATPase activity in saline environment (Alvarez-Pizarro et al. 2009). Also, phospholipid, sterol, and fatty acid requirement for activation of proton translocating activity by PM ATPase has been illustrated (Sandstrom and Cleland 1989; Cook et al. 1993; Yu et al. 1999). It is worth to mention that changes in Na^+ content and Na^+/K^+ ratio in leaves of various cultivars were associated with the enhanced PM H^+-ATPase activity under salt stress, as H^+-ATPase is known to energize the PM for Na^+ exclusion by Na^+/H^+ antiporters (Mansour 2014; Mansour et al. 2015), and also polyunsaturated fatty acids have been reported to modulate ion channels activities (Gutla et al. 2012). More detail for the role of the PM lipids (i.e., phospholipids, glycolipids, sterols, fatty acids) in salt adaptation in various plant species under saline conditions is addressed in the review by Mansour et al. (2015). Furthermore, Barkla et al. (2018) recently provide evidence that triacylglycerol might similarly play an important role in response to salinity as either an energy reserve for Na^+ accumulation and/or driving membrane biosynthesis for cell expansion in epidermal bladder cells of *Mesembryanthemum crystallinum*. It is obvious that alterations in the PM lipid composition under high salinity may have advantage for modulating the transport systems and hence ion homeostasis and subsequently salt tolerance. Besides the PM lipids role in maintaining membrane physical and chemical properties, regulating membrane transport systems, and activating membrane bound transcription factors (Seo et al.

2010; Mansour et al. 2015), production of biologically active lipid molecules has been shown to play an important role in salt stress response and tolerance (Ruelland et al. 2014; Okazaki and Saito 2014). Studies report that PM lipids have been implied as key signaling molecules in plant responses to saline conditions (Ruelland et al. 2014; Okazaki and Saito 2014). These specifically include phosphoinositides and phosphatidic acid, which have been proven to be responsive to salt stress and have role in salt acclimation (Meringer et al. 2016; Meijer et al. 2017; Barkla et al. 2018). The investigation of Shen et al. (2011) illustrates that in rice suspension-cultured, phospholipase Dα mediated H^{+}-ATPase activity and transcription and hence is involved in salt tolerance. It has been therefore reported that plants overexpressing phospholipase genes frequently display tolerance to high salinity (Ruelland et al. 2014). The correlation between phospholipases and salt adaptation comes from the fact that the action of phospholipases and/or lipid kinases results in phosphatidic acid generation, which is implied as a key signaling molecule in plant responses to saline conditions (Ruelland et al. 2014; Mansour et al. 2015). Lipid signaling pathways have been also demonstrated to be involved in the regulation of the level of proline and ROS under salt stress (Ruelland et al. 2014). Similarly, induction of phosphoinositides signaling pathways under salinity has been reported by Barkla et al. (2018), which is suggested to be involved in enhanced membrane trafficking during salinity stress. On light of the above, PM lipids remodeling and functions are obviously having a key role in plant salt adaptation.

Proteomics of the PM has been indicated as one of the important cellular tolerance mechanisms to salt imposition (Zamani et al. 2010, Mansour 2013, 2014). For example, the pepper *CaPIMP1* gene encodes a PM protein involving in plant stress tolerance in adverse environmental conditions (Hong and Hwang 2009). Moreover, qualitative and quantitative differences in the PM proteins of species/genotypes contrasting in salt sensitivity have reported under salt stress (Kerkeb et al. 2001; Salama et al. 2007; Zamani et al. 2010; Kosová et al. 2013; Mansour 2013, 2014). In agreement, Kosová et al. (2013) report that proteins involved in stress response reveal differences in protein structure, activity, and level between tolerant plants and sensitive ones. These salt-responsive PM proteins have been found to improve plant salt tolerance under high salinity. In addition, several studies report that salt-responsive proteins produced in salt-tolerant species/cultivars improve their salt tolerance and hence play an important role in salt acclimation to saline conditions (Kosová et al. 2013; Fercha 2017). Among salt-responsive PM proteins, we mention their functions ascribed as protective proteins, transport proteins, and sensory proteins (Table 8.3). It has been shown that overexpression of the genes encoding these proteins contribute to enhancing tolerance to high salinity (Osakabe et al. 2013; Mansour 2014; Wang et al. 2014). For the involvement of the PM transport proteins in plant adaptation to saline environments, the readers are referred to the review by Mansour (2014). The PM protective proteins such as osmotin (Salama et al. 2007) is hypothesized to regulate plant responses to different environmental stresses through its involvement as an osmoprotectant and a cell signal pathway modulator (Abdin et al. 2011). Consistently, Xu et al. (1996) indicate that osmotin and LEA proteins contribute to alleviating salt stress by protecting

Table 8.3 Representative plasma membrane proteins evolved in response to saline conditions and contribute to plant salt tolerance

Species	Protein	Function in salt tolerance	References
Oryza sativa	Protein kinases	Signal transduction	Vialaret et al. (2014)
Kandelia candel	14-3-3-like protein	H^+-ATPase regulator	Wang et al. (2014)
Arabidopsis thaliana	Na^+/H^+ antiporter	Na^+ extrusion	Mansour (2014)
Porteresia coarctata	Cellulose synthase	Cell wall synthesis	Sengupta and Majumder (2009)
Oryza sativa	14-3-3 protein	H^+-ATPase regulator	Malakshah et al. (2007)
Oryza sativa	OsRPK1	Signal transduction	Cheng et al. (2009)
Puccinella tenuiflora	PutPMP3	Plasma membrane hyperpolarization reserving	Chang-Qing et al. (2008)
Musa paradisiaca	Dehydrin	Plasma membrane protectant	Shekhawat et al. (2011)
Fragaria ananassa	Osmotin	Plasma membrane protectant	Mansour et al. (2015)
Arabidopsis thaliana	AtLTL1	Releasing fatty acids from plasma membrane, and fatty acids have a role as signaling molecules	Naranjo et al. (2006)
Nicotiana tabacum	14-3-3 protein	ABA signaling pathway in salt adaption through regulating ABA biosynthesis, downstream of ABA signaling, stomatal closure, antioxidant defense	Zhang et al. (2018)
Oryza sativa	LEA (HVA7) protein	Protecting macromolecules	Xu et al. (1996)
Kandelia candel	14-3-3-like protein	Plasma membrane H^+-ATPase regulator and involved in signal transduction	Wang et al. (2014)

macromolecules from damages caused by ion toxicity. In support, HVA7, a LEA from barley, when transferred to rice, confers water and salt stress tolerance (Xu et al. 1996). Another possible role of salinity-induced changes in the PM polypeptides is that these new proteins may modulate specific molecular interactions between lipids and proteins and between proteins themselves (Mansour 2013; Mansour et al. 2015). Such effects play diverse roles in subcellular targeting, protein-protein interactions, and signaling, thus regulating the PM transport systems and in turn plant ion homeostasis under saline conditions (Mansour et al. 2015). Furthermore, some of the PM responsive proteins have been found to participate in signal transduction (Ghosh and Xu 2014) and also act as chaperone proteins that repair and protect the PM under stress conditions (Malakshah et al. 2007; Nouri and Komatsu 2010).

The finding of Chang-Qing et al. (2008) demonstrated that salt stress induced two PM protein three genes (PutPMP3) from *Puccinellia tenuiflora*, which have been found to be involved in tolerance to salinity. Second, Malakshah et al. (2007) similarly identified eight PM proteins that responded to salt stress including 14-3-3 proteins, which are well known as H^+-ATPase regulators and also suggested to be involved in pH regulation in cells under salt stress conditions. Further support is the result of Wang et al. (2014) who report that one upregulated protein under salinity is 14-3-3-like protein, which was also proposed to function as PM H^+-ATPase regulator and involved in signal transduction. Moreover, one PM protein that was upregulated under salt stress is cellulose synthase which was proposed to contribute to adaptation of wild halophytic rice to high salinity (Sengupta and Majumder 2009). Cheng et al. (2009) identified 18 salt-responsive PM proteins in rice roots, which were involved in membrane stabilization, ion homeostasis, and signal transduction. The receptor proteins identified on the PM are known to sense various environmental stimuli and transduce them to downstream intracellular signaling networks (Cheng et al. 2009). It is important to mention that the receptor proteins located on the PM are suggested as potential candidate genes for genetic engineering of improved stress-tolerant crops (Osakabe et al. 2013; Valmonte et al. 2014). Further evidence emphasizing the prime role of the PM proteins in salt tolerance is provided by the finding that overexpression of *AtLTL1* (Li-tolerant lipase 1), a salt-induced gene encoding a GDSL-motif lipase, increases salt tolerance in transgenic *Arabidopsis* plants (Naranjo et al. 2006). AtLTL1 is a protein anchored to the PM and has been suggested to act by releasing fatty acids from membrane lipids. The importance of this protein comes from the fact that fatty acids are essential signaling molecules in plants that are involved in mechanisms of defense against biotic and abiotic stresses (Naranjo et al. 2006). Among the different groups, group 2 of LEA proteins known as dehydrins are particularly interesting and were shown to enhance plant tolerance to various stresses (Hanin et al. 2011). Brini et al. (2007) showed that the expression of the wheat dehydrin DHN-5 in *Arabidopsis* led to an increase in salt and osmotic stress tolerance. In addition, overexpression of a novel dehydrin gene, *MusaDHN-1*, contributes affirmatively to salt stress tolerance in banana (Shekhawat et al. 2011). Dehydrin has been reported to present in vicinity of the PM (Rorat 2006), bind and protect membranes, and also protect against oxidative stress under various stresses (Graether and Boddington 2014). It is also important to note that not all proteins produced in saline conditions are correlated with stress tolerance. For instance, Sobhanian et al. (2011) demonstrate that some of the salt response proteins are probably part of a general stress response and may not be related to salt tolerance. Other reports indicate also that polypeptide pattern of the PM from wheat-sensitive and wheat-tolerant roots (Mansour 1998, 2000) and from sugar beet roots (Yahya et al. 1995) did not change significantly under salinity. Further, the PM total proteins from roots of canola-tolerant and canola-sensitive cultivars did not significantly change in response to salt imposition (Zamani et al. 2010). Moreover, the PM total proteins of the halophyte *Spartina patens* callus were decreased in response to NaCl stress (Wu et al. 2005). This apparent discrepancy could be attributed to the fact that not all salt-responsive proteins are related to salinity adaptation and may be merely a detrimental effect of salinity (Mansour et al. 2015). The aforementioned discussion, however, implies clear evidence for the involvement of salt-responsive PM proteins in adaptation to salinity.

8.6 Cell Wall Modifications and Salt Tolerance

The plant cell wall determines cell size and shape through the mechanical control of cell expansion, which regulates tissue and organ morphology (Le Gall et al. 2015). It also provides stability and protection to the plant and is one of the first layers of the stress stimuli perception. The primary cell wall is composed of cellulose microfibrils, different hemicelluloses, and non-cellulosic neutral polysaccharides embedded in pectin matrix, cross-linked with structural proteins (Le Gall et al. 2015). As the availability of the cell wall ROS determines this cross-linking, it reinforces the strength and rigidity of the cell walls and can be a key component of the plant response to environmental stresses (Le Gall et al. 2015). Since the cell wall is rigid to provide protection and support of cell and also extensible to allow cell growth through its high intracellular turgor pressure, regulation between wall rigidity and flexibility is crucial to regulate the differential growth under stress conditions (Cosgrove 2018). Wang and Ruan (2013) also emphasize this situation as they indicate that to maintain the plant growth, cell wall is either loosened or tightened in response to salinity, which obviously illustrates the complexity of cell wall under abiotic stress. One cell wall constituent playing a central role in modulating cell wall extensibility and thus controlling cell enlargement and expansion is its proteins (Cosgrove 2018). For instance, interaction between hemicellulosic polymers and cellulose fibrils can be modulated by two groups of enzymes, expansins (EXP) and xyloglucan endotransglucosylases/hydrolases (XTH) (Tenheken 2015). The secondary cell walls are formed by massive deposition of cellulose and hemicelluloses within the primary walls. The secondary walls are produced when the cell stops expanding and it is more rigid and can accumulate lignin and suberin (Byrt et al. 2018). Role of cell wall chemical composition and architecture in tolerance to salt stress is supported by the finding that Na^+ displaces Ca^{2+} from its pectin cross-linking, which triggers the cell wall rigidification, the impact which is proposed to be recognized by proteins such as wall-associated kinases (WAKs), feronia, and arabinogalactan proteins (Byrt et al. 2018). More elaboration on the roles of cell wall composition remodeling in plant salt adaptation is discussed in the following sections. Based on the below discussion, cell wall mechanisms for adaptation to salinity are evident although additional mechanisms may be revealed by further future research.

Plant cell wall is a dynamic and crucial component of the plant cell response to abiotic stress, and it is the first barrier between the plant cell and its surrounding environment. It is thus suggested that different stress reception mechanisms most probably localized in the cell wall (Zagorchev et al. 2014), and hence it may be considered as the first line of defense at cellular level against abiotic stress. This cell wall essential role in stress sensing and signal transduction is also confirmed by Le Gall et al. (2015). As salt stress reduces root growth of most crop plants, roots of salt-tolerant plants change the cell wall composition, transport processes, cell size and shape, and root architecture by modulating metabolism, gene expression, and protein activity (Byrt et al. 2018). One effect of high Na^+ on the cell wall is that accumulation of Na^+ in the apoplast increases the interaction between Na^+ and negatively charged sites within cell wall polymers changing their chemical properties and also causes alkalinization of the apoplast resulting in growth reduction (Byrt et al. 2018). It is known that Ca^{2+} cross-links some pectin molecules, while under high salinity Na^+ displaces Ca^{2+} from these binding sites and reduces pectin cross-linking, which subsequently slows down cell elongation (Byrt et al. 2018). This is proposed as a mechanism that contributes to root elongation reduction under high salinity, and therefore externally applied Ca^{2+} ameliorates these effects (LaHaye and Epstein 1971). One way Ca^{2+} ameliorates salt toxicity was through maintaining the composition of the cell wall: increasing the pectin level and thus root growth restoration (An et al. 2014a). The ameliorative effect of Ca^{2+} was more conspicuous in the salt-tolerant soybean cultivar than the salt-sensitive one (An et al. 2014b). Under high Na^+, the chemical composition of the cell wall is changed, and these changes are varied among different cell types and species, which are suggested to reflect different strategies to cope with salt stress (Byrt et al. 2018). This is because compositional variation in different cell wall polymers influences Na^+ binding, K^+ passage, or the binding and passage of Ca^{2+} in roots. Alterations in the cell wall structure in response to saline conditions also include the regulation of cell wall-modifying proteins to allow turgor and drive cellular expansion (Sasidharan et al. 2011). Also, changes in cell wall enzyme activities and contents, which contribute to cell wall adjustment and tolerating salt stress, have been reported (Byrt et al. 2018). These cell wall enzymes include endo-β-1,4-glucanases, EXP, and XTH. It is reported that biosynthesis of cellulose and xyloglucan is induced in most tolerant species which was associated with an upregulation of the expression of genes encoding EXP and XTH (Tenheken 2015). It seems that these modifications might be a part of a long-term cell wall remodeling strategy for the plant adaptation to high salinity.

Cell wall constituents such as structural proteins and hemicellulosic and phenolic residues can be cross-linked by the action of cell wall peroxidases (Tenheken 2015). Higher transcript levels for peroxidase in cowpea were attributed to the inhibition of root growth under salinity (Maia et al. 2013). When one isoform (*SWPA4*) for peroxidase genes from sweet potato was overexpressed in tobacco plants, highly elevated peroxidase enzyme activity in the cell wall was found to result in transgenic plants which were more tolerant against salt (Kim et al. 2008). In these works, peroxidase is involved in modifying the cell wall and subsequently enhances salt tolerance, which highlights the role of the cell wall in salt tolerance mechanism. Beside the increase in peroxidase activity, if ROS formation is occurred in the same tissue, cross-linking of cell wall components might strengthen the mechanical properties of the wall (Tenheken 2015), which might be beneficial for plant cells to better withstand the changes in turgor pressure of osmotic stress. Moreover, treatment of maize root with NaCl increased XTH which may cleave and reform the bonds between xyloglucan chains to regulate cell wall rigidity (Choi et al. 2011). It is also reported that enhanced salt tolerance was associated with the overexpression of *XTH* gene in tomato (Choi et al. 2011). The importance of that comes from the fact that the loosening of cell wall polysaccharides is very important under salt stress (as a reduction in the rate of cell elongation is one consequence of salt stress) to maintain the possibility for cells and organs to expand, and hence differential transcription of genes regulating the rate of cell wall loosening and stiffening is consistently observed in response to salt stress (Byrt et al. 2018). That is, transcriptional changes occur for genes linked to cell wall extensibility and lignin and suberin synthesis, those associated with the modification of cell wall polysaccharides such as XTH. There are many types of root cell wall proteins such as phosphatase proteins, WAK proteins, hydroxyproline-rich glycoproteins, arabinogalactan proteins, glycine-rich proteins, and proline-rich proteins. The roles of these proteins include salt stress sensing, maintenance of cell wall integrity, and transport of Na^+ via vesicle trafficking between the apoplast and the vacuole in saline conditions. In addition, it is worth mentioning that one beneficial role of ROS, such as hydroxyl radicals, is based on their capability to cleave sugar bonds in plant polysaccharides, which causes a cell wall loosening similar to the action of classical loosening enzymes like expansins or xyloglucan-modifying enzymes (Tenheken 2015). Also, apoplastic ROS and Ca^{2+} are implicated in rapid systemic signaling and are involved in the relationship between cell wall changes and plasma membrane function in response to salt stress (Tenheken 2015). Moreover, ROS production in the apoplast is required also for lignin formation. Based on the above discussion, cell wall architecture appears to be a crucial player in the response of plants to salt stress.

Another cell wall constituent which plays a role in salt adaptation is pectin. Changes in cell wall pectin content have been reported in response to high salinity, and higher content is associated with increased tolerance to salinity (An et al. 2014b). Although greater content of pectin can bind a large amount of water molecules and provide a hydrated medium around plasma membranes, this beneficial effect of pectin is argued by Byrt et al. (2018). In addition, some type of pectin (in particular rhamnogalacturonan II) is important in root cell elongation in higher plants, and supporting to that, overexpressing genes controlling programmed cell death in rice transgenes modulates pectin deposition and enhanced salt tolerance relative to control plants (Hoang et al. 2015). On the other hand, lower pectin content might cause a reduced flow of water into cells and hence slow down cell elongation and transpiration rates (Byrt et al. 2018). In addition to cell wall pectin, salt imposition also changes synthesis and structure of cellulose, and the function of proteins involved in stabilizing cellulose synthesis complexes has been shown to be important in coping salt stress, and plants deficient in cellulose are therefore more sensitive to saline conditions (Endler et al. 2015). Another modification on the walls of specific cells is that suberization of the endodermis and/or developing a second diffusion barrier, exodermis, occurs in response to salt stress (Byrt et al. 2018), and these walls have massive amounts of lignin, suberin, cell wall proteins, and carbohydrates. These modifications have been suggested to influence the movement of solutes and water and help prevent water loss and the entry of excessive Na^+ and Cl^- into cells or the xylem (Byrt et al. 2018), which contributes to tolerance to salt stress.

8.7 Conclusion and Future Directions

The physiological and biochemical mechanisms of salt tolerance can be summarized as follows: (1) ion homeostasis and compartmentalization through primarily activation of the PM and tonoplast transport systems; (2) biosynthesis of osmoprotectants and compatible solutes and their contribution to osmotic adjustment and protective functions; (3) activation of antioxidant enzymes and synthesis of antioxidant compounds for ROS detoxification, signaling, and protection; (4) PM lipids and proteins remodeling, thereby maintaining membrane integrity, transport system activity, and signaling; and (5) cell wall modifications and in order to regulate cell enlargement and expansion, cell wall integrity, and salt stress sensing signaling transduction. It is important to mention that information on some issues of the discussed adaptive cellular mechanisms of salt tolerance is not consistent and conclusive, and therefore more realistic detailed studies are necessary and required for unmasking the disagreement and full assessment of cellular traits associated with plant salt tolerance. For example, although the cell wall plays an important role in salt tolerance, the cell wall modifications are different in glycophytes and halophytes. In glycophytes, there is an increase in primary and secondary cell walls, whereas in halophytes only the secondary cell wall structure is altered (Le Gall et al. 2015). We believe that a large array of mechanisms work in tandem with each other to tolerate the impacts associated with high salinity and eventually result in salt-tolerant plants that can properly continue their life cycle and yield. Our contention is based on the fact that plant response to salinity is complex and involves multiple genes implicated in distinct or overlapping regulatory pathways. Thus, adopting one cellular mechanism and engineering of a single downstream effector gene may be efficient in some circumstances but usually might have limited success when one considers the field combined conditions and our ultimate goal, the increasing yield. This is because most works are typically being studied under controlled growth conditions. However, the field environment is very different and often involving simultaneous exposure of plants to more than one abiotic and/or biotic stress condition. Alternatively, a variety of traits or strategies overlapping in a complementary orchestrate to control cell function and development under stress conditions has to be adopted. Developing salt-tolerant genotypes must be therefore generated by combining different strategies involved in enhancing tolerance to salinity, i.e., transgene pyramiding must be considered as a promising feasible strategy to improve salt tolerance and productivity in crop plants. Also, testing these physiological traits under field conditions will shed light on their exact contribution to salt tolerance mechanism and in turn in increasing plant yield.

References

AbdElgawad H, Zinta G, Hegab MM, Pandey R, Asard H, Abuelsoud W (2016) High salinity induces different oxidative stress and antioxidant responses in maize seedlings organs. Front Plant Sci 7:276

Abdin MZ, Kiran U, Alam A (2011) Analysis of osmotin, a PR protein as metabolic modulator in plants. Bioinformatics 5:336–340

Acosta-Motos JR, Ortuno MF, Bernal-Vicente A, Diaz-Vivancos P, Sanchez-Blanco MJ, Antonio Hernandez J (2017) Plant responses to salt stress: adaptive mechanisms. Agronomy 7:18

Afzal Z, Howton TC, Sun Y, Mukhtar MS (2016) The roles of aquaporins in plant stress responses. J Dev Biol 4:9

Aghaei KA, Ehsanpour A, Komatsu S (2009) Potato responds to salt stress by increased activity of antioxidant enzymes. J Integr Plant Biol 51:1095–1103

Ali A, Yun D (2017) Salt stress tolerance; what do we learn from halophytes? J Plant Biol 60:431–439

Ali A, Park HC, Aman R, Ali Z, Yun DJ (2013) Role of HKT1 in *Thellungiella salsuginea*, a model extremophile plant. Plant Signal Behav 8:e25196

Almeida DM, Oliveira MM, Saibo NJM (2017) Regulation of Na^+ and K^+ homeostasis in plants: towards improved salt stress tolerance in crop plants. Gene Mol Biol 40:326–345

Alvarez-Pizarro JC, Gomes-Filho E, de Lacerda CF, Alencar NM, Prisco JT (2009) Salt- induced changes on H^+-ATPase activity, sterol and phospholipid content and lipid peroxidation of root plasma membrane from dwarf-cashew activity and lipid composition of plasma membrane vesicles isolated from roots (*Anacardium occidentale* L.) seedlings. Plant Growth Regul 59:125–135

Amjad M, Akhtar SS, Yang A, Akhtar J, Jacobsen SE (2015) Antioxidative response of quinoa exposed to iso-osmotic, ionic and non-ionic salt stress. J Agron Crop Sci 201:452–460

An P, Li X, Zheng Y, Eneji AE, Inanaga S (2014a) Calcium effects on root cell wall composition and ion contents in two soybean cultivars under salinity stress. Can J Plant Sci 94:733–740

An P, Li X, Zheng Y, Matsuura A, Abe J, Eneji AE, Inanagas S (2014b) Effects of NaCl on root growth and cell wall composition of two soybean cultivars with contrasting salt tolerance. J Agron Crop Sci 200:212–218

Anjaneyulu E, Reddy PS, Sunita MS, Kavi Kishor PB, Meriga B (2014) Salt tolerance and activity of antioxidative enzymes of transgenic finger millet overexpressing a vacuolar H^+-pyrophosphatase gene (*SbVPPase*) from *Sorghum bicolor*. J Plant Physiol 171:789–798

Annunziata MG, Ciarmiello LF, Woodrow P, Maximova E, Fuggi A, Carillo P (2017) Durum wheat roots adapt to salinity remodeling the cellular content of nitrogen metabolites and sucrose. Front Plant Sci 7:2035

Araujo SS, Beebe S, Crespi M, Delbreil B, Gonzalez EM, Gruber V, Lejeune-Henaut I, Link W, Monteros MJ, Prats E, Rao I, Vadez V, Patto MCV (2015) Abiotic stress responses in legumes: strategies used to cope with environmental challenges. Crit Rev Plant Sci 34:237–280

Ashraf M, Foolad MR (2007) Roles of glycinebetaine and proline in improving abiotic stress resistance. Environ Exp Bot 59:206–216

Assaha DVM, Mekawy AMM, Liu L, Noori MS, Kokulan KS, Ueda A, Nagaoka T, Saneoka H (2017a) Na^+ retention in the root is a key adaptive mechanism to low and high salinity in the glycophyte, *Talinum paniculatum* (Jacq.) Gaertn. (Portulacaceae). J Agron Crop Sci 203:56–67

Assaha DVM, Ueda A, Saneoka H, Al-Yahyai R, Yaish MW (2017b) The role of Na^+ and K^+ transporters in salt stress adaptation in glycophytes. Front Plant Sci 8:509

Bao A, Du B, Touil L, Kang P, Wang Q, Wang S (2016) Co-expression of tonoplast Cation/H^+ antiporter and H^+-pyrophosphatase from xerophyte *Zygophyllum xanthoxylum* improves alfalfa plant growth under salinity, drought and field conditions. Plant Biotechnol J 14:964–975

Barkla BJ, Rhodes T, Tran KT, Wijesinghege C, Larkin JC, Dassanayake M (2018) Making epidermal bladder cells bigger: developmental- and salinity-induced endopolyploidy in a model halophyte. Plant Physiol 177:615–632

Bazrafshan AH, Ehsanzadeh P (2016) Evidence for differential lipid peroxidation and antioxidant enzyme activities in *Sesamum indicum* L. genotypes under NaCl salinity. J Agric Sci Technol 18:207–222

Bhaedwaj R, Sharma I, Kanwar M, Sharma R, Handa N, Kaur H (2013) Aquaporins: role under salt stress in plants. In: Ahmad P, Azooz MM, Prasad MNV (eds) Ecophysiology and responses of plants under salt stress. Springer, New York, pp 213–248

Bhutta WM (2011) Antioxidant activity of enzymatic system of two different wheat (*Triticum aestivum* L.) cultivars growing under salt stress. Plant Soil Environ 57:101–107

Bose J, Xie Y, Shen W, Shabala S (2013) Haem oxygenase modifies salinity tolerance in *Arabidopsis* by controlling K^+ retention via regulation of the plasma membrane H^+-ATPase and by altering SOS1 transcript levels in root. J Exp Bot 64:471–481

Bose J, Rodrigo-Moreno A, Shabala S (2014) ROS homeostasis in halophytes in the context of salinity stress tolerance. J Exp Bot 65:1241–1257

Bose J, Rodrigo-Moreno A, Lai D, Xie Y, Shen W, Shabala S (2015) Rapid regulation of the plasma membrane H^+-ATPase activity is essential to salinity tolerance in two halophyte species, *Atriplex lentiformis* and *Chenopodium quinoa*. Ann Bot 115:481–494

Brini F, Hanin M, Lumrerous V, Amara I, Khoudi H, Masmoudi K (2007) Overexpression of wheat dehydrin DHN-5 enhances tolerance to salt and osmotic stress in *Arabidopsis thaliana*. Plant Cell Rep 26:2017–2026

Byrt CS, Munns R, Burton RA, Gilliham M, Wege S (2018) Root cell wall solutions for crop plants in saline soils. Plant Sci 269:47–55

Chang-Qing Z, Shunsaku N, Shenkui L (2008) Characterization of two plasma membrane protein 3 genes (*PutPMP3*) from the alkali grass, *Puccinellia tenuiflora*, and functional comparison of the rice homologues, OsLti6a/b from rice. BMB Rep 41:448–454

Chen Q, Zhang M, Shen S (2011) Effect of salt on malondialdehyde and antioxidant enzymes in seedling roots of Jerusalem artichoke (*Helianthus tuberosus* L.). Acta Physiol Plant 33:273–278

Cheng Y, Qi Y, Zhu Q, Chen X, Wang N, Zhao X (2009) New changes in the plasma-membrane-associated proteome of rice roots under salt stress. Proteomics 9:3100–3114

Choi JY, Seo YS, Kim SJ, Kim WT, Shin JS (2011) Constitutive expression of CaXTH3 a hot pepper xyloglucan endotransglucosylase/hydrolase, enhanced tolerance to salt and drought stresses without phenotypic defects in tomato plants (*Solanum lycopersicum* cv. Dotaerang). Plant Cell Rep 30:867–877

Cook DT, Ros R, Burden RS, James CS (1993) A comparison of the influence of sterols on the specific activity of the H^+-ATPase in isolated plasma membrane vesicles from oat, rye and rice shoots. Physiol Plant 88:397–402

Cosgrove DJ (2018) Diffuse growth of plant cell walls. Plant Physiol 176:16–27

Cuin TA, Shabala S (2008) Compatible solutes mitigate damaging effects of salt stress by reducing the impact of stress-induced reactive oxygen species. Plant Signal Behav 3:207–208

Cuin TA, Tian Y, Betts SA, Chalmandrier R, Shabala S (2009) Ionic relations and osmotic adjustment in durum and bread wheat under saline conditions. Funct Plant Biol 36:1110–1119

D'Amelia L, Dell'Aversana E, Woodrow P, Ciarmiello LF, Carillo P (2018) Metabolomics for crop improvement against salinity stress. In: Kumar V, Wani SH, Suprasanna P, Tran LP (eds) Salinity responses and tolerance in plants, vol 2. Springer, Cham, pp 267–288

Daneshmand F, Arvin MJ, Kalantari KM (2010) Physiological responses to NaCl stress in three wild species of potato in vitro. Acta Physiol Plant 32:91–101

Das SK, Patra JK, Thatoi H (2016) Antioxidative response to abiotic and biotic stresses in mangrove plants: a review. Int Rev Hydrobiol 101:3–19

Diray-Arce J, Gul B, Khan MA, Nielsen B (2016) Halophyte transcriptomics: understanding mechanisms of salinity tolerance. In: Khan MA, Ozturk M, Ahmed MZ (eds) Halophytes for food security in dry lands. Academic, London, pp 157–175

Endler A, Kesten C, Schneider R, Zhang Y, Ivakov T, Froehlich A, Persson S (2015) A mechanism for sustained cellulose synthesis during salt stress. Cell 162:1353–1364

Falleh H, Jalleli I, Ksouri R, Boulaaba M, Guyot S, Magné C, Abdelly C (2012) Effect of salt treatment on phenolic compounds and antioxidant activity of two *Mesembryanthemum edule* provenances. Plant Physiol Biochem 52:1–8

Fan Y, Zhu M, Shabala S, Li CD, Johnson P, Zhou MX (2014) Antioxidant activity in salt-stressed barley leaves: evaluating time- and age-dependence and suitability for the use as a biochemical marker in breeding programs. J Agron Crop Sci 200:261–272

Fan W, Denga G, Wang H, Zhang H, Zhang P (2015) Elevated compartmentalization of Na^+ into vacuoles improves salt and cold stress tolerance in sweet potato (*Ipomoea batatas*). Physiol Plant 154:560–571

Fercha A (2017) Understanding seed priming in wheat: a proteomics approach. Lambert Academic, Saarbrücken

Flowers TJ (2004) Improving crop salt tolerance. J Exp Bot 55:307–319

Flowers TJ, Colmer TD (2008) Salinity tolerance in halophytes. New Phytol 179:945–963

Flowers TJ, Colmer TD (2015) Plant salt tolerance: adaptations in halophytes. Ann Bot 115:327–331

Flowers TJ, Munns R, Colmer TD (2015) Sodium chloride toxicity and the cellular basis of salt tolerance in halophytes. Ann Bot 115:419–431

Foyer CH, Noctor G (2005) Oxidant and antioxidant signalling in plants: a re-evaluation of the concept of oxidative stress in a physiological context. Plant Cell Environ 28:1056–1071

Garg AK, Kim J, Owens TG, Ranwala AP, Choi YD, Kochian LV, Wu RJ (2002) Trehalose accumulation in rice plants confers high tolerance levels to different abiotic stresses. PNAS 99:15898–15903

Ghosh D, Xu J (2014) Abiotic stress responses in plant roots: a proteomics perspective. Front Plant Sci 5:6

Gill SS, Tuteja N (2010) Reactive oxygen species and antioxidant machinery in abiotic stress tolerance in crop plants. Plant Physiol Biochem 48:909–930

Gilliham M, Able JA, Roy SJ (2017) Translating knowledge about abiotic stress tolerance to breeding programmes. Plant J 90:898–917

Gouiaa S, Khoudi H (2015) Co-expression of vacuolar Na^+/H^+ antiporter and H^+-pyrophosphatase with an IRES-mediated dicistronic vector improves salinity tolerance and enhances potassium biofortification of tomato. Phytochemistry 117:537–546

Graether SP, Boddington KF (2014) Disorder and function: a review of the dehydrin protein family. Front Plant Sci 5:576

Gutla PV, Boccaccio A, De Angeli A (2012) Modulation of plant TPC channels by polyunsaturated fatty acids. J Exp Bot 63:6187–6197

Han Y, Yin S, Huang L, Wu X, Zeng J, Xiaohui L, Qiu L, Munns R, Chen Z, Zhang GP (2018) A sodium transporter HvHKT1;1 confers salt tolerance in barley via regulating tissue and cell ion homeostasis. Plant Cell Physiol 59:1976–1989

Hanin M, Brini F, Ebel C, Toda Y, Masmoudi K (2011) Plant dehydrins and stress tolerance, versatile proteins for complex mechanisms. Plant Signal Behav 6:1503–1509

Hanin M, Ebel C, Ngom M, Laplaze L, Masmoudi K (2016) New insights on plant salt tolerance mechanisms and their potential use for breeding. Front Plant Sci 7:1787

Hasanuzzaman M, Nahar K, Rahman A (2017) Approaches to enhance salt stress tolerance in wheat. In: Wanyera R, Owuoche O (eds) Wheat improvement, management and utilization. InTech, Croatia, pp 151–187

Hasegawa PM, Bressan RA, Zhu JK, Bohnert HJ (2000) Plant cellular and molecular responses to high salinity. Annu Rev Plant Physiol Plant Mol Biol 51:463–499

Hernández JA, Jiménez A, Mullineaux PM, Sevilla F (2000) Tolerance of pea (*Pisum sativum* L.) to long-term salt stress is associated with induction of antioxidant defenses. Plant Cell Environ 23:853–862

Hoang TM, Moghaddam L, Williams B, Dale J, Mundree SG (2015) Development of salinity tolerance in rice by constitutive-overexpression of genes involved in the regulation of programmed cell death. Front Plant Sci 6:175

Hong JK, Hwang BK (2009) The promoter of the pepper pathogen-induced membrane protein gene CaPIMP1 mediates environmental stress responses in plants. Planta 229:249–259

Horei T, Karahara I, Katsuhara M (2012) Salinity tolerance mechanisms in glycophytes and overview central focus on rice plants. Rice 5:11–18

Hou Q, Ufer G, Bartels D (2016) Lipid signaling in plant responses to abiotic stress. Plant Cell Environ 39:1029–1048

Hu W, Yuan Q, Wang Y, Cai R, Deng X (2012) Overexpression of a wheat aquaporin gene, TaAQP8, enhances salt tolerance in transgenic tobacco. Plant Cell Physiol 53:2127–2141

Hu L, Xiang L, Li S, Zou Z, Hu X (2016) Beneficial role of spermidine in chlorophyll metabolism and D1 protein content in tomato seedlings under salinity–alkalinity stress. Physiol Plant 156:468–477

Jaarsma R, de Boer AH (2018) Salinity tolerance of two potato cultivars (*Solanum tuberosum*) correlates with differences in vacuolar transport activity. Front Pant Sci 9:737

Janicki-Russak M (2011) Plant plasma membrane H^+-ATPase in adaptation of plants to abiotic stresses. In: Shanker AK, Venkateswarlu B (eds) Abiotic stress response in plants – physiological, biochemical and genetic perspectives. InTech, Croatia, pp 179–218

Kamiab F, Talaie A, Khezri M, Javanshah A (2014) Exogenous application of free polyamines enhances salt tolerance of pistachio (*Pistacia vera* L.) seedlings. Plant Growth Regul 72:257–268

Kavitha PG, Miller AJ, Mathew MK, Maathuis FJM (2012) Rice cultivars with differing salt tolerance contain similar cation channels in their root cells. J Exp Bot 63:3289–3296

Kerkeb L, Donaire JP, Rodriguez-Rosale MP (2001) Plasma membrane H^+-ATPase activity is involved in adaptation of tomato to NaCl. Physiol Plant 111:483–490

Khan MS, Ahmad D, Khan MA (2015a) Utilization of genes encoding osmoprotectants in transgenic plants for enhanced abiotic stress tolerance. Electron J Biotechnol 8:257–266

Khan K, Agarwal P, Shanware A, Sane VA (2015b) Heterologous expression of two *Jatropha* aquaporins imparts drought and salt tolerance and improves seed viability in transgenic *Arabidopsis thaliana*. PLoS One 10:e0128866

Kiani-Pouya A (2015) Changes in activities of antioxidant enzymes and photosynthetic attributes in triticale (×Triticosecale Wittmack) genotypes in response to long-term salt stress at two distinct growth stages. Acta Physiol Plant 37:1–11

Kibinza S, Bazin J, Bailly C, Corbineau F, El-Maarouf-Bouteau H (2011) Catalase is a key enzyme in seed recovery from ageing during priming. Plant Sci 181:309–315

Kim YH, Kim CY, Song WK, Park DS, Kwon SY, Lee HS (2008) Overexpression of sweet potato swpa4 peroxidase results in increased hydrogen peroxide production and enhances stress tolerance in tobacco. Planta 227:867–881

Kirch HH, Vera-Estrella R, Golldack D, Quigley F, Michalowski CB, Barkla BJ, Bohnert HJ (2000) Expression of water channel proteins in *Mesembryanthemum crystallinum*. Plant Physiol 123:111–124

Kosová K, Prášil IT, Vítámvás P (2013) Protein contribution to plant salinity response and tolerance acquisition. Int J Mol Sci 14:6757–6789

Koyro H, Ahmad P, Geissler N (2012) Abiotic stress responses in plants: an overview. In: Ahmad P, Prasad MNV (eds) Environmental adaptations and stress tolerance of plants in the era of climate change. Springer, Berlin, pp 1–28

Kumar V, Shriram V, Kavi Kishor PB, Jawali N, Shitole MG (2010) Enhanced proline accumulation and salt stress tolerance of transgenic *indica* rice by over-expressing *P5CSF129A* gene. Plant Biotechnol Rep 4:37–48

LaHaye PA, Epstein E (1971) Calcium and salt toleration by bean plants. Physiol Plant 25:213–218

Lai S, Lai M, Lee R, Chen Y, Yen HE (2014) Transgenic Arabidopsis expressing osmolyte glycine betaine synthesizing enzymes from halophilic methanogen promote tolerance to drought and salt stress. Plant Mol Biol 85:429–441

Le Gall H, Philippe F, Domon J, Gillet F, Pelloux J, Rayon C (2015) Cell wall metabolism in response to abiotic stress. Plants 4:112–166

Lee G, Carrowa RN, Duncana RR, Eitemanb ME, Rieger MW (2008) Synthesis of organic osmolytes and salt tolerance mechanisms in *Paspalum vaginatum*. Eviron Exp Bot 63:19–27
Lin K, Pu SF (2010) Tissue- and genotype-specific ascorbate peroxidase expression in sweet potato in response to salt stress. Biol Plant 54:664–670
Lin F, Liu Z, Shi Q, Wang X, Wei M, Yang F (2012) Exogenous nitric oxide (NO) increased antioxidant capacity of cucumber hypocotyl and radicle under salt stress. Sci Hortic 142:118–127
Lin K, Jwo W, Chandrika NNP, Wu T, Lai M, Wang C, Hong C (2016) A rice mutant defective in antioxidant-defense system and sodium homeostasis possesses increased sensitivity to salt stress. Biol Plant 60:86–94
López-Gómez E, Sanjuán MA, Diaz-Vivancos P, Mataix-Beneyto J, García-Legaz MF, Hernández JA (2007) Effect of salinity and rootstocks on antioxidant systems of loquat plants (*Eriobotrya japonica* Lindl.): response to supplementary boron addition. Environ Exp Bot 160:151–158
Lopez-Perez L, Martinez-Ballesta MC, Maurel C, Carvajal M (2009) Changes in plasma membrane lipids, aquaporins and proton pump of broccoli roots, as an adaptation mechanism to salinity. Phytochemistry 70:492–500
Maia JM, Voigt EL, Ferreira-Silva SLA, Fontenele V, Macedo CEC, Silveira JAG (2013) Differences in cowpea root growth triggered by salinity and dehydration are associated with oxidative modulation involving types I and III peroxidases and apoplastic ascorbate. J Plant Growth Regul 32:376–387
Malakshah SN, Rezaei MH, Heidari M (2007) Proteomics reveals new salt responsive proteins associated with rice plasma membrane. Biosci Biotechnol Biochem 71:2144–2154
Mallik S, Nayak M, Sahu BB, Panigrahi AK, Shaw BP (2011) Response of antioxidant enzymes to high NaCl concentration in different salt-tolerant plants. Biol Plant 55:191–195
Maksimovic JD, Zhang J, Zeng F, Zivanovic B, Shabala L, Zhou M, Shabala S (2013) Linking oxidative and salinity stress tolerance in barley: can root antioxidant enzyme activity be used as a measure of stress tolerance? Plant Soil 365:141–155
Mansour MMF (1998) Protection of plasma membrane of onion epidermal cells by glycine betaine and proline against NaCl stress. Plant Physiol Biochem 36:767–772
Mansour MMF (2000) Nitrogen containing compounds and adaptation of plants to salinity stress. Biol Plant 43:491–500
Mansour MMF (2013) Plasma membrane permeability as an indicator of salt tolerance in plants. Biol Plant 57:1–10
Mansour MMF (2014) The plasma membrane transport systems and adaptation to salinity. J Plant Physiol 171:1787–1800
Mansour MMF, Ali EF (2017a) Glycinebetaine in saline conditions: an assessment of the current state of knowledge. Acta Physiol Plant 39:56
Mansour MMF, Ali EF (2017b) Evaluation of proline functions in saline conditions. Phytochemistry 140:52–68
Mansour MMF, Salama KHA (2004) Cellular basis of salinity tolerance in plant. Environ Exp Bot 52:113–122
Mansour MMF, Stadelmann EJ (1994) NaCl-induced changes in protoplasmic characteristics of *Hordeum vulgare* cultivars differing in salt tolerance. Physiol Plant 91:389–394
Mansour MMF, Lee-Stadelmann OY, Stadelmann EJ (1993) Solute potential and cytoplasmic viscosity in *Triticum aestivum* and *Hordeum vulgare* under salt stress. A comparison of salt resistant and salt sensitive lines and cultivars. J Plant Physiol 142:623–628
Mansour MMF, Salama KHA, Allam HYH (2015) Role of the plasma membrane in saline conditions: lipids and proteins. Bot Rev 81:416–451
Mansour MMF, Ali EF, Salama KHA (2019) Does seed priming play a role in regulating reactive oxygen species under saline conditions? In: Hasanuzzaman M, Fotopoulos V, Nahar K, Fujita M (eds) Reactive oxygen, nitrogen and sulfur species in plants: production, metabolism, signaling and defense mechanisms. Wiley and Sons, New Jersey, USA, 437–488

Martinez-Ballesta MC, Silva C, Lopez-Berenguer C, Cabaoero FJ, Carvajal M (2006) plant aquaporins: new perspectives on water and nutrient uptake in saline environment. Plant Biol 57:1–10

Maswada HF, Djanaguiraman M, Prasad PVV (2018) Response of photosynthetic performance, water relations and osmotic adjustment to salinity acclimation in two wheat cultivars. Acta Physiol Plant 40:105

Meijer HJG, van Himbergen JAJ, Musgrave A, Munnik T (2017) Acclimation to osmotic stress-activated lipid signaling pathways in *Chlamydomonas*. Phytochemistry 135:64–72

Meringer MV, Villasuso AL, Margutti MP, Usorach J, Racagni GE (2016) Saline and osmotic stresses stimulate PLD/diacylglycerol kinase activities and increase the level of phosphatidic acid and proline in barley roots. Environ Exp Bot 128:69–78

Mittler R, Vanderauwera S, Suzuki N, Miller G, Tognetti VB, Gollery M, van Breusegem F (2011) ROS signaling: the new wave? Trends Plant Sci 16:300–309

Mittova V, Tal M, Volokita M, Guy M (2002) Salt stress induces up-regulation of an efficient chloroplast antioxidant system in the salt-tolerant wild tomato species *Lycopersicon pennellii* but not in the cultivated species. Physiol Plant 115:393–400

Mittova V, Tal M, Volokita M, Guy M (2003) Up-regulation of the leaf mitochondrial and peroxisomal antioxidative systems in response to salt-induced oxidative stress in the wild salt-tolerant tomato species *Lycopersicon pennellii*. Plant Cell Environ 26:845–856

Mottaleb SA, Rodriguez-Navarro A, Haro R (2013) Knockouts of *Physcomitrella patens* CHX1 and CHX2 transporters reveal high complexity of potassium homeostasis. Plant Cell Physiol 54:1455–1468

Munns R, Termaat A (1986) Whole plant response to salinity. Aust J Plant Physiol 13:143–160

Munns R, Tester M (2008) Mechanisms of salinity tolerance. Annu Rev Plant Biol 59:651–681

Munns R, James RA, Xu B, Athman A, Conn SJ, Jordans C, Byrt CS, Hare RA, Tyerman SD, Tester M, Plett D, Gilliham M (2012) Wheat grain yield on saline soils is improved by an ancestral Na^+ transporter gene. Nat Biotechnol 30:360–364

Nahar K, Hasanuzzaman M, Rahman A, Alam MM, Mahmud JA, Suzuki T, Fujita M (2016) Polyamines confer salt tolerance in mung bean (*Vigna radiata* L.) by reducing sodium uptake, improving nutrient homeostasis, antioxidant defense, and methylglyoxal detoxification systems. Front Plant Sci 7:1104

Naranjo MA, Forment J, Roldan M (2006) Overexpression of *Arabidopsis thaliana* LTL1, a salt-induced gene encoding a GDSL-motif lipase, increases salt tolerance in yeast and transgenic plants. Plant Cell Environ 29:1890–1900

Nikalje GC, Variyar PS, Joshi MV, Nikam TD, Suprasanna P (2018) Temporal and spatial changes in ion homeostasis, antioxidant defense and accumulation of flavonoids and glycolipid in a halophyte *Sesuvium portulacastrum* (L.). PLoS One 13:e0193394

Niu X, Xiong F, Liu J, Sui Y, Zeng Z, Lu B, Liu Y (2014) Co-expression of ApGSMT and ApDMT promotes biosynthesis of glycine betaine in rice (*Oryza sativa* L.) and enhances salt and cold tolerance. Environ Exp Bot 104:16–25

Noreen Z, Ashraf M, Aisha N (2010) Salt-induced regulation of some key antioxidant enzymes and physio-biochemical phenomena in five diverse cultivars of turnip (*Brassica rapa* L.). J Agron Crop Sci 196:273–285

Nounjan N, Chansongkrow P, Charoensawan V, Siangliw JL, Toojinda T, Chadchawan S, Theerakulpisut P (2018) High performance of photosynthesis and osmotic adjustment are associated with salt tolerance ability in rice carrying drought tolerance QTL: physiological and co-expression network analysis. Front Plant Sci 9:1135

Nouri MZ, Komatsu S (2010) Comparative analysis of soybean plasma membrane proteins under osmotic stress using gel-based and LC MS/MS-based proteomics approaches. Proteomics 10:1930–1945

Okazaki Y, Saito K (2014) Roles of lipids as signaling molecules and mitigators during stress response in plants. Plant J 79:584–596

Osakabe Y, Yamaguchi-Shinozaki K, Shinozaki K, Tran LP (2013) Sensing the environment: key roles of membrane-localized kinases in plant perception and response to abiotic stress. J Exp Bot 64:445–458

Petridis A, Therios I, Samouris G, Tananaki C (2012) Salinity-induced changes in phenolic compounds in leaves and roots of four olive cultivars (*Olea europaea* L.) and their relationship to antioxidant activity. Environ Exp Bot 79:37–43

Plett DC, Moller IS (2010) Na^+ transport in glycophytic plants: what we know and would like to know. Plant Cell Eviron 33:612–626

Pushpalatha G, Subrahmanyam D, Sreenu K, Ram T, Subbarao LV (2013) Effect of salt stress on seedling growth and antioxidant enzymes in two contrasting rice introgression lines. Indian J Plant Physiol 18:360–366

Queiros F, Rodrignes JA, Almeida JM, Almeida DPF, Fidalgo F (2011) Differential responses of the antioxidant defence system and ultrastructure in a salt-adapted potato cell line. Plant Physiol Biochem 49:1410–1419

Rady MOA, Semida WM, Abd El-Mageed TA, Hemidad KA, Rady MM (2018) Up-regulation of antioxidative defense systems by glycine betaine foliar application in onion plants confer tolerance to salinity stress. Sci Hortic 240:614–622

Rahman R, Nahar K, Hasanuzzaman M, Fujita M (2016) Calcium supplementation improves Na^+/K^+ ratio, antioxidant defense and glyoxalase systems in salt-stressed rice seedlings. Front Plant Sci 7:609

Rangani J, Parida AK, Panda A, Kumari A (2016) coordinated changes in antioxidative enzymes protect the photosynthetic machinery from salinity induced oxidative damage and confer salt tolerance in an extreme halophyte *Salvadora persica* L. Front Plant Sci 7:50

Rasool S, Ahmad A, Siddiqi TO, Ahmad P (2013) Changes in growth, lipid peroxidation and some key antioxidant enzymes in chickpea genotypes under salt stress. Acta Physiol Plant 35:1039–1050

Rios JJ, Martinez-Ballesta MC, Ruiz JM, Blasco B, Carvajal M (2017) Silicon-mediated improvement in plant salinity tolerance: the role of aquaporins. Front Plant Sci 8:948

Radic S, Stefanic PP, Lepedus H, Roje M (2013) Salt tolerance of *Centaurea ragusina* L. is associated with efficient osmotic adjustment and increased antioxidative capacity. Environ Exp Bot 87:39–48

Rorat T (2006) Plant dehydrins: tissue location, structure and function. Cell Mol Biol Lett 11:536–556

Roy S, Chakraborty U (2015) Cross-generic studies with rice indicate that ion homeostasis and antioxidant defense is associated with superior salinity tolerance in *Cynodon dactylon* (L.) Pers. Indian J Plant Physiol 20:14–22

Rubio MC, Bustos-Sanmaded P, Clemente RM, Becana M (2009) Effects of salt stress on the expression of antioxidant genes and proteins in the model legume *Lotus japonicus*. New Phytol 181:851–859

Ruelland E, Kravets V, Derevyanchuk M (2014) Role of phospholipid signaling in plant environmental responses. Environ Exp Bot 114:129–143

Russell NJ (1989) Functions of lipids: structural roles and membrane functions. In: Ratledge C, Wilkinson SC (eds) Microbial lipids. Academic, London, pp 279–365

Sah SK, Reddy KR, Li J (2016) Abscisic acid and abiotic stress tolerance in crop plants. Front Plant Sci 7:571

Salama KHA, Mansour MMF, Ali FZM (2007) NaCl-induced changes in plasma membrane lipids and proteins of *Zea mays* L. cultivars differing in their response to salinity. Acta Physiol Plant 29:351–359

Salama KHA, Mansour MMF, Hassan NS (2011) Choline priming improves salt tolerance in wheat (*Triticum aestivum* L.). Aust J Basic Appl Sci 5:126–132

Salama KHA, Mansour MMF, Al-Malawi HA (2015) Glycinebetaine priming improves salt tolerance of wheat. Biologia 70:1334–1339

Sandstrom RP, Cleland RE (1989) Comparison of the lipid composition of oat root and coleoptiles plasma membranes. Plant Physiol 90:1207–1213

Sanoubar R, Cellini A, Veroni AM, Spinelli F, Masia A, Antisari LV, Orsini F, Gianquinto G (2016) Salinity thresholds and genotypic variability of cabbage (*Brassica oleracea* L.) grown under saline stress. J Sci Food Agric 96:319–330

Sarkar RK, Mahata KR, Singh DP (2013) Differential responses of antioxidant system and photosynthetic characteristics in four rice cultivars differing in sensitivity to sodium chloride stress. Acta Physiol Plant 35:2915–2926

Sasidharan R, Voesenek LA, Pierik R (2011) Cell wall modifying proteins mediate plant acclimatization to biotic and abiotic stresses. Crit Rev Plant Sci 30:548–562

Sengupta S, Majumder AL (2009) Insight into the salt tolerance factors of a wild halophytic rice, *Porteresia coarctata*: a physiological and proteomic approach. Planta 229:911–929

Seo DC, Kim MJ, Song JS, Kim HJ, Park CM (2010) Proteolytic processing of an Arabidopsis membrane-bound NAC transcription factor is triggered by cold-induced changes in membrane fluidity. Biochem J 427:359–367

Shabala S (2013) Learning from halophytes: physiological basis and strategies to improve stress tolerance in crops. Ann Bot 112:1209–1221

Shabala S, Mackay A (2011) Ion transport in halophytes. Adv Bot Res 57:152–199

Shabala L, Zhang J, Pottosin I, Bose J, Zhu M, Fuglsang AT, Velarde-Buendia A, Massart A, Hill CB, Roessner U, Bacic A, Wu H, Azzarello E, Pandolfi C, Zhou M, Poschenrieder C, Mancuso S, Shabala S (2016) Cell-type-specific H^+-ATPase activity in root tissues enables K^+ retention and mediates acclimation of barley (*Hordeum vulgare*) to salinity stress. Plant Physiol 172:2445–2458

Shabaz M, Mushtaq Z, Andaz F, Masood A (2013) Does proline application ameliorate adverse effects of salt stress on growth, ions and photosynthetic ability of eggplant (*Solanum melongena* L.)? Sci Hortic 164:507–511

Shekhawat UKS, Srinivas L, Ganapathi TR (2011) Musa DHN-1, a novel multiple stress-inducible SK3-type dehydrin gene, contributes affirmatively to drought-and salt-stress tolerance in banana. Planta 234:915–932

Shen P, Wang R, Zhang W (2011) Rice phospholipase Dα is involved in salt tolerance by the mediation of H^+-ATPase activity and transcription. J Integr Plant Biol 53:289–299

Silva P, Gerós H (2009) Regulation by salt of vacuolar H^+-ATPase and H^+-pyrophosphatase activities and Na^+/H^+ exchange. Plant Signal Behav 4:718–726

Silva P, Facanha AR, Tavares RM, Gerós H (2010) Role of tonoplast proton pumps and Na^+/H^+ antiport system in salt tolerance of *Populus euphratica* Oliv. J Plant Growth Regul 29:23–34

Sobhanian H, Aghaei K, Komatsu S (2011) Changes in the plant proteome resulting from salt stress: toward the creation of salt-tolerant crops? J Proteomics 7:1323–1337

Sorkheh K, Shiran B, Rouhi V, Khodambashi M, Sofo A (2012) Salt stress induction of some key antioxidant enzymes and metabolites in eight Iranian wild almond species. Acta Physiol Plant 34:203–213

Souid A, Gabriele M, Longo V, Pucci L, Bellani L, Smaoui A, Abdelly C, Hamed K (2016) Salt tolerance of the halophyte Limonium delicatulum is more associated with antioxidant enzyme activities than phenolic compounds. Funct Plant Biol 43:607–619

Suga S, Komatsu S, Maeshima M (2002) Aquaporin isoforms responsive to salt and water stresses and phytohormones in radish seedlings. Plant Cell Physiol 43:1229–1237

Sunarpi HT, Motoda J, Kubo M, Yang H, Yoda K, Horie R, Chan WY, Leung HY, Hattori K, Konomi M, Osumi M (2005) Enhanced salt tolerance mediated by AtHKT1 transporter-induced Na unloading from xylem vessels to xylem parenchyma cells. Plant J 44:928–938

Tanou G, Filippou P, Belghazi M, Job D, Fotopoulos V, Molassiotis A (2012) Oxidative and nitrosative-based signaling and associated post-translational modifications orchestrate the acclimation of citrus plants to salinity stress. Plant J 72:585–599

Tarchoune I, Sgherri C, Izzo R, Lachaal M, Querghi Z, Navari-Izzo F (2010) Antioxidative responses of *Ocimum basilicum* to sodium chloride or sodium sulphate salinization. Plant Physiol Biochem 48:772–777

Tenheken R (2015) Cell wall remodeling under abiotic stress. Front Plant Sci 5:771

Termaat A, Pasioura JB, Munns R (1985) Shoot turgor does not limit growth of NaCl affected wheat and barley. Plant Physiol 77:869–872
Turan S, Cornish K, Kumar S (2012) Salinity tolerance in plants: breeding and genetic engineering. Aust J Crop Sci 6:1337–1348
Turkan I, Demiral T (2009) Recent developments in understanding salinity tolerance. Environ Exp Bot 67:2–9
Valmonte GR, Arthur K, Higgins CM (2014) Calcium-dependent protein kinases in plants: Evolution, expression and function. Plant Cell Physiol 55:551–569
Vialaret J, Pietro M, Hem S, Maural C, Rossignal M, Santoni V (2014) Phosphorylation dynamics of membrane proteins from Arabidopsis roots submitted to salt stress. Proteomics 14:1058–1070
Wang L, Ruan YL (2013) Regulation of cell division and expansion by sugar and auxin signaling. Front Plant Sci 4:163
Wang WB, Kim YH, Lee HS, Kim KY, Deng XP, Kwak SS (2009) Analysis of antioxidant enzyme activity during germination of alfalfa under salt and drought stresses. Plant Physiol Biochem 47:570–577
Wang L, Liu X, Liang M (2014) Proteomic analysis of salt-responsive proteins in the leaves of mangrove *Kandelia candel* during short-term stress. PLoS One 9:e83141
Wang N, Qi H, Qiao W, Shi J, Xu Q, Zhou H, Yan G, Huang Q (2017) Cotton (*Gossypium hirsutum* L.) genotypes with contrasting K^+/Na^+ ion homeostasis: implications for salinity tolerance. Acta Physiol Plant 39:77
Wang W, Tian F, Hao Q, Han Y, Li Q, Wang X, Wang W, Wang Y, Wang W (2018) Change in growth, lipid peroxidation and some key antioxidants enzymes in chick pea genotypes under salt stress. Acta Physiol Plant 40:39
Wu J, Seliskar DM, Gallagher JL (2005) The response of plasma membrane lipid composition in callus of the halophyte, *Spartina patens*, to salinity stress. Am J Bot 92:852–858
Wu H, Shabala L, Liu X, Azzarello E, Zhou M, Pandolfi C, Chen Z, Bose J, Mancuso S, Shabala S (2015a) Linking salinity stress tolerance with tissue-specific Na^+ sequestration in wheat roots. Front Plant Sci 6:71
Wu H, Shabala L, Zhou H, Shabala S (2015b) Chloroplast-generated ROS dominate NaCl-induced K^+ efflux in wheat leaf mesophyll. Plant Signal Behav 10(5):e1013793
Xu D, Duan X, Wang B, Hong B, Hot T, Wu R (1996) Expression of a late embryogenesis abundant protein gene, HVA1, from barley confers tolerance to water deficit and salt stress in transgenic rice. Plant Physiol 110:249–257
Xu Y, Hu W, Liu J, Zhang J, Jia C, Miao H (2014) A banana gene, *MaPIP1;1*, is involved in tolerance to drought and salt stresses. BMC Plant Biol 12:188
Yahya A, Liljenberg C, Nilsson R, Lindberg S, Banas A (1995) Effects of pH and minerals nutrition supply on lipid composition and protein pattern of plasma membranes from sugar beet roots. J Plant Physiol 146:81–87
Yamada S, Katsuhara M, Kelly WB, Michalowski CB, Bohnert HJ (1995) A family of transcripts encoding water channel proteins: tissue-specific expression in the common ice plant. Plant Cell 7:1129–1142
Yao S, Lan H, Zhang F (2010) Variation of seed heteromorphism in Chenopodium album and the effect of salinity stress on the descendants. Ann Bot 105:1015–1025
Yousuf M, Fariduddin Q, Khan T, Hayat S (2017) Epibrassinolide reverses the stress generated by combination of excess aluminum and salt in two wheat cultivars through altered proline metabolism and antioxidants. S Afr J Bot 112:391–398
Yu BJ, Gong HM, Liu YL (1999) Effects of exogenous fatty acids on H^+-ATPase and lipid composition of plasma membrane vesicles isolated from roots of barley seedlings under salt stress. J Plant Physiol 155:646–651
Zagorchev L, Kamenova P, Odjakova M (2014) The role of plant cell wall proteins in response to salt stress. Sci World J 2014:764089

Zamani B, Bybordi A, Khorshidi S, Nezami T (2010) Effects of NaCl salinity levels on lipids and proteins of canola (*Brassica napus* L.) cultivars. Adv Environ Biol 4:397–403

Zaoui S, Gautier H, Bancel D, Chaabani G, Hanen W, Lachaal M, Karray-Bouraoui N (2016) Antioxidant pool optimization in *Carthamus tinctorius* L. leaves under different NaCl levels and treatment durations. Acta Physiol Plant 38:187

Zhang Y, Zhao H, Zhou S, He Y, Luo Q, Zhang F, Qiu D, Feng J, Wei Q, Chen L, Chen M, Chang J, Yang G, He G (2018) Expression of *TaGF14b*, a 14-3-3 adaptor protein gene from wheat, enhances drought and salt tolerance in transgenic tobacco. Planta 248:117–137

Zheng J, Zhao L, Wu C, Shen B, Zhu A (2015) Exogenous proline reduces NaCl-induced damage by mediating ionic and osmotic adjustment and enhancing antioxidant defense in *Eurya emarginata*. Acta Physiol Plant 37:181

Zorb C, Senbayram M, Peiter E (2014) Potassium in agriculture – status and perspectives. J Plant Physiol 171:656–669

Chapter 9
Microorganisms Aiding Existence and Efficiency of Plants in Saline Environment: What We Know and What to Expect

Usha Chakraborty, Swarnendu Roy, and Bishwanath Chakraborty

Abstract Salinity poses a major threat to agriculture and hence to food production globally. Though a number of plants are halophytes, surviving in saline environment and, in some cases, requiring a saline soil for growth, none of the cultivated crops are halophytes. Microorganisms, being ubiquitous, are present everywhere, and a large number of salt-tolerant microbes have been found to be associated with halophytes. Many of them are involved in providing tolerance to the plants against salt stress through various mechanisms such as adjusting osmotic balance, ion homeostasis, enhancing antioxidant machinery for scavenging the toxic reactive oxygen species and production of hormones, etc. The use of beneficial microorganisms which have the potential for plant growth promotion as well as salt stress alleviation is gaining momentum. The most commonly used are the plant growth-promoting bacteria of genus *Bacillus*, as well as some arbuscular mycorrhizal fungi, particularly belonging to the genus *Glomus*. This approach would provide a cost-efficient, eco-friendly means of salt stress alleviation linked with plant growth promotion, which in turn could lead to sustainable agriculture under changing climatic conditions.

Keywords Salinity · Halophytes · Salt-tolerant microbes · Rhizosphere · PGPR · Mycorrhiza

9.1 Introduction

At this juncture of the existence of man and other life forms on earth, climate change poses the greatest threat to their survival. For mankind, the problem of climate change is of extreme importance because global food production depends on

U. Chakraborty (✉) · S. Roy
Department of Botany, Plant Biochemistry Laboratory, University of North Bengal, Siliguri, West Bengal, India

B. Chakraborty
Department of Biological Sciences, Aliah University, Kolkata, West Bengal, India

B. Giri, A. Varma (eds.), *Microorganisms in Saline Environments: Strategies and Functions*, Soil Biology 56, https://doi.org/10.1007/978-3-030-18975-4_9

favourable climatic conditions and several abiotic factors such as temperature, water availability, salinity, heavy metals, etc. cause stresses which in turn lead to metabolic disorders. Excess salt concentration in soil and water resources declines agricultural productivity, turns fertile fields to marginal lands and leads to their abandonment. The Food and Agriculture Organization estimates that salinity has affected more than 6% of land area (Ilangumaran and Smith 2017). Though most of this land is not cultivable, it is estimated that approximately 45 million ha of irrigated land (20% of total) and 32 million ha under dryland agriculture (about 2% of total) has been affected (Munns and Tester 2008).

Soil is considered as saline when its electrical conductivity (EC) exceeds 4 dSm^{-1} which is an equivalent of 40 mM NaCl. Besides sodium chloride (NaCl), other salts, such as sodium sulphate (Na_2SO_4), sodium nitrate ($NaNO_3$), magnesium sulphate ($MgSO_4$), magnesium chloride ($MgCl_2$), potassium sulphate (K_2SO_4), calcium carbonate ($CaCO_3$), etc., are also present in saline soil out of which NaCl causes serious problems for higher plants. Increase in these salt limits leads to two major stresses for the plants: osmotic stress and ionic stress (Vaishnav et al. 2016). The immediate impact of salinity, which is observed within minutes of the stress, is osmotic imbalance brought about by the hypertonic conditions, and ion toxicity which occurs over several hours to days and weeks is the result of toxic ions (Na^+ and Cl^-) accumulating in the cells. These in turn affect hormonal status, transpiration, photosynthesis, translocation of nutrients and other metabolic processes (Munns 2002). Physiological processes in crops negatively affected by high concentrations of salt in soil include seed germination, growth, flowering and fruit set, as well as decrease in yield and the quality in crops (Arora et al. 2008). Considering the negative impacts of salinity, the development of salt-tolerant plants has been a much-desired scientific goal, which has only been met with limited success (Hahm et al. 2017), and only a few major genetic determinants of salt tolerance have been identified (Munns and Tester 2008; Schubert et al. 2009).

However, as an alternative to breeding and genetic manipulation, tolerance to salinity has also been shown to be improved by the application of salt-tolerant microorganisms. In this context, beneficial microorganisms, such as plant growth-promoting rhizobacteria (PGPR), or arbuscular mycorrhizal fungi (AMF) from the rhizosphere of plants have been investigated for their potential to alleviate salt stress. These studies have shown that the majority of PGPR, including strains of *Azospirillum*, *Bacillus*, *Pseudomonas*, *Rhizobium* and *Ochrobactrum* species, increase the salt tolerance of their host plants (Hamdia et al. 2004; Ahmad et al. 2013; Bharti et al. 2013; Chakraborty et al. 2013) and, in addition, that PGPR can stimulate plant growth through a variety of mechanisms, including the fixation of atmospheric nitrogen, solubilization of phosphate and production of phytohormones [e.g. indole-3-acetic acid (IAA), gibberellin, cytokinin and abscisic acid], 1-aminocyclopropane-1-carboxylate (ACC) deaminase and exopolysaccharide (EPS) (Chakraborty et al. 2009). It is further believed that beneficial soil microorganisms enhance soil-water-plant relations through intricate signalling mechanisms that are not yet fully understood. Dodd and Perez-Alfocea (2012) have opined that the ability of soil microorganisms to modulate phytohormonal signalling and trigger several other mechanisms to work in an integrated fashion contributes to enhanced stress tolerance in plants.

9.2 Salt Tolerance in Halophytes

It is now established that there is a continuum of tolerance to Na^+ and Cl^{-1} in their environment, ranging from the very sensitive (Flowers et al. 2010b) to the tolerant (English and Colmer 2013). There are probably less than 500 species that can grow in salt concentrations of the 'average' seawater and hence are very rare (Flowers et al. 2010a). This is quite surprising since at the evolutionary time when plants started colonizing land, low frequency of occurrence of halophytes, seawater contained substantial salt concentrations (about 30 g kg^{-1}; Railsback et al. 1990). It would therefore seem, from the low frequency of halophytes, that salt tolerance is not a fundamental trait in terrestrial plants, but rather, it appeared later during their evolution (Flowers and Colmer 2015). Halophytes thus represent a very small group of gifted plants that possess the inherent ability to grow, reproduce and sustain in salt-rich regions of the earth, the regions that remain practically uninhabitable by the majority of earth flora. This quality of the halophytes has earned them a special status, and comparative studies with salt-avoiding plants have revealed the contribution of anatomical adaptations and modification in seed germination, etc. in making them highly adjustable to saline environments. One of the most suitable ways to define halophytes is done on the basis of their ability to complete life cycle in a salt concentration of a minimum of 200 mM NaCl in natural environments (Flowers et al. 1986). The requirement of salts and the type of salt (NaCl, KCl, $MgSO_4$) for the optimal growth and development of halophytes may vary, and also the halophytes can be grown in laboratory conditions completely without salt or at a very low concentration of salts. This implies that though the halophytes survive and complete their life cycle in saline environments, the requirement of salt for their survival and reproduction is not totally exclusive. However, few halophytes thus exists, viz. *Halogeton glomeratus* and *Salicornia* spp., which comes close to the definition of 'obligate halophytes'—the plants that necessarily require the presence of high amount of salts in their vicinity for growth and development (Flowers et al. 1977). The halophytes are adapted to saline environments, the physiological and molecular framework of which has been slowly and gradually tuned on during the evolutionary courses. This therefore brought forward the variations in adaptation mechanism among the halophytic plants (Ungar 1991). On the basis of their adaptation mechanism, the halophytes can be broadly classified into euhalophytes or true halophytes (the plants that accumulate high amount of salts in tissue), pseudohalophytes or salt-avoiders (the plants that restrict the uptake of salts) and crinohalophytes or salt-excretors (the plants that exude out the excess salt in their tissue with the help of salt glands or bladders) (Glenn and Brown 1999).

During the course of evolution, halophytic plants have developed several physiological and biochemical mechanisms which enable them to survive in soils with high salt concentration. Some of these are ion homeostasis and compartmentalization, ion transport and uptake, accumulation of osmoprotectants and compatible solutes, enhancement of antioxidative enzyme activity and synthesis of antioxidant compounds, hormone modulation and synthesis of polyamines (Gupta and Huang 2014). All of these do not work together and the mechanism in each plant may vary.

Sarcocornia natalensis is an extremely tolerant succulent halophytic plant that depends upon the massive inorganic ion accumulation that drives the influx of water to maintain the turgor pressure of the cells, thereby assisting the plants to metabolize in saline environments (Naidoo and Rughunanan 1990). *Salicornia europaea* which is one of the most tolerant euhalophytes also relies upon the uptake of Na^+ for osmoregulation. In *S. europaea* the excess amounts of Na^+ were sequestrated into the vacuoles of shoot endodermis which was accomplished at the cellular level due to the enhanced activity of ion transporters like SeNHX1, SeVHA-A and SeVP1 (Lv et al. 2012). Even several miRNAs were identified in *S. europaea* in both shoot and root that contribute to salt tolerance of this species by regulating the genes and transcription factors involved in lignin biosynthesis and carbon and nitrogen metabolism (Feng et al. 2015).

Roy and Chakraborty (2018) investigated the mechanisms of salinity tolerance in *Cynodon dactylon*, a wild perennial chloridoid grass having a wide range of ecological distribution which is generally adaptable to varying degrees of salinity stress. They suggested, based on their experiments, that salt exclusion mechanism present exclusively in the wild grass was one of the major factors contributing to its tolerance. Accumulation of osmolytes like proline and glycine betaine, as well as increased CdSOS1 and CdNHX expression, also pointed to their role in salinity tolerance. In *Suaeda maritima*, the process of Na^+ sequestration and exclusion both seems to be important in conferring salt tolerance. In-depth transcriptomic analysis of this plant has revealed an overexpression of wide array of genes, most of which are transcription factors related to cell wall synthesis, carbohydrate metabolism, ion and redox homeostasis and phosphoinositide and hormone signalling (Gharat et al. 2016). *Suaeda nudiflora*—a succulent halophyte of the Chenopodiaceae family—grows comfortably at high salt concentrations, though at germination phase it is somewhat sensitive to salinity. The study of salinity tolerance of this species during the early stages of growth has revealed the role of antioxidative enzymes like superoxide dismutase and peroxidase in conferring salinity tolerance (Cherian and Reddy 1999). Functional characterization of *Paspalum vaginatum* by yeast cDNA expression library technique revealed the overexpression of at least 18 salinity tolerance genes that have been predicted to be involved mostly with increasing photosynthetic efficiency, maintaining oxidative balance, post-translational modifications and phospholipid biosynthesis (Chen et al. 2016). In the halophyte *Atriplex nummularia*, the concentration of Na^+ ion has been found to be higher in comparison to that of in roots. The shoot Na^+ ions have been found to be channelized into the cell vacuoles that help to maintain a low level of the toxic ion in cytoplasm for the antioxidative enzymes to function properly (de Araujo et al. 2006). Studies on the extreme halophyte *Salvadora persica* revealed the role of accumulation of proline and reducing sugars along with anatomical adaptations in the form of an increase in thickness of hypodermal layer and pith area in shoot and increased epidermal layer with decrease in the diameter of xylem vessels in roots as an aid to salinity tolerance (Parida et al. 2016). Proteomic analysis in a dicotyledonous halophyte of the family Zygophyllaceae—*Tangut nitraria*—showed the involvement of 71 proteins in salt tolerance that were mainly involved with redox homeostasis by scavenging of reactive oxygen species (ROS), signal transduction, membrane transport for ion homeostasis, carbohydrate metabolism, etc. (Cheng et al. 2015).

Salinity results in the formation of reactive oxygen species (ROS) in all plants and antioxidant metabolism, including antioxidative enzymes and nonenzymatic compounds, play critical parts in detoxifying ROS (Gupta and Huang 2014). It has been shown by several authors that salinity tolerance is positively correlated with the activity of antioxidative enzymes, such as superoxide dismutase (SOD), catalase (CAT), glutathione peroxidise (GPX), ascorbate peroxidase (APX) and glutathione reductase (GR), and with the accumulation of nonenzymatic antioxidant compounds (Asada 1999; Gupta et al. 2005; Chakraborty and Pradhan 2012, 2013). Gill et al. (2013) and Tuteja et al. (2013) have reported the functioning of few helicase proteins (e.g. DESD-box helicase and OsSUV3 dual helicase) in conferring plant salinity tolerance by improving/maintaining photosynthesis and antioxidant machinery.

9.3 Overview of Salt-Tolerant Microbes and Their Mechanism of Tolerance

The microorganisms that inhabit the saline environments are known as ‘halophilic’ or ‘salt-loving’. They can be differentiated into ‘slight’, ‘moderate’ and ‘extreme’ halophiles depending on the range of NaCl at which they can sustain their optimum growth. The range of NaCl required for the growth of ‘slight halophiles’ is 2–5%, for ‘moderate halophiles’ being 5–20% and for ‘extreme halophiles’ being more than 20% (Vreeland 1987). The microorganisms can also be categorized into obligate and non-obligate halophiles. The former includes mainly the halophilic archaea that can grow up to a salt concentration in excess of 30% but fails to grow if the salt concentration falls below 10–12% (Grant and Larsen 1989). Non-obligate halophiles include the bacteria, algae and fungi that can survive in a broad range of salt concentrations from 0 to 30% NaCl, which are also characterized by the accumulation of organic solutes (Litchfield 1998). The halophiles survive the adverse impacts of saline habitats by employing mainly two mechanisms: ‘salt-in-cytoplasm’ and ‘organic-osmolyte’ (Kunte et al. 2002).

The ‘salt-in-cytoplasm’ mechanism deals with the increase in accumulation of K^+ and Cl^- ions in the bacterial cells under saline environment. Though the enhanced salt concentration is not ideally suited for the proper functioning of cellular enzymes, the cell adapts to the situation by including majority of acidic amino acids, thus assisting the enzymatic machinery to function actively in such saturating salt concentrations (Oren 2008). Siglioccolo et al. (2011), after analysing the three-dimensional properties of halophilic and non-halophilic proteins, propounded that the adaptation of proteins in high salt concentrations depends upon the weakening of the core hydrophobic interactions with simultaneous counterbalancing in the presence of salts preventing the loss of function of proteins. Though the salt-in-cytoplasm mechanism is energetically more economical as compared to the accumulation of osmolytes, this is not widely used by the halophilic microorganisms (Oren 1999). This strategy of salt tolerance is known from the extreme halophilic archaean members of the family: *Halobacteriaceae* like *Halobacterium salinarum*, *Haloarcula marismortui*, *Natronomonas pharaonis* and *Halquadratum walsbyi* (Kennedy et al. 2001; Baliga et al. 2004; Falb et al. 2005; Bolhuis et al. 2006). Apart from archaea, this strategy is also employed by bacteria such as anaerobic fermentative *Halanaerobiales* and aerobic members like *Salinibacter ruber* (Oren 2008).

The 'organic-osmolyte' mechanism of salt tolerance is evident in many bacteria and methanogenic archaea (Kunte 2006). These halophiles adapt to extreme environments by accumulating a wide range of osmolytes. Nearly 30 years back, it was realized that the accumulation of osmolytes in the members of *Halomonadaceae* played an important role in balancing the osmotic pressure imparted by the saline medium (Wohlfarth et al. 1990). In the studies with *Deleya halophila*, *Halomonas elongata* and *Flavobacterium halmephilum*, it was concluded that the accumulation of osmolytes in the form of an abundant amino acid—ectoine, glycine betaine and an unidentified sugar—was sufficient enough to maintain the osmotic pressure of the bacterial cells. Lai et al. (2000) reported the accumulation of glycine betaine as osmoprotectant and maintenance of high molar concentrations of K^+ in the cells under saline conditions in the halophilic archaea *Methanohalophilus portucalensis*. The accumulation of glutamine and glutamate in the moderate halophilic bacteria *Halobacillus halophilus* was shown to maintain the osmotic balance along with the accumulation of proline. The accumulation of glutamine was also shown to be dependent on chloride uptake as the transcription rate of glutamine synthetase was dependent on it (Saum and Muller 2008). The fungi that are mostly neglected are also present in hypersaline environments, and halophilic fungi like *Hortaea werneckii*, *Wallemia ichthyophaga* and *Aureobasidium pullulans* mostly accumulate osmolytes in the form of polyols such as glycerol, arabitol and mannitol to maintain the osmotic balance and low levels of salt inside their cells (Gunde-Cimerman et al. 2009; Ma et al. 2010).

Also the cellular membranes of extreme archaeal halophiles account for increased salinity tolerance. The membranes of several archaea like *Halobacterium halobium* and *Halobacterium salinarum* are reported with a high proportion of a specific diacidic phospholipid designated as archaetidylglycerol methylphosphate which confers stability at NaCl concentrations up to 5 M (Tenchov et al. 2006).

9.4 Association of Salt-Tolerant Microbes with Halophytes

Recently the association of microbes with halophytes that harbour the rhizosphere and root interiors of halophytic plants has garnered special attention of the scientific community. Several researches in the past decade have nevertheless acknowledged the role of these microbes, whatever the magnitude may be, in assisting the halophytes to grow in saline environments with ease. This halophytic microbiome adds to a new dimension in the understanding of salt tolerance mechanism in halophytes. Halophytic microbiome that remains associated with rhizosphere, endosphere and phyllosphere may include the members of prokaryotes like bacteria and archaea and also eukaryotes like fungi (Ruppel et al. 2013). It is now a widely accepted fact that the microbes existing in close proximity of plants growing in environmentally challenged habitats have something to do with enhancing the chances of the survival of the plants (Lau and Lennon 2012). Quite necessarily the interest in search for novel microorganisms associated with halophytic plants has thus increased in leaps and bounds in the past decade. Coastal ecosystem consisting of shrubs growing in the coastal area is considered as a low-nutrient environment due to the sparseness of nitrogen and phosphorus (Frosini et al. 2012). These coastal dune areas of the world represent a unique flora as a result of variations in salinity, anaerobicity and temperature fluctuations during different seasons (Bibi et al. 2018). These conditions make this habitat unusual and thus may have been selective for the presence of truly unique microorganisms having a rich biodiversity. Hence, under such conditions, it is highly probable that microorganisms associated with these coastal dune plants play a role in their survival and perform diverse biological functions in plants (Del Vecchio et al. 2015).

Jha et al. (2011) investigated the presence of plant-microbe interaction in the extremely halophytic plant *Salicornia brachiata*. Many diazotrophic bacteria were eventually identified as *Brachybacterium saurashtrense*, *Zhihengliuella* sp., *Brevibacterium casei*, *Haererehalobacter* sp., *Halomonas* sp., *Vibrio* sp., *Cronobacter sakazakii*, *Pseudomonas* spp., *Rhizobium radiobacter* and *Mesorhizobium* sp. These bacteria were also reported to possess plant growth-promoting activities like production of indole-3-acetic acid (IAA), solubilization of rock phosphate and presence of 1-aminocyclopropane-1-carboxylic acid (ACC) deaminase activity. Yuan et al. (2016) reported the presence of several bacterial members of γ-proteobacteria, α-proteobacteria, *Bacteroidetes* and *Verrucomicrobia* from the rhizo- and endosphere of halophyte *Suaeda salsa*. These included eight halotolerant genera, viz. *Microbulbifer*, *Pelagibius*, *Halomonas*, *Marinoscillum*, *Fulvivirga*, *Haloferula*, *Pelagicoccus* and *Marinobacter*. Also, fungal genera were reported to inhabit the rhizo- and endosphere of *Suaeda salsa*, viz. *Beauveria*, *Monosporascus*, *Leptosphaeria* and *Retroconis*, all of them belonging to *Ascomycota*. Several halotolerant bacteria have been isolated from *Arthrocnemum indicum*, viz. *Klebsiella* sp., *Pseudomonas* sp., *Agrobacterium* sp. and *Ochrobactrum* sp. that were also shown to possess plant growth-promoting activity and ability to fix atmospheric nitrogen in non-host plants (Sharma et al. 2016). Navarro-Torre et al. (2017) reported the presence of at least eight bacterial strains from the phyllosphere of a halophytic plant *Arthrocnemum macrostachyum*. Among these, three strains, viz. *Bacillus alcalophilus* EA1, *Bacillus thuringiensis* EA3 and *Gracilibacillus saliphilus* EA8, were found to possess plant growth-promoting activities like phosphate solubilization, IAA production and siderophore production along with enzymatic activities like chitinase, pectinase, amylase and protease under saline environments. The presence of these attributes in these three strains made them highly desirable candidates to promote plant growth under saline conditions in consortium. Bibi et al. (2018) reported the association of several bacteria with the halophyte *Salsola imbricata* belonging to the major classes of *Actinobacteria, Firmicutes*, β-*Proteobacteria* and γ-*Proteobacteria*, some of them having antagonistic activity against fungal plant pathogens like *Phytophthora capsici* and *Pythium ultimum.* Among the identified bacteria, *Nocardioides luteus*, *Nocardiopsis dassonvillei*, *Bordetella avium*, *Pseudomonas boreopolis* and *Arthrobacter subterraneus* were found to be associated with the rhizosphere; *Bacillus licheniformis*, *Streptomyces geysiriensis*, *Bacillus sonorensis* and *Streptomyces violaceochromogenes* with root endosphere; and *Bacillus tequilensis*, *Agromyces indicus* and *Bacillus halosaccharovorans* with leaf endosphere.

A significant number of fungi are also reported to remain associated with the halophytic plants. Indira and Srinivasan (2017) have isolated a large number of fungal endophytes from the different parts of *Suaeda maritima* and *Suaeda monoica*, viz. *Acremonium* sp., *Alternaria* spp., *Aspergillus* spp., *Cladosporium* sp., *Phoma* sp., *Ulocladium* sp., *Meyerozyma* sp., etc.

Bacteria isolated from extreme environments such as deserts and oceans have been shown to induce salt tolerance in crop plants. Zerrouk et al. (2016) reported that *Pseudomonas fluorescens* strain isolated from date palm rhizosphere in Saharan region promoted root growth in maize (*Zea mays*) seedlings under salt stress. Inoculation of wheat plants with salt-tolerant bacteria has shown their ability to promote plant growth and enhance salt tolerance (Chakraborty et al. 2013; Upadhyay and Singh 2015; Singh and Jha 2016a, b).

There are also reports of the association of arbuscular mycorrhizal fungi (AMF) with halophytes naturally in saline environments (Carvalho et al. 2001; Hilderbrandt et al. 2001; Harisnaut et al. 2003; Yamato et al. 2008). Aliasgharzadeh et al. (2001) observed that the most predominant species of AMF in the severely saline soils of the Tabriz plains [with an electrical conductivity of 162 dS m^{-1}] were *Glomus intraradices*, *G. versiform* and *G. etunicatum*. Other studies have also shown that the AMF most commonly observed in saline soils are *Glomus* spp. (Ho 1987; Wang et al. 2004). Molecular biological techniques revealed that approximately 80% of these spores belonged to one single species, *Glomus geosporum* (Wilde et al. 2009), though it is not yet established that this fungus confers salt tolerance on plants (Evelin et al. 2009). However, this finding does not necessarily indicate that *Glomus geosporum* confers salt tolerance on plants and enables them to grow under salt stress conditions. Porras-Soriano et al. (2009) tested the efficacy of three species of AMF—*Glomus mosseae*, *G. intraradices* and *G. claroideum*—to alleviate salt stress in olive trees under nursery conditions. Of the three, *G. mosseae* was the most efficient in terms of olive tree performance and particularly in the protection offered against the detrimental effects of salinity. These findings suggest that the capability of AMF in protecting plants from the detrimental effects of salt stress may depend on the behaviour of each species. A list of some of the microbes associated with halophytes have been provided in Table 9.1.

Table 9.1 Salt-tolerant microorganisms isolated from the different plant parts of halophytes depicting the diversity of halophyte-associated microorganisms

Name of the halophyte	Family	Name of the major microorganisms isolated	Parts with which associated	References
Arthrocnemum macrostachyum	Amaranthaceae	*Bacillus alcalophilus*, *Bacillus thuringiensis*, *Oceanobacillus* sp., *Gracilibacillus saliphilus*	Aerial parts	Navarro-Torre et al. (2017)
Carex scabrifolia	Cyperaceae	*Gynuella sunshinyii* gen. nov., sp. nov.	Rhizosphere	Chung et al. (2015)
Halocnemum strobilaceum	Amaranthaceae	*Fodinicurvata* sp., *Halomonas* sp., *Mesorhizobium* sp., *Salegentibacter* sp.	Rhizosphere	Li et al. (2018)
Halocnemum strobilaceum	Amaranthaceae	*Bacillus* sp., *Pseudomonas* sp., *Isoptericola* sp., *Marinobacter* sp., *Halomonas* sp., *Staphylococcus* sp.	Rhizosphere	Marasco et al. (2016)
Halostachys caspica (accepted name: *Halostachys belangeriana*)	Amaranthaceae	*Citrobacter* sp., *Exiguobacterium* sp., *Pseudomonas* sp.	Rhizosphere	Li et al. (2018)
Kalidium foliatum	Amaranthaceae	*Gracilimonas* sp., *Jiangella* sp., *Marinimicrobium* sp., *Planococcus* sp., *Pontibacter* sp.	Rhizosphere	Li et al. (2018)
Limonium gmelinii	Plumbaginaceae	*Albidovulum* sp., *Bauldia* sp., *Deferrisoma* sp., *Geminicoccus* sp., *Thiohalomonas* sp.	Rhizosphere	Li et al. (2018)
Limonium sinense	Plumbaginaceae	*Bacillus* spp., *Pseudomonas* spp., *Klebsiella* spp., *Serratia rubidaea*, *Arthrobacter* sp., *Streptomyces pactum*, *Isoptericola dokdonensis*, *Microbacterium paraoxydans*	Endophytic on root, stem and leaves	Qin et al. (2018)
Limonium sinense	Plumbaginaceae	*Bacillus* spp., *Pseudomonas* spp., *Klebsiella* spp., *Serratia* sp., *Arthrobacter* sp., *Streptomyces* sp., *Isoptericola* sp., *Microbacterium* sp.	Endophytic on root, stem and leaves	Qin et al. (2014)
Lycium barbarum	Solanaceae Elaeagnaceae	*Glomus* spp., *Septoglomus* spp.,	Rhizosphere	Liu et al. (2017b)

(continued)

Table 9.1 (continued)

Name of the halophyte	Family	Name of the major microorganisms isolated	Parts with which associated	References
Elaeagnus angustifolia		*Acaulospora* spp., *Funneliformis coronatum*, *Scutellospora* spp., *Ambispora callosa*, *Diversispora* spp., *Claroideoglomus* spp., *Pacispora* spp., *Rhizophagus clarus*		
Lycium ruthenicum	Solanaceae	*Gimesia* sp., *Pelagibius* sp.	Rhizosphere	Li et al. (2018)
Messerschmidia sibirica (accepted name: *Tournefortia sibirica*)	Boraginaceae	*Paracoccus*, *Aeromicrobium*, *Devosia*, *Flavobacterium*, *Actinoplanes*, *Mycobacterium*, *Arthrobacter*, *Blastococcus*, *Nocardioides* and *Microbacterium*	Rhizosphere	Tian and Zhang (2017)
		Bacillus, *Serratia*, *Kocuria*, *Massilia* and *Curtobacterium*	Endophytic in stem	
		Rhizobium, *Novosphingobium*, *Sphingopyxis*, *Hydrogenophaga*, *Herbiconiux*, *Streptomyces*, *Hyphomicrobium*, *Steroidobacter*, *Bradyrhizobium* and *Variibacter*	Endophytic in roots	
		Methylobacterium, *Aureimonas*, *Sphingomonas*, *Providencia* and *Pseudomonas*	Endophytic in leaves	
Salicornia europaea	Amaranthaceae	*Salinicola* sp., *Pseudomonas* sp., *Serratia* sp., *Bacillus* spp., *Streptomyces* spp., *Microbacterium oxydans*	Endophytic rhizosphere	Szymanska et al. (2016)
		Ensifer sp., *Serratia* sp.	Rhizosphere	
Salsola imbricata	Amaranthaceae	*Nocardioides luteus*, *Nocardiopsis dassonvillei*, *Bordetella avium*, *Pseudomonas boreopolis*, *Arthrobacter subterraneus*	Rhizosphere	Bibi et al. (2018)
			Endophytic in roots	
			Endophytic in leaves	

(continued)

Table 9.1 (continued)

Name of the halophyte	Family	Name of the major microorganisms isolated	Parts with which associated	References
		Bacillus licheniformis, *Arthrobacter subterraneus*, *Streptomyces* spp., *Bacillus sonorensis* *Bacillus tequilensis*, *Nocardioides luteus*, *Agromyces indicus*, *Bacillus halosaccharovorans*		
Suaeda fruticosa	Amaranthaceae	*Gracilibacillus* spp., *Staphylococcus* spp., *Virgibacillus salarius*, *Salinicoccus sesuvii*, *Bacillus* spp., *Zhihengliuella flava*, *Brevibacterium halotolerans*, *Oceanobacillus picturae*, *Exiguobacterium mexicanum*, *Pseudomonas japonica*, *Bacillus* spp., *Exiguobacterium* spp., *Staphylococcus* spp., *Salinicoccus* spp., *Oceanobacillus* spp., *Arthrobacter* spp., *Pseudomonas rhizosphaerae*, *Halomonas nanhaiensis*	Rhizosphere Phytoplane	Aslam and Ali (2018)
Suaeda salsa	Amaranthaceae	*Microbulbifer* sp., *Pelagibius* sp., *Halomonas* sp., *Marinoscillum* sp., *Fulvivirga* sp., *Haloferula* sp., *Pelagicoccus* sp., *Marinobacter* sp., *Beauveria* sp., *Monosporascus* sp., *Leptosphaeria* sp., *Retroconis* sp.	Rhizosphere Endophytic in roots Rhizosphere and endophytic in roots	Yuan et al. (2016)

9.5 How Do Microbes Induce Tolerance in Plants?

Most recently the interest in the study of dynamics of plant-microbe interaction has grown exponentially as it has been postulated to enhance sustainable agriculture by replacing the conventional practices. Many bacteria (plant growth-promoting microbes/rhizobacteria, PGPM or PGPR) and fungi associated with the plants have been associated with plant growth promotion and amelioration of biotic and abiotic stresses by regulating plant hormones, nutrition acquisition, siderophore production and enhancing antioxidative enzymes (Kumar and Verma 2018). In this context, the role of endophytes—the microbes that harbour the interior of a healthy living plant—also provides the plants with inherent mechanisms to tolerate biotic and abiotic stresses. These mechanisms include the maintenance of plant growth under stressed condition by increasing the root biomass, nutrient uptake, nitrogen fixation and maintenance of water level inside the plant tissues by osmotic adjustments, controlling stomatal movements, balancing the redox environment and so on (Lata et al. 2018).

Root hydraulic conductance under salt stress is affected by mycorrhizae, and PGPR can affect root hydraulic conductance (L). It was reported that bean plants (*Phaseolus vulgaris*) with mycorrhiza (*Glomus intraradices* BEG 123) had a greater osmotic root hydraulic conductance under saline (3.1 dS m^{-1}) stress (Aroca et al. 2007). It was proposed that mycorrhizal roots increased active solute transport as a mechanism to sustain water flow across the root. Although regulation of root hydraulic properties by AM symbiosis was strongly correlated with the regulation of aquaporin (PvPIP2) protein abundance and phosphorylation state (Aroca et al. 2007), different *PIP* genes were differentially regulated under salinity stress. Similarly, maize (*Zea mays*) plants that were inoculated with the PGPR *Bacillus megaterium* (Marulanda et al. 2010) and then exposed to salinity (2.59 dS m^{-1}) had increased length compared with uninoculated plants, which were correlated with increased root expression of two (of the six studied) ZmPIP isoforms (Dodd and Perez-Alfocea 2012).

The role of mycorrhizal fungi in plant protection under salt stress has been correlated with increased uptake of nutrients, accumulation of an osmoregulator, an increase in photosynthetic rate and water-use efficiency. Several mechanisms by which mycorrhizal colonization can alleviate salt stress in host plant have been discussed by Evelin et al. (2009). It has been reported that AMF-inoculated plants grow better than noninoculated plants under salt stress (Giri et al. 2003; Sannazzaro et al. 2007; Zuccarini and Okurowska 2008). *Cucurbita pepo* plants following successful colonization with *Glomus intraradices* under salinity stress conditions improved their growth, yield, quality of fruits, water status as well as nutrient content which was attributed to AMF-mediated enhanced nutrient acquisition, especially better P nutrition (Sharifi et al. 2007). Mycorrhizal dependency increases with increasing salt concentrations (Giri and Mukerji 2004) and varies with the fungal isolates and plant species (Tian et al. 2004).

Mycorrhizal colonization in plants can increase the P concentration by enhancing its uptake by external fungal hyphae which allows them to explore more soil volume than non-mycorrhizal plants. It has been suggested that improved P nutrition in AM-inoculated legume plants may improve their growth rate, increase antioxidant production and enhance nodulation and nitrogen fixation (Garg and Manchanda 2008). Higher accumulation of N was recorded in shoots of *Sesbania grandiflora* colonized with AMF than non-mycorrhizal control plants. It has been reported that improved N nutrition may help to reduce the toxic effect of Na ions by reducing its uptake which may indirectly help in maintaining the chlorophyll content (Giri and Mukerji 2004).

Induced systemic tolerance (IST) to abiotic stresses in plants has been found to be conferred by the PGPR by several mechanisms. The term 'IST' was coined to describe the physical and chemical changes imparted by microorganisms in plants in response to abiotic stresses and is generally associated with enhanced tolerance of plants under stress (Vurukonda et al. 2016). PGPR-elicited mechanisms under abiotic stresses include the production of growth hormones like cytokinins leading to ABA accumulation in plants; production of antioxidative enzymes aiding ROS scavenging; production of ACC (1-aminocyclopropane-1-carboxylic acid) deaminase which facilitates the degradation of ACC, precursor of ethylene, thereby releasing plant stress and promoting normal plant growth; and production of volatiles that is associated with the downregulation of root ion transporters like HKT1, thereby lowering the amount of Na^+ ion in plants under salinity stress (Yang et al. 2009). IST imparted by the bacterium *Achromobacter piechaudii* in the tomato plants was shown to enhance salt tolerance of tomato plants by mainly reducing the amount of ethylene accumulation under stress by degrading the precursor of ethylene biosynthesis through the activity of ACC deaminase enzyme secreted by the bacterium in the rhizosphere of the plants under stress (Mayak et al. 2004). Singh and Jha (2016b) also demonstrated the activity of a bacterium—*Serratia marcescens* CDP-13—isolated from the rhizosphere of *Capparis decidua* in imparting systemic resistance and tolerance to wheat plants under salinity stress. The application of *S. marcescens* enhanced the growth of wheat plants even in NaCl concentrations of 150–200 mM. The enhanced tolerance could be attributed to the efficiency of the bacterial isolate to stimulate the accumulation of osmolytes and modulate the activity of antioxidative enzymes like superoxide dismutase, peroxidase and catalase.

Salt stress, as other abiotic stresses, induces the formation of reactive oxygen species (ROS), which can cause severe oxidative damage to plants. Plants are protected from oxidative damage through a set of antioxidant systems which involve a variety of antioxidative enzymes, including superoxide dismutase (SOD), dehydroascorbate reductase (DHAR), glutathione reductase (GR), APX, CAT and GPX (Debez et al. 2001; Caverzan et al. 2012).

Transcriptome profiling of genes in *Arabidopsis thaliana* was done to evaluate the differential expression of genes stimulated by the application of *Bacillus amyloliquefaciens* to the former under salinity stress (Liu et al. 2017a). Transcriptome profiling of plants under NaCl stress revealed the differential expression of 1288 genes of which 1024 were upregulated, which mostly included the genes associated with photosynthesis, ROS scavenging, production of trehalose and proline, Na^+ translocation, jasmonate and ABA-dependent pathway. Single inoculation and co-inoculation of bacteria like *Azospirillum brasilense* and *Rhizobium tropici* has also been shown to stimulate IST in wheat plants treated with NaCl. The wheat plants treated with these bacteria showed an upregulation of the genes like APX1, CAT1, SOD2 and SOD4 in leaves and APX2 in roots clearly indicating the role of antioxidative enzymes in maintaining the redox balance of the plants under stress (Fukami et al. 2017). Also the accumulation of proline and malondialdehyde decreased indicating a better stress management of the plants following application along with the downregulation of genes associated with pathogenesis. Some of the mechanisms through which microbes induce tolerance are represented schematically in Fig. 9.1.

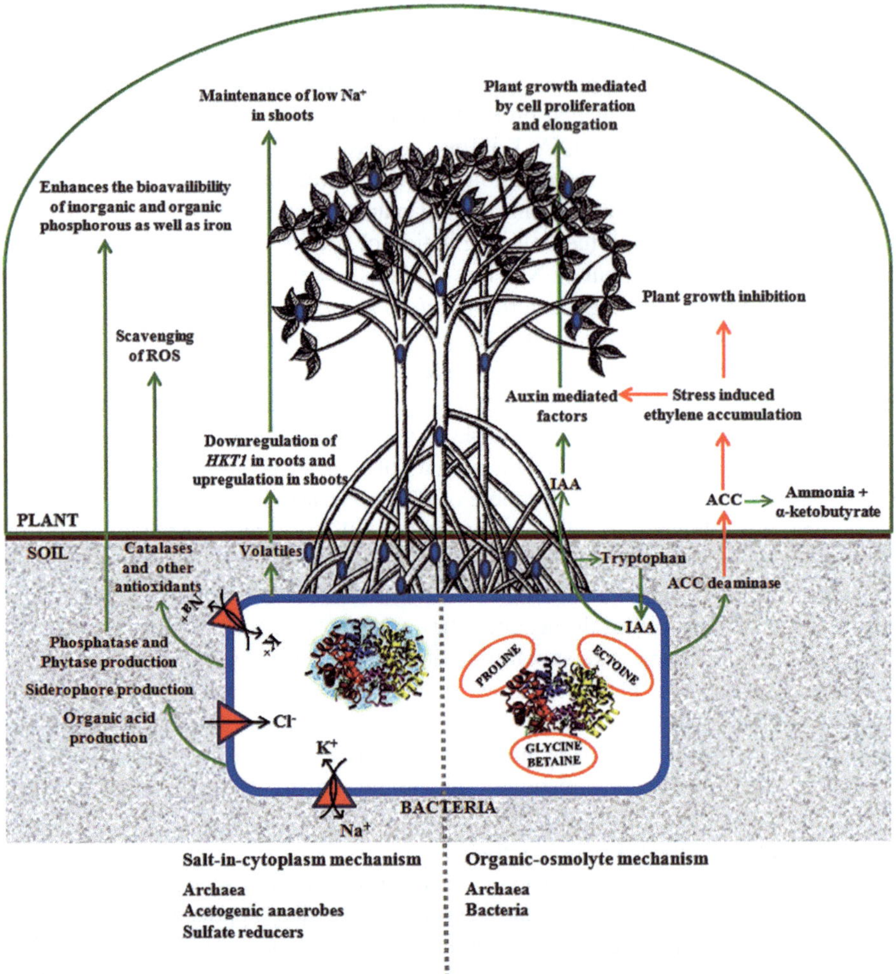

Fig. 9.1 Halophyte-microbe interaction: depiction of the mechanisms of salt tolerance in the rhizospheric, endophytic (root, stem and leaves) and phyllospheric bacteria along with the induced tolerance and growth promotion conferred by these bacteria to the halophytes. Bacteria (shown in blue dots) harbouring the saline environments maintain their normal biochemical machinery mainly by two mechanisms: salt-in-cytoplasm characterized by an increased uptake of K^+ and Cl^- and maintaining the hydration shell of the enzymes by employing more acidic amino acids towards the outside. Organic-osmolyte mechanism relies more upon the accumulation of osmolytes like proline, ectoine, glycine betaine, trehalose, etc. which act as osmoprotectants by protecting the enzyme functions

9.6 Alleviation of Salt Stress Through Beneficial Microbes: The Next Step for Sustainable Agriculture

As a natural phenomenon, the halophytes and its associated microorganisms are generally salt-tolerant, and some of the microorganisms are also able to confer the plants with several benefits for survival and sustenance. These microorganisms possessing plant growth-promoting attributes can be utilized for conferring induced systemic tolerance towards a wide range of abiotic stresses in non-host plants. The bacteria associated with halophytes most often possess growth-promoting attributes like production of growth hormones, viz. auxin, cytokinin and gibberellin; production of volatile organic compounds; production of siderophore; potentiality to fix atmospheric nitrogen; solubilization of organic and inorganic phosphate; and so on (Numan et al. 2018).

Muthezhilan et al. (2012) reported the isolation of 46 bacterial strains from the rhizosphere of *Ipomoea* sp. growing in the sand dunes of Chennai coastal area (India). Out of the isolated strains, 18 strains were able to produce indole-3-acetic acid (IAA), and 6 strains were able to significantly solubilize inorganic phosphate. The in vivo potentiality of the phosphate-solubilizing strains was evaluated on black gram and green gram, and it was found that a strain identified as *Pseudomonas* sp. AMET1148 was the most potent one as a plant growth promoter. Similarly, Goswami et al. (2014) isolated 85 strains from the coastal region of Little Rann of Kutch (India), which is a saline desert, and tested the isolates for plant growth-promoting attributes like IAA production, phosphate solubilization, siderophore production, ammonia, HCN, chitinase and antagonistic activity against fungal pathogens. Out of all the screened isolates, *Bacillus licheniformis* strain A2 was found to be the most potent and was thus selected for the evaluation of its growth promotion activity in a non-host plant *Arachis hypogea* under saline conditions. The results indicated that the bacterial strain was able to increase biomass, root length and shoot length of the plants under saline conditions. Also, several bacterial strains isolated from the coastal areas of Sundarbans were used as bioinoculants along with farmyard manure (FYM) for the field trials on rice and lady's finger which successfully enhanced the plant height and crop yield (Barua et al. 2012). These bacterial strains isolated from the forest bed and agricultural fields were identified as *Agrobacterium* sp. and *Bacillus* spp. and were found to possess several plant growth-promoting attributes like IAA production, phosphate solubilization, acetylene reductase activity and nitrogen fixation.

The application of halotolerant bacteria like *Enterobacter* sp. (MN17) and *Bacillus* sp. (MN54) has been found to increase the salt tolerance capacity in the facultative halophyte *Chenopodium quinoa* by reducing the Na^+ ion uptake, thus promoting ionic and osmotic balance (Yang et al. 2016). These bacteria possessed several plant growth-promoting activities like ACC deaminase activity, secretion of exopolysaccharide and IAA production; also the ability of these two strains to colonize the rhizosphere profoundly conferred better salinity tolerance to the plants.

Five salt-tolerant bacteria isolated from the rhizosphere of halophytes like *Atriplex leucoclada*, *Haloxylon salicornicum*, *Lespedeza bicolor*, *Suaeda fruticosa* and *Salicornia virginica* with plant growth-promoting activity were assessed for their growth-promoting ability on maize under salinity stress (Ullah and Bano 2015). Among the strains identified as *Bacillus* sp., *Bacillus pumilus* NB5, *Arthrobacter aurescens*, *Arthrobacter pascens* and *Bacillus pumilus* BSH-4 isolated, respectively, from the halophytic plants, *Bacillus* sp. and *A. Pascens* were the most potent ones in terms of increasing length and fresh and dry mass of root and shoot of maize plants. These two strains possessed phosphate-solubilizing activity and siderophore- and bacteriocin-producing activity and also conferred better tolerance to maize plants under salinity stress by enhancing osmolyte accumulation and antioxidative enzymes. A salt-tolerant bacterium from a halophytic desert shrub *Haloxylon ammodendron* identified as *Pseudomonas* sp. M30–35 was inoculated to the rhizosphere of perennial ryegrass to determine the ability of the bacteria to confer drought and salinity tolerance to the plants along with plant growth promotion (He et al. 2018). It was found that the bacterium was able to significantly increase the growth and salt tolerance of the plants at the same time as evident from the increase in shoot fresh and dry weights, chlorophyll content, catalase activity, soluble sugar and proline contents. The bacterium was able to maintain the ionic balance by maintaining a high K^+ content and a low Na^+ content in the plant tissues, at a same time decreasing the malondialdehyde (MDA) content.

Despite the straightforward approach of screening plant growth promoting attributes in salt-tolerant bacteria isolated from the rhizosphere, interior of plant tissue and surrounding soil and water of halophytes; there has been another strategy for identifying salt-tolerant plant growth-promoting bacteria that completely depends on the inverse methodology and has been reported to be effective in a similar fashion. This methodology depends upon the isolation of bacteria from the vicinity of the non-halophytes at first, and the screening for the salt tolerance and plant growth promotion of the isolated strains is performed afterwards. Chakraborty et al. (2011) reported the isolation of nine bacterial strains that were highly salt-tolerant and could grow in nutrient agar medium supplemented with 10% NaCl. Out of the nine strains, a bacterial strain was known to possess significant plant growth-promoting activity in terms of phosphate solubilization, IAA production, protease production and siderophore production, which was later identified as *Bacillus cereus*. The application of the bacterium to the rhizosphere of crop plants grown under NaCl stress revealed that the isolate could impart induced systemic tolerance to the plants by increasing the production of antioxidative enzymes and defence-related enzymes in the crop plants. Chakraborty et al. (2013) also isolated and identified two halotolerant bacteria—*Bacillus safensis* and *Ochrobactrum pseudogrignonense* from the rhizosphere of wheat and blady grass, respectively, which could promote growth of plants as well as alleviate salinity through mechanisms such as enhanced activities of antioxidative enzymes, accumulation of osmolytes and antioxidants, phosphate solubilization, hormone production, etc. Similarly, Habib et al. (2016) isolated 15 bacterial isolates from the crop fields and screened them for ACC deaminase activity, nitrogen fixation, phosphate solubilization, IAA production and salt-tolerant attributes. Two potent isolates, UPMR2 and UPMR18, identified as *Bacillus megaterium* and *Enterobacter* sp., were evaluated for their efficacy of growth promotion in okra. The studies revealed that these two bacteria could increase germination percentage and chlorophyll content and enhance antioxidative enzymes in the okra plants under salinity stress in comparison to control plants. Shi-Ying et al. (2018) reported the isolation of at least 305 bacterial strains from the soil collected from the paddy fields in Taoyuan, China. Out of the tested strains, only 9.3% of the bacterial strains were highly salt-tolerant which could survive in a NaCl concentration of 150 mg/L and belonged to the *Bacillales*, *Actinomycetales*, *Rhizobiales* and *Oceanospirillales*. Some of the strains could successfully improve the growth of rice plants under NaCl stress due to their ability to produce 1-aminocyclopropane-1-carboxycarboxylate deaminase (ACC deaminase), IAA and siderophores and at the same time induce the accumulation of proline in the stressed plants. Table 9.2 depicts some of the salt-tolerant microbes which have been used for stress alleviation in different plants.

9.7 Conclusion

Salinity is already a major problem globally due to shrinking usable land, increasing soil salinity and excess use of fertilizers, and with human habitation spreading more and more towards coastal areas, the problem, in the future, will be compounded. While halophytes can thrive in this environment, most of the cultivated crops would be unable to sustain growth in a saline environment, and even if they do, their productivity would be greatly reduced. However, under such circumstances, the resilience of plants makes them enhance their tolerance mechanisms, providing them with the ability to withstand increasing salinity. Over and above, it is now clear that there are a large group of microorganisms in such rhizosphere which can afford protection to the plants against the stress. Thus, an important area of research in agriculture is now focused on studies leading to the understanding of how microorganisms provide increased tolerance to the plants. Exploitation of microorganisms which have other beneficial properties such as plant growth promotion and nutrient mobilization, in addition to protection against abiotic stresses, is proving to be a highly efficient, cost-effective, eco-friendly method of sustainable agriculture. Present status of research and their outcomes in this context have been discussed above which will pave the way for the future so that we can still maintain the productivity of crops to feed the ever-growing masses, under various stresses to which the plants are already exposed and will be exposed in the coming years.

Table 9.2 Salt-tolerant bacteria isolated from halophytes or non-halophytes used to ameliorate salinity stress and plant growth promotion in plants

Name of the salt-tolerant microorganism	Isolated from	Plant growth-promoting attributes	References
Arthrobacter sp., *Bacillus megaterium*	Rhizosphere of *Tamarix chinensis*, *Suaeda salsa* and *Zoysia sinica*	IAA production, phosphate solubilization	Fan et al. (2016)
Azospirillum halopreferens	Rhizosphere of *Leptochloa fusca*	Nitrogen fixation	Bashan et al. (2000)
Bacillus aquimaris	Salt-affected soils	IAA production	Li and Jiang (2017)
Bacillus licheniformis	Rhizosphere of *Suaeda fruticosa*	Ammonia, siderophore, chitinase production and antagonistic activity	Goswami et al. (2014)
Bacillus safensis	Wheat rhizosphere	Phosphate solubilization, siderophore production, enhancing antioxidative enzyme activities, ACC deaminase activity, accumulation of osmolytes, antagonistic activity	Chakraborty et al. (2013)
Bacillus sp.	Rhizosphere of maize plants	Phosphate solubilization, siderophore production, ACC deaminase activity, antagonistic activity	Kaur et al. (2018)
Bacillus sp. and *Arthrobacter pascens*	Rhizospheric soil of mixed halophytic flora	Phosphate solubilization, bacteriocin and siderophore production	Ullah and Bano (2015)
Curtobacterium flaccumfaciens, *Ensifer garamanticus*	Endophytic species from *Hordeum secalinum* and *Plantago winteri*	ACC deaminase activity, auxin production, nitrogen fixation	Cardinale et al. (2015)
Kocuria erythromyxa, *Staphylococcus kloosii*	Rhizosphere of plants growing in high-salt region of Upper Coruh Valley, Turkey	Phosphate solubilization, catalase production	Karlidag et al. (2013)
Ochrobactrum pseudogrignonense	Rhizosphere of *Imperata cylindrica*	Phosphate solubilization, siderophore production, enhancing antioxidative enzyme activities, ACC deaminase activity, accumulation of osmolytes, antagonistic activity	Chakraborty et al. (2013)
Pseudomonas sp.	Rhizosphere of *Haloxylon ammodendron*	Phosphate solubilization, auxin biosynthesis, oxidative stress reliever	He et al. (2018)
Staphylococcus jettensis, *Zhihengliuella flava*, *Bacillus megaterium*	Rhizosphere and phytoplane of *Suaeda fruticosa*	ACC deaminase activity, IAA production, phosphate solubilization	Aslam and Ali (2018)

References

Ahmad M, Zahir ZA, Nazli F, Akram F, Arshad M, Khalid M (2013) Effectiveness of halo-tolerant, auxin producing *Pseudomonas* and *Rhizobium* strains to improve osmotic stress tolerance in mung bean (*Vigna radiata* L.). Braz J Microbiol 44:1341–1348

Aliasgharzadeh N, Saleh Rastin N, Towfighi H, Alizadeh A (2001) Occurrence of arbuscular mycorrhizal fungi in saline soils of the Tabriz Plain of Iran in relation to some physical and chemical properties of soil. Mycorrhiza 11:119–122

Aroca R, Porcel R, Ruiz-Lozano JM (2007) How does arbuscular mycorrhizal symbiosis regulate root hydraulic properties and plasma membrane aquaporins in *Phaseolus vulgaris* under drought, cold or salinity stresses? New Phytol 173:808–816

Arora N, Bhardwaj R, Sharma P, Arora HK (2008) Effects of 28-homobrassinolide on growth, lipid peroxidation and antioxidative enzyme activities in seedlings of *Zea mays* L. under salinity stress. Acta Physiol Plant 30:833–839

Asada K (1999) The water-water cycle in chloroplasts: scavenging of active oxygens and dissipation of excess photons. Annu Rev Plant Biol 50:601–639

Aslam F, Ali B (2018) Halotolerant bacterial diversity associated with *Suaeda fruticosa* (L.) Forssk. improved growth of maize under salinity stress. Agronomy 8:131. https://doi.org/10.3390/agronomy8080131

Baliga NS, Bonneau R, Facciotti MT, Pan M, Glusman G, Deutsch EW, Shannon P, Chiu Y, Sting Weng R, Richie Gan R, Hung P, Date SV, Marcotte E, Hood L, Ng WV (2004) Genome sequence of *Haloarcula marismortui*: a halophilic archaeon from the Dead sea. Genome Res 14:2221–2234

Barua S, Tripathi S, Chakraborty A, Ghosh S, Chakrabarti K (2012) Characterization and crop production efficiency of diazotrophic bacterial isolates from coastal saline soils. Microbiol Res 167:95–102

Bashan Y, Moreno M, Troyo E (2000) Growth promotion of the seawater-irrigated oilseed halophyte *Salicornia bigelovii* inoculated with mangrove rhizosphere bacteria and halotolerant *Azospirillum* spp. Biol Fertil Soils 32:265–272

Bharti N, Yadav D, Barnawal D, Maji D, Kalra A (2013) *Exiguobacterium oxidotolerans*, a halotolerant plant growth promoting rhizobacteria, improves yield and content of secondary metabolites in *Bacopa monnieri* (L.) Pennell under primary and secondary salt stress. World J Microbiol Biotechnol 29:379–387

Bibi F, Strobel GA, Naseer MI, Yasir M, Al-Ghamdi AAK, Azhar EI (2018) Microbial flora associated with the halophyte – *Salsola imbricate* and its biotechnical potential. Front Microbiol 9:65. https://doi.org/10.3389/fmicb2018.00065

Bolhuis H, Palm P, Wende A, Falb M, Rampp M, Rodriguez-Valera F, Pfeiffer F, Oesterhelt D (2006) The genome of the square archaeon "*Haloquadratum walsbyi*": life at the limits of water activity. BMC Genet 7:169

Cardinale M, Ratering S, Suarez C, Montoya AMZ, Geissler-Plaum R, Schnell S (2015) Paradox of plant growth promotion potential of rhizobacteria and their actual promotion effect on growth of barley (*Hordeum vulgare* L.) under salt stress. Microbiol Res 181:22–32

Carvalho LM, Correia PH, Martins-Loucao A (2001) Arbuscular mycorrhizal fungal propagules in a salt marsh. Mycorrhiza 14:165–170

Caverzan A, Passaia G, Rosa SB, Ribeiro CW, Lazzarotto F, Margis-Pinheiro M (2012) Plant responses to stresses: role of ascorbate peroxidase in the antioxidant protection. Genet Mol Biol 35:1011–1019

Chakraborty U, Pradhan B (2012) Oxidative stress in five wheat varieties (*Triticum aestivum* L.) exposed to water stress and study of their antioxidant enzyme defense system, water stress responsive metabolites and H_2O_2 accumulation. Braz J Plant Physiol 24:117–130

Chakraborty U, Pradhan B (2013) Drought stress-induced oxidative stress and antioxidative responses in four wheat (*Triticum aestivum* L.) varieties. Arch Agron Soil Sci 58:617–630

Chakraborty U, Chakraborty BN, Basnet M, Chakraborty AP (2009) Evaluation of *Ochrobactrum anthropi* TRS-2 and its talc based formulation for enhancement of growth of tea plants and management of brown root rot disease. J Appl Microbiol 107:625–634
Chakraborty U, Roy S, Chakraborty AP, Dey P, Chakraborty BN (2011) Plant growth promotion and amelioration of salinity stress in crop plants by a salt-tolerant bacterium. Rec Res Sci Technol 3:61–70
Chakraborty U, Chakraborty BN, Chakraborty AP, Dey PL (2013) Water stress amelioration and plant growth promotion in wheat plants by osmotic stress tolerant bacteria. World J Microbiol Biotechnol 29:789–803
Chen Y, Chen C, Tan Z, Liu J, Zhuang L, Yang Z, Huang B (2016) Functional identification and characterization of genes cloned from halophyte seashore *Paspalum* conferring salinity and cadmium tolerance. Front Plant Sci 7:102
Cheng T, Chen J, Zhang J, Shi S, Zhou Y, Lu L, Wang P, Jiang Z, Yang J, Zhang S, Shi J (2015) Physiological and proteomic analyses of leaves from the halophyte *Tangut nitraria* reveals diverse response pathways critical for high salinity tolerance. Front Plant Sci 6:30
Cherian S, Reddy MP (1999) Salt tolerance in the halophyte *Suaeda nudiflora* Moq.: effect of NaCl on growth, ion accumulation and oxidative enzymes. Indian J Plant Physiol 5:32–37
Chung EJ, Park JE, Jeon CO, Chung YR (2015) *Gynuella sunshinyii* gen. nov., sp. nov., an antifungal rhizobacterium isolated from a halophyte, *Carex scabrifolia* Steud. Int J Syst Evol Microbiol 65:1038–1043
de Araujo SAM, Silveira JAG, Almeida TD, Rocha IMA, Morais DL, Viegas RA (2006) Salinity tolerance of halophyte *Atriplex nummularia* L. grown under increasing NaCl levels. J Agric Environ Eng 10:848–854
Debez A, Chaibi W, Bouzid S (2001) Effect of NaCl and growth regulators on germination of *Atriplex halimus* L. Cahiers Agric 10:135–138
Del Vecchio S, Prisco I, Acosta AT, Starrisci A (2015) Changes in plant species composition of coastal dune habitats over a 20 year period. AoB Plants 7:plv018. https://doi.org/10.1093/aobpla/plv018
Dodd IC, Perez-Alfocea F (2012) Microbial amelioration of crop salinity stress. J Exp Bot 63:3415–3428
English JP, Colmer TD (2013) Tolerance of extreme salinity in two stem-succulent halophytes (*Tecticornia* species). Funct Plant Biol 40:897–912
Evelin H, Kapoor R, Giri B (2009) Arbuscular mycorrhizal fungi in alleviation of salt stress: a review. Ann Bot 104:1263–1280
Falb M, Pfeiffer F, Palm P, Rodewald K, Hickmann V, Tittor J, Oesterhelt D (2005) Living with two extremes: conclusions from the genome sequence of *Natronomonas pharaonis*. Genome Res 15:1336–1343
Fan P, Chen D, He Y, Zhou Q, Tian Y, Gao L (2016) Alleviating salt stress in tomato seedlings using *Arthrobacter* and *Bacillus megaterium* isolated from the rhizosphere of wild plants grown on saline-alkaline lands. Int J Phytoremediation 18:1113–1121
Feng J, Wang J, Fan P, Jia W, Nie L, Jiang P, Chen X, Lv S, Wan L, Chang S, Li S, Li Y (2015) High-throughput deep sequencing reveals that microRNAs play important roles in salt tolerance of euhalophyte *Salicornia europaea*. BMC Plant Biol 15:63
Flowers TJ, Colmer TD (2015) Plant salt tolerance: adaptations in halophytes. Ann Bot 115:327–331
Flowers TJ, Troke PF, Yeo AR (1977) The mechanism of salt tolerance in halophytes. Annu Rev Plant Physiol 28:89–121
Flowers TJ, Hajibagheri MA, Clipson NJW (1986) Halophytes. Q Rev Biol 61:313–337
Flowers TJ, Galal HK, Bromham L (2010a) Evolution of halophytes: multiple origins of salt tolerance in land plants. Funct Plant Biol 37:604–612
Flowers TJ, Gaur PM, Gowda CLL, Krisnamurthy L, Samineni S, Siddique KH, Turner NC, Vadez V, Varshney RK, Colmer TD (2010b) Salt sensitivity in chickpea. Plant Cell Environ 33:490–509

Frosini S, Lardicci C, Balestri E (2012) Global change and response of coastal dune plants to the combined effects of increased sand accretion (burial) and nutrient availability. PLoS One 7: e47561. https://doi.org/10.1371/journal.pone.0047561

Fukami J, de la Osa C, Ollero FJ, Megías M, Hungria M (2017) Co-inoculation of maize with *Azospirillum brasilense* and *Rhizobium tropici* as a strategy to mitigate salinity stress. Funct Plant Biol 45:328–339

Garg N, Manchanda G (2008) Effect of arbuscular mycorrhizal inoculation of salt-induced nodule senescence in *Cajanus cajan* (pigeon pea). J Plant Gr Reg 27:115–124

Gharat SA, Parmar S, Tambat S, Vasudevan M, Shaw BP (2016) Transcriptome analysis of the response to NaCl in *Suaeda maritima* provides an insight into salt tolerance mechanisms in halophytes. PLoS One 11(9):e0163485

Gill SS, Tajrishi M, Madan M, Tuteja N (2013) A DESDbox helicase functions in salinity stress tolerance by improving photosynthesis and antioxidant machinery in rice (*Oryza sativa* L. cv. PB1). Plant Mol Biol 82:1–22

Giri B, Mukerji KG (2004) Mycorrhizal inoculant alleviates salt stress in *Sesbania aegyptiaca* and *Sesbania grandiflora* under field conditions: evidence for reduced sodium and improved magnesium uptake. Mycorrhiza 14:307–312

Giri B, Kapoor R, Mukerji KG (2003) Influence of arbuscular mycorrhizal fungi and salinity on growth, biomass and mineral nutrition of *Acacia auriculiformis*. Biol Fertil Soils 38:170–175

Glenn EP, Brown JJ (1999) Salt tolerance and crop potential of halophytes. Crit Rev Plant Sci 18:227–255

Goswami D, Dhandhukia P, Patel P, Thakker JN (2014) Screening of PGPR from saline desert of Kutch: Growth promotion in *Arachis hypogea* by *Bacillus licheniformis* A2. Microbiol Res 169:66–75

Grant WD, Larsen H (1989) Group III. Extremely halophilic archaeobacteria order Halobacteriales ord. nov. In: Staley JT, Bryant MP, Pfennig N, Holt JG (eds) Bergey's manual of systematic bacteriology. Williams and Wilkins, Baltimore, pp 2216–2233

Gunde-Cimerman N, Ramos J, Plemenitaš A (2009) Halotolerant and halophilic fungi. Mycol Res 113:1231–1241

Gupta B, Huang B (2014) Mechanism of salinity tolerance in plants: physiological, biochemical, and molecular characterization. Int J Genom 701596, 18 p. https://doi.org/10.1155/2014/701596

Gupta KJ, Stoimenova M, Kaiser WM (2005) In higher plants, only root mitochondria, but not leaf mitochondria reduce nitrite to NO, in vitro and in situ. J Exp Bot 56:2601–2609

Habib SH, Kausar H, Saud HM (2016) Plant growth-promoting rhizobacteria enhance salinity stress tolerance in okra through ROS-scavenging enzymes. Bio Med Res Intern 6284547, 10 p. https://doi.org/10.1155/2016/6284547

Hahm MS, Son JS, Hwang YJ, Kwon DK, Sa-Youl Ghim SY (2017) Alleviation of salt stress in pepper (*Capsicum annum* L.) plants by plant growth-promoting rhizobacteria. J Microbiol Biotechnol 27:1790–1797

Hamdia ABE, Shaddad MAK, Doaa MM (2004) Mechanisms of salt tolerance and interactive effects of *Azospirillum brasilense* inoculation on maize cultivars grown under salt stress conditions. Plant Growth Regul 44:165–174

Harisnaut P, Poonsopa D, Roengmongkol K, Charoensataporn R (2003) Salinity effects on antioxidant enzymes in mulberry cultivar. Sci Asia 29:109–113

He AL, Niu SQ, Zhao Q, Li YS, Gou JY, Gao HJ, Suo SZ, Zhang JL (2018) Induced salt tolerance of perennial ryegrass by a novel bacterium strain from the rhizosphere of a desert shrub *Haloxylon ammodendron*. Int J Mol Sci 19:469

Hilderbrandt U, Janetta K, Ouziad F, Renne B, Nawrath K, Bothe H (2001) Arbuscular mycorrhizal colonization of halophytes in Central European salt marshes. Mycorrhiza 10:175–183

Ho I (1987) Vesicular-arbuscular mycorrhizae of halophytic grasses in the Alvord desert of Oregon. Northwest Sci 61:148–151

Ilangumaran G, Smith DL (2017) Plant growth promoting rhizobacteria in amelioration of salinity stress: a systems biology perspective. Front Plant Sci 8:1768. https://doi.org/10.3389/fpls.2017.01768

Indira K, Srinivasan M (2017) Diversity and ecological distribution of endophytic fungi associated with salt marsh plants. Ind J Geo Mar Sci 46:612–623

Jha B, Gontia I, Hartmann A (2011) The roots of the halophyte *Salicornia brachiata* are a source of new halotolerant diazotrophic bacteria with plant growth-promoting potential. Plant Soil 356:265–277

Karlidag H, Yildirim E, Turan M, Pehluvan M, Donmez F (2013) Plant growth-promoting rhizobacteria mitigate deleterious effects of salt stress on strawberry plants (*Fragaria x ananassa*). Hort Sci 48:563–567

Kaur A, Devi SR, Vyas P (2018) Stress-tolerant antagonistic plant growth-promoting rhizobacteria from *Zea mays*. J Plant Prot Res 58:115–123

Kennedy SP, Ng WV, Salzberg SL, Hood L, DasSarma S (2001) Understanding the adaptation of *Halobacterium* species NRC-1 to its extreme environment through computational analysis of its genome sequence. Genome Res 11:1641–1650

Kumar A, Verma JP (2018) Does plant-microbe interaction confer stress tolerance in plants: a review? Microbiol Res 207:41–52

Kunte HJ (2006) Osmoregulation in bacteria: Compatible solute accumulation and osmosensing. Environ Chem 3:94–99

Kunte HJ, Trüper HG, Stan-Lotter H (2002) Halophilic microorganisms. In: Horneck G, Baumstark-Khan C (eds) Astrobiology. Springer, Berlin

Lai M-C, Hong T-Y, Gunsalus RP (2000) Glycine betaine transport in the obligate halophilic archaeon *Methanohalophilus portucalensis*. J Bacteriol 182:5020–5024

Lata R, Chowdhury S, Gond SK, White JF Jr (2018) Induction of abiotic stress tolerance in plants by endophytic microbes. Lett Appl Microbiol 66:268–276

Lau JA, Lennon JT (2012) Rapid responses of soil microorganisms improve plant fitness in novel environments. Proc Natl Acad Sci 109:14058–14062

Li HQ, Jiang XW (2017) Inoculation with plant growth-promoting bacteria (PGPB) improves salt tolerance of maize seedling. Russ J Plant Physiol 64:235–241

Li Y, Kong Y, Teng D, Zhang X, He X, Zhang Y, Lv G (2018) Rhizobacterial communities of five co-occurring desert halophytes. Peer J 6:e5508

Litchfield CD (1998) Survival strategies for microorganisms in hypersaline environments and their relevance to life on early Mars. Meteorit Planet Sci 33:813–819

Liu S, Hao H, Lu X, Zhao X, Wang Y, Zhang Y, Xie Z, Wang R (2017a) Transcriptome profiling of genes involved in induced systemic salt tolerance conferred by *Bacillus amyloliquefaciens* FZB42 in *Arabidopsis thaliana*. Sci Rep 7:10795

Liu H, Wang Y, Tang M (2017b) Arbuscular mycorrhizal fungi diversity associated with two halophytes *Lycium barbarum* L. and *Elaeagnus angustifolia* L. in Ningxia, China. Arch Agron Soil Sci 63:796–806

Lv S, Jiang P, Chen X, Fan P, Wang X, Li Y (2012) Multiple compartmentalization of sodium conferred salt tolerance in *Salicornia europaea*. Plant Physiol Biochem 51:47–52

Ma Y, Galinski EA, Grant WD, Oren A, Ventosa A (2010) Halophiles 2010: life in saline environments. Appl Environ Microbiol 76:6971–6981

Marasco R, Mapelli F, Rolli E, Mosqueira_santillan MJ, Fusi M, Bariselli P, Reddy M, Cherif A, Tsiamis G, Borin S, Daffonchio D (2016) *Salicornia strobilacea* (synonym of *Halocnemum strobilaceum*) grown under different tidal regimes selects rhizosphere bacteria capable of promoting plant growth. Front Microbiol 7:1286

Marulanda A, Azcón R, Chaumont F, Ruiz-Lozano JM, Aroca R (2010) Regulation of plasma membrane aquaporins by inoculation with a *Bacillus megaterium* strain in maize (*Zea mays* L) plants under unstressed and salt-stressed conditions. Planta 232:533–543

Mayak S, Tirosh T, Glick BR (2004) Plant growth-promoting bacteria confer resistance in tomato plants to salt stress. Plant Physiol Biochem 42:565–572

Munns R (2002) Comparative physiology of salt and water stress. Plant Cell Environ 25:239–250
Munns R, Tester M (2008) Mechanisms of salinity tolerance. Annu Rev Plant Biol 59:651–681. https://doi.org/10.1146/annurev.arplant.59.032607.092911
Muthezhilan R, Sindhuja BS, Hussain AJ, Jayaprakashvel M (2012) Efficiency of plant growth promoting rhizobacteria isolated from sand dunes of Chennai coastal area. Pak J Biol Sci 15:795–799
Naidoo G, Rughunanan R (1990) Salt tolerance in the succulent, coastal halophyte, *Sarcocornia natalensis*. J Exp Bot 41:497–502
Navarro-Torre S, Barcia-Piedras JM, Mateos-Naranjo E, Redondo-Gomez S, Camacho M, Caviedes MA, Pajuelo E, Rodríguez-Llorente ID (2017) Assessing the role of endophytic bacteria in the halophyte *Arthrocnemum macrostachyum* salt tolerance. Plant Biol 19:249–256
Numan M, Bashir S, Khan Y, Mumtaz R, Shinwari ZK, Khan AL, Khan A, AL-Harrasi A (2018) Plant growth promoting bacteria as an alternative strategy for salt tolerance in plants: a review. Microbiol Res 209:21–32
Oren A (1999) Bioenergetic aspects of halophilism. Microbiol Mol Biol Rev 63:334–348
Oren A (2008) Microbial life at high salt concentrations: phylogenetic and metabolic diversity. Saline Syst 4:2
Parida AK, Veerabathini SK, Kumari A, Agarwal PK (2016) Physiological, anatomical and metabolic implications of salt tolerance in the halophyte *Salvadora persica* under hydroponic culture condition. Front Plant Sci 7:351
Porras-Soriano A, Soriano-Martin ML, Porras-Piedra A, Azcón R (2009) Arbuscular mycorrhizal fungi increased growth, nutrient uptake and tolerance to salinity in olive trees under nursery conditions. J Plant Physiol 166:1350–1359
Qin S, Zhang Y-J, Yuan B, Xu P-Y, Xing K, Wang J, Jiang J-H (2014) Isolation of ACC deaminase-producing habitat-adapted symbiotic bacteria associated with halophyte *Limonium sinense* (Girard) Kuntze and evaluating their plant growth-promoting activity under salt stress. Plant Soil 374:753–766
Qin S, Feng W-W, Zhang Y-J, Wang T-T, Xiong Y-W, Xing K (2018) Diversity of bacterial microbiota of coastal halophyte *Limonium sinense* and amelioration of salinity stress damage by symbiotic plant growth-promoting actinobacterium *Glutamicibacter halophytocola* KLBMP 5180. Appl Environ Microbiol 84:e01533–e01518
Railsback LB, Ackerly SC, Anderson TF, Cisne JL (1990) Paleontological and isotope evidence for warm saline deep waters in Ordovician oceans. Nature 343:156–159
Roy S, Chakraborty U (2018) Role of sodium ion transporters and osmotic adjustments in stress alleviation of *Cynodon dactylon* under NaCl treatment: a parallel investigation with rice. Protoplasma 255:175–191
Ruppel S, Franken P, Witzel K (2013) Properties of the halophyte microbiome and their implications for plant salt tolerance. Funct Plant Biol 40:940–951
Sannazzaro AI, Echeverria M, Albertó EO, Ruiz OA, Menéndez AB (2007) Modulation of polyamine balance in *Lotus glaber* by salinity and arbuscular mycorrhiza. Plant Physiol Biochem 45:39–46
Saum SH, Muller V (2008) Regulation of osmoadaptation in the moderate halophile *Halobacillus halophilus*: chloride, glutamate and switching osmolyte strategies. Saline Syst 4:4
Schubert S, Neubert A, Schierholt A, Sumer A, Zorb C (2009) Development of salt-resistant maize hybrids: the combination of physiological strategies using conventional breeding methods. Plant Sci 177:196–202
Sharifi M, Ghorbanli M, Ebrahimzadeh H (2007) Improved growth of salinity-stressed soybean after inoculation with salt pretreated mycorrhizal fungi. J Plant Physiol 164:1144–1151
Sharma S, Kulkarni J, Jha B (2016) Halotolerant rhizobacteria promote growth and enhance salinity tolerance in peanut. Front Microbiol 7:1600
Shi-Ying Z, Cong F, Yong-Xia W, Yun-Sheng X, Wei X, Xiao-Long C (2018) Salt-tolerant and plant growth-promoting bacteria isolated from high-yield paddy soil. Can J Microbiol 64:968–978. [Epub ahead of print]

Siglioccolo A, Paiardini A, Piscitelli M, Pascarella S (2011) Structural adaptation of extreme halophilic proteins through decrease of conserved hydrophobic contact surface. BMC Struct Biol 11:50
Singh RP, Jha PN (2016a) Alleviation of salinity-induced damage on wheat plant by an ACC deaminase-producing halophilic bacterium *Serratia* sp. SL-12 isolated from a salt lake. Symbiosis 69:101–111. https://doi.org/10.1007/s13199-016-0387-x
Singh RP, Jha PN (2016b) The multifarious PGPR *Serratia marcescens* CDP-13 augments induced systemic resistance and enhanced salinity tolerance of wheat (*Triticum aestivum* L.). PLoS One 11(6):e0155026
Szymanska S, Płociniczak T, Piotrowska-Seget Z, Hrynkiewicza K (2016) Endophytic and rhizosphere bacteria associated with the roots of the halophyte *Salicornia europaea* L. – community structure and metabolic potential. Microbiol Res 192:37–51
Tenchov B, Vescio EM, Sprott GD, Zeidel ML, Mathai JC (2006) Salt tolerance of archaeal extremely halophilic lipid membranes. J Biol Chem 281:10016–10023
Tian XY, Zhang CS (2017) Illumina-based analysis of endophytic and rhizosphere bacterial diversity of the coastal halophyte *Messerschmidia sibirica*. Front Microbiol 8:2288
Tian CY, Feng G, Li XL, Zhang FS (2004) Different effects of arbuscular mycorrhizal fungal isolates from saline or non-saline on salinity tolerance of plants. Appl Soil Ecol 26:143–148
Tuteja N, Sahoo RK, Garg B, Tuteja R (2013) OsSUV3 dual helicase functions in salinity stress tolerance by maintaining photosynthesis and antioxidant machinery in rice (*Oryza sativa* L. cv. IR64). Plant J 76:115–127
Ullah S, Bano A (2015) Isolation of plant-growth-promoting rhizobacteria from rhizospheric soil of halophytes and their impact on maize (*Zea mays* L.) under induced soil salinity. Can J Microbiol 61:307–313
Ungar I (1991) Ecophysiology of vascular halophytes. CRC, Boca Raton
Upadhyay SK, Singh DP (2015) Effect of salt-tolerant plant growth promoting rhizobacteria on wheat plants and soil health in a saline environment. Plant Biol 17:288–293. https://doi.org/10.1111/plb.12173
Vaishnav A, Varma A, Tuteja N, Choudhary DK (2016) PGPR-mediated amelioration of crops under salt stress. In: Choudhary DK, Varma A, Tuteja N (eds) Plant-microbe interaction: an approach to sustainable agriculture. Springer, Singapore, pp 205–226. https://doi.org/10.1007/978-981-10-2854-0_10
Vreeland RH (1987) Mechanisms of halotolerance in microorganisms. Crit Rev Microbiol 14:311–356
Vurukonda SSKP, Vardharajula S, Shrivastava M, SkZ A (2016) Enhancement of drought stress tolerance in crops by plant growth promoting rhizobacteria. Microbiol Res 184:13–24
Wang FY, Liu RJ, Lin XG, Zhou JM (2004) Arbuscular mycorrhizal status of wild plants in saline-alkaline soils of the Yellow River Delta. Mycorrhiza 14:133–137
Wilde P, Manal A, Stodden M, Sieverding E, Hilderbrandt U, Bothe H (2009) Biodiversity of arbuscular mycorrhizal fungi in roots and soils of two salt marshes. Environ Microbiol 11:1548–1561
Wohlfarth A, Severin J, Galinski EA (1990) The spectrum of compatible solutes in heterotrophic halophilic eubacteria of the family *Halomonadaceae*. J Gen Microbiol 136:705–712
Yamato M, Ikeda S, Iwase K (2008) Community of arbuscular mycorrhizal fungi in coastal vegetation on Okinawa Island and effect of the isolated fungi on growth of sorghum under salt-treated conditions. Mycorrhiza 18:241–249
Yang J, Kloepper JW, Ryu CM (2009) Rhizosphere bacteria help plants tolerate abiotic stress. Trends Plant Sci 14:1–4
Yang A, Akhtar SS, Iqbal S, Amjad M, Naveed M, Zahir ZA, Jacobsen S-E (2016) Enhancing salt tolerance in quinoa by halotolerant bacterial inoculation. Funct Plant Biol 43:632–642
Yuan Z, Druzhinina IS, Labbe J, Redman R, Qin Y, Rodriguez R, Zhang C, Tuskan GA, Lin F (2016) Specialized microbiome of a halophyte and its role in helping non-host plants to withstand salinity. Sci Rep 6:32467. https://doi.org/10.1038/srep32467

Zerrouk IZ, Benchabane KL, Yokawa K, Ludwig-Muller J, Baluska F (2016) A *Pseudomonas* strain isolated from date-palm rhizospheres improves root growth and promotes root formation in maize exposed to salt and aluminum stress. J Plant Physiol 191:111–119. https://doi.org/10.1016/j.jplph.2015.12.009

Zuccarini P, Okurowska P (2008) Effects of mycorrhizal colonization and fertilization on growth and photosynthesis of sweet basil under salt stress. J Plant Nutr 31:497–513

Chapter 10
Effect of Salinity on Physiological Processes in Plants

Okon Godwin Okon

Abstract Soil salinity is amongst the most severe globally threatening abiotic stress problems that affects approximately 20% of irrigated lands and reduces crop yields and productivity significantly. The process of soil salinization can be majorly attributed to natural contributors such as weathering of parental rocks, oceanic salt deposition and groundwater table fluctuations, while the common anthropogenic factors contributing to salinization include irrigational water and the incessant and disproportionate usage of inorganic fertilizers and pesticides. In saline environs, soil salinity affects plants in several ways and induces dissimilar but distinctive stresses such as osmotic, ionic and oxidative stress. As a consequence of excessive accumulation and toxicity of excessive Na^+ and Cl^- in the plants, there is a negative effect on almost every known physiological and biochemical metabolic pathway in plants which includes inhibition of photosynthetic activities, membrane disorder, production of toxic metabolites, production of ROS and attenuated water and nutrient acquisition leading to physiological drought, chlorosis and necrosis and cell and subsequently whole plant death as a result of Na^+ and Cl^- intrusion on many physiological processes in plants. Plants grown in saline environments device a series of adaptive measures to acclimatize to saline conditions; this includes morphological, physiological and biochemical variations. These variations comprise of upsurges in the photosynthetic pigment composition, root/canopy quotient as well as leaf anatomy and ion compartmentalization that ultimately leads to reduction in leaf ion toxicity, thus preserving the water status, limiting water loss and, hence, protecting the process of photosynthesis. This chapter thus discusses the effects of salinity on physiological processes of plants. It also provides some future viewpoints which will widen scientific understanding of exactly how soil salinity distresses the physiological physiognomies of plants.

Keywords Ionic toxicity · Nutrient disparity · Osmotic stress · Oxidative stress · Salt tolerance · Soil salinization

---

O. G. Okon (✉)
Department of Botany, Akwa Ibom State University, Ikot Akpaden, Akwa Ibom, Nigeria

B. Giri, A. Varma (eds.), *Microorganisms in Saline Environments: Strategies and Functions*, Soil Biology 56, https://doi.org/10.1007/978-3-030-18975-4_10

10.1 Introduction

Plants are faced by a wide-ranging assortment of environmental stresses which includes drought, temperature (heat and cold), waterlogging, UV stress, alkalinity, salinity, etc. Soil salinization is a worldwide problem that affects approximately 20% of irrigated land as well as coastal regions and reduces crop yields significantly (Qadir et al. 2014; Negrao et al. 2017). Salinization in water used for irrigation and abundance in agricultural soils constitutes one of the numerous major abiotic constrictions on agriculture globally, and the situation has degenerated in a span of about 20 years as a result of the upsurge in demand and irrigational necessities in semi and arid areas as observed in the Mediterranean (Colla et al. 2010; Cirillo et al. 2016; Jose et al. 2017). Soil salinization distresses nearby 800 million ha of agriculturally arable lands worldwide (Munns and Tester 2008). A soil solution at an EC (electrical conductivity) of about 4 dS m^{-1} which is equal to about 40 mM of NaCl and producing an osmotic force reaching almost 0.2 MPa resulting in the meaningful decrease of the growth and productivity of most crop plants is regarded as saline (Munns and Tester 2008). Soil solution contains various important soluble salts and is regarded as the best media for plant growth and development (Saxena et al. 2017). The absorption and translocation of these soluble salts is carried out by roots of plants to various plant parts where they are required for numerous metabolic activities.

One of the major problems facing crop production globally is the excess accumulation of soluble salts in agriculturally viable soils (Nasim 2010; Ebrahim 2014). The excessive content of water-soluble salts, basically sodium (Na), chloride (Cl), calcium (Ca), potassium (K) and magnesium (Mg), is referred to as salinity (US Department of Agriculture 2008). This definition of salinity is driven from the ECe that significantly reduces the yield of most crops. Under natural conditions, the salts in soil contaminated by sea water are usually mixtures of several cations and anions. It is difficult to ascertain the specific ion effect from the study of such condition (Nayer et al. 2013). With the additional environmental issues of global warming persisting severely too, soil salinity has been projected to affect at least half of the globally arable lands (Gamalero et al. 2009).

Aside from plants, a whole community of soil microorganisms is also destroyed as well as favourable soil desirable properties (Zhu 2002; Yuan et al. 2007). A typical example of soils affected by salts usually shows excess presence of ions of salts, namely, Na^{+2}, Na^{+}/K^{+}, Mg^{+2}/Ca^{+2} and Cl^{-}/NO^{-3} (Zhu 2002). In a hypersaline soil or water, the other ion ingredients may be B, Sr^{2+}, SiO_2, Mo, Ba^{2+} and Al^{3+} (Hu et al. 2000). The use of groundwater that possesses very high contents of salts for irrigational purposes, increased temperatures resulting in high rate of evaporation as well as excessive addition of fertilizers to the soil all contribute to the severe cases of soil salinity (Cantrell and Linderman 2001; Ebrahim 2014). There are about three different types of categories for salt-affected soils, namely, saline, saline-alkali and alkaline soils depending on total EC concentration of water-soluble paste of salts retained in them, soil pH as well as the exchangeable sodium percentage of the soil

(Rasool et al. 2012; Parvaiz et al. 2014). Although NaCl appears to be the principal salt present in soils mostly affected by salts, other salts such as Na_2SO_4, $MgCl_2$, KCl and $NaHCO_3$ are also frequently found in considerable amounts in such soils (Bie et al. 2004).

10.2 Effects of Salinity on Plants

The injurious effects of soil salinity on plant growth and agricultural yield are enormous (Tester and Davenport 2003), resulting in huge damages in crop yield and productivity (Giri et al. 2003; Mathur et al. 2007). Salinity affects plant through various means such as:

1. Severe drought physiologically to plants as a result of limitations of water availability induced by the reduced soil solution osmotic potential. In order to reverse this effect, plant adjusts to maintaining a very low internal osmotic potential so as to prevent outward movement of water from the plant roots into the soil (Feng et al. 2002; Jahromi et al. 2008; Evelin et al. 2009).
2. Excessive toxicity of Na^+ and Cl^- ions in the cell results in the interference with plasma membrane integrity, destruction of organelles in cells and inhibition of photosynthetic activities, protein synthesis, enzyme structures and respiration (Juniper and Abbott 1993; Feng et al. 2002; Evelin et al. 2009).
3. Nutrient disproportion because of the exclusion of essential mineral nutrients and excessive uptake and/or upward transport of ions of Na^+ and Cl^- to plant aerial part which leads to essential mineral nutrients ion deficits (Adiku et al. 2001) (Fig. 10.1).

10.2.1 Effects of Na^+ and Cl^- Toxicity

Osmotic and ionic stresses are the foremost effects imposed on plants by soil salinization. Excessive incursion of salt ions may cause excessive build-up of Na^+ and Cl^- in the cytosol and lower the apoplastic water potential (Shan 2009). Damages osmotically can happen in plants, as salinity in the soil increases, as a consequence of the build-up of excess salt ions. For most plant species, Na^+ is considered to be more toxic than Cl^- (Munns and Tester 2008). Once Na^+ is taken up by the roots, it can be rapidly translocated to shoots via the xylem where it accumulates as water evaporates. Additionally, recirculation of Na^+ back to roots is limited, suggesting that Na^+ transport is somewhat unidirectional ensuing in accumulation in shoots and foliage (Tester and Davenport 2003).

Na^+ ions overt damage is concomitant with the excessive uptake by roots and accumulation of Na^+ in leaves, ensuing in necrosis and shortening the lifetime of specific leaves, eventually resulting in the reduction in net crop yield and

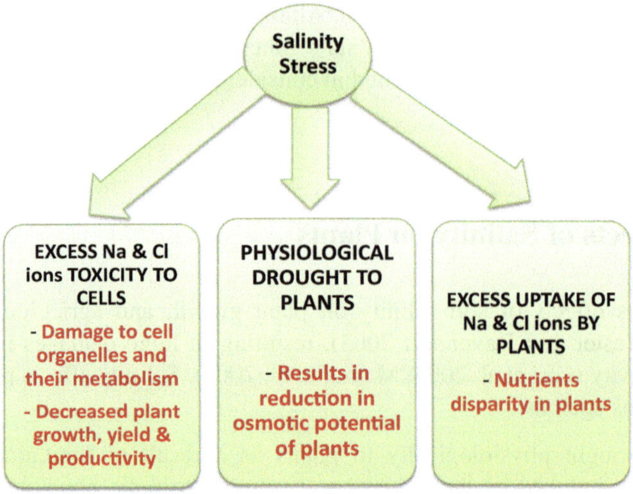

Fig. 10.1 Effect of salinity stress on plants. Salinity stress results in physiological drought to plants as a result of water deficit, disparity in macro- and micronutrient uptake and composition as a result of excessive toxicity due to Na and Cl ion accumulation thereby leading to damage of cell organelles and disruption of essential metabolic activities, ultimately resulting in poor plant growth and reduces plant productivity

productivity (Munns 1993). Plants need to maintain turgor pressure and water uptake. Without proper intracellular compartmentalization (i.e. vacuole storage), excess of Na^+ is toxic in plants due to osmotic disproportions. Moreover, high soil Na^+ can also cause deficiencies of other nutrients by meddling with ion uptake and transporters. K^+ homeostasis is essential for cellular functions, and this can be disrupted due to antagonism of Na^+ for K^+ binding sites. K^+ plays a very important role in the activation of about 50 enzymes in plant cells; however, this important role played by K^+ cannot be easily replaced by substituting K^+ with Na^+; therefore enzyme activity can be inhibited in the presence of excess Na^+ (Bhandal and Malik 1988). Similarly, the processes of binding the tRNA to ribosomes requires a very high concentration of K^+; therefore, competition of K^+ binding sites by Na^+ can cause severe damage in plant cells (Shan 2009).

The chloride ion (Cl^-) is an essential micronutrient for higher plants. It is involved in oxygen evolution reactions in photosynthesis (Olesen and Andreasson 2003), maintaining electrical charge across membranes (Läuchli and Lüttge 2002) and adjusting osmotic potential in vacuoles and the cytosol. Toxicity and inhibition of photosynthesis may occur if Cl^- is present in excess. In some plant species, namely, grapevine, citrus and soybean, only small amounts of Na^+ reach the leaves, whereas Cl^- can continue to accumulate in the leaves; therefore, Cl^- is considered a more toxic component in the aforementioned plants (Munns and Tester 2008; Shan 2009).

10.3 Salinity Effects on Plant Growth, Development and Biomass Yield

At the early stage of salinity stress, the reduction of plant growth observed is as a result of the osmotic stress caused by soil salinization, while excessive uptake and accumulation of Na^+ in plant leaves is responsible for the plant growth reduction observed in the later stage (Munns 2005; Munns and Tester 2008; Rahnama et al. 2010). Reduction in percentage germination/seedling emergence, shoot length, leaf area, petiole length and internode length has been reported as well.

10.3.1 Shoot Length

Salt stress significantly affects the shoot length of plants; this has been demonstrated by the work of Amira and Abdul (2010) who reported significant differences, whether an increase or decrease for *Vicia faba* grown under saline conditions. To further corroborate to these findings, Taffouo et al. (2009) on cowpea (*Vigna unguiculata* L.), Mathur et al. (2006) on their research on moth bean (*Vigna aconitifolia* L.), Kapoor and Srivastava (2010) on *Vigna mungo* L. and Jamil et al. (2007) on radish plant (*Raphanus sativus* L.), they all reported that increased soil salinity levels developed a significant decline in the shoot lengths of the plants.

10.3.2 Number of Leaves

Several researchers have confirmed that salt stress significantly affects the leaf numbers of plants; they include Karen et al. (2002), with their study on *Cirer arietinum* L., and Raul et al. (2003); they carried out a research which involves leaves of *Phaseolus acutifolius* L. (tepary bean) as well as *Phaseolus filiformis* L. (wild bean) and *Vigna unguiculata* L. (cowpea). They observed that the number of leaves in these plants was reduced as a result of NaCl treatment when they compared the NaCl treated plants to the control. The observations made by these researchers can be further explained that mature leaves of plants grown in saline conditions accumulate excess NaCl in their cytoplasm and cell walls. Simultaneously, since the sap of the vacuole cannot accommodate more salts than it's supposed to, thus, there is a decrease in the concentration of salt inside the cells, resulting in early senescence and leaf abscission (Munns 2002).

10.3.3 Leaf Area

Previous studies have shown that salt stress significantly affected the leaf area of plants. Such studies include the reports of Amira and Abdul (2010) who reported significant reduction in leaf area in *Vicia faba* grown under salinity stress. Mathur et al. (2006) also reported that the stress of the moth bean plant (*Vigna aconitifolia* L.) and the reduction in the plant leaf area was a response to the upsurge in NaCl concentrations. Other researchers such as Zhao et al. (2007) also made similar observations while studying *Avena sativa* L. (oat); similarly, Yilmaz and Kina (2008), also research on *Fragaria xananssa* (L.) and also made similar observations. They all reported that leaf area reduction is a major effect of the exposure of plants to high soil salinity caused by NaCl. The remarkable decrease in the leaf area in salinity treatment plants in the reports stated above is due to the upsurge in absorption and accumulation of Na^+ and Cl^- in the leaves. This effect could be clarified by the negative interruption effect of salt on photosynthesis that ultimately leads to the diminished growth of plant, leaf growth to avoid escape of water via transpiration and chlorophyll content (Netondo et al. 2004).

10.3.4 Biomass

The extent of plant growth and the above-ground biomass they generate are very important parameters to access when a plant is grown under stressed conditions. The above-ground plant parts or aerial parts usually experience or show remarkable reduction as a result of salt stress. This can be observed by the general reduction in the plant fresh (FW) and dry weight (DW) of all the plant parts (Jose et al. 2017). Beltrano et al. (2013) corroborate these findings in their work on *Capsicum annum*; they observed that shoot and root DWt declined as soil salinity increased. According to the findings of Yasar and Esra (2012), significant decreases were observed in the fresh and dry total plant weights (shoot and roots) depending on the increasing irrigation water salinity. Other supporting researches include the works by Rui et al. (2009) on *Bruguiera gymnorrhiza* L., Jamil et al. (2007) who worked on *Raphanus sativus* L. (radish plants), similarly, He et al. (2008) reported similar observation made while working on *Kyllinigia peruviana* L. and then Memon et al. (2010) reported same on *Brassica campestris* L.

To further expound on this, many researches have suggested that plants stressed by soil salinity usually activates the synthesis of 1-aminocyclopropane-1-carboxilic acid (ACC) which is a consequence of specific uptake and build-up of Cl^-; 1-aminocyclopropane-1-carboxilic acid (ACC) is then converted to the highly efficient hormone ethylene which will be released in a high quantity enough to activate the abscission of leaves as reported in several plants including citrus (Tudela and Primo-Millo 1992; Dodd 2005; Jose et al. 2017).

Several researchers have reported that salt and osmotic stress stimulated 1-aminocyclopropane-1-carboxilic acid (ACC) to be converted to the highly efficient hormone ethylene in the halophyte plants such as *Allenrolfea occidentalis* (Chrominski et al. 1988; Jose et al. 2017). Several other researches have also backed up the fact that under salt-induced osmotic stress, early leaf death and subsequent abscission are a result of the uptake and accumulation of toxic Na^+ and Cl^- ions in the leaves as observed and reported in tomato irrigated with high concentrations of salt. ACC is responsible for the declines in chlorophyll contents of salinized plants and is also implicated in the early onset of oxidative damage. Under saline conditions, there is a progressive decrease in leaf contents of cytokinin (CK) and indol-3-acetic acid (IAA) which favours the accumulation and actions of abscisic acid (ABA) which induces leaf senescence (Albacete et al. 2008; Ghanem et al. 2008; Jose et al. 2017).

10.4 Effect of Salinity on Photosynthetic Pigments and Photosynthetic Process

Salinity causes a reduction in photosynthetic pigment concentrations due to reduction in water potential. NaCl prevents the uptake of N which is a major element required by plants and plays a very vital role in the make-up of the structure of chlorophyll (Kaya et al. 2009); NaCl also impedes the uptake of Mg which plays an important role in the biosynthesis of chlorophyll, as well as deterring the specific activities of enzymes which are needed in photosynthetic pigments production (Murkute et al. 2006; Sheng et al. 2008; Abdel-Lateef and Chaoxing 2011).

Soil salinity reduces photosynthetic capability of plants as it affects the photosynthetic apparatus and concentration of plant's photosynthetic pigments (Doganlar et al. 2010; Ayala-Astorga and Alcaraz-Melendez 2010; Perveen et al. 2010; Akram and Ashraf 2011), and it also hinders chlorophyll biosynthesis by subduing its precursors (Santos 2004) and lowers uptake of mineral nutrient such as magnesium (Giri and Mukerji 2004; Murkute et al. 2006; Sheng et al. 2008). Diffusion limitations of CO_2 used in the process of photosynthesis is hindered by soil salinity as well (Flexas et al. 2007) and a remarkable decrease in the of photosynthetic pigments contents (Delfine et al. 1999; Ashraf and Harris 2013). This have been demonstrated in spinach that the excessive salt accumulation inhibits photosynthesis as reported by Di Martino et al. (1999), primarily by limiting the CO_2 stomatal and mesophyll diffusion and conductance (Delfine et al. 1998) and plunging the content of chlorophyll, where light absorbance is significantly affected (Ashraf and Harris 2013; Alvino et al. 2000).

Similar observation was made in radish plant, where research has it that approximately 80% of inhibited growth at high salt stress levels results from light capture limitations resulting from the reduced leaf area. Reduced conductance of the stomata then accounts for the outstanding 20% of reduced growth due to salt stress (Marcelis and Van Hooijdonk 1999). Thus, soil salinity ultimately hinders the whole photosynthetic process of plants by reduction in plant leaf area which inhibits the process of photosynthesis thereby limiting plant ability to grow and generate biomass (Yeo 2007; Rui and Ricardo 2017).

This corresponds with the findings of several researchers like Jing et al. (2015) who observed that the total photosynthetic pigments when the plant *Suaeda aralocaspica* was exposed to high levels of salinity was drastically lowered; destruction of the photosynthetic apparatus (chloroplast) was responsible for this reduction. Amira and Abdul (2010) while working in *Vicia faba* also reported decline in chlorophyll 'a', 'b' and total photosynthetic pigments. Exposing the plant *Hordeum vulgare* L. to 0, 120 and 240 mM of NaCl also resulted in the decline in chlorophyll 'a', 'b' and total photosynthetic pigments as reported by Tort and Turkyilmaz (2004). Other researches to back up this observation includes the works of Taffouo et al. (2009, 2010) on *Vigna unguiculata* L. and *Vigna subterranean* L., Lee et al. (2004) on *Paspalum vaginatum* L., Turan et al. (2007) on *Phaseolus vulgaris* L., Siler et al. (2007) on *Centaurium erythraea* L. and Cheruth et al. (2008) on *Catharanthus roseus* L.; all these researchers in their work reported the decline in total photosynthetic pigments caused by NaCl.

The photosystem II (PS II) is a pigment system that plays a very vital part in the process of photosynthesis; however, salinity alters the structure and activity of PS II (Al-Taweel et al. 2007; Abdeshahian et al. 2010; Saleem et al. 2011) and to a greater extent obstructs the activities of this reaction centre (Ruiz-Lozano et al. 2012). When there is excessive energy in the photosystem II (PS II) and plant fails to disperse the excess energy, this can damage the photosystem (Ruiz-Lozano et al. 2012). Under stress conditions, the photosynthetic activity of the plant is often affected as a result of the adjustment in the level and the activity of various photosynthetic enzymes such as Rubisco (Aragao et al. 2005; Valentine et al. 2006), sucrose phosphate synthase and fructose-1,6-bisphosphatase (Ghosh et al. 2001) and PEP carboxylase (Abdel-Lateef 2008).

10.5 Relative Membrane Permeability

Soil salinity results in increased electrolyte permeability of plasma membranes which could be attributed to the efflux of K^+ due to excessive Na^+ uptake. Salinity causes injury to cellular membranes which has been reported by several researchers such as Mumtaz et al. (2013), in their research on *Cucumis sativus* that, at high salt stress levels, there was an enhancement of electrolyte leakage when compared to the non-salinized plants or control. Equally, Mumtaz et al. (2013) again working on salt-sensitive cultivars of *Cucumis* observed increased leakage of electrolytes than on the cultivars tolerant to salt. Several authors have also made these observations in cucumber, namely, Kaya et al. (2001); in rice, Lutts et al. (1996); in tomato, Atilla (2014); and in sugar beet, Ghoulam et al. (2002). A major effect of environmental stress (i.e., salt, drought) on plant is membrane modification as a result of cell membrane perturbation function or total dysfunction. Changes in membrane leakage and injury can be measured by the extent of electrolyte leakage (EL) in tissues (Atilla 2014).

10.6 Plant Water Status Under Salinity

Soil salinity creates a difficult water acquisition condition for plants which faces difficult water uptake as the salinity lowers water potential, consequently, decreasing plant growth and yield. Therefore, sustaining water homeostasis and the normal functioning of photosynthetic apparatus is essential for mitigating the impact of salinity on plant development and productivity. Soil salinity reduces root hydraulic conductivity of plants (Martínez-Ballesta et al. 2003); however, the precise mechanism by which salinity reduces hydraulic conductivity of plant roots is not clear; some way it is validated with the functioning of aquaporins or their abundance in the membrane (Ruiz-Lozano et al. 2012).

Plant-water use and relations elucidate the conduct of plants in relations of how they regulate water use within their cells; this is important because it has vital effect on the different water-dependent physiological and metabolic activities within the plants that control the quality and quantity of general plant growth, yield and productivity (Jose et al. 2017). Plants essentially grow and exist on two dissimilar environments, namely, soil and atmosphere, which are usually dissimilar in their manner of coordination and function. The atmosphere is practically continuous in its evaporative demand; however, rain water supply and that from irrigation may not make up enough water in the soil for uptake. A significant portion of absorbed water by plant roots from soil is subsequently lost to the atmosphere via transpiration. In this situation, a continuous water column is created in the xylem, leading to continuous water flow from soil to the roots to the leaves (Jose et al. 2017).

This upward water take-up by plants is relational to force driving its take-up and is also oppositely proportionate to oppositional force met via the soil-plant system. Transpiration, however, produces the internal conditions by which water is taken up by the plant root system, a transpirational pull and conveyed to the plant canopy.

Consequently, as water is lost via the process of transpiration, there is a drastic drop in the pressure of the leaf water potential and xylem (Jose et al. 2017). The opening and closing of the stomata controls the process of transpiration; however, if the stomata is continuously open and there is a constant loss of water via transpiration without enough replacement or replenishment of this loss water in the plants from the soil, there will be dehydration in the plants, and plant leaves will subsequently lose its turgidity. Many environmental and climatic factors influence the amount of water loss and dehydration in plants: the irrigational water quality, plant canopy nature and propagative growth age and stage of the plant (Jose et al. 2017). In a well-hydrated plant, plant leaf cooling can be achieved via transpiration.

Under water deficit conditions, plants resist transpiration by closing their stomata, and as such the plant is forced to transpire via the cuticles and canopy. This leads to the gradual rise in the plant temperature resulting in resistance of the stomata to transpiration and gradual escalation in the respiration of plant tissues. Consequently, this condition results in limiting the photosynthetic capabilities of the plant thereby resulting in decreased growth and subsequently plant productivity (Jose et al. 2017).

The presence of abundant salts in the root zone rhizosphere of plants gives rise to a decline in the osmotic potential and thereby lessens the water availability for take-up by the plant via the root zone (Franco et al. 2011; Niu and Rodriguez 2006). Ali and Hassan (2014) reported that under salt stress because of inadequate water uptake, relative water content (RWC) was significantly decreased in relation to salinity in chamomile herb. Shou-Jun et al. (2014) also reported that under saline condition, leaf relative turgidities in plants reduced throughout the experiment. As salinity stress was fulfilled, the significant decreases in leaf relative water content and leaf turgidity occurred at all salinity levels. Plants growing under normal conditions maintain relatively higher leaf turgidity compared to plants grown under saline conditions.

Additionally, in soil solutions where salts are existent, there are always incidences of imbalances of nutrients, thus, deterring the uptake of essentials by the roots. Several other limitations caused by salt stress include impeded photosynthesis, reduction of enzymatic activities, alteration and fluctuations of hormones and alterations of metabolism (Munns 2002; Acosta-Motos et al. 2014). In all, the above-listed alterations and effects result in poor plant growth and general loss in yield and productivity (Bhatnagar-Mathur et al. 2007).

10.7 Antioxidant Enzyme Activities Under Salt Stress

Under salt stress environments, there is always reactive oxygen species (ROS) being produced. Free radicals, namely, hydroxyl radical ($^{\cdot}OH$) and superoxide anion ($O2^{-}$), are amongst the ROS produced, while single oxygen (O_2) and hydrogen peroxide (H_2O_2) make up the non-radical molecules. The unavoidable seepage of electrons onto O_2 results in the formation of reactive oxygen species in plants. This would be the effect of the electron transport to chloroplasts, plasma membranes and mitochondria or as a by-product of different metabolic pathways in numerous cellular partitions (Sharma et al. 2012).

The site of production of reactive oxygen species in cytosol includes the mitochondria, chloroplasts as well as apoplastic spaces of plant cells where ROS basically O_2^{-}, H_2O_2 and a hydroxyl ion OH^{-} are produced (Mittler 2002; Abbasi et al. 2014, 2016).

Injury to membranes usually occurs as a result of the rise in the formation of reactive oxygen species (Shalata et al. 2001). At a very high concentration, all reactive oxygen species are exceedingly damaging to organisms. ‘Oxidative stress’ results when the level of tolerance of a particular cell is lower than the ROS level or level of ROS formation. The boosted formation and increased ROS level in the cell when plants come under several environmental stresses as severe as salinity, this ROS can adversely impede cell activities directly by triggering proteins oxidation as well as lipids peroxidation thereby averting the activity of enzymes, which ultimately leads to cell deactivation and death (Sharma et al. 2012).

It is a well-known fact that plants have both enzymatic and non-enzymatic machineries designed for rummaging reactive oxygen species (ROS). These enzymatic antioxidants comprise of cycle such as ascorbate peroxidase (APX), dehydroascorbate reductase (DHAR), superoxide dismutase (SOD), guaiacol peroxidase (GPX), catalase (CAT), monodehydroascorbate reductase (MDHAR), glutathione reductase (GR) and ascorbate glutathione (AsA-GSH). The non-enzymic antioxidant present in the cell includes ascorbate (AsA), glutathione (GSH), carotenoids, tocopherols as well as phenolics (Sharma et al. 2012).

Essential to plant tolerance under stressed environments, plants put in place a system of antioxidant defence which helps them cleanse excessive ROS in plant systems; amongst such non-enzymatic antioxidant combinations are carotenoids in addition tocopherols, while peroxidase (POD), catalase (CAT) and superoxide dismutase (SOD) make up the enzymatic antioxidant components (Ali et al. 2011; Abbasi et al. 2014, 2016).

This can be corroborated with previous researches. Hernandez et al. (1995) reported in their research with *Pisum sativum* (pea) which is a salt-tolerant cultivar of pea plants recorded series of elevated chloroplast CuZn-SOD as well as mitochondrial Mn-SOD and ascorbate peroxidase more than their more salt-sensitive *Pisum sativum* cultivar. They also observed excessive production of ascorbate peroxidase (APX) and glutathione reductase (GSH); all these show mitigation and tolerance properties against oxidative stress, improving water use in wheat (Sairam et al. 1998).

Cavalcanti et al. (2004), while researching on cowpea (*Vigna radiata* L.), concluded that the well-organized antioxidant system SOD-APX-CAT to some extent does not always take part in the improvement of plants' salt-tolerance. Elevated actions of APX, CAT, SOD, glutathione reductase (GR), relatively minor H_2O_2, O_2^- and thiobarbituric acid reactive contents with relation to the salt-tolerant and salt-sensitive *Zea mays* genotypes at different salt stress levels have been reported by Kholova et al. (2010) and Abbasi et al. (2016). In comparison with other severe abiotic stresses, salinity stress also induces oxidative stress in plants (Abdel 2011; Abdel and Miransari 2014).

10.7.1 Salinity and Abscisic Acid Signalling

Abscisic acid (ABA) is a plant hormone, which acts as a stress signal and thus enables plants to adjust and cope with various environmental stresses such as salinity and drought stress (Keskin et al. 2010; Javid et al. 2011; Miransari et al. 2014). There is usually an upsurge in ABA concentrations in plants subjected to soil salinity conditions. The endogenous production of ABA usually results mostly from deficit of water and not particularly from the effect of salt stress, hence linked to soil or leaf water potential as the case may be (Javid et al. 2011).

10.8 Effect of Salinity on Mineral Nutrition in Plants

The soil solution, having abundance of Na^+ and Cl^- ions in it, presents a difficult uptake issue for vital ions such as Ca^{+2}, P, K^+, Mg^{+2} and N because Na^+ and Cl^- ions compete and alter the ideal salt ratios in the soil solution, therefore affecting plant nutrient acquisition and restricting plant growth and biomass (Saxena et al. 2017). It is now a well-known and established fact that mineral nutrients and salt relations give rise to substantial nutrient instability (Azeem and Ahmad 2011). When there are excess Na^+ and Cl^- ions in the soil, the plant is likely to take up these ions and accumulate them, and in return they hinder the take-up of other necessary macro and micro minerals including Ca^{2+}, K^+ and Mn^{2+} (Karimi et al. 2005; Abbasi et al. 2016).

In many plant species, the excessive take-up of these toxic ions (Na^+ and Cl^-) which in turn limits needed nutrient (K^+, Mg^{2+} and Ca^{2+}) significantly leads to meagre growth, development and yield (Abbasi et al. 2016). This is further buttressed by the researches of Shabala et al. (2006) who worked on wheat plant, Akram et al. (2007) on their work on sunflower and Mahmood (2011) on beans; they all reported that salt stress induces poor plant growth in the above-mentioned crops as a result of antagonistic relations of Na^+ and Cl^- ions with essential nutrients such as Ca, K or N.

Studies on leaves and roots of canola (*Brassica napus*) cultivars have revealed that soil salinity limits the uptake and build-up of K^+ and Ca^{2+} (Ulfat et al. 2007; Ashraf and Ali 2008). Hajiboland et al. (2010) observed the negative effects of excessive salts on tomato cultivars. Certain processes of metabolism in plants are deleteriously affected by elevated Na^+/K^+ ratio (Dogan et al. 2010; Abbasi et al. 2016). However, certain major elements like N, K and Mg mediate many processes of metabolism and thus can be said to play a vital role in the mitigation of the undesirable effects of soil salinity (Giri and Mukerji 2004; Evelin et al. 2009, 2012).

10.8.1 Nitrogen

Nitrogen (N) is a major mineral element that is needed in large quantities by plants. N forms an integral part of numerous components of cell and plants, namely, nucleic acids and amino acids. If a plant or plant tissue experiences deficiency in N, yellowing of leaves (chlorosis) is observed followed by remarked decline in growth. Salinity interferes with the N metabolism stages such as uptake and reduction of NO_3^- and protein synthesis (Frechill et al. 2001; Evelin et al. 2009). Several experiments have reported negative influence of soil salinity on the uptake of N; this includes the work of Cantrell and Linderman (2001) on the plants and parts of onions and lettuce, Silveira et al. (2001) on cowpea and Colla et al. (2008) on zucchini plants who all recorded reduction of N composition with increase in salinity.

10.8.2 Phosphorus

Phosphorus (P) is a very important major element which plays a very crucial role in many structures of plant cells, namely, phospholipid components that make up the plant cell membranes, as well as sugar-phosphate intermediary of metabolic processes of respiration and photosynthesis. Due to the deleterious effects of soil salinity, P deficiency may arise which will lead to poor and diminutive growth, slender and willowy plant stems, leaf coloration changes to dark green and subsequently senescence and abscission of older leaves (Evelin et al. 2012). Evelin et al. (2009, 2011) in their researches on fenugreek plant tissues stated that lower contents of P resulted in reduced and stunted growth, dark green coloration of the leaves, production of slender stems and death of the old plants parts as a result of soil salinity (Evelin et al. 2012; Miransari 2017).

10.8.3 Potassium and Sodium Relations

Potassium (K) is very vital macronutrient which plays a vital role in plant metabolism. It is required for the activation of a range of enzymatic reactions such as during the formation of pyruvate, stomatal activities, synthesis of protein during tRNA bonding to ribosomes and most importantly sustaining vacuolar and cell turgor pressure (Giri et al. 2007; Evelin et al. 2012).

Jaarsma et al. (2013) recorded high accumulation of Na^+ in *Solanum tuberosum* under high salinity. Studies carried out by Evelin et al. (2011) revealed that NaCl treatment reduced the level of K^+, an antagonist of Na^+. Porcel et al. (2016) reported increased accumulation of Na^+ in rice root than shoot tissues due to a decreased root-to-shoot distribution of Na^+ and suggested to be one of the key strategies adopted by plants to adapt to growth under salt stress conditions.

The functioning of the major elemental transporter found in the root plasma membranes like the K^+ selective ion channels is impeded by rise in the level of soil Na^+ ions found in the soil solution; this retards the growth of plants roots as well as affects the soil structure (Porcel et al. 2012).

Many metabolic processes depend on the presence of K^+; however, Na^+ is a major antagonist of the essential mineral K^+. Since both ions share related properties physiologically, Na^+ usually seeks to replace K^+ in the cytoplasm as both compete for binding site therefore impeding certain metabolic activities that depends on K^+ (Giri et al. 2007; Hajiboland 2013). The higher K^+/Na^+ ratio results in the sustenance of essential K-facilitated enzymatic activities as well as the efflux of Na^+ thereby induction of ionic balance in plant cell cytoplasm (Evelin et al. 2009).

10.8.4 Calcium

Calcium (Ca) is an essential macronutrient that has some very vital functions in the plant system including composition of the structure of cell walls, the regulation transport across membrane and maintenance of ion selectivity and integrity maintenance of plant membranes (Maathuis 2009; Evelin et al. 2012). Garg and Manchanda (2009) reported decrease on Ca^{2+} of *Cajanus cajan* under salt stress. Evelin et al. (2011) also recorded similar reduction on *Acacia auriculiformis.*

10.8.5 Magnesium

Magnesium (Mg) is a very important macronutrient, and it is the central element that constitutes the molecule of chlorophyll (Evelin et al. 2012). Giri et al. (2003) stated also that NaCl imparted negative effects on Mg^{2+} uptake *Acacia nilotica.* This finding is corroborated with the works of Sheng et al. (2008) on *Zea mays*, Abdel-Lateef and Chaoxing (2011) on tomato leaves and Evelin et al. (2011) on fenugreek plants who all reported a decrease in Mg under saline treatments. Their discoveries suggest that salt interferes with the processes of synthesis of chlorophyll in plants (Giri and Mukerji 2004). Photosynthetic proficiency and plant growth is subsequently achieved by rise in the uptake of Mg^{2+} resulting in the upsurge in photosynthetic pigments (Evelin et al. 2009).

10.8.6 Iron, Manganese and Zinc

These are essential micronutrients which are required by plants in small quantities. Iron (Fe) takes part in various important processes in plant N_2 fixation and respiration as well as being a key component of cytochromes plus nonheme iron proteins which takes part in photosynthesis. Activities of some decarboxylases, kinases, dehydrogenases, oxidases and peroxidases greatly depend on the presence of Manganese (Mn). Certain cation-initiated enzymes as well as photosynthetic O_2 evolution depend greatly on manganese, while Zn is a component of alcohol dehydrogenase, carbonic anhydrase, glutamic dehydrogenase, etc. Many enzymes require zinc ions (Zn^{2+}) for their activity, and zinc may be required for chlorophyll biosynthesis in some plants. Recently, in most of the salinity- and plant mineral nutrition-related researches, little or no attention has been paid to the micronutrients. The availability of most of these micronutrients depends on soil solution's pH and

EC and the type of binding sites on the organic plus inorganic particle surfaces. The mobility of Cu, Fe, Mn and Zn^{2+} in the soil solution is low. In saline soils, the solubility of these micronutrients also decreases (Evelin et al. 2009).

10.8.7 Chloride

The essentiality of Cl^- as a micronutrient is still debatable. Cl^- is needed by plants for the purpose of evolution of O_2 as a photosynthetic reaction. In regions with extremely high saline, concentration of Cl^- ion can hinder plant growth as well as become very toxic to plants (Xu et al. 2000; Evelin et al. 2009). Evelin et al. (2011) also recorded similar increase in Cl^- on *Acacia auriculiformis*. This is in line with the work of Cantrell and Linderman (2001) who reported increased Cl^- uptake in lettuce. Root cells take up Cl^- from the soil solution through H^+/Cl^- symporters, as well as the anion channels in saline environments. In order to get to the aerial plant part, Cl^- navigates the root by a symplastic pathway and therefore released into the stele with the help of anion channels. The rise and accumulation of Cl^- in plants is very toxic and causes severe losses to agriculture in saline areas (Xu et al. 2000; White and Broadley 2001).

10.9 Salinity Effects on Proximate Composition of Plants

Recent studies have revealed the deleterious effects of salinity on the proximate composition of plants. Okon (2018) recorded that the proximate composition of *Cucurbita maxima* and *Telfairia occidentalis* such as carbohydrate, caloric value and fibre was significantly reduced with salinity, while ash, protein and lipids increased in saline treatments than nonsaline treatments. Reduction in the proximate composition of plants could be as a result of the deleterious effect of salinity. Increase in ash, protein and lipids is also reported by Uddin et al. (2017); they reported increase in ash, crude protein and lipids in *Clinacanthus nutans* with increasing salinity. This finding corroborates results of a study by Ali et al. (2014) on *Portulaca oleracea*, *Hibiscus sabdariffa* and *Sorghum bicolor* which support the previous results stating increase in ash, crude protein and lipids under salinity. Uzun et al. (2013) and Kapoor and Srivastava (2010) also reported that increasing salinity level tended to enhance the crude protein synthesis on bitter vetch and *Vigna mungo* L., respectively. The increase in crude lipid content in response to salinity may be because plants have some tolerance level at these salinity levels or its accumulation of compatible solutes (Uddin et al. 2017). These results fit with those previously reported on *P. oleracea* by Teixeira and Carvalho (2009) and *Oenothera biennis* seeds by Heuer et al. (2002). They observed increase in the total fat content of plants exposed to moderate saline environment. Kekere (2014) reported increase in ash content of *Arachis hypogaea*. In saline environments, excess ion accumulation of Na^+ and Cl^- accounts for high ash content (Kekere 2014).

10.10 Biochemical Alterations in Plants Under Salinity Stress

In plants, under severe environmental stress such as salinity stress, the increase and over-assembly of different categories of several organic solutes takes place (Abdel et al. 2009). These compatible organic solutes are known to be non-toxic at very significant cellular concentrations as well as possess very low molecular weight. They perform the vital role of detoxifying the cell of ROS and maintenance of osmotic adjustment, stabilization of proteins and enzymes and importantly membrane integrity preservation (Abdel and Chaoxing 2014).

Additionally, some of these solutes are referred to as osmo-protectants because they protect cellular constituents from damage as a result of physiological drought. These organic solutes include proline, polyols, trehalose, soluble sugars and quaternary ammonium compounds (QACs) such as proline, betaine, alanine betaine, glycine betaine, pipecolate betaine and hydroxy proline betaine (Hamdia and Shaddad 2010; Abdel and Miransari 2014).

10.10.1 Proline

Proline is a major plant cell osmo-protectant. Several experiments conducted under saline stressed conditions have reported the accumulation of proline (Abdel-Lateef 2010, Abdel 2011; Hameed et al. 2014). Under severe soil salinity conditions, cell membrane and protein stabilization is one of the advantages of proline accumulation. Proline has been reported to sometimes perform the duty of a salt stress injury sensor (Abdel-Lateef et al. 2009).

Other researches such as that of Hassanein et al. (2009) reported that the accumulation of proline is an adjustment strategy to cope and tolerate the adverse effect of salt stress. Meanwhile, other reports suggest that salt-sensitive plants accumulated more proline at various salinity levels as a response and adaptive measure to salinity (Jahromi et al. 2008; Sheng et al. 2011; Evelin and Kapoor 2013; Abdel and Chaoxing 2014).

10.10.2 Betaines

Once betaines are formed, they aren't easily metabolized (Kapoor et al. 2013). Betaines are derived from amino acids and are regarded as quaternary ammonium compounds. Under severe soil salinity conditions, cell membrane structure, enzymes and protein stabilization is one of the functions of betaine accumulation (Hajiboland 2013).

10.11 Changes in Morphology and Anatomy of Plants Due to Salt Stress

Besides physiological and biochemical adaptations, plants adapt to saline conditions by changes in the anatomical structure of leaves and roots plus changes in morphology (Ola et al. 2012; Saule et al. 2013). Morphology and anatomy of leaf and roots are affected by salt stress. Salinity induced reduction in almost all morphological and anatomical variables (Ceccoli et al. 2011). High salinity mostly causes anatomical alterations such as reduction of stomata number (Çavişoglu et al. 2007), decrease of length of bundle, xylem rows, number of vessels and increase in both spongy and palisade tissue (Hussein et al. 2012; Saule et al. 2013).

10.12 Impact of Salinity Stress on Ultrastructure of Plants

High level of water-soluble salts in soil may result in the alteration of cell wall thickness, increase in the frequency of plasmodesmata, vacuolization of cytoplasm, shrinkage of protoplasm, widening of apoplastic space in the middle of cell wall as well as cell membrane, disruption of thylakoid and thylakoid membrane, swelling and decrease in total number of thylakoids, degeneration of chloroplast membrane, amassing of plastoglobules, stretching of cristae as well as denser matrix in mitochondria and dissipation of mitochondrial membrane potential, aggregation of chromatin in nucleus and induction of ovule abortion, thereby hampering the plant physiology and metabolism (Evelin et al. 2009, 2013; Saxena et al. 2017).

Salinity leads to structural and ultrastructural effects, especially in plant species that are salt sensitive. Some of the injuries occur at the early stage of salt stress, such as clumping of the photosynthetic apparatus chloroplast and the resulting granal and fret partition swellings or complete reorganization of the entire photosynthetic apparatus components thylakoid and grana (Ashok et al. 2012). Other structural modifications such as the mitochondria density increase which contributes to the particle regularity of the enzyme ATPase which portrays that there may be energy demand at modest salinity levels.

10.13 Conclusion and Future Perspective

Soil salinity is one of the weightiest environmental abiotic challenges limiting plant growth and productivity. The root system initially is the first to perceive salt stress, thereby limiting water available to the plants resulting in osmotic stress and subsequently physiological drought. Ionic toxicity is one of the long-term effects of salt stress; this happens as a result of excess uptake and accumulation of Na^+ and Cl^- ions which antagonizes the uptake of relevant nutrients which leads to nutrient disproportions in the cytosol. Salinity stress induces osmotic, ionic, oxidative and nutritional disorders, therefore altering the major metabolic processes of plants (Saxena et al. 2017). One of the machineries of plant salt tolerance adjustment is the production of ROS. Amongst the compatible organic osmolytes produced by plants, proline (Pro) and glycine betaines (GB) are the most essential as effective. Salt-tolerant plant species accumulate Pro after being subjected to salinity stress condition. GB performs and acts like an osmolyte and shields the photosynthetic mechanism PSII during salinity (Jose et al. 2017). Conclusively, it is suggested that more researches should be carried at the molecular level to determine the molecular approaches that could help to genetically engineer salt-tolerant plants. This will go a long way to provide some future insights and will contribute to our understanding of plant salt tolerance mechanisms and the extent of technical know-how required to further produce salt-tolerant species of crops.

References

Abbasi GH, Akhtar J, Haq MA, Ali S, Chen ZH, Malik W (2014) Exogenous potassium differentially mitigates salt stress in tolerant and sensitive maize hybrids. Pak J Bot 46:135–146

Abbasi H, Moazzam J, Anwar H, Shafaqat A, Rafiq A, Zafar M, Parveen D (2016) Salt stress manifestation on plants, mechanism of salt tolerance and potassium role in alleviating it: a review. Zemdirbyste-Agriculture 103(2):229–238

Abdel LAA (2011) Ameliorative effect of calcium chloride on growth, antioxidant enzymes, protein patterns and some metabolic activities of canola (*Brassica napus* L) under seawater stress. J Plant Nutr 34:1303–1320

Abdel LAA, Chaoxing H (2014) Does the inoculation with *Glomus mosseae* improve salt tolerance in pepper plants. J Plant Growth Regul 10:23–31

Abdel L, Miransari M (2014) The role of arbuscular mycorrhizal fungi in alleviation of salt stress. Springer, New York, pp 23–33

Abdel-Lateef A (2008) Phosphoenolpyruvate carboxylase activity of wheat and maize seedlings subjected to salt stress. Aust J Basic Appl Sci 2:37–41

Abdel-Lateef AA (2010) Changes of antioxidative enzymes in salinity tolerance among different wheat cultivars. Cereal Res Commun 38:43–55

Abdel-Lateef AHA, Chaoxing H (2011) Effect of arbuscular mycorrhizal fungi on growth, mineral nutrition, antioxidant enzymes activity and fruit yield of tomato grown under salinity stress. Sci Hortic 127:228–233

Abdel-Lateef AA, Shaddad KAM, Ismail MA, Abu AFM (2009) Benzyladenine can alleviate saline injury of two roselle (*Hibiscus sabdariffa*) cultivars via equilibration of cytosolutes including anthocyanins. Int J Agric Biol 11:151–157

Abdeshahian M, Nabipour M, Meskarbashee M (2010) Chlorophyll fluorescence as criterion for the diagnosis salt stress in wheat (*Triticum aestivum*) plants. Int J Chem Biol Eng 4:184–186

Acosta-Motos JR, Álvarez S, Barba-Espín G, Hernández JA, Sánchez-Blanco MJ (2014) Salts and nutrients present in regenerated waters induce changes in water relations, antioxidative metabolism, ion accumulation and restricted ion uptake in *Myrtus communis* L. plants. Plant Physiol Biochem 85:41–50

Adiku G, Renger M, Wessolek G, Facklam M, Hech-Bischoltz C (2001) Simulation of dry matter production and seed yield of common beans under varying soil water and salinity conditions. Agric Water Manag 47:55–68

Akram MS, Ashraf M (2011) Exogenous application of potassium dihydrogen phosphate can alleviate the adverse effects of salt stress on sunflower (*Helianthus annuus* L.). J Plant Nutr 34:1041–1057

Akram MS, Athar HUR, Ashraf M (2007) Improving growth and yield of sunflower (*Helianthus annuus* L.) by foliar application of potassium hydroxide (KOH) under salt stress. Pak J Bot 39:769–776

Albacete A, Ghanem ME, Martínez-Andújar C, Acosta M, Sánchez-Bravo J, Martínez V, Lutts S, Dodd IC, Pérez-Alfocea F (2008) Hormonal changes in relation to biomass partitioning and shoot growth impairment in salinized tomato (*Solanum lycopersicum* L.) plants. J Exp Bot 59:4119–4131

Ali EF, Hassan FAS (2014) Alleviatory effects of salt stress by mycorrhizal fungi and gibberellic acid on chamomile plant. Int J Sci Res 3(11):109–111

Ali S, Zeng F, Cai S, Qiu B, Zhang G (2011) The interaction of salinity and chromium in the influence of barley growth and oxidative stress. Plant Soil Environ 57:153–159

Ali AS, Mohamed BF, Dreyling G (2014) Salt tolerance and effects of salinity on some agricultural crops in the Sudan. J For Prod Ind 3(2):56–65

Al-Taweel K, Iwaki T, Yabuta Y et al (2007) A bacterial transgene for catalase protects translation of d1 protein during exposure of salt-stressed tobacco leaves to strong light. Plant Physiol 145:258–265

Alvino A, D'Andria R, Delfine S, Lavini A, Zanetti P (2000) Effect of water and salinity stress on radiation absorption and efficiency in sunflower. Ital J Agron 4:53–60

Amira MS, Abdul Q (2010) Effect of salt stress on plant growth and metabolism of bean plant *Vicia faba* (L.). J Saudi Soc Agric Sci 10:7–15

Aragao MEF, Guedes MM, Otoch MLO et al (2005) Differential responses of ribulose-1, 5-bisphosphate carboxylase/oxygenase activities of two *Vigna unguiculata* cultivars to salt stress. Braz J Plant Physiol 17:207–212

Ashok A, Nisha K, Karishma N, Anju T, Gupta KK (2012) Arbuscular mycorrhizal symbiosis and alleviation of salinity stress. J Appl Nat Sci 4(1):144–155

Ashraf M, Ali Q (2008) Relative membrane permeability and activities of some antioxidant enzymes as the key determinants of salt tolerance in canola (*Brassica napus* L.). Environ Exp Bot 63:266–273

Ashraf M, Harris PJC (2013) Photosynthesis under stressful environments: an overview. Photosynthetica 51:163–190

Atilla LT (2014) Influence of foliarly applied different triazole compounds on growth, nutrition, and antioxidant enzyme activities in tomato (*Solanum lycopersicum* L.) under Salt Stress. Aust J Crop Sci 8(1):71–79

Ayala-Astorga GI, Alcaraz-Melendez L (2010) Salinity effects on protein content, lipid peroxidation, pigments and praline in *Paulownia imperialis* (Siebold and Zaccarini) and *Paulownia fortunei* (Seemann and Hemsley) plant grown in vitro. Electron J Biomed 13:1–15

Azeem M, Ahmad R (2011) Foliar application of some essential minerals on tomato (*Lycopersicon esculentum*) plant grown under two different salinity regimes. Pak J Bot 43:1513–1520

Beltrano J, Ruscitti M, Arango MC, Ronco M (2013) Effects of arbuscular mycorrhiza inoculation on plant growth, biological and physiological parameters and mineral nutrition in pepper grown under different salinity and P levels. J Soil Sci Plant Nutr 13(1):123–141

Bhandal IS, Malik CP (1988) Potassium estimation, uptake, and it's role in the physiology and metabolism of flowering plants. Int Rev Cytol 110:205–254

Bhatnagar-Mathur P, Devi MJ, Reddy DS, Lavanya M, Vadez V, Serraj R, Yamaguchi-Shinozaki-K, Sharma KK (2007) Stress inducible expression of At DREB1A in transgenic peanut (*Arachis hypogaea* L.) increases transpiration efficiency under water limiting conditions. Plant Cell Rep 26:2071–2082

Bie Z, Ito T, Shinohara Y (2004) Effects of sodium sulfate and sodium bicarbonate on the growth, gas exchange and mineral composition of lettuce. Sci Hortic 99:215–224

Cantrell IC, Linderman RG (2001) Preinoculation of lettuce and onion with VA mycorrhizal fungi reduces deleterious effects of soil salinity. Plant Soil 233:269–281

Cavalcanti FR, Oliveira JTA, Martins-Miranda AS, Viegas RA, Silveira JAG (2004) Superoxide dismutase, catalase and peroxidase activities do not confer protection against oxidative damage in salt-stressed cowpea leaves. New Phytol 163(3):563–571

Çavişoglu KS, Kiliç S, Kabar K (2007) Some morphological and anatomical observations during alleviation of salinity (NaCl) stress on seed germination and seedling growth of barley by polyamines. Acta Physiol Plant 29:551–557

Ceccoli G, Ramos JC, Ortega LI, Acosta JM, Perreta MG (2011) Salinity induced anatomical and morphological changes in *Chloris gayana* Kunth roots. Biocell 35(1):9–17

Cheruth J, Ragupathi G, Ashot K, Paramasivam M, Beemarao S, Rajaram P (2008) Interactive effects of triadimefon and salt stress on antioxidative status and ajmalicine accumulation in *Catharanthus roseus*. Acta Physiol Plant 30(3):287–292

Chrominski A, Bhat RB, Weber DJ, Smith BN (1988) Osmotic stress-dependent conversion of 1-aminocyclopropane-1-carboxylic acid (ACC) to ethylene in the halophyte, *Allenrolfea occidentalis*. Environ Exp Bot 28:171–174

Cirillo C, Rouphael Y, Caputo R, Raimondi G, Sifola MI, De Pascale S (2016) Effects of high salinity and the exogenous of an osmolyte on growth, photosynthesis and mineral composition in two ornamental shrubs. J Hortic Sci Biotechnol 91:14–22

Colla G, Rouphael Y, Cardarelli M, Tullio M, Rivera C, Rea E (2008) Alleviation of salt stress by arbuscular mycorrhizal in zucchini plants grown at low and high phosphorus concentration. Biol Fertil Soils 44:501–509

Colla G, Rouphael Y, Leonardi C, Bie Z (2010) Role of grafting in vegetable crops grown under saline conditions. Sci Hortic 127:147–155

Delfine S, Alvino A, Zacchini M, Loreto F (1998) Consequences of salt stress on conductance to CO_2 diffusion, Rubisco characteristics and anatomy of spinach leaves. Funct Plant Biol 25:395–402

Delfine S, Alvino A, Villani MC, Loreto F (1999) Restrictions to carbon dioxide conductance and photosynthesis in spinach leaves recovering from salt stress. Plant Physiol 119:1101–1106

Di Martino C, Delfine S, Alvino A, Loret F (1999) Photorespiration rate in spinach leaves under moderate NaCl stress. Photosynthetica 36:233–242

Dodd IC (2005) Root-to-shoot signalling: assessing the roles of 'up' in the up and down world of long-distance signalling in planta. Plant Soil 74:257–275

Dogan M, Tipirdamaz R, Demir Y (2010) Salt resistance of tomato species grown in sand culture. Plant Soil Environ 56:499–507

Doganlar ZB, Demir K, Basak H et al (2010) Effects of salt stress on pigment and total soluble protein contents of three different tomato cultivars. Afr J Agric Res 5:2056–2065

Ebrahim KE (2014) Role of arbuscular mycorrhizal fungi in fighting soil salinity. Published Ph.D. Thesis, Royal Holloway-University of London, London

Evelin H, Kapoor R (2013) Arbuscular mycorrhizal symbiosis modulates antioxidant response in salt-stressed *Trigonella foenum-graecum* Plants. Mycorrhiza 5:45–77

Evelin H, Kapoor R, Giri B (2009) Arbuscular mycorrhizal fungi in alleviation of salt stress: a review. Ann Bot 104:1263–1280

Evelin H, Giri B, Kapoor R (2011) Contribution of Glomus intraradices inoculation to nutrient acquisition and mitigation of ionic imbalance in NaCl-stressed *Trigonella foenum-graecum.* Mycorrhiza 22:1–15

Evelin H, Giri B, Kapoor R (2012) Contribution of *Glomus intraradices* inoculation to nutrient acquisition and mitigation of ionic imbalance in NaCl-stressed *Trigonella foenum-graecum.* Mycorrhiza 22:203–217

Evelin H, Giri B, Kapoor R (2013) Ultrastructural evidence for AMF mediated salt stress mitigation in *Trigonella foenumgraecum.* Mycorrhiza 23:71–86

Feng G, Zhang FS, Li X, Tian CY, Tang C, Rengel Z (2002) Improved tolerance of maize plants to salt stress by arbuscular mycorrhiza is related to higher accumulation of soluble sugars in roots. Mycorrhiza 12:185–190

Flexas J, Diaz-Espejo A, Galmés J, Kaldenhoff R, Medrano H, Ribas-Carbo M (2007) Rapid variations of mesophyll conductance in response to changes in CO_2 concentration around leaves. Plant Cell Environ 30:1284–1298

Frechill S, Lasa B, Ibarretxe L, Lamsfus C, Aparicio TP (2001) Pea response to saline stress is affected by the source of nitrogen nutrition (ammonium or nitrate). Plant Growth Regul 35:171–179

Franco JA, Bañón S, Vicente MJ, Miralles J, Martínez-Sánchez JJ (2011) Root development in horticultural plants grown under abiotic stress conditions—a review. J Hortic Sci Biotechnol 86:543–556

Gamalero E, Lingua G, Berta G, Glick BR (2009) Beneficial role of plant growth promoting bacteria and arbuscular mycorrhizal fungi on plant responses to heavy metal stress. Can J Microbiol 55(5):501–514

Garg N, Manchanda G (2009) Role of arbuscular mycorrhizae in the alleviation of ionic, osmotic and oxidative stresses induced by salinity in *Cajanus cajan* (L.) Millsp. (Pigeonpea). J Agron Crop Sci 195:110–123

Ghanem ME, Albacete A, Martínez-Andújar C, Acosta M, Romero-Aranda R, Dodd IC, Lutts S, Pérez-Alfocea F (2008) Hormonal changes during salinity-induced leaf senescence in tomato (*Solanum lycopersicum* L.). J Exp Bot 59:3039–3050

Ghosh S, Bagchi S, Majumder AL (2001) Chloroplast fructose-1,6-bisphosphatase from *Oryza* differs in salt tolerance property from the porteresia enzyme and is protected by osmolytes. Plant Sci 160:1171–1181

Ghoulam C, Foursy A, Fares K (2002) Effects of salt stress on growth, inorganic ions and proline accumulation in relation to osmotic adjustment in five sugar beet cultivars. Environ Exp Bot 47:39–50

Giri B, Mukerji KG (2004) Mycorrhizal inoculant alleviates salt stress in *Sesbania aegyptiaca* and *Sesbania grandiflora* under field conditions: evidence for reduced sodium and improved magnesium uptake. Mycorrhiza 14:307–312

Giri B, Kapoor R, Mukerji KG (2003) Influence of arbuscular mycorrhizal fungi and salinity on growth, biomass and mineral nutrition of *Acacia auriculiformis.* Biol Fertil Soils 38:170–175

Giri B, Kapoor R, Mukerji KG (2007) Improved tolerance of *Acacia nilotica* to salt stress by arbuscular mycorrhiza, *Glomus fasciculatum*, maybe partly related to elevated K^+/Na^+ ratio in root and shoot tissues. Microb Ecol 54:753–760

Hajiboland R (2013) Role of arbuscular mycorrhiza in amelioration of salinity. In: Ahmad P et al (eds) Salt stress in plants: signalling, omics and adaptations. Springer, New York, pp 301–354

Hajiboland R, Aliasgharzadeh N, Laiegh SF, Poschenreider C (2010) Colonization with arbuscular mycorrhizal fungi improves salinity tolerance of tomato (*Solanum lycopersicum* L.) plants. Plant Soil 331:313–327

Hamdia MA, Shaddad MAK (2010) Salt tolerance of crop plants. Review. J Stress Physiol Biochem 6:64–90

Hameed A, Dilfuza E, Abd-Allah EF, Hashem A, Kumar A, Ahmad P (2014) Salinity stress and arbuscular mycorrhizal symbiosis in plants. In: Miransari M (ed) Use of microbes for the alleviation of soil stresses. Springer, New York, p 159

Hassanein RA, Hassanein AA, Haider AS, Hashem HA (2009) Improving salt tolerance of *Zea mays* l. plants by presoaking their grains in glycine betaine. Aust J Basic Appl Sci 3:928–942
He E, Ikhajiagba B, Bamidele JF, Ogic-odia E (2008) Salinity effects on young healthy seedling of Kyllingia peruviana collected from Escravos, Delta state. Glob J Environ Res 2(2):74–88
Hernandez JA, Olmos E, Corpas FJ, Sevilla F, Del Rio LA (1995) Salt-induced oxidative stress in chloroplasts of pea plants. Plant Sci 105:151–167
Heuer B, Yaniv Z, Ravina I (2002) Effect of late salinization of chia (*Salvia hispanica*), stock (*Matthiola tricuspidata*) and evening primrose (*Oenothera biennis*) on their oil content and quality. Ind Crop Prod 15(2):163–167
Hu YC, Schnyder H, Schmidhalter U (2000) Carbohydrate deposition and partitioning in elongating leaves of wheat under saline soil conditions. Aust J Plant Physiol 27:363–370
Hussein MM, Abo-Leila BH, Metwally SA, Leithy SZ (2012) Anatomical structure of jatropha leaves affected by proline and salinity conditions. J Appl Sci Res 8(1):491–496
Jaarsma R, Rozemarijn SM, Albertus H (2013) Effect of salt stress on growth, Na^+ accumulation and proline metabolism in potato (*Solanum tuberosum*) cultivars. PLoS One 8(3):60–183
Jahromi F, Aroca R, Porcel R, Ruiz-Lozano JM (2008) Influence of salinity on the in vitro development of *Glomus intraradices* and on the in vivo physiological and molecular responses of mycorrhizal lettuce plants. Microb Ecol 55:45–53
Jamil M, Rehman SU, Lee KJ, Kim JM, Rha HK (2007) Salinity reduced growth Ps2 photochemistry and chlorophyll content in radish. Sci Agric (Piracicaba, Braz.) 64(2):111–118
Javid MJ, Sorooshzadeh A, Moradi F, Sanavy SAMM, Allahdadi I (2011) The role of phytohormones in alleviating salt stress in crop plants. Aust J Crop Sci 5:726–734
Jing C, Xiu YL, Ling C, Jia JX, Hai YL (2015) Effects of salinity on the growth, physiology and relevant gene expression of an annual halophyte grown from heteromorphic seeds. AoB Plants 7:112
Jose RA, Maria FO, Agustina BV, Pedro DV, Maria JS, Jose AH (2017) Plant responses to salt stress: adaptive mechanisms. Agronomy 7(18):2–38
Juniper S, Abbott LK (1993) Vesicular-arbuscular mycorrhizas and soil salinity. Mycorrhiza 4:45–57
Kapoor K, Srivastava A (2010) Assessment of salinity tolerance of *Vinga Mungo* var. Pu-19 using ex vitro and in vitro methods. Asian J Biotechnol 2(2):73–85
Kapoor R, Evelin H, Mathur P, Giri B (2013) Arbuscular mycorrhiza: approaches for abiotic stress tolerance in crop plants for sustainable agriculture. In: Tuteja N, Gill SS (eds) Plant acclimation to environmental stress. Springer, LLC, Berlin, pp 359–401
Karen W, Anthony RY, Timothy JF (2002) Effects of salinity and ozone, individually and in combination on growth and ion contents of two chickpea (*Cicer aritinum* L.) varieties. Environ Pollut 120(2):397–403
Karimi G, Ghorbanli M, Heidari H, Khavarinejad RA, Assareh MH (2005) The effects of NaCl on growth, water relations, osmolytes and ion content in Kochia prostrate. Biologia Plantarum 49:301–304
Kaya C, Higgs D, Kirnak H (2001) Effects of supplementary phosphorus and potassium on physiological development and mineral nutrition of cucumber and pepper cultivars grown at high salinity (NaCl). J Plant Nutr 24:1457–1471
Kaya C, Ashraf M, Sonmez O, Aydemir S, Tuna AL, Cullu MA (2009) The influence of arbuscular mycorrhizal colonisation on key growth parameters and fruit yield of pepper plants grown at high salinity. Sci Hortic 121:1–6
Kekere O (2014) Growth, yield and seed nutritional composition of groundnut (*Arachis hypogaea* LINN) under elevated level of soil salinity. Mol Soil Biol 5(5):31–38
Keskin BC, Sarikaya AT, Yuksel B, Memon AR (2010) Abscisic acid regulated gene expression in bread wheat. Aust J Crop Sci 4:617–625
Kholova J, Sairam RK, Meena RC (2010) Osmolytes and metal ions accumulation, oxidative stress and antioxidant enzymes activity as determinants of salinity stress tolerance in maize genotypes. Acta Physiol Plant 32(3):477–486

Läuchli A, Lüttge U (2002) Salinity: environment-plants-molecules. Kluwer Academic, Dordrecht, p 14
Lee G, Carrow RN, Duncan RR (2004) Photosynthetic responses to salinity stress of halophytic seashore paspalum ecotypes. Plant Sci 166(6):1417–1425
Lutts S, Kinet JM, Bouharmont J (1996) NaCl-induced senescence in leaves of rice (*Oryza sativa* L.) cultivars differing in salinity resistance. Ann Bot 78:389–398
Maathuis FJM (2009) Physiological functions of mineral macro nutrients. Curr Opin Plant Biol 12:250–258
Mahmood K (2011) Salinity tolerance in barley (*Hordeum vulgare* L.): effects of varying NaCl, K^+/Na^+ and $NaHCO_3$ levels on cultivars differing in tolerance. Pak J Bot 43:1651–1654
Marcelis LFM, Van Hooijdonk J (1999) Effect of salinity on growth, water use and nutrient use in radish (*Raphanus sativus* L.). Plant Soil 215:57–64
Martínez-Ballesta MC, Aparicio F, Pallás V et al (2003) Influence of saline stress on root hydraulic conductance and PIP expression in *Arabidopsis*. J Plant Physiol 160:689–697
Mathur N, Singh J, Bohra S, Bohra A, Vyas A (2006) Biomass production, productivity and physiological changes in moth bean genotypes at different salinity levels. Am J Plant Physiol 1(2):210–213
Mathur N, Singh J, Bohra S, Vyas A (2007) Arbuscular mycorrhizal status of medicinal halophytes in saline areas of Indian Thar Desert. Int J Soil Sci 2:119–127
Memon SA, Hou X, Wang LJ (2010) Morphological analysis of salt stress response of Pak Choi. EJEAFChe 9(1):248–254
Miransari M (2017) Arbuscular mycorrhizal fungi and heavy metal tolerance in plants. In: Wu QS (ed) Arbuscular mycorrhizas and stress tolerance of plants. Springer, Singapore, pp 148–150
Miransari M, Abrishamchib A, Khoshbakht K, Niknam V (2014) Plant hormones as signals in arbuscular mycorrhizal symbiosis. Crit Rev Biotechnol 34(2):123–133
Mittler R (2002) Oxidative stress, antioxidants and stress tolerance. Trends Plant Sci 7:405–410
Mumtaz KM, Ruqaya S, Al-Mas'oudi M, Al-Said F, Khan I (2013) Salinity effects on growth, electrolyte leakage, chlorophyll content and lipid peroxidation in cucumber (*Cucumis sativus* L.). Int Conf Food Agric Sci 55:30–32
Munns R (1993) Physiological processes limiting plant growth in saline soils: some dogmas and hypotheses. Plant Cell Environ 16:15–24
Munns R (2002) Comparative physiology of salt and water stress. Plant Cell Environ 25 (2):239–250
Munns R (2005) Genes and salt tolerance: bringing them together. New Phytol 167:645–663
Munns R, Tester M (2008) Mechanisms of salinity tolerance. Annu Rev Plant Biol 59:651–681
Murkute AA, Sharma S, Singh SK (2006) Studies on salt stress tolerance of citrus rootstock genotypes with arbuscular mycorrhizal fungi. Hortic Sci 33:70–76
Nasim G (2010) The role of arbuscular mycorrhizae in inducing resistance in drought and salinity stress in crops. In: Ashraf M, Ozturk M, MSA A (eds) Plant adaptation and phytoremediation, vol 1. Springer, Berlin, pp 119–141
Nayer AKKS, Sara TA, Mohammad P (2013) Comparative study of different salts (sodium chloride, sodium sulfate, potassium chloride, and potassium sulfate) on growth of forage species. J Plant Nutr 36(2):214–230
Negrao S, Schmockel SM, Tester M (2017) Evaluating physiological responses of plants to salinity stress. Ann Bot 119:1–11
Netondo GW, Onyango JC, Beck E (2004) Crop physiology and metabolism sorghum and salinity II – gas exchange and chlorophyll fluorescence of sorghum under salt stress. Crop Sci 44 (3):806–811
Niu G, Rodriguez DS (2006) Relative salt tolerance of five herbaceous perennials. Hortic Sci 41:1493–1497
Okon OG (2018) Effects of arbuscular mycorrhizal fungi inoculation on the growth and physiology of some vegetables grown in saline soil amended with poultry manure. Unpublished Ph.D. Thesis, University of Uyo, Akwa Ibom State

Ola HAE, Reham EF, Eisa SS, Habib SA (2012) Morpho-anatomical changes in salt stressed kallar Grass (*Leptochloa fusca* L. Kunth). Res J Agric Biol Sci 8(2):158–166
Olesen K, Andreasson LE (2003) The function of the chloride ion in photosynthetic oxygen evolution. Biochemistry 42(7):2025–2035
Parvaiz A, Munir O, Satyawati S, Salih G (2014) Effect of sodium carbonate-induced salinity–alkalinity on some key osmoprotectants, protein profile, antioxidant enzymes, and lipid peroxidation in two mulberry (*Morus alba* L.) cultivars. J Plant Interact 9(1):460–467
Perveen S, Shahbaz M, Ashraf M (2010) Regulation in gas exchange and quantum yield of photosystem II (PSII) in salt-stressed and non-stressed wheat plants raised from seed treated with triacontanol. Pak J Bot 42:3073–3081
Porcel R, Aroca R, Ruiz-Lozano JM (2012) Salinity stress alleviation using arbuscular mycorrhizal fungi. A review. Agron Sustain Dev 32:181–200
Porcel R, Aroca R, Azcon R et al (2016) Regulation of cation transporter genes by the arbuscular mycorrhizal symbiosis in rice plants subjected to salinity suggests improved salt tolerance due to reduced Na+ root-to-shoot distribution. Mycorrhiza 26:673
Qadir M, Quillerou E, Nangia V et al (2014) Economics of salt-induced land degradation and restoration. Nat Resour Forum 38:282–295
Rahnama A, James RA, Poustini K, Munns R (2010) Stomatal conductance as a screen for osmotic stress tolerance in durum wheat growing in saline soil. Funct Plant Biol 37(3):255–263
Rasool S, Hameed A, Azooz MM, Rehman M, Siddiqi TO, Ahmad P (2012) Salt stress: causes, types and responses of plants. In: Ahmad P, Azooz MM, Prasad MNV (eds) Ecophysiology and responses of plants under salt stress. Springer, New York, pp 1–24
Raul L, Andres O, Armado L, Bernardo M, Enrique T (2003) Response to salinity of three grain legumes for potential cultivation in arid areas (plant nutrition). Soil Sci Plant Nutr 49 (3):329–336
Rui MAM, Ricardo PS (2017) Soil salinity: effect on vegetable crop growth. Management practices to prevent and mitigate soil salinization. Horticulturae 3(30):1–13
Rui L, Wei S, Mu-xiang C, Cheng-jun J, Min W, Bo-ping Y (2009) Leaf anatomical changes of *Burguiera gymnorrhiza* seedlings under salt stress. J Trop Subtrop Bot 17(2):169–175
Ruiz-Lozano JM, Porcel R, Azcón C, Aroca R (2012) Regulation by arbuscular mycorrhizae of the integrated physiological response to salinity in plants: new challenges in physiological and molecular studies. J Exp Bot 63(11):4033–4044
Sairam RK, Deshmukh PS, Saxena DC (1998) Role of antioxidant system in wheat genotypes tolerance to water stress. Biol Plant 41:387–394
Saleem A, Ashraf M, Akram NA (2011) Salt (NaCl) induced modulation in some key physio-biochemical attributes in Okra (*Abelmoschus esculentus* L.). J Agron Crop Sci 197:202–213
Santos CV (2004) Regulation of chlorophyll biosynthesis and degradation by salt stress in sunflower leaves. Sci Hortic 103:93–99
Saule A, Akmaral N, Subhash M, Aygul A, Saule K, Saule A, Asil N, Anar Z, Saltanat A, Ravilya A, Tamara L (2013) The effect of salinity on growth and anatomical attributes of barley seedling (*Hordeum vulgare* L.). Afr J Biotechnol 12(18):2366–2377
Saxena B, Giri B, Shukla K (2017) Arbuscular mycorrhizal fungi and tolerance of salt stress in plants. In: Wu QS (ed) Arbuscular mycorrhizas and stress tolerance of plants. Springer, Singapore, pp 68–73
Shabala S, Demidchik V, Shabala L, Cuin TA, Smith CJ, Miller AJ, Davies JM, Newman IA (2006) Extracellular Ca^{2+} ameliorates NaCl-induced K^{+} loss from Arabidopsis root and leaf cells by controlling plasma membrane K+-permeable channels. Plant Physiol 141(4):1653–1665
Shalata A, Mittova V, Volokita M, Guy M, Tal M (2001) Response of cultivated tomato and its wild salt tolerant relative *Lycopersicon pennellii* to salt-dependent oxidative stress: the root antioxidative system. Physiol Plant 112(4):487–494
Shan SW (2009) Enhanced phytoremediation of salt-impacted soils using plant growth-promoting rhizobacteria (PGPR). Published Master's Thesis, University of Waterloo, Waterloo

Sharma P, Jha AB, Dubey RS, Pessarakli M (2012) Reactive oxygen species, oxidative damage, and antioxidant defense mechanism in plants under stressful conditions. A review. J Bot 2012:1–26

Sheng M, Tang M, Chan H, Yang B, Zhang F, Huang Y (2008) Influence of arbuscular mycorrhizae on photosynthesis and water status of maize plants under salt stress. Mycorrhiza 18:287–296

Sheng M, Tang M, Zhang F, Huang Y (2011) Influence of arbuscular mycorrhiza on organic solutes in maize leaves under salt stress. Mycorrhiza 21:423–430

Shou-Jun Y, Zhong-Lan Z, Yuan-Xia X, Zhi-Fen Z, Shu-Yi S (2014) Arbuscular mycorrhizal fungi increase salt tolerance of apple seedlings. Bot Stud 55(70):4–7

Siler B, Misic D, Filipovic B, Popovic Z, Cvetic T, Mijovic A (2007) Effects of salinity on in vitro growth and photosynthesis of common centaury (*Centaurium erythraea* Rafn.). Achiev Biol Sci 59(2):129–134

Silveira JAG, Melo ARB, Viegas RA, Oliveira JTA (2001) Salinity induced effects on nitrogen assimilation related to growth in cowpea plants. Environ Exp Bot 46:171–179

Taffouo VD, Kouamou JK, Ngalangue LMT, Ndjeudji BAN, Akoa A (2009) Effects of salinity stress on growth, ions partitioning and yield of some cowpea (*Vigna ungiuculata* L., walp) cultivars. Int J Bot 5(2):135–143

Taffouo VD, Wamba OF, Yombi E, Nono GV, Akoe A (2010) Growth, yield, water status and ionic distribution response of three Bambara groundnut (*Vigna subterranean* (L.) verdc.) landraces grown under saline conditions. Int J Bot 6(1):53–58

Teixeira M, Carvalho IS (2009) Effects of salt stress on purslane (*Portulaca oleracea*) nutrition. Ann Appl Biol 154:77–86

Tester M, Davenport R (2003) Na^+ tolerance and Na^+ transport in higher plants. Ann Bot 91:503–527

Tort N, Turkyilmaz B (2004) A physiological investigation on the mechanisms of salinity tolerance in some barley culture forms. JFS 27:1–16

Tudela D, Primo-Millo E (1992) l-Aminocyclopropane-1-carboxylic acid transported from roots to shoots promotes leaf abscission in *Cleopatra mandarin* (*Citrus reshni* Hort. ex Tan.) seedlings rehydrated after water stress. Plant Physiol 100:131–137

Turan MA, Kalkat V, Taban S (2007) Salinity-induced stomatal resistance, proline, chlorophyll and ion concentrations of bean. Int J Agric Res 2(5):483–488

Uddin MK, Shamsuzzaman S, Zi LQ, Mohdselamat MH (2017) Effects of salinity on growth, antioxidant contents and proximate compositions of Sabah snake grass (*Clinacanthus Nutans* (Burm. F.) Lindau). Bangladesh J Bot 46(1):263–269

Ulfat M, Athar HR, Ashraf M, Akram NA, Jamil A (2007) Appraisal of physiological and biochemical selection criteria for evaluation of salt tolerance in canola (*Brassica napus* L.). Pak J Bot 39:1593–1608

US Department of Agriculture (2008) Research database. Bibliography on salt tolerance. In: George E, Brown Fr (eds) Salinity lab. US Department of Agriculture. Serv. Riverside, CA. http://www.Ars.usda.gov/Services/docs.htm?docid=8908. Retrieved 26 Aug 2008

Uzun S, Uzun O, Kaplan M, Ilbas AI (2013) Response of bitter vetch lines to salt stress. Bulgarian J Agric Sci 19(5):1061–1067

Valentine AJ, Mortimer PE, Lintnaar M et al (2006) Drought responses of arbuscular mycorrhizal grapevines. Symbiosis 41(3):127–133

White PJ, Broadley MR (2001) Chloride in soils and its uptake and movement within the plant: a review. Ann Bot 88:967–988

Xu G, Magen H, Tarchitzky J, Kafkaki U (2000) Advances in chloride nutrition. Adv Agron 68:96–150

Yeo AR (2007) Salinity. In: Yeo AR, Flowers TJ (eds) Plant solute transport. Blackwell, Oxford, pp 340–365

Yilmaz H, Kina A (2008) The influence of Nacl salinity on some vegetative and chemical changes of strawberries (*Fragaria xananassa* L.). Afr J Biotechnol 7(18):3299–3305

Yuan BC, Li ZZ, Liu H, Gao M, Zhang YY (2007) Microbial biomass and activity in salt affected soils under arid conditions. Appl Soil Ecol 35:319–328

Zhao GQ, Ma BL, Ren CZ (2007) Growth, gas exchange, chlorophyll fluorescence and ion content of naked oat in response to salinity. Crop Sci 47(1):123–131

Zhu JK (2002) Regulation of ion homeostasis under salt stress. Curr Opin Plant Biol 6:441–445

Part III
Plant-Microbe Interactions in Saline Environments

Chapter 11
Influence of Native Arbuscular Mycorrhizal Fungi and *Pseudomonas fluorescens* on *Tamarix* Shrubs Under Different Salinity Levels

Karima Bencherif, Yolande Dalpé, and Anissa Lounès-Hadj Sahraoui

Abstract *Tamarix articulata* and *Tamarix gallica* are salt-tolerant shrubs naturally associated with arbuscular mycorrhizal fungi (AMF). A greenhouse experiment was conducted to evaluate the mycorrhizal dependency of two *Tamarix* species using *Rhizophagus irregularis* (DAOM 197198) as arbuscular mycorrhizal inoculant. *T. articulata* mycorrhizal dependency reached twice the one found for *T. gallica*. Based on those results, a second greenhouse experiment aimed to compare the effectiveness of native AMF inoculum originating from saline soil of Algerian arid and semiarid areas, while combined or not with a native endophytic bacteria *Pseudomonas fluorescens*, with commercial AMF inoculum (Symbivit) on *T. articulata* growth under three soil salinity levels, non-saline (0.6 ds m^{-1}), moderately saline (2.33 ds m^{-1}), and saline soils (7.52 ds m^{-1}). Root mycorrhizal rates were improved by the co-inoculation with native AMF and *P. fluorescens* and with the native inoculum more than commercial inoculum in saline and moderately saline soils. They were about four folds higher with the co-inoculum and two folds with native AMF inoculum. Improvements of shoot biomasses were about 1.4-fold higher with native AMF inoculum and co-inoculum as compared to commercial inoculum in moderately saline soil. The positive effect on plant growth of the co-inoculation in moderately saline soil was associated with enhancement of phosphorus and nitrogen contents in all soil salinity level. These findings highlight the efficiency of native co-inoculation (AMF/*P. fluorescens*) in improving plant growth under saline conditions and suggest potential use of native inoculum for increasing *T. articulata* plantation in disturbed soils.

K. Bencherif (✉) · A. Lounès-Hadj Sahraoui
Faculté des sciences de la Nature et de la vie, Université de Djelfa, Djelfa, Algeria

Unité de Chimie Environnementale et Interactions sur le Vivant (UCEIV), Université du Littoral Côte d'Opale, Calais, France
e-mail: lounes@univ-littoral.fr

Y. Dalpé
Agriculture et agroalimentaire Canada, Centre de recherche et développement d'Ottawa, Ottawa, Ontario, Canada

B. Giri, A. Varma (eds.), *Microorganisms in Saline Environments: Strategies and Functions*, Soil Biology 56, https://doi.org/10.1007/978-3-030-18975-4_11

Keywords AMF inoculation · Nutrient uptake · Tamarix growth · Mycorrhizal dependency · Soil salinity · *Pseudomonas fluorescens*

11.1 Introduction

Soil salinity is increasing steadily in many parts of the world and affecting negatively plant production (Badda et al. 2014). Increasing salt concentration in soil decreased the plant ability to absorb water and adversely affects metabolic process, osmotic balance, nutrient absorbance, hydraulic conductivity, stomatal conductance, net photosynthetic rate, and intercellular CO_2 concentration (Al-Karaki et al. 2001; Liu et al. 2016). In this respect, biological processes such as arbuscular mycorrhizal fungi (Kohl et al. 2016) and bacterial application (Frey-Klett et al. 2007) to alleviate salt stress would be a better option (Badda et al. 2014). AMF and symbiotic bacteria closely interact with roots, improving plant growth and health (Dey et al. 2006; Smith and Read 2008; Bona et al. 2015), enhancing biocontrol efficiency, abiotic stress management, and phytoremediation (Ramasamy et al. 2011). AMF improve plant mineral nutrition, hormone biosynthesis, and consequently plant metabolism (Lingua et al. 2012; Baslam et al. 2013; Bona et al. 2015). The impact of AMF on host plants depends on the mycorrhizal potential of the fungal inoculant, on the plant host mycorrhizal dependency, as well as on the soil and environmental conditions in which the association occurs (Hoeksma et al. 2010; Johnson 2010; Sikes et al. 2014). *Pseudomonas* bacteria are known to stimulate plant growth through biosynthesis of plant hormones, reduction of stress ethylene produced by plants, nitrogen fixation, and phosphate solubilization (Bonfante and Anca 2009). *Pseudomonas* bacteria may also promote plant growth by improving mycorrhizal symbiosis establishment and functioning (Dey et al. 2006), by supporting the spore germination and mycelia extension of AMF (Frey-Klett et al. 2007; Ramasamy et al. 2011).

Tamarix articulata and *Tamarix gallica* are indigenous shrubs in saline soils, mainly in arid and semiarid Algerian areas (Belarouci 1991). In Algeria, during the last decade, several extensive restoration projects expanded using *Tamarix* species alone or associated with other species such as cypress (*Cypresus* spp.) (Ergr 2013). A neglected aspect of those restoration efforts is the role played by mycorrhizal symbiosis and root-inhabiting soil microorganisms in the revegetation of disturbed areas (Smith and Read 2008). *Tamarix* are mycotrophic plants naturally associated with the arbuscular mycorrhizal fungi (AMF) (Beauchamp et al. 2005; Chaudhry et al. 2013; Bencherif et al. 2015).

There is an increasing interest to utilize beneficial soil biota as a tool to enhance plant nutrition and productivity, which can be deliberately introduced into agroecosystems through inoculation in order to add desired functions or to enhance an already existing one (Kohl et al. 2016). It is well reported that AMF inoculation associated with symbiotic bacteria can improve plant growth (Johnson 2010; Xun et al. 2015; Liu et al. 2016) and positively influence the physiology of plants

(Ramasamy et al. 2011). This emphasizes the importance of testing the effectiveness of different AMF strains to develop inoculum suitable for *Tamarix* seedling production with the aim to revegetate the stressed soil areas (Varma 1999; Miransari 2011). Several investigations using *Tamarix ramosissima* have already been carried out under greenhouse conditions using inoculated plants (Beauchamp et al. 2005; Meinhardt and Gehring 2012; Taniguchi et al. 2015), but limited investigations concerned *Tamarix articulata* and *Tamarix gallica*, two species frequently encountered under steppic Algerian environments.

The first investigation on the determination of microorganisms associated to *T. articulata* rhizosphere revealed an interesting AMF biodiversity under adverse saline environmental conditions. Moreover, it was showed that even mycorrhizal colonization rate of *T. articulata* decreased when salinity level increased, and it was not completely inhibited and remained possible (Bencherif et al. 2015). In the present study, greenhouse pot culture of *T. articulata* and *T. gallica* were undertaken in order to (1) compare the mycorrhizal dependency of the two studied species, to (2) evaluate the benefit of AMF inoculation on *T. articulata* growth and mineral nutrition under soil salinity conditions, and at last (3) to compare the efficiency of commercial and native inoculum while tested alone or in combination with the endophytic bacteria, *Pseudomonas fluorescens*.

11.2 Material and Methods

11.2.1 Mycorrhizal Dependency

T. articulata and *T. gallica* cuttings were sampled from mature shrubs (10–20 years) devoid of any apparent infection. Cuttings were stored at 4 °C for 15 days. The cutting base was sterilized with hydrogen peroxide (10%) solution for 5 min and then rinsed in sterile distilled water to remove any trace of chemical that could interfere with seedling development. Plantation was elaborated in greenhouse at 25 °C. Ten replicates for each treatment per plant were done. Substrate was composed of a mixture of black peat compost, resinous bark, and wood (1:1:1) with a pH of 6.3 and a water retention capacity of 820 mL L^{-1}.

In a first experiment, the mycorrhizal dependency (MD) of *T. articulata* and *T. gallica* inoculated with commercial inoculum *Rhizophagus irregularis* (DAOM 197198) (MYKE PRO, Premier Tech, Canada), as granules with a concentration of 500 propagules/g were evaluated according to the method described by Hetrick et al. (1992). Mycorrhizal dependency (MD) is evaluated as follows:

$$\text{MD} = \big[(\text{Dry weight of mycorrhizal plants} - \text{Dry weight of non-mycorrhizal plants}) \\ /\text{Dry weight of non-mycorrhizal plants}\big] \times 100.$$

11.2.2 Production of Native Inoculum

Tamarix native rhizosphere soil was harvested from naturally grown shrubs located in the experimental arid and semiarid Algerian areas, with salinity level about 0.6 ds m^{-1}, 2.33 ds m^{-1} and 7.52 ds m^{-1}. One volume of soil from each prospected location was mixed with one volume of sterilized peat soil in pot culture (300 mL) to trap native AMF on alfalfa (*Medicago sativa*) roots according to Fortin et al. (2008). Seeds of alfalfa were previously sterilized in ethanol 90° for 5 min and rinsed with sterile distilled water. Five pots per location were placed under growth chamber conditions with 16 h photoperiod at 15 °C and watered twice a week. After 4 months, greenhouse conditions became 25 °C with 12 h photoperiod. After 10 months, aerial parts were cut, roots were mixed with soil, and new seeds were planted for 12 months. The inoculum produced after 22 months consisted of a mixture of different native AMF spores, mycorrhizal alfalfa roots, hyphae, and the inert substrate. After the trapping period, spores were extracted from the soils using the wet-sieving method described by Gerdemann and Nicolson (1963). Spores were identified on the basis of their morphological characteristics (Blaszkowski 2012). Morphological identification of spores was done on three repetitions in time. Alfalfa root colonization rate was evaluated according to McGonigle et al. (1990). On an interval of 6 months, spores were extracted to control and identify the species in the native inoculum, and root colonization rate was evaluated. During inoculum production, the same AMF species were observed.

11.2.3 Pseudomonas fluorescens *Propagation*

P. fluorescens endophytic bacteria were originally isolated from the *Tamarix* roots and identified according to the methods described by Fages and Mulard (1988). The concentration of the bacterial suspension used for inoculation was 3.3×10^8 cell mL^{-1}, and 30 mL of the suspension was added to each pot culture (Fages and Mulard 1988; Akram 2008).

11.2.4 Mycorrhizal Inoculation Treatments of T. articulata Seedlings

The second experiment involved three different AMF inoculation treatments and three soil salinity levels. Three soils harvested from arid and semiarid areas were used in this experiment. They are classified according to their salinity levels as non-saline (0.6 ± 0.14 ds m^{-1}) (NS), moderately saline (2.32 ± 0.4 ds m^{-1}) (MS), and saline soils (7.52 ± 3.01 ds m^{-1}) (S). Based on the results of the first experiment, *T. articulata* was used as unique host plant. Cuttings were stratified at 4 °C for

15 days and planted into pots filled with substrate consisting of autoclaved sand and naturel sampled soil (1:1). Control substrate and sand were sterilized twice (120 °C and 1 bar for 30 min). Three types of inoculum were used: (1) commercial inoculum (Symbivit®; Inoculum plus, France) consisting of a mix of six *Glomus* sp. on natural clay carriers and natural ingredients supporting mycorrhiza (hamates, ground minerals, extract from sea organisms), naturally degradable granules of a water-retaining gel, with an average of 50 spores/g; (2) native inoculum made of trap culture inoculum containing an average of 40 spores/g and 0.68 g/g of AMF colonized root fragments; and (3) co-inoculum with native AMF plus native *P. fluorescens* containing an average of 40 spores/g supplemented with 30 mL of a bacterial solution (3.3×10^8 cell mL^{-1}) diluted in $MgSO_4$ (0.01 M). The inoculation methodology described by Akram (2008) modified was used to inoculate *Tamarix* seedlings. Whatever the inoculum type, each pot (capacity of 500 mL) was supplemented with 50 g of inoculum. Non-inoculated plants were supplemented with the equivalent amount of autoclaved inoculum, in order to provide plants with the same organic substances as inoculated plants. The inoculum was placed in the layer beneath the cutting in the pot cultures to facilitate fungal colonization of plant roots. Seedlings were incubated in controlled conditions at 15 °C night/day, 10–14 h photoperiod, and 70% humidity. Five replicates were employed for each treatment.

11.2.5 Determination of Root and Shoot Dry Weight

After 24 weeks of culture, shoots and roots were harvested and then dried in a hot air oven at 70 °C for 72 h until constant weight. Three plants were used as biological replicates for each treatment.

11.2.6 Determination of Phosphorus and Nitrogen Shoot Contents

Dried shoots (1 g) were digested in a mixture of boiling sulfuric acid and hydrogen peroxide. Nitrogen content was determined using Kjeldahl method (Muñoz-Huerta et al. 2013). Phosphorus was extracted by HCl acid and evaluated by spectrophotometric method using molybdenum sulfate (Drogba 2011).

11.2.7 Estimation of Mycorrhizal Rate

All events related to AMF root colonization, namely, internal hyphae, arbuscules, and vesicles, were visualized after 10 and 24 weeks of culture by microscopic observation on trypan blue-stained roots according to Phillips and Hayman (1970) method modified by Dalpé and Seguin (2013). Root colonization rate was calculated using the grid line intersect method (McGonigle et al. 1990). Two thousand and two hundred fifty root fragments from 50 plants were observed under an optical microscope (×100 magnification). The total mycorrhization rate (containing arbuscules, vesicles, and hyphae) was estimated.

11.2.8 Isolation, Quantification, and Identification of AMF Spores

AMF spore extraction was performed on 24-weeks plant cultures using a wet-sieving method (Gerdemann and Nicolson 1963). Spore morphotypes were described and species tentatively identified using original descriptions of species, identification keys (Blaszkowski 2012), and specialized websites: http://www.zor.zut.edu.pl/Glomeromycota/index.html and http://invam.caf.wvu.edu.

11.3 Results

11.3.1 Mycorrhizal Dependency of T. articulata *and* T. gallica

Evaluation of mycorrhizal dependency of *T. articulata* and *T. gallica* revealed that *T. articulata* was much more responsive to AMF *Rhizophagus irregularis* than *T. gallica* by a factor of 1.7 (Table 11.1).

11.3.2 AMF Trap Culture Evaluation Under Different Soil Salinity Levels

The number of spores isolated from *Medicago sativa* trap cultures after 22 months was about 46, 32, and 16 spores/g of soil in moderately, saline, and non-saline soils,

Table 11.1 Mycorrhizal dependency of *T. articulata* and *T. gallica*

Tamarix species	Mycorrhizal responsiveness index
T. articulata	$30.09^{a} \pm 0.5$
T. gallica	$17.53^{b} \pm 0.12$

Data are presented as mean ± standard deviation. Means were obtained from ten replicates for each plant species ($n = 10$)
Different letters indicate significant difference between the species according to the LSD test ($p < 0.05$)

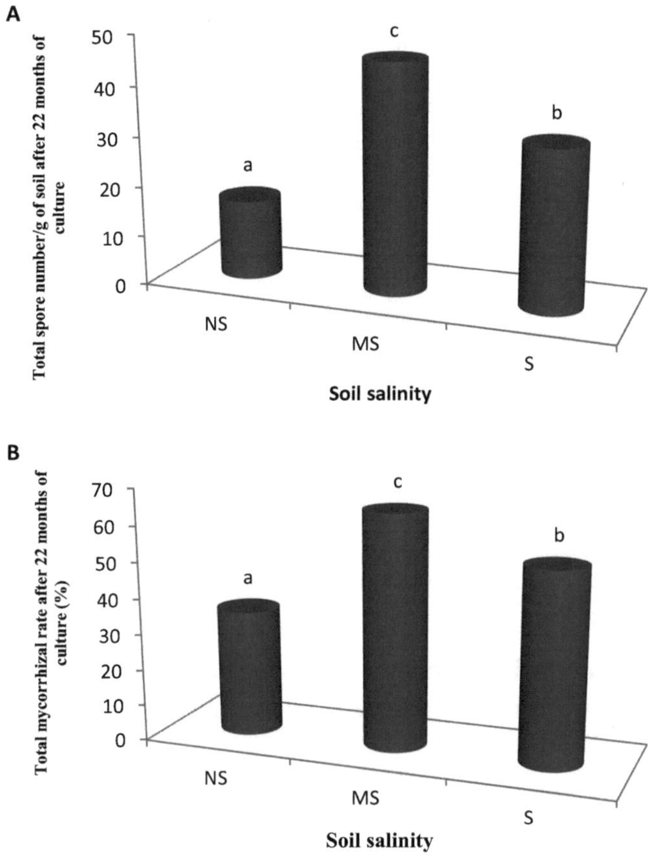

Fig. 11.1 Inoculum richness of soil extracted from *T. articulata* rhizosphere in sampled sites using trap culture with *Medicago sativa* after 22 months of culture. (**a**) Spore number isolated from 22 months old pot cultures. (**b**) AMF root colonization rate of *Medicago sativa* after 22 months of culture. Data are presented as mean ± standard deviation. Means were obtained from five replicates for each soil. Different letters indicate significant difference between the different soils according to the LSD test ($p < 0.05$). *NS* non-saline, *MS* moderately saline soil, *S* saline soils

respectively (Fig. 11.1a). *Medicago sativa* root colonization rates followed the same pattern (Fig. 11.1b). Moderately saline soils showed the highest root colonization rate (65%), and the lowest rate was found in the non-saline soil (35%). The root colonization rate in the most saline soils was about 53%. Native AMF species were *Septoglomus constrictum*, *Funneliformis geosporum*, *F. mosseae*, *F. coronatum*, *F. caledonium*, *Rhizophagus fasciculatus*, *R. diaphanous*, and *Gigaspora gigantea* (Table 11.2). *S. constrictum* dominated in non-saline and moderately saline soils, while *F. geosporum* was abundant in saline soil.

Table 11.2 Evaluation of AMF trap culture under soil salinity levels

Soil types	AMF species	Spores number g^{-1} of soil
Non-saline	*S. constrictum*	7 ± 2
	F. geosporum	5 ± 1
	F. mosseae	2 ± 1
Moderately saline	*S. constrictum*	13 ± 1
	F. geosporum	11 ± 1
	F. mosseae	6 ± 1
	F. caledonium	2 ± 1
	O. diaphana	2 ± 1
	F. coronatum	3 ± 1
	Gi. gigantea	4 ± 1
	R. fasciculatum	3 ± 1
Saline	*S. constrictum*	6 ± 2
	F. geosporum	13 ± 4
	F. mosseae	8 ± 3
	F. caledonium	4 ± 1

Data are presented as mean of AMF isolated spores after 22 weeks of alfa culture. Means for each species were obtained from five replicates per each soil

11.3.3 Impact of Mycorrhizal Inoculation on T. articulata *Root Colonization Rate Under Different Soil Salinity Levels*

T. articulata root colonization rates after 10 and 24 weeks of culture were presented in Fig. 11.3. After 10 weeks of plant culture, mycorrhizal rates were 1.7- and 1.5-fold higher when the soil was inoculated with native AMF and co-inoculated with AMF plus *P. fluorescens*, respectively, as compared to non-inoculated soil. In contrast, in the presence of commercial AMF inoculum, AMF roots were not colonized in saline soil with low root colonization rate in non-saline and moderately saline soils (Fig. 11.2). After 24 weeks, evaluation of AMF root colonization rate indicates that use of native inoculum and co-inoculum increased significantly total arbuscular mycorrhizal rates comparatively with non-inoculated plants in non-saline and saline soil. They were about 2- and 17-fold higher for the co-inoculation, respectively, in non-saline and saline soils and about 1-fold and 15-fold higher, respectively, in non-saline and saline soil with native AMF inoculum. With commercial inoculum, AMF mycorrhizal rate was 6.6-fold higher than at 10 weeks for saline soil and about 33.4-fold higher in moderately saline soil.

A positive correlation was recorded between root colonization rate and shoot biomass in moderately saline soil for all studied inoculum and for co-inoculum in saline soils (Table 11.4). For non-inoculated soil, a positive correlation was recorded between root colonization rate and plant biomasses on non-saline and moderately saline soils. Root colonization rate recorded positive correlation with spore number

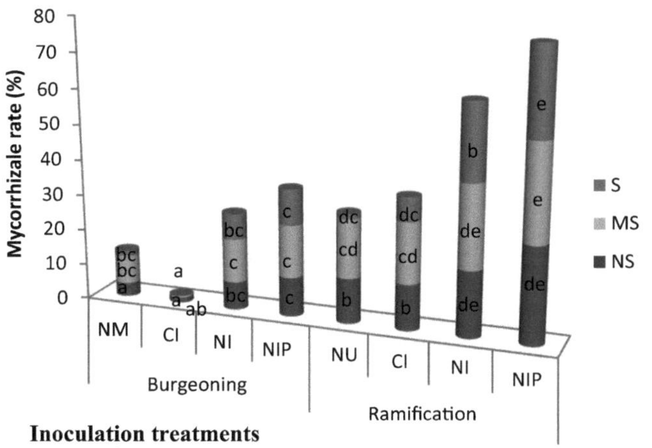

Fig. 11.2 Root colonization rates of *T. articulata* at 10 and 24 weeks for each treatment (*NM* non-mycorrhizal, *CI* commercial inoculum, *NI* native inoculum, *NIP* native inoculum plus *Pseudomonas fluorescens*) in each soil (*NS* non-saline soil, *MS* moderately saline soil, *S* saline soil). Data are means based on five replicates ±standard error (SE), and different letters indicates significant difference among treatments ($p < 0.05$, Tukey HSD test)

on commercial inoculum on saline soils, with native inoculum on moderately saline soils, and with co-inoculum on non-saline and saline soils (Table 11.4).

11.3.4 Impact of Mycorrhizal Inoculation on AMF Spore Abundance and Biodiversity in T. articulata *Rhizosphere Under Different Soil Salinity Levels*

AMF spores isolated from *T. articulata* rhizosphere cultured in pot during 24 weeks under different soil salinity levels were identified and quantified (Table 11.3). They belong to *Funneliformis*, *Septoglomus*, *Rhizophagus*, *Oehlia*, and *Gigaspora* genera. Eight species were identified as *F. geosporum*, *F. mosseae*, *F. caledonium*, *F. coronatum*, *S. constrictum*, *Gi. gigantea*, *Oehlia diaphanous*, *R. fasciculatum*, and one non-identified glomoid spore species. While *S. constrictum* was the most abundant species in non-inoculated soils regardless of soil salinity level, *F. geosporum* and *F. mosseae* were the most abundant species in co-inoculated non-saline, moderately saline, and saline soils. With AMF native and commercial inoculums, *F. geosporum* was the most abundant species in the three-studied soil. *Gi. gigantea* was not recorded in saline soils.

Table 11.3 AMF spore species according to inoculation treatments through soil salinity levels

AMF species	Non-saline soils				Moderately saline soils				Saline soils			
	NM	CI	NI	NIP	NM	CI	NI	NIP	NM	CI	NI	NIP
Funneliformis geosporum	14 ± 1	22 ± 2	27 ± 5	28 ± 5	15 ± 3	30 ± 5	29 ± 6	30 ± 6	16 ± 4	45 ± 7	30 ± 4	31 ± 2
Funneliformis mosseae	21 ± 2	18 ± 1	26 ± 2	25 ± 2	13 ± 2	30 ± 7	25 ± 1	27 ± 3	16 ± 6	26 ± 5	39 ± 4	42 ± 6
Funneliformis caledonium	2 ± 1	4 ± 1	5 ± 1	5 ± 1	nd	4 ± 2	2 ± 1	3 ± 1	nd	nd	nd	nd
Funneliformis coronatum	Nd	3 ± 1	nd	1 ± 2	nd	20 ± 3	7 ± 1	6 ± 1	12 ± 1	2 ± 2	17 ± 5	18 ± 2
Septoglomus constrictum	26 ± 5	13 ±2	17±2	17 ± 7	17 ± 2	23 ± 1	28 ± 4	26 ± 5	25 ± 8	31 ± 6	18 ± 3	23 ± 2
Gigaspora gigantea	6 ± 1	nd	3 ± 1	4 ± 2	8 ± 1	3 ± 1	1 ± 1	1 ± 1	nd	nd	nd	nd
Oehlia diaphana	Nd	1 ±1	4 ± 1	3 ± 1	3 ± 1	1 ± 1	2 ± 1	3 ± 2	nd	5 ± 1	nd	nd
Rhizophagus fasciculatum	1 ± 1	1 ±1	4 ± 1	3 ± 2	5 ± 2	3 ± 1	nd	nd	nd	nd	nd	nd
Glomus sp1	Nd	2 ±1	2 ± 1	2 ± 1	4 ± 1	1 ± 1	nd	nd	8 ± 1	4 ± 1	3 ± 2	3 ± 2
Total AMF spore number/g of soil	70	64	88	88	65	115	94	96	77	113	107	117

Data are presented as mean of AMF isolated spores after 24 weeks of seedling culture. Means for each treatment were obtained from five replicates. *NM* non-mycorrhizal, *CI* commercial inoculum, *NI* native inoculum, *NIP* native inoculum plus *Pseudomonas fluorescens*, *nd* not detected

11.3.5 *Impact of Mycorrhizal Inoculation on* T. articulata *Growth Under Different Soil Salinity Levels*

Impact of inoculation by the three different mycorrhizal inoculum types (commercial inoculum, AMF native inoculum, and co-inoculum with AMF plus *P. fluorescens*) on *T. articulata* growth is presented in Fig. 11.3a, b. Whereas no effects of inoculation on root and shoot biomasses were observed with the commercial inoculum in moderately and saline soils, a significant increase was recorded in non-saline soil in comparison with non-inoculated non-saline soil. In contrast, positive effects on shoot biomasses were observed with the AMF native inoculum and co-inoculum on moderately saline soils. Shoot biomasses were about 1.07 and 1.08 g, respectively, with native inoculum and co-inoculum in moderately saline soils, while it was

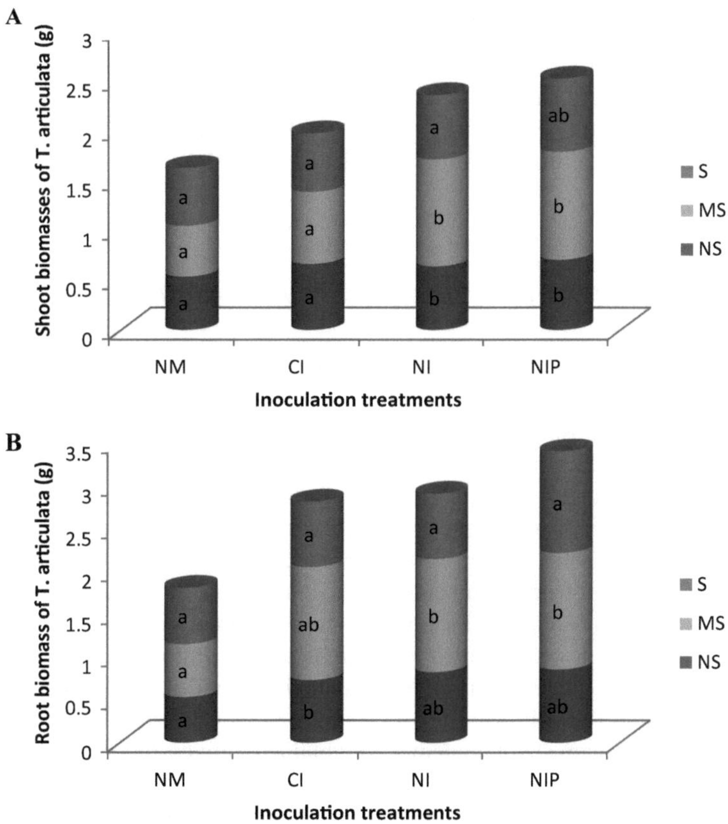

Fig. 11.3 *T. articulata* shoot (**a**) and root (**b**) biomasses after 24 weeks of culture in a greenhouse. *NM* non-mycorrhizal plants. Data are means based on three replicates ±standard error (SE), and different letters indicate significant difference among treatments ($p < 0.05$, Tukey HSD test). *CI* commercial inoculum, *NI* native inoculum, *NIP* native inoculum plus *Pseudomonas*, *NS* non-saline, *MS* moderately saline soil, *S* saline soils

Table 11.4 Correlation between soil salinity and total mycorrhizal rate at ramification with different measured parameters under saline soil conditions

Parameters	CI		NI		NIP	
	AMF	S	AMF	S	AMF	S
Mycorrhizal rate at burgeoning	–	−0.96**	–	0.43*	–	0.69*
Mycorrhizal rate at ramification	–	−0.63**	–	0.48**	–	0.6**
Shoots biomasses	0.32 ns	−0.61	0.56**	0.58*	0.69**	0.54**
Roots biomasses	0.56*	−0.74**	0.67*	0.67*	0.97*	0.56**
Phosphorus content	0.34 ns	−0.64*	0.65*	0.58*	0.53**	0.86*
Nitrogen content	0.23 ns	−0.25 ns	0.72*	0.32 ns	0.5*	0.65*
Total spore number	0.17 ns	0.38 ns	0.55*	0.7*	0.71**	0.54*

CI commercial inoculum, *NI* native inoculum, *NIP* co-inoculum (native inoculum plus *Pseudomonas fluorescens*). Correlation was established according to Pearson coefficient test; $*p < 0.05$, $**p < 0.01$, *ns* not significant

about 0.7 g in non-inoculated plants and with commercial inoculum (Fig. 11.3a). Moreover, AMF native inoculum and co-inoculum improved *T. articulata* root biomass in moderately saline soil. It was two folds higher than in non-inoculated moderately saline soil (Fig. 11.3b), while AMF inoculation showed significant positive effect on root biomass at $p < 0.001$ (Table 11.4).

11.3.6 Phosphorus and Nitrogen Shoot Contents

The impact of inoculation treatment under three soil salinity levels on *T. articulata* shoot phosphorus content is showed in Fig. 11.4a. Soil salinity significantly reduced *T. articulata* shoot phosphorus content in non-inoculated soils. However, addition of native inoculum and co-inoculum enhanced phosphorus content except for native inoculum in moderately saline soil. In the presence of the native co-inoculum, phosphorus content was about one fold higher in non-saline and moderately saline soils and about two folds higher in saline soils comparatively with non-inoculated soils. With addition of native inoculum, phosphorus shoot content was enhanced about one fold in non-saline and saline soils as compared with non-inoculated soil.

Positive correlation was recorded between root colonization rates and phosphorus contents except for saline soil in non-inoculated soil and in the presence of commercial inoculum (Table 11.4).

Increases in nitrogen shoot content were obtained with co-inoculum as compared with non-inoculated soils whatever the salinity soil levels (Fig. 11.4b). They were about one fold higher in non-saline and saline soil and about 3.8-fold higher in moderately saline soils. In addition, native co-inoculum and inoculum increased shoot nitrogen content more than commercial inoculum in non-saline and moderately saline soils and for saline soil only with co-inoculum.

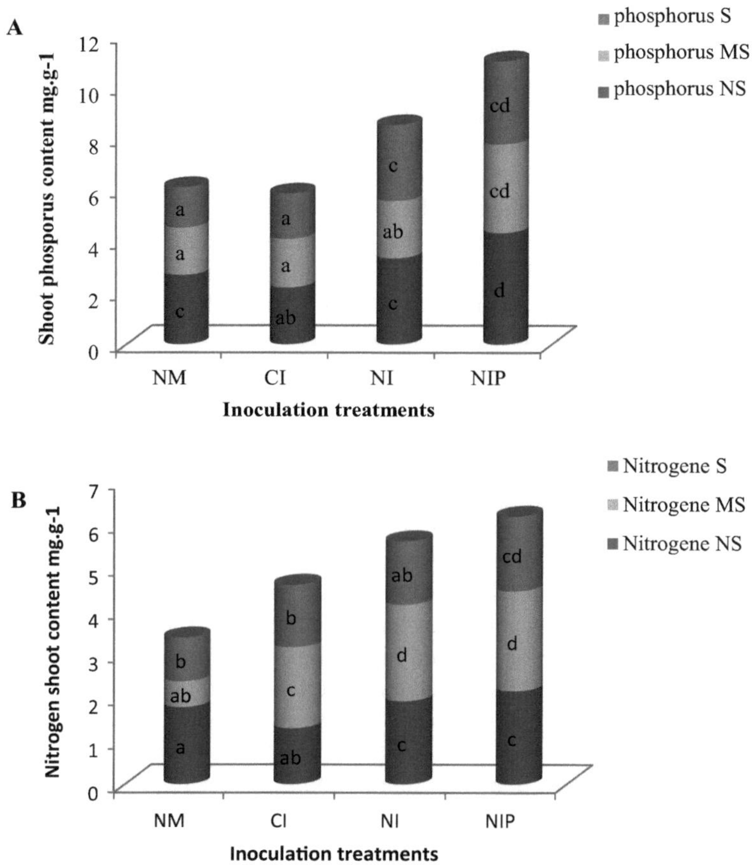

Fig. 11.4 Phosphorus (**a**) and nitrogen (**b**) shoot contents after 24 weeks of *T. articulata* growth under the four inoculum treatments in saline soils. Data are means based on five replicates ±standard error (SE), and different letters indicate significant difference among treatments ($p < 0.05$, Turkey HSD test). *NM* non-mycorrhizal, *CI* commercial inoculum, *NI* native inoculum, *NIP* native inoculum plus *Pseudomonas*, *NS* non-saline soil, *MS* moderately saline soil, *S* saline soil

11.4 Discussion

In the first part of the present study, it was revealed that *T. articulata* and *T. gallica* were dependent on AMF (*R. irregularis*) symbiosis. The mycorrhizal dependency of *T. articulata* was almost two times higher than for *T. gallica* (30 vs. 17%). These results agree with an earlier study of Mathr and Vyas (2000), which reported that *T. articulata* shows tremendous dependency on AMF (*Gigaspora margarita* and *Glomus fasciculatum*) for biomass production and nutrient uptake in saline soils. Giri (2017) also observed that the dependence of *Gliricidia sepium* on AMF increased with increasing levels of salinity, which disturbed water potential and reduced soil nutrient availability to plants. Renuka et al. (2012) demonstrated that mycorrhizal dependency varied according to the AMF species used as inoculum in *Acacia melanoxylon*. Decklerck et al. (1995) explained that mycorrhizal dependency varied greatly according to plant species and plant genotype (Sangabriel-conde et al. 2014). Moreover, plant dependency to mycorrhizal colonization can affect plant response and plant benefits from mycorrhizae (Dalpé and Monreal 2004). This result conducted us to select *T. articulata* to pursue experimentations on growth and salt tolerance of *Tamarix* experiment in the current study.

In the second part of this study, the performance of the three kinds of AMF inoculum (one commercial AMF inoculum and two native inoculums: one was composed of AMF alone, and the second contains AMF and *P. fluorescens*) was tested with respect to *T. articulata* growth and mineral nutrition under different soil salinity levels. Firstly, evaluation of *T. articulata* root colonization rates at two time cultures (10 and 24 weeks) showed that no AMF mycorrhizal structure were observed at 10 weeks on *T. articulata* roots with commercial inoculum in saline soils. At 24 weeks with commercial inoculum, AMF root colonization rate was significantly apparent in all studied soils. This indicated that commercial inoculum took more time to establish compared to other inoculums especially on saline soils. In non-inoculated soils, AMF root colonization was the lowest in saline soils both at 10 and 24 weeks of culture. This could be due to a direct effect of salt on spore germination, hyphal spread, and/or mycorrhizae formation (Garg and Pandey 2015). In saline soils, co-inoculation allowed higher root colonization rate both at 10 and 24 weeks as compared with non-inoculated soil and commercial inoculum. These findings are in accordance with those of Garg and Pandey (2015) who observed higher root colonization rate in *Cajanus cajan* L. with native inoculum when compared to commercial one in salt conditions. Likewise, Bona et al. (2015) observed high AMF root colonization rate in strawberry culture inoculated with a mixture of AMF and *Pseudomonas* ssp. The difference in mycorrhizal rate between both native inoculum and co-inoculum as compared with commercial inoculum must be due to the characteristics of AMF species (Evelin et al. 2009) and to their specific adaptation to environmental conditions (Symanczik et al. 2015). Moreover, compatibility between host plant and inoculum may be influenced also by root exudates. The same result was reported by Estrada et al. (2013) who showed an increased colonization ability of *Asteriscus maritimus* inoculated with *R. irregularis* as native isolate under saline conditions as compared to *R. irregularis* in commercial inoculum. Our results suggest that AMF species in native AMF inoculum recognized more easily *T. articulata* root exudates than AMF species in commercial inoculum.

AMF spore isolation from the *T. articulata* pots culture after 24 weeks of culture indicated that all AMF species belong to Glomeromycota phylum. *F. geosporum* and *F. mosseae* were the most abundant species in inoculated soils, while *S. constrictum* dominate in non-inoculated soils. Some species of *Rhizophagus* and *Gigaspora* genera were also detected. Such fungal diversity occurs frequently in natural ecosystems where many AMF can colonize the same root system (Sharma et al. 2009). *F. geosporum* predominance may indicate a better adaptation of this species to saline soils. In fact, *F. geosporum* has already been described in *T. ramosissima* salt marsh (Taniguchi et al. 2015) and in field saline soils in the rhizosphere of *T. articulata*

(Chaudhry et al. 2013; Bencherif et al. 2015). No significant differences on AMF spore number were detected between the different inoculation treatments. However, a tendency to get higher spore content in the saline soils especially under the co-inoculum was noticed. Such a result suggested that the addition of *P. fluorescens* to the native AMF inoculum can potentially stimulate AMF sporulation in saline soil. This is in accordance with the study of Panneerselvam et al. (2012), which reported that combined inoculation of *G. mosseae* with *Pseudomonas putida* improved significantly AMF colonization and spore proliferation in addition to enhancement of the growth promotion in guava seedlings.

An interesting result found in the present study is the ability of the native microorganisms to enhance plant growth in saline stressful condition. Indeed, a positive effect of native inoculum and co-inoculum was observed on *T. articulata* shoot and root biomasses as compared with commercial inoculum especially in moderately saline soil for the co-inoculum. Native inoculum enhanced positively nitrogen shoot content similarly to shoot and root biomasses in moderately saline soil. This is in agreement with Ortiz et al. (2015) and Bona et al. (2015) studies which showed that native inoculum and *Pseudomonas* sp. increased positively *Trifolium repens* and strawberry growth. In fact, *Pseudomonas* being a good auxin and cytokinin producer may solubilize phosphate and stimulate AMF root colonization and promote root growth (Frey-Klett et al. 2007). Positive correlation was recorded between root colonization rates and *T. articulata* biomasses. The synergistic interaction between AMF and bacteria such as *Rhizobia*, *Azospirillum*, and *Pseudomonas* can stimulate plant growth (Sanon et al. 2009; Miransari 2011; Vyas and Vyas 2014). Likewise, Estrada et al. (2013) recorded an increased growth improvement of *Asteriscus maritimus* with native inoculum based on *R. irregularis* under saline conditions as compared to commercial inoculum of the same species. In contrast, nonsignificant difference was recorded between commercial, co-inoculum, and native inoculum in the non-saline soils on *T. articulata* growth except for root biomass with commercial inoculum. This finding underlined the importance of arbuscular mycorrhizal association for plant survival and growth in arid and semiarid environments (Dupponois et al. 2011; Garg and Pandey 2015).

The main mechanism for enhanced *T. articulata* growth under inoculation conditions seems to be improvement in nutrient uptake. Positive correlation was recorded between AMF colonization rate and *T. articulata* growth and phosphorus and nitrogen content in non-saline, saline, and moderately saline soils with co-inoculum and native inoculum. In contrast, commercial inoculum showed a positive correlation only in moderately saline soil between root mycorrhizal rate, shoot and root biomasses, and phosphorus content, while no significant correlation was recorded in saline soils. Non-saline soil shows positive correlation for phosphorus shoots content with commercial inoculum. Indeed, co-inoculation enhanced nitrogen and phosphorus shoot contents under non-saline, moderately saline, and saline soils for phosphorus and on moderately saline soil for nitrogen. These findings are in accordance with previous data which demonstrated that AMF improve plant phosphorus uptake (Hamel 2004) and that this effect is enhanced by the co-inoculation of AMF with *Pseudomonas* spp. (Artursson et al. 2006) under condition of low soil phosphorus and nitrogen level. In the same way, Vyas and Vyas (2014) recorded that plant phosphorus content in *capsicum mathania* cultiver was observed upon dual inoculation (*Glomus deserticola* plus *Azospirillum*) rather than single inoculation. On the other hand, soil salinity significantly reduces phosphorus shoot content in non-inoculated plants. This result could be explained by the fact that phosphate ions precipitate with Ca^{+2}, Mg^{+2}, and Zn^{+2} ions in saline soils and became unavailable to plants (Evelin et al. 2009). The detrimental salinity effect can be attenuated through AMF inoculation (Evelin et al. 2009). In our study, co-inoculation of *T. articulata* by AMF and *Pseudomonas* increased phosphorus shoot content approximately by two folds compared to the commercial AMF inoculum. Co-inoculation can activate the production of phosphatases by the mycorrhizal fungus and increase the production, by mycorrhizal hyphae, of organic acids that mineralize phosphorus (Varma 1999).

In the same way, our findings showed that inoculation treatment enhanced nitrogen shoot content. Same result was described with Cypress (*Cupressus atlantica*) in Morocco inoculated in greenhouse conditions (Dupponois et al. 2011). This can be attributed to increased nitrogen availability by AMF inoculum (Hamel 2004; Garg and Pandey 2015). In saline soils, plants increasingly accumulate Na^+, which directly damages membrane systems and disturbs N-reductase activity (Evelin et al. 2009). To counteract this detrimental effect, the AMF inoculation tends to enhance K^+ and decreased Na^+ by preselecting nutrients and preventing Na^+ ion accumulation (Garg and Pandey 2015). Likewise, AMF have been reported to directly take up both organic and inorganic nitrogen from the soil solution and transfer it to their host plants (Ferrol and Pérez-Tienda 2009).

11.5 Conclusion

Application of native AMF inoculum and AMF/*P. fluorescens* co-inoculum significantly enhanced growth and affected mineral uptake of *T. articulata* on saline soil. Native AMF inoculum single or combined with native *P. fluorescens* conferred higher salt tolerance to *T. articulata* than AMF commercial inoculum. Although the mechanisms are not well understood, this study supports the hypothesis that *P. fluorescens* and AMF induce changes in plant physiology and create favorable microenvironments. This can be regarded as an efficient biotechnology to improve plant development under stressed conditions of arid and semiarid areas such as soil salinity. However, the interaction of non-native isolates with the local microflora needs to be explored. However, it is imperative to understand the nature of the interaction between native beneficial endophytic bacteria and native AMF in order to exploit and manage the co-inoculation for obtaining sustainable plant development, it will also be interesting in the future to produce co-inoculum originated from arid and semiarid areas for *Tamarix* species and other shrubs to increase plant establishment and growth and phosphorus and nitrogen uptake. Instead, it must be useful to enlarge this experimentation to several sites and diversified studies on abiotic stresses to confirm the universality of the co-inoculation efficiency.

Acknowledgments This study was supported by the Algerian Ministry of Higher Education and Scientific Research. We thank the technical staff of NFRI (National Forest Research Institute) for their assistance in soil sampling and the Science of Nature and Life Faculty of Djelfa University (Algeria) for the facilities provided.

References

Akram A (2008) Elicitation de la résistance systémique induite chez la tomate et le concombre et activation de la voie de la lipoxygénase par des rhizobactéries non-pathogènes. Dissertation, Université de Liège

Al-Karaki GN, Hammad R, Russan M (2001) Response of two tomato cultivars differing in salt stress. Mycorrhiza 11:43–47

Artursson V, Finlay R, Jansson J (2006) Interactions between arbuscular mycorrhizal fungi and bacteria and their potential for stimulating plant growth. Environ Microbiol 8:1–10

Badda N, Aggarwal A, Kadian N, Sharma N (2014) Influence of arbuscular mycorrhizal fungi and different salinity levels on growth enhancement and nutrient uptake of *Gossypium arboreum* L. Kavaka 43:14–21

Baslam M, Esteban R, Garcia-Plazaola JI, Giocoechea N (2013) Effectiveness of arbuscular mycorrhizal fungi (AMF) for inducing the accumulation of major carotenoids, chlorophylls and tocopherol in green and red leaf lettuces. Appl Microbiol Biotechnol 97:3119–3128

Beauchamp V, Stromberg JC, Stutz J (2005) Interaction between *Tamarix ramosissima* (Saltcedar), *Populus fremontii* (Cottonwood), and mycorrhizal fungi: effects on seedling growth and plant species coexistence. Plant Soil 275:221–231

Belarouci LN (1991) Les reboisements en Algérie et leurs perspectives d'avenir. V1 et V2 OPU, Alger, p 641

Bencherif K, Boutekrabt A, Fontaine J, Laruelle F, Dalpè Y, Lounes-Hadj Sahraoui A (2015) Impact of soil salinity on arbuscular mycorrhizal biodivesity and microflora biomass associated to *Tamarix articulata* Vahll rhizosphere in arid and semiarid Algerian areas. Sci Total Environ 533:488–494

Blaszkowski J (2012) Glomeromycota. IB Publisher Polish Academy of Sciences, Poland, p 274

Bona E, Lingua G, Manassero P, Cantamessa S (2015) AM fungi and PGPR pseudomonas increase flowering, fruit producing, and vitamin content in strawberry grown at low nitrogen and phosphorus levels. Mycorrhiza 25:181–193

Bonfante P, Anca IA (2009) Plants, mycorrhizal fungi and bacteria: a network of interactions. Ann Rev Microbiol 63:363–383

Chaudhry MS, Saeed M, Nasim FUH (2013) Soil chemical heterogeneity may affect the diversity of arbuscular mycorrhizal fungi in the rhizosphere of *Tamarix aphylla* under arid climate. An Stiint Univ AI I Cuza Iasi S Iia Biol Veg 59:53–63

Dalpé Y, Monreal M (2004) Arbuscular mycorrhizal inoculum to support sustainable cropping systems. Symposium great plains inoculant from proceeding crop management. Crop Manag 3:301–309

Dalpé Y, Seguin SM (2013) Microwave-assisted technology for the clearing and staining of arbuscular mycorrhizal fungi in roots. Mycorrhiza 23:333–340

Decklerck S, Plenchette C, Strullu DG (1995) Mycorrhizal dependency of banana (*Musa acuminata*, AAA group) cultivar. Plant Soil 176:183–187

Dey R, Pal KK, Tilak KVBR (2006) Interaction of plant growth promoting rhizobacteria and AM fungi: present status and future prospects. In: Podila GK, Varma A (eds) Basic research applications of mycorrhizae. Kent UK, Anshan, pp 83–110

Drogba SA (2011) Propriétés physico-chimique et fonctionnelles des tubercules et des amidons d'igname (Discorea) Envireonnment, p 154

Dupponois R, Hafidi M, Ndoye I, Galiana A, Dreyfus B, Prin Y (2011) Gestion et valorisation des ressources microbiennes des sols pour une revégétalisation durable des milieux sahéliens. Le projet majeur africain de la Grande Muraille Verte, p 12

ERGR (2013) Entreprise régional du génie rurale. Rapport de travaux des projet ERGR Atlas Djelfa (ex SAFA) Bilan d'activités et rapports d'évaluation 2011–2013

Estrada B, Aroca R, Azcon Aguilar C, Barea JM, Ruiz-Lozano JM (2013) Native arbuscular mycorrhizal fungi isolated from a saline habitat improved maize antioxidant systems and plant tolerance to salinity. Plant Sci 201–202:42–51

Evelin H, Kapoor R, Giri B (2009) Arbuscular mycorrhizal fungi in alleviation of salt stress: a review. Ann Bot. 104:1263–1280

Fages J, Mulard D (1988) Isolement de bacteries rhizospheriques et effet de leur inoculation en pots chez Zea mays. Agronomie 4:309–314

Ferrol N, Pérez-Tienda J (2009) Coordinated nutrient exchange in arbuscular mycorrhiza interface. In: Azcon-Aguilar et al (eds) Mycorrhizas: functional processes and ecological impact. Springer, Berlin, pp 73–87

Fortin JA, Plenchette C, Piché Y (2008) Les mycorhizes: la nouvelle révolution verte. In: Multi mondes-Quea (eds), p 126

Frey-Klett P, Garbaye J, Tarkka M (2007) The mycorrhiza helper bacteria revisited. New Phytol 176:22–36

Garg N, Pandey R (2015) Effectiveness of native and exotic arbuscular mycorrhizal fungi on nutrient uptake and ion homeostasis in salt-stressed *Cajanus cajan* L. (Millsp.) genotypes. Mycorrhiza 25:165–180

Gerdemann JW, Nicolson TH (1963) Spores of mycorrhizal endogone species extracted from soil by wet sieving and decanting. Trans Br Mycol Soc 46:235–244

Giri B (2017) Mycorrhizal dependency and growth response of *Gliricidia sepium* (Jacq.) Kunth ex Walp. under saline condition. Plant Sci Today 4(4):154–160

Hamel C (2004) Impact of arbuscular mycorrhizal fungi on N and P cycling in the root zone. Can J Soil Sci 84:383–395

Hetrick BAD, Wilson GWT, Cox TS (1992) Mycorrhizal dependence of modern wheat varieties, landraces, and ancestors. Can J Bot 70:2032–2040

Hoeksma JD, Chaudhary VB, Gehring CA (2010) A meta-analysis of context-dependency in plant response to inoculation with mycorhizal fungi. Ecol Lett 13:394–407

Johnson NC (2010) Resources stoichiometry elucidates the structure and function of arbuscular mycorrhizas across scales. New Phytol 185:631–647

Kohl L, Lukasiewicz CE, VanderHeijden MGA (2016) Establishment and effectiveness of inoculated arbuscular mycorrhizal fungi in agricultural soil. Plant Cell Environ 39:136–146

Lingua G, Bona E, Todeschini V, Cattanea C, Marsano F, Berta G, Cavaletto M (2012) Effects of heavy metals and arbuscular mycorrhiza on the leaf proteome of a selected poplar clone: a time course analysis. Conf Proc PLoS One 7:e38662

Liu S, Guo X, Feng G, Maimaitiaili B, Fan J, He X (2016) Indigenous arbuscular mycorrhizal fungi can alleviate salt stress and promote growth of cotton and maize in saline fields. Plant Soil 398:195–206

Mathr N, Vyas A (2000) Mycorrhizal dependency of *Tamarix aphylla* in saline areas of Thar Desert. J Nat 25:105–110

McGonigle TP, Miller MH, Evans DG, Fairchild GL, Swan JA (1990) A new method which gives an objective measure of colonization of roots by vesicular arbuscular mycorrhizal fungi. New Phytol 115:495–501

Meinhardt KA, Gehring CA (2012) Disrupting mycorrhizal mutualisms: a potential mechanism by which exotic tamarisk outcompetes native cottonwoods. Ecol Appl 22:532–549

Miransari M (2011) Hyperaccumulators, arbuscular mycorrhizal fungi and stress of heavy metals. Biotechnol Adv 29:645–653

Muñoz-Huerta RF, Guevara-Gonzalez RG, Contreras-Medina LM, Torres-Pacheco I, Prado-Olivarez J, Ocampo-Velazquez RV (2013) A review of methods for sensing the nitrogen status in plants: advantages, disadvantages and recent advances. Sensors 13:10823–10843

Ortiz N, Armada E, Duque E, Roldan A, Azcon R (2015) Contribution of arbuscular mycorrhizal fungi and/or bacteria to enhancing plant drought tolerance under natural soil conditions: effectiveness of autochthonous or allochthonous strains. J Plant Physiol 174:87–96

Panneerselvam P, Mohandas SS, Saritha B, Upreti KK, Poovarasan, Monnappa A, Sulladmath VV (2012) *Glomus mosseae* associated bacteria and their influence on stimulation of mycorrhizal colonization, sporulation, and growth promotion in guava (*Psidium guajava* L.) seedlings. Biol Agric Hortic 28:267–279

Phillips JM, Hayman DS (1970) Improved procedures for clearing roots and staining parasitic and vesicular-arbuscular mycorrhizal fungi for rapid assessment of infection. Trans Br Mycol Soc 55:158–161

Ramasamy K, Joe MM, Kim K, Lee S, Shagol C, Rangasamy A, Chung J, Islam MR, Sa T (2011) Synergistic effects of arbuscular mycorrhizal fungi and plant growth promoting rhizobacteria for sustainable agricultural production. Korean J Soil Sci Fert 44:637–649

Renuka G, Rao MS, Praveen Kumar V, Ramesh M, Sanditi Ram R (2012) Arbuscular mycorrhizal dependency of *Acacia melanoxylon* R.Br. Proc Natl Acad Sci India Sect B: Biol Sci 82:441–446

Sangabriel-Conde W, Negrete-Yankelevich S, Maldonado-Mendoza IE, Trejo-Aguilar D (2014) Native maize landraces from Los Tuxtlas, Mexico show varying mycorrhizal dependency for P uptake. Biol Fertil Soils 50:405–414

Sanon A, Andrianjaka Z, Prin Y, Bally R, ThioulousnJ CG, Duponnois R (2009) Rhizosphere microbiota interferes with plant-plant interactions. Plant Soil 321:259–278

Sharma D, Kapoor R, Bhatnagar AK (2009) Differential growth response of *Curculigo orchioides* to native arbuscular mycorrhizal fungal (AMF) communities varying in number and fungal components. Eur J Soil Biol 45:328–333

Sikes BA, Maherali H, Klironomos JN (2014) Mycorrhizal fungal growth responds to soil characteristics, but not host plant identity, during a primary lacustrine dune. Mycorrhiza 24:219–226

Smith SE, Read DJ (2008) Mycorrhizal symbiosis. Academic, San Diego

Symanczik S, Courty PE, Bolleri T, Wiemken A, Al-Yahya MN (2015) Impact of water regimes on an experimental community of four desert arbuscular mycorrhizal fungal (AMF) species, as affected by the introduction of a non-native AMF species. Mycorrhiza 25:639–647

Taniguchi T, Acharya K, Imada S, Iwanaga FYN (2015) Arbuscular mycorrhizal colonization of *Tamarix ramosissima* along a salinity gradient in the southwestern United States. Landsc Ecol Eng 11:221–225

Varma A (1999) Fungi in arid and semi-arid soils. In: Hock B (ed) Mycorrhiza structure, function, molecular biology and biotechnology, 2nd edn. Springer, Berlin, pp 521–556

Vyas M, Vyas A (2014) Field response of *Capsicum annuum* dually inoculated with AM fungi and PGPR in Western Rajasthan. Int J Res Stud Biosci 2:21–26

Xun F, Xie B, Liu S, Guo C (2015) Effect of plant growth-promoting bacteria (PGPR) and arbuscular mycorrhizal fungi (AMF) inoculation on Oats in saline-alkali soil contaminated by petroleum to enhance phytoremediation. Environ Sci Pollut Res Int 22:598–608

Chapter 12
Salt-Tolerant Microbes: Isolation and Adaptation

Mohammad Javad Zarea

Abstract One of the major obstacles limiting agriculture in arid and semiarid areas of the world is salinity. Arid and semiarid areas are affected by salinity of soil and water. Global changes aggress this issue and trigger the agricultural area into unproductive. Increasing salinity of soil and water decreases yield of importance of economic crops. Several strategies have been assumed to be used in order to deal with soil salinity. Moreover, using beneficial microorganisms is of great interest. This article review doesn't aim to repeat the beneficial role of plant growth-promoting bacteria in agriculture since, in this regard, many valuable article reviews have been published previously. The major aim and point of this review are to discuss the strategies used for the isolation of the beneficial saline-adapted bacteria for crop production under salinity condition. Isolation procedures of the different groups of beneficial bacteria including rhizobacteria, rhizobia, and phosphorus solubilizing bacteria have been reviewed in this chapter.

Keywords Salinity · Salt stress · Crop production · Plant growth-promoting rhizobacteria · Salt adaptation

12.1 Introduction

Salinity of soil and water are characteristics of arid and semiarid areas. Higher concentration of soluble salts results in salinity of soil or irrigation water. Low leaching of soluble soils drastically accumulates soluble soils in the soil moisture of the root zone. Higher concentrations of soluble salts decrease water uptake by the plant. In addition, absorption of some important nutrients like potassium, nitrogen,

M. J. Zarea (✉)
Faculty of Agriculture, Department of Agronomy and Plant Breeding, Ilam University, Ilam, Iran
e-mail: mj.zarea@ilam.ac.ir

B. Giri, A. Varma (eds.), *Microorganisms in Saline Environments: Strategies and Functions*, Soil Biology 56, https://doi.org/10.1007/978-3-030-18975-4_12

and phosphorus is restricted by salinity. Toxic elements such as sodium and chloride ions are accumulated in plant tissue. High osmotic pressures resulting from soluble salts restrict uptake of water by the plant (Zarea et al. 2013a, b, 2014). Developing and using salt-tolerant crops has been of interest but with little achievement (Munns and Tester 2008).

Beneficial microorganisms involve diverse mechanisms in order to promote plant growth and development of the host plant. Enhancing nutrient uptake and assimilation, phosphorus solubilization, biological nitrogen fixation, and modulating abiotic stress are diverse mechanisms by which beneficial microorganisms enhance growth and development of plant (Ilangumaran and Smith 2017).

Salinity of soil is a major problem in semiarid and, especially, arid areas of the world limiting crop productivity. Use of beneficial microorganisms has been considered for crop production under salinity stress condition (Zarea et al. 2012, 2013a, b, 2014). It has been assumed that adaptation of plants to their local environment might be driven from association relationships with microbes (Rodriguez and Redman 2008). Use of plant growth-promoting bacteria in saline soil can confer salinity tolerance to plant, which results in promising strategy in sustainable crop production under saline areas.

12.2 Plant Growth-Promoting Rhizobacteria and Salt Adaptation

Free-living beneficial bacteria occurring in the plant rhizosphere with beneficial activities and effects on plant are known as plant growth-promoting rhizobacteria (PGPR). Some of PGPR colonize the interior root and tissue of the host plant. A considerable number of PGPR are physically in contact with the plant root, and some of them don't colonize the root surface (Gray and Smith 2005). Molecular and physiological studies have indicated that bacteria respond to salinity in different ways. Bacteria involve several mechanisms by which develop various adaptations to salinity of soil and water. Accumulation of compatible solutes (Whatmore et al. 1990; Hagemann 2011), production of exopolysaccharides (Sandhya et al. 2009), changing in bacterial membrane fatty acid composition (Klein et al. 1999), expressions of salt-responsive genes (Diby et al. 2005), repressed or overexpressed in stress-related proteins (Paul et al. 2006; Steil et al. 2003; Weber and Jung 2002; Liu et al. 2005; Georg et al. 2009; Paul and Lade 2014), and activity of Na+ pump (Oren 2006). It has been considered that bacterial strains isolated from saline soil are more tolerant to salt than those isolated from non-saline area (Reinhold et al. 1986, 1988; Zarea et al. 2012).

Different methods have been developed for the isolation of PGPR. Beneficial bacteria are isolated from rhizosphere, root, and interior tissues of the plant and then screened for plant growth-promoting properties. Plant growth-promoting characteristics of these bacteria are assayed in vitro, in greenhouse, and then under field condition.

12.3 Key Notes for the Isolation of Salt-Tolerant and Halotolerant Bacteria

It should be noted that concentration of NaCl of the medium for the screen of salt-tolerant and halotolerant bacteria of phosphate-solubilizing bacteria, rhizobia, and diazotrophs is a key factor. It seems that isolation of bacteria based on common media and then, are screened for salinity tolerance are not suitable procedure for the screen of halotolerant bacteria under laboratory since some halotolerant bacteria need salinity condition for the optimum growth and this procedure doesn't allow to eliminate salinity-sensitive ones. For example, Zhu et al. (2011) found better growth of phosphate-solubilizing halophilic bacterium *Kushneria* sp. YCWA18 at 6% (w/v) NaCl. Some diazotrophic bacteria with salinity-tolerant properties need NaCl in media for better enrichment and optimal growth. For example, Sharma et al. (2016) and Jha et al. (2012) used NFb medium supplemented up to 4% concentration of NaCl for the isolation of halotolerant rhizobacteria and halotolerant diazotrophic bacteria. Loganathan and Nair (2004) used LGI medium supplemented with 250 mM NaCl for the isolation of salt-tolerant *Swaminathania salitolerans*. Xiang et al. (2011) used NBRIP medium plate containing tricalcium phosphate with 10% NaCl for the isolation of halotolerant phosphate-solubilizing bacteria.

12.4 Phosphorus Solubilizing Microorganism Isolation

Phosphorus solubilizing microorganisms (PSM) are the most important plant growth-promoting bacteria found in soil and rhizosphere. These microorganisms including certain group of fungi and bacteria play an important role in availability of phosphorus to plants and involve in P dynamic in soil (Richardson 2007). Microorganisms involved in phosphorus acquisition include mycorrhizal fungi and PSMs. Bacteria belonging to the genera *Agrobacterium*, *Alcaligenes*, *Azotobacter*, *Bacillus*, *Bradyrhizobium*, *Chromobacterium*, enterobacteria, *Erwinia*, *Escherichia*, *Flavobacterium*, *Micrococcus*, *Salmonella*, *Serratia*, *Streptomyces*, *Thiobacillus*, *Pseudomonas*, *and Rhizobium* have been mentioned as the most powerful P solubilizers (Whitelaw 2000; Zhao and Lin 2001). Strains of these bacterial genera through several mechanisms, including acidification, exchange reactions, and polymeric substances formation, convert the insoluble phosphorus to soluble forms (Delvasto et al. 2006).

The Pikovskaya (PVK) medium first described by Pikovskaya is used for screening of efficient PSB (Pikovskaya 1948). Modified $Ca_3(PO_4)_2$ culture medium was then introduced by Nautiyal (1999) as an efficient microbiological growth medium for screening of PSM. Nautiyal (1999), from comparison of three media including National Botanical Research Institute's phosphate growth medium (NBRIP), National Botanical Research Institute's phosphate growth medium devoid of yeast extract (NBRIY), and Pikovskaya (PVK) medium, concluded that NBRIP medium

Table 12.1 Composition of media and chemical reagents for the isolation of phosphorous solubilizing microorganisms

Ingredients	The modified Pikovskaya (PVK) (% per liter) (Son et al. 2006)	Pikovskaya (PVK) (g L^{-1}) (Pikovskaya 1948)	Research Institute's phosphate growth medium (NBRIP) (g L^{-1}) (Nautiyal 1999)	National Botanical Research Institute's phosphate growth medium devoid of yeast extract (NBRIY) medium contained (g L^{-1}) (Nautiyal 1999)	Modified $Ca_3(PO_4)_2$ culture medium (g L^{-1}) (Zhu et al. (2011)
Glucose	1.0	–	10	10	10
Dextrose	–	10	–	–	–
$Ca_3(PO4)^{2-}$	0.5	10	5	5	10
$(NH_4)_2SO_4$	0.05	0.5	0.1	0.5	0.5
NaCl[a]	0.02	–	–	0.2	30
$CaCl_2 \cdot 2H_2O$	0.01	–	–	–	–
$MgSO_4 \cdot 7H_2O$	0.01	0.1	0.25	0.1	0.3
$MgCl_2 \cdot 6H_2O$	–	–	5	–	–
$MnSO_4 \cdot 7H_2O$	0.05	0.0001	–	0.002	0.03
$FeSO_4 \cdot 7H_2O$	0.05	0.0001	–	0.002	0.03
KCl	0.02	0.20	2	0.2	0.3
Yeast extract	0.05	0.5	–	–	–
Agar-agar	NR[b]	18	NR	NR	20
pH	7.5	7.0	7.0	7.0	7.0–7.5

[a]Adding higher-level concentrations of NaCl more than its recommended seems to facilitate screening for the isolation of salt-tolerant and halotolerant PSB
[b]Not reported

was the best among three tested media for the screening of the most efficient phosphate solubilizers (Table 12.1). Later PVK was modified by Son et al. (2006) for the isolation of salt- and pH-tolerant *Pantoea agglomerans* R-42 (Son et al. 2006). In this media, the tricalcium phosphate ($Ca_3(PO_4)_2$) at the concentration rate of 0.5% (wv^{-1}) is added as source of insoluble phosphate. The tricalcium phosphate is prepared separately and, after autoclaving, aseptically mixed with other ingredients. It seems that cultural medium used for the isolation of phosphate-solubilizing halophilic bacterium must be added with NaCl to eliminate those PSBs that cannot survive at higher concentration of NaCl. For example, Zhu et al. (2011) found better growth of phosphate-solubilizing halophilic bacterium *Kushneria* sp. YCWA18 at 6% (w/v) NaCl.

For the isolation of salt-tolerant PSBs, the following procedure can be followed: collecting soil sample from plant root rhizosphere or saline soil. Serial dilution of the soil sample must be prepared. For the preparation of serial dilution of the soil sample, 1 g of soil sample is dissolved in 10–50 mL of sterile water containing 0.85% NaCl (w/v). Dissolved soil is kept shaken at 240 rpm for 30 min (Han et al. 2014). For isolation of PSB from roots, adhering soil must be removed from root by washing with sterile water. Roots are then macerated. Serial dilution of the roots is then prepared. The bacterial isolates can be obtained by plating a serial of tenfold dilutions of root or soil samples onto a standard NBRIP growth agar mediums (Nautiyal 1999) or modified Pikovskaya (PVK) medium (Son et al. 2006) supplemented with desired concentrations of NaCl. Adding NaCl in the medium seems to play an important role for the screening of those PSB that show salinity tolerance. Xiang et al. (2011) used NBRIP medium plate containing tricalcium phosphate with 10% NaCl for the isolation of halotolerant phosphate-solubilizing bacteria. Zhu et al. (2011) screened the phosphate-solubilizing halophilic bacterium. These researchers used the modified $Ca_3(PO4)_2$ culture containing 30 g NaCl (Nautiyal 1999). Phosphate-solubilizing halophilic bacterium *Kushneria* sp. YCWA18 with the ability to withstand 6% NaCl was isolated. These researchers reported that the fastest growth of *Kushneria* sp. YCWA18 is observed at 6% NaCl (w/v), 28 °C, and pH 7.0–8.0, indicating the need of these strain to NaCl. An aliquot (10–100 μL) cultured overnight from decimal dilutions must be inoculated on agar plate and be incubated at 28–30 °C for 2–3 days. Bacterial isolates which show clear halo are selected for the purification as halotolerant phosphate-solubilizing bacteria.

12.5 Rhizobia Isolation

Biological nitrogen fixation plays a vital role in N supply to the biosphere. Plant from family of leguminous through their leguminous certain group of bacteria known as rhizobia fix atmospheric nitrogen. 21 Tg of atmospheric N annually is biologically fixed through rhizobia-legume symbiosis (Herridge et al. 2008). Responses of rhizobial population to the environmental factors are not the same, and there are differences in their tolerance to environmental limiting factors (Wei et al. 2008). Although most agricultural legumes cannot withstand salinity, strains of *Rhizobium* have the ability to survive and grow at higher salinity (Singleton et al. 1982). Higher pH and salinity are prevalent in semiarid and especially in arid areas. The accumulation of soluble salts in these areas results in increasing pH value of soil. Shamseldin and Werner (2005) reported that biological N_2 fixation and nodule formation are more restricted by salinity than alkalinity stress. It might be also considered that native rhizobia even salt-sensitive species may be superior in that exogenous salt-resistant rhizobia (Ventorino et al. 2012).

Rhizobia create symbiotic relationship with almost 100 agricultural legumes. Microbial population and microbial activity are low under saline soils. Soil salinity condition may directly influence symbiont. Salinity also may directly impact on symbiont interaction between legume and *Rhizobium* spp. (Parker et al. 1977). However, tolerance of many strains of *Rhizobium* to salinity concentration is higher than most agricultural legumes (Singleton et al. 1982), but also salinity tolerance between different rhizobial strains is different (Laranjo and Oliveira 2011; Moussaid et al. 2015). For example, while *Sinorhizobium meliloti* growth is not affected by the higher concentration of NaCl (at 0.8 M NaCl), 0.2 M NaCl inhibits *Mesorhizobium mediterraneum* and *Rhizobium gallicum* growth (Abdelmoumen et al. 1999). *Sinorhizobium meliloti*, *S. fredii*, and *Rhizobium tropici* were reported to grow at higher concentration of salt, 300 mmol L^1 (Graham and Parker 1964; Sauvage et al. 1983; Yelton et al. 1983; Graham 1992). Rhizobia involve several mechanisms in order to withstand salinity including accumulation of low molecular weight organic solutes or compatible solutes, ectoines, glycine betaine, proline polyamines, and inorganic cations such as K and regulation of the enzymatic antioxidant system (Santos and da Costa 2002; Domínguez-Ferreras et al. 2006; Pérez-Montaño et al. 2016).

Singleton and Bohlool (1984) reported that nodule initiation is very sensitive to NaCl and even low concentration of salt can affect nodule development. Singleton and Bohlool (1984) elucidated that sensitivity of nodule initiation to NaCl is related to the sensitivity of root infection sites and was not related to the rhizobial survival. El-Sheikh and Wood (1995) reported that nodulation formation severely reduced (50%) salinity (34.2 mm NaCl) in soybean. These researchers also elucidate that N_2 fixation rate by the salt-tolerant strain was more than that fixed by salt-sensitive strain under salinity condition. In addition, effect of salinity on *Rhizobium*-legume symbiosis may indirectly relate to the available photosynthates and plant growth since salinity decreases plant growth and photosynthesis rate (Seeman and Critchley 1985; Brignoli and Lauteri 1991).

Based on the reported studies, it should be noted that tolerance of the legume host itself is also an important factor to ensure success of forming effective symbioses under high salinity concentration (Craig et al. 1991). It has been shown that rhizobial symbiosis can affect the salt tolerance of the host plant. Rhizobial inoculation improved salt tolerance in alfalfa and soybean under salinity stress (Bertrand et al. 2015; Qu et al. 2016). Salt-tolerant rhizobial strain compared to the salt-sensitive *Rhizobium* strain seems to be more effective in enhancing salt tolerance of the host plant (Bertrand et al. 2015). It can be concluded that screening for the salt-tolerant rhizobia is important for growing agricultural legumes in saline soil. However, salinity tolerance of the host plant (legumes) also must be consecrated. Rhizobia are different in salinity tolerance (Table 12.2). The salt-tolerant strain of rhizobia seems to be better able to enhance salinity tolerance and better improve growth of the host plant than salt-sensitive species from non-saline conditions.

Isolation of *Rhizobium* from plant root has been shown in Fig. 12.1. Roots of plant must be carefully uprooted, and then root systems are washed to remove all the adhesive soil particles from them and are then surface sterilized. Nodules exhibit

Table 12.2 Salinity tolerance of strains of *Rhizobium* to salinity

Bradyrhizobium strain RJS9-2	0.35 M NaCl	Dong et al. (2017)
Rhizobium strain ANG4	Up to 850 mM NaCl	Lal and Khanna (1995)
Rhizobium strain ANG5	Up to 850 mM NaCl	Lal and Khanna (1995)
Rhizobium leguminosarum TAL 271	2 % (w/v) NaCl	Helemish and El-Gammal (1987)
Rhizobium halophytocola	7.5% (w/v) NaCl	Bibi et al. (2011)
Gynuella sunshinyii	7% (w/v) NaCl	Chung et al. (2015)
Rhizobium fredii USDA 20X	0.34 M NaCl	El-Sheikh and Wood (1995)
Rhizobium fredii P220	Up to 0.4 M NaCl	Fujihara and Yoneyama (1993)

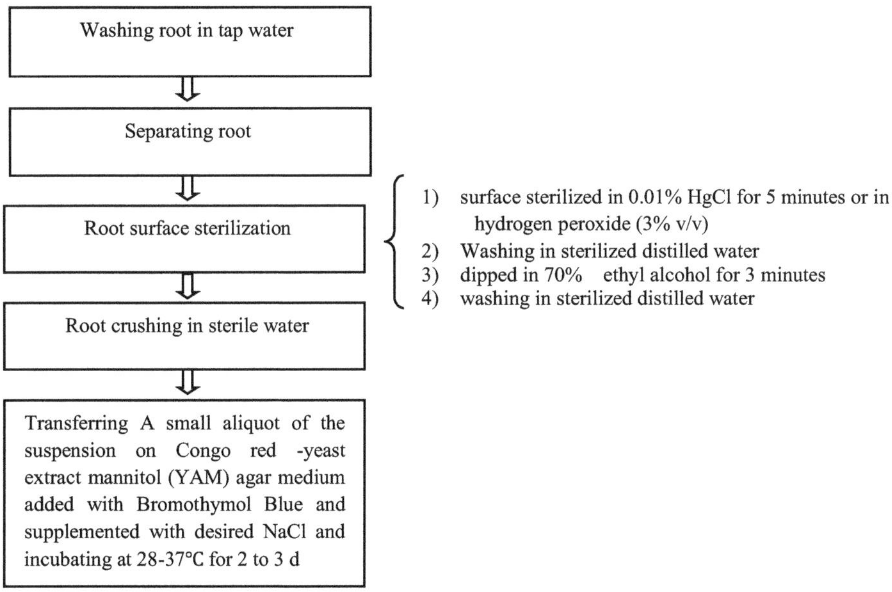

Fig. 12.1 Diagram representing all the steps applied to isolate rhizobia

different color. Depending on the state of pigment, present in the nodules, color of nodules is varied from brown to pink. Pink nodules are healthier. Hydrogen peroxides and mercuric chloride are sterilizing agent used for sterilizing of nodules for 4–5 min. After sterilizing, nodules must be repeatedly washed with sterile distilled water, seven times. Sterilized roots are dipped in 70% ethyl alcohol for 3 min and washed in sterilized distilled water. Nodules are crushing in sterile water, and water serial dilutions of nodular extract are prepared. Nodular extract is prepared from 1g of nodular extract in 10 ml of disinfected distilled water. Dilutions of nodular extract are then inoculated onto Yeast Mannitol Agar (YMA), YMA+bromothymol blue (BTB), YMA+ Congo red (CR), or YMA+BTB+CR (Table 12.3). Supplementing these media with NaCl helps to eliminate salt-sensitive strains. Chen et al. (2000) isolated halotolerant rhizobia from root nodules of *Canavalia rosea* on mannitol

Table 12.3 Composition of media and chemical reagents for the isolation of *Rhizobium*

Yeast mannitol broth (YMB)	Ingredients
Mannitol	1–10.0 g
K_2HPO_4	0.5 g
$MgSO_4 \cdot 7H_2O$	0.2 g
NaCl[a]	0.1 g
Yeast extract	0.5 g
Distilled water	1.0 L
pH	6.8
Yeast Mannitol Agar (YMA)	
YMB	1 L
Agar	15 g
YMA+bromothymol blue (BTB)[b]	
YMA	1 L
BTB (0.5 g/100 mL ethanol)	5 mL
YMA+ Congo Red (CR)[c]	
YMA	1 L
CR (0.25 g/100 mL)	10 mL

[a]Desired NaCl

[b]Turning BTB into yellow and blue indicates fast-growing and slow-growing rhizobia with acid and alkaline reaction, respectively

[c]Rhizobia strains don't adsorb CR or absorb it in a little quantity. Those strains that adsorb CR appear in faintly pink to white color

(YEM) agar containing 0.0025% (w/v) red and reported that all isolates grew at salt concentrations as high as 3%.

12.6 Isolation of Halotolerant Plant Growth-Promoting Diazotrophs (HPGPR)

Diazotrophic bacteria or nitrogen-fixing bacteria have been reported to be isolated from different geographic regions and various plant using different formulations of N-free semisolid media. These bacteria have been reported that they were isolated from the rhizosphere or roots of various plants. Diazotrophic bacteria are present in the rhizosphere or interior of plant. *Azospirillum*, *Azotobacter*, *Beijerinckia*, *Derxia*, *Herbaspirillum*, *Sphingomonas*, *Gluconacetobacter diazotrophicus*, *Burkholderia*, and *Azoarcus are of importance* and are well-known diazotrophic bacteria. These bacteria through direct and indirect mechanism lead plant to better performances under nonstress and, especially, under stress condition. These bacteria may be halophilic bacteria.

During the last decade, various media were recommended for the isolation of diazotrophic bacteria. At first silica gel plates or N-free agar recommended by Fred and Waksman's (1928) and Winogradsky's (1949) was used for the isolation of

Table 12.4 Composition of media and chemical reagents for the isolation of nitrogen-fixing bacteria (Hartmann and Baldani 2006)

DL-malic acid	5.0 g
K_2HPO_4	0.5 g
$MgSO_4 \cdot 7H_2O$	0.2 g
NaCl[a]	0.1 g
$CaCl_2 \cdot 2H_2O$	0.02 g
Micronutrient solution	2 mL
Bromothymol blue solution (0.5% in 0.2 N KOH)	2 mL
Fe(III) EDTA (1.64%)	4.0 mL
Vitamin solution	1.0 mL
Distilled water	1.0 L
pH	6.8
Micronutrient solution	
$CuSO_4 \cdot 5H_2O$	0.4 g
$ZnSO_4 \cdot 7H_2O$	0.12 g
H_3BO_3	1.4 g
$Na_2MoO_4 \cdot 2H_2O$	1.0 g
$MnSO_4 \cdot H_2O$	1.5 g
Distilled water	1.0 L
Vitamin solution	
Biotin	10 mg
Pyridoxol HCL	20 mg
Distilled water	0.1 L

[a]Desired NaCl

these bacteria. These media allowed few microorganisms to grow. Then N-free semisolid medium recommended by Döbereiner and Day (1976) was vastly used for the isolation of diazotrophic bacteria (Table 12.4). This introduced media allowed for the isolation of many diazotrophic bacteria (Döbereiner and Pedrosa 1987). N-free semisolid medium is a medium suitable for the growth of the diazotrophic bacteria that their nitrogenase is very sensitive to O_2 (Dalton and Postgate 1968). Various N-free semisolid-based media have been used to isolate nonsymbiotic nitrogen-fixing bacteria of diazotrophs. Table 12.5 shows several media for the isolation of diazotrophic bacteria belonging to the genus *Azospirillum.* LGI has been recommended for the isolation of *Azospirillum amazonense* (Baldani 1984); FAM for *Azospirillum amazonense* (Magalhães and Dobereiner 1984); JNFb for *Herbaspirillum seropedicae* (Baldani et al. 1992), *Herbaspirillum rubrisubalbicans* (Baldani et al. 1996), and *Herbaspirillum frisingense* (Kirchhof et al. 2001); JNFb for *Sphingomonas* spp. (Videira et al. 2009); LGI-P for *Gluconacetobacter* (Cavalcante and Döbereiner 1988) and diazotrophicus (Reis et al. 1994); JMV for *Burkholderia kururiensis* (Baldani et al. 2000); BAz for diazotrophic *Burkholderia* (Estrada-de-los-Santos et al. 2001); and BAc for diazotrophic *Burkholderia* (Estrada-de-los-Santos et al. 2001). Some of the isolated diazotrophs had salinity tolerance to NaCl, and some are halotolerant diazotrophic bacteria, *Swaminathania salitolerans* (Loganathan and Nair 2004), *Azospirillum*

Table 12.5 The media used for the isolation of nonsymbiotic diazotrophs of the genus *Azospirillum*

NFb medium	*Azospirillum brasilense and A. lipoferum*	Döbereiner and Day (1976)
	Azospirillum doebereinerae	Eckert et al. (2001)
	Azoarcus olearius	Chen et al. (2013)
	Azospirillum melinis	Peng et al. (2006)
	Azoarcus olearius	Chen et al. (2013)
Nutrient agar (NA)	*Azospirillum fermentarium*	Lin et al. (2013)
NFB and LGI	*Azospirillum melinis*	Peng et al. (2006)
Semisolid MPSS medium with a freshly prepared FeS suspension	*Azospirillum thiophilum*	Lavrinenko et al. (2010)
Mineral salts medium (MSM) supplemented with sodium acetate	*Azospirillum humicireducens*	Zhou et al. (2012)
M medium without biotin	*Azospirillum canadense*	Mehnaz et al. (2007a, b)
M medium without biotin	*Azospirillum zeae*	Mehnaz et al. (2007a, b)
Modified N-free malate media with 1.2 g of NaCl	*Azospirillum halopraeferens*	Reinhold et al. (1987)

brasilense strain NH (Nabti et al. 2007), and *Salicornia brachiata* (Jha et al. 2012). *Azospirillum halopraeferens* could grow on N-free semisolid medium in the presence of 0.25% NaCl (Reinhold et al. 1987). *Bacillus alkalidiazotrophicus* also could tolerate 1.2 M Na+ (Sorokin et al. 2001).

Very little research concerning the isolation of PGPR and diazotrophic bacteria has been done. Halophytes have been mentioned to be extremely salt-tolerant. These bacteria commonly grow and survive in environments with high concentrations of soil. A few bacterial diazotrophs have also been detected in saline areas in association with plants, and their contribution to nitrogen fixation has remained unknown. Most of diazotrophs have been isolated from temperate and tropical areas. Therefore, acknowledgment of the salt-tolerant diazotrophs and their specific contribution of nitrogen fixation under saline areas are not well known. It seems that rhizosphere of halophytic plants under saline areas must be an ideal source for isolation of salt-tolerant nitrogen-fixing bacteria that could improve the growth of various crops in areas faced with salinity stress. These groups of bacteria can be isolated from soil, rhizosphere, and root of plant. The following procedure can be followed for the isolation of halotolerant and salt-tolerant diazotrophic bacteria.

Halotolerant can be isolated from the rhizosphere of plant and interior of plant root (endophytic). Rhizosphere soil collected from the roots is suspended in sterile saline solution (0.85% NaCl, Tween 80) (Çelen and Mehmet 2004). 1–10 g of soil samples is mixed with 9–90 mL of saline solution. After preparing appropriate dilutions, 1 mL of each dilutions is added to 10 mL vial of N-free semisolid medium (NFB) containing various or desired NaCl (Table 12.4). Different formulations of

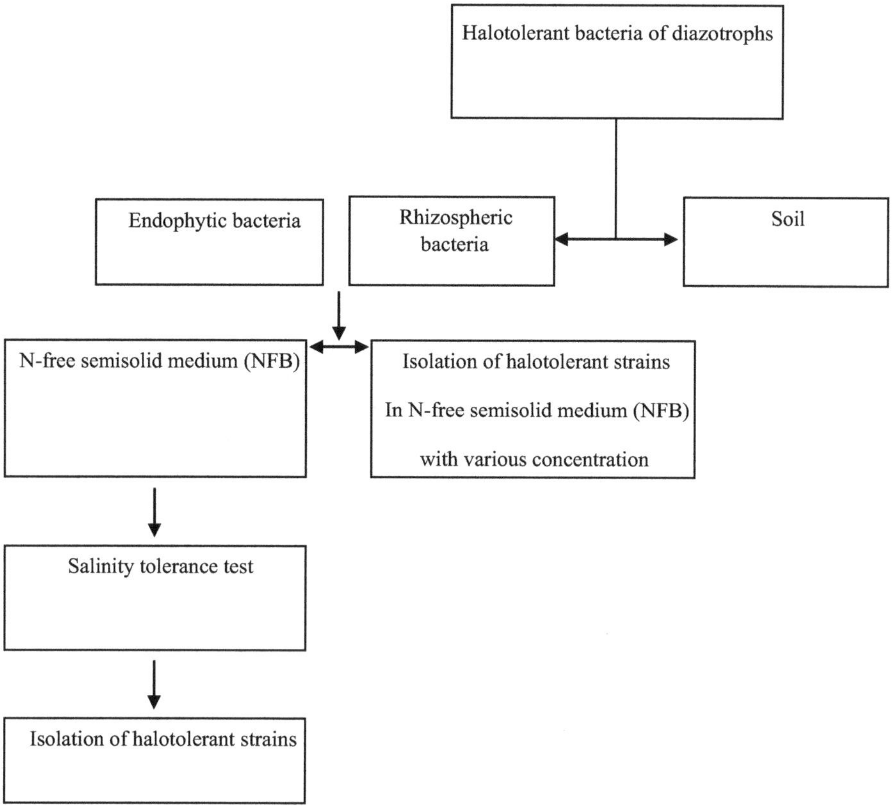

Fig. 12.2 Diagram showing the steps applied to isolate diazotrophs

N-free semisolid media are used to isolate nonsymbiotic nitrogen-fixing bacteria including NFb, LGI, FAM, JNFb, LGI-P, JMV, BAz, and BAz media (Baldani et al. 2014). Vial of NFB must be incubated for 7 days at 28–30 °C. Pellicle formation is usually observed 2–3 days after inoculation (Baldani et al. 2014).

Two methods can be used for the isolation of the salt-tolerant strains of diazotrophic bacteria in N-free semisolid medium (NFB) with various concentration of NaCl, using NFb media with various salinity or desired salinity to eliminate non-salinity-tolerant bacteria or using standard NFb without salinity, purification of all isolates, and then screening of the strains for salinity tolerance (Fig. 12.2). The first procedure seems to be suitable since many non-salinity-tolerant bacteria will be eliminated. In addition some diazotrophic bacteria with salinity-tolerant properties need NaCl in media for better enrichment and optimal growth. For example, Sharma et al. (2016) and Jha et al. (2012) used NFb medium supplemented up to 4% concentration of NaCl for the isolation of halotolerant rhizobacteria and halotolerant diazotrophic bacteria. Loganathan and Nair (2004) used LGI medium supplemented with 250 mM NaCl for the isolation of salt-tolerant *Swaminathania salitolerans*.

12.7 Conclusions

One of the sustainable major problems in agricultural crops under arid and semiarid areas of the world is salinity. Salinity limits crop production in these areas and decreases yield of importance of economic crops. *B. licheniformis* (Goswami et al. 2014), *Brachybacterium* sp. (Jha et al. 2012; Shukla et al. 2012), *Halobacillus* sp. (Ramadoss et al. 2013), *Exiguobacterium oxidotolerans* (Bharti et al. 2013), and *Pseudomonas* sp. (Rosenberg 1983; Jha et al. 2012; Egamberdieva et al. 2015) are examples of halotolerant rhizobacteria with ability to confer salinity tolerance to crops. Several strategies have emerged in order to cope with salinity stress. Beneficial bacteria that can thrive under such hazardous environment could be used as an option for the crop production under saline environment. However, procedure isolation of beneficial halotolerant bacteria to be used in crop production under saline areas is not well considerable. In this regard, it seems that common methods used for the isolation of plant growth-promoting bacteria of halotolerant bacteria must be standardized and revised.

References

Abdelmoumen H, Filali-Maltouf A, Neyra M, Belabed A, El Idrissi MM (1999) Effect of high salts concentrations on the growth of rhizobia and responses to added osmotica. J Appl Microbiol 86:889–898

Baldani JI, Baldani VLD, Sampaio MJAM, Döbereiner J (1984) A fourth *Azospirillum* species from cereal roots. An Acad Bras Ci 56:265

Baldani VLD, Baldani JI, Olivares FL, Döbereiner J (1992) Identification and ecology of *Herbaspirillum seropedicae* and the closely related *Pseudomonas rubrisubalbicans*. Symbiosis 19:65–73

Baldani JI, Pot TB, Kirchhof G, Falsen E, Baldani VLD, Olivares FL, Hoste B, Kersters K, Hartmann A, Gillis M, Döbereiner J (1996) Emended description of *Herbaspirillum*, a mild plant pathogen, as *Herbaspirillum rubrisubalbicans* comb. nov.; and classification of a group of clinical isolates (EF group 1) as *Herbaspirillum* species 3. Int J Syst Bacteriol 46:802–810

Baldani VLD, Baldani JI, Döbereiner J (2000) Inoculation of rice plants with the endophytic diazotrophs *Herbapirillum seropedicae* and *Burkholderia* spp. Biol Fertil Soils 30:485–491

Baldani JI, Reis VM, Videira SS, Boddey LH, Baldani VLD (2014) The art of isolating nitrogen-fixing bacteria from non-leguminous plants using N-free semi-solid media: a practical guide for microbiologists. Plant Soil 384:413–431

Bertrand A, Dhont C, Bipfubusa M, Chalifour FP, Drouin P, Beauchamp CJ (2015) Improving salt stress responses of the symbiosis in alfalfa using salt-tolerant cultivar and *rhizobial* strain. Appl Soil Ecol 87:108–117

Bharti N, Yadav D, Barnawal D, Maji D, Kalra A (2013) Exiguobacterium oxidotolerans, a halotolerant plant growth promoting rhizobacteria, improves yield and content of secondary metabolites in *Bacopa monnieri* (L.) Pennell under primary and secondary salt stress. World J Microbiol Biotechnol 29:379–387

Bibi F, Chung EJ, Yoon HS, Song GC, Jeon CO, Chung YR (2011) *Haloferula luteola* sp nov., an endophytic bacterium isolated from the root of a halophyte, *Rosa rugosa*, and emended description of the genus *Haloferula*. Int J Syst Evol Microbiol 61:1837–1841

Brignoli E, Lauteri M (1991) Effects of salinity on stomatal conductance, photosynthetic capacity, and carbon isotope discrimination of salt-tolerant (*Gossypium hirsutum* L.) and salt-sensitive (*Phaseolus vulgaris* L.) C, non-halophytes. Plant Physiol 95:638–635

Cavalcante VA, Döbereiner J (1988) A new acid-tolerant nitrogen-fixing bacterium associated with sugarcane. Plant Soil 108:23–31

Çelen E, Mehmet AK (2004) Isolation and characterization of aerobic denitrifiers from agriculture soil. Turk J Biol 28:9–14

Chen WM, Lee TM, Lan CC, Cheng CP (2000) Characterization of halotolerant rhizobia isolated from root nodules of *Canavalia rosea* from seaside areas. FEMS Microbiol Ecol 34:9–16

Chen M-H, Sheu S-Y, James EK, Young C-C, Chen W-M (2013) *Azoarcus olearius* sp. nov., a nitrogen-fixing bacterium isolated from oil-contaminated soil. Int J Syst Evol Microbiol 63:3755–3761

Chung EJ, Park JA, Jeon CO, Chung YR (2015) *Gynuella sunshinyii* gen. nov., sp nov., an antifungal *rhizobacterium* isolated from a halophyte, *Carex scabrifolia* Steud. Int J Syst Evol Microbiol 65:1038–1043

Craig GF, Atkins CA, Bell DT (1991) Effect of salinity on growth of Rhizobium and their infectivity and effectiveness on two species of Acacia. Plant Soil 133:253–262

Dalton H, Postgate JR (1968) Effect of oxygen on growth of *Azotobacter chroococcum* in batch and continuous cultures. J Gen Microbiol 54:463–473

Delvasto P, Valverde A, Igual JM, Munoz JA, Gonzalez F, Blazquez ML, Garcia-Balboa C (2006) Diversity and activity of phosphate bioleaching bacteria from a high phosphorous iron-ore. Soil Biol Biochem 38:2645–2654

Diby P, Sarma YR, Srinivasan V, Anandaraj M (2005) Pseudomonas fluorescens mediated vigour in black pepper (*piper nigrum* L.) under greenhouse cultivation. Ann Microbiol 55:171–174

Döbereiner J, Day JM (1976) Associative symbiosis in tropical grasses: characterization of micro-organisms and dinitrogen fixing sites. In: Newton WE, Nyman CJN (eds) Proceedings of 1st international symposium on nitrogen fixation. Washington State University Press, Washington, pp 518–538

Döbereiner J, Pedrosa FO (1987) Nitrogen-fixing bacteria in nonleguminous crop plants, Brock/ Springer series in contemporary bioscience. Springer, New York

Domínguez-Ferreras A, Pérez-Arnedo R, Becker A, Olivares J, Soto MJ, Sanjuán J (2006) Transcriptome profiling reveals the importance of plasmid pSymB for osmoadaptation of *Sinorhizobium meliloti*. J Bacteriol 188:7617–7625

Dong R, Zhang J, Huan H, Bai C, Chen Z, Liu G (2017) High salt tolerance of a *Bradyrhizobium* strain and its promotion of the growth of *Stylosanthes guianensis*. Int J Mol Sci 18:1625

Eckert B, Weber OB, Kirchhof G, Halbritter A, Stoffels M, Hartmann A (2001) *Azospirillum doebereinerae* sp. nov., a nitrogen fixing bacterium associated with the C4-grass Miscanthus. Int J Syst Evol Microbiol 51:17–26

Egamberdieva D, Lindström LL, Räsänen La K (2015) A synergistic interaction between salt-tolerant *Pseudomonas* and *Mesorhizobium* strains improves growth and symbiotic performance of liquorice (*Glycyrrhiza uralensis* Fish.) under salt stress. Appl Microbiol Biotechnol 100:2829–2841

El-Sheikh EAE, Wood M (1995) Nodulation and N_2 fixation by soybean inoculated with salt-tolerant rhizobia or salt-sensitive Bradyrhizobium in saline soil. Soil Biol Biochem 27:657–661

Estrada-de-los-Santos P, Bustillos-Cristales R, Caballero-Mellado J (2001) *Burkholderia*, a genus rich in plant-associated nitrogen fixers with wide environmental and geographic distribution. Appl Environ Microbiol 67:2790–2798

Fred EB, Waksman SA (1928) Laboratory manual of general microbiology. McGraw-Hill Book, London

Fujihara S, Yoneyama T (1993) Effects of pH and osmotic stress on cellular polyamine contents in the soybean rhizobia *Rhizobium fredii* P220 and *Bradyrhizobium japonicum* A1017. Appl Environ Microbiol 59:1104–1109

Georg J, Voss B, Scholz I, Mitschke J, Wilde A, Hess W (2009) Evidence for a major role of antisense RNAs in cyanobacterial gene regulation. Mol Syst Biol 5:305

Goswami D, Dhandhukia P, Patel P, Thakker JN (2014) Screening of PGPR from saline desert of Kutch: growth promotion in *Arachis hypogaea* by *Bacillus licheniformis* A2. Microbiol Res 169:66–75

Graham PH (1992) Stress tolerance in *Rhizobium* and *Bradyrhizobium* and nodulation under adverse soil conditions. Can J Microbiol 38:475–484

Graham PH, Parker CA (1964) Diagnostic features in the characterization of the root nodule bacteria of legumes. Plant Soil 20:383–396

Gray EJ, Smith DL (2005) Intracellular and extracellular PGPR: commonalities and distinctions in the plant-bacterium signaling processes. Soil Biol Biochem 37:395–412

Hagemann M (2011) Molecular biology of cyanobacterial salt acclimation. FEMS Microbiol Rev 35:87–123

Han Y, Wang Ch, Li X, Cao X, Cao A, Zhao N (2014) Isolation and identification of saline tolerance phosphate-solubilizing bacteria derived from salt-affected soils and their mechanisms of P-solubilizing. In: Zhang TC et al (eds) Proceedings of the 2012 international conference on applied biotechnology (ICAB 2012). Lecture notes in electrical engineering, vol 250. Springer, Berlin. https://doi.org/10.1007/978-3-642-37922-2_135

Hartmann A, Baldani JI (2006) The Genus *Azospirillum*. In: Rosemberg E, Schleifer K-H, Stackerbrandt E (eds) The prokaryotes. Springer, New York

Helemish FA, El-Gammal SMA (1987) Salt and pH tolerance of *Rhizobium leguminosarum* TAL-271. Zbl Microbiol 142:211

Herridge DF, Peoples MB, Boddey RM (2008) Global inputs of biological nitrogen fixation to agricultural systems. Plant Soil 311:1–18

Ilangumaran G, Smith DL (2017) Plant growth promoting rhizobacteria in amelioration of salinity stress: a systems biology perspective. Front Plant Sci 8:1768

Jha B, Gontia I, Hartmann A (2012) The roots of the halophyte *Salicornia brachiata* are a source of new halotolerant diazotrophic bacteria with plant growth-promoting potential. Plant Soil 356:265–277

Kirchhof G, Eckert B, Stoffels M, Baldani JI, Reis VM, Hartmann A (2001) *Herbaspirillum frisingense* sp. nov., a new nitrogen-fixing bacterial species that occurs in C4-fiber plants. Int J Syst Evol Microbiol 51:157–168

Klein W, Weber MH, Marahiel MA (1999) Cold shock response of *Bacillus subtilis*: isoleucine-dependent switch in the fatty acid branching pattern for membrane adaptation to low temperatures. J Bacteriol 181:5341–5349

Lal B, Khanna S (1995) Selection of salt tolerant *Rhizobium* isolates of *Acacia nilotica*. World J Microbiol Biotechnol 10:637–639

Laranjo M, Oliveira S (2011) Tolerance of Mesorhizobium type strains to different environmental stresses. Anton Leeuw 99:651–662

Lavrinenko K, Chernousova E, Gridneva E et al (2010) *Azospirillum thiophilum* sp. nov., a novel diazotrophic bacterium isolated from a sulfide spring. Int J Syst Evol Microbiol 60:2832–2837

Lin S-H, Liu Y-C, Hameed A, Hsu YH, Lai WA, Shen FT, Young CC (2013) *Azospirillum fermentarium* sp. nov., a novel nitrogen fixing species isolated from a fermenter. Int J Syst Evol Microbiol 63:3762–3768. (in Taiwan)

Liu Y, Gao W, Wang Y, Wu L, Liu X, Yan T et al (2005) Transcriptome analysis of *Shewanella oneidensis* MR-1 in response to elevated salt conditions. J Bacteriol 187:2501–2507

Loganathan P, Nair S (2004) *Swaminathania salitolerans* gen. nov., sp. nov., a salt-tolerant, nitrogen-fixing and phosphate solubilizing bacterium from wild rice (*Porteresia coarctata* Tateoka). Int J Syst Evol Microbiol 54:1185–1190

Magalhães FMM, Döbereiner J (1984) Occurrence of *Azospirillum amazonense* in some Amazonian ecosystems. Rev Microbiol 4:246–252

Mehnaz S, Weselowski B, Lazarovits G (2007a) *Azospirillum canadense* sp. nov., a nitrogen-fixing bacterium isolated from corn rhizosphere. Int J Syst Evol Microbiol 57:620–624

Mehnaz S, Weselowski B, Lazarovits G (2007b) *Azospirillum zeae* sp. nov., a diazotrophic bacterium isolated from rhizosphere soil of *Zea mays*. Int J Syst Evol Microbiol 57:2805–2809

Moussaid S, Domínguez-Ferreras A, Muñoz S, Aurag J, Sanjuán J (2015) Increased trehalose biosynthesis improves *Mesorhizobium ciceri* growth and symbiosis establishment in saline conditions. Symbiosis 67:103–111

Munns R, Tester M (2008) Mechanisms of salinity tolerance. Annu Rev Plant Biol 59:651–681

Nabti E, Sahnoune M, Adjrad S, Van Dommelen A, Ghoul M, Schmid M, Hartmann A (2007) A halophilic and osmotolerant *Azospirillum brasilense* strain from Algerian soil restores wheat growth under saline conditions. Eng Life Sci 7:354–360

Nautiyal CS (1999) An efficient microbiological growth medium for screening phosphate solubilizing microorganisms. FEMS Microbiol Lett 170:265–270

Oren A (2006) The order Halobacteriales. In: Dworkin M, Falkow S, Rosenberg E, Schleifer K-H, Stackebrandt E (eds) The prokaryotes, vol 3, 3rd edn, pp 113–164

Parker CA, Trinick MR, Chatel DL (1977) Rhizobia as soil and rhizosphere inhabitants. In: Hardy RWF, Gibson AH (eds) A treatise on dinitrogen fixation, vol 4. Wiley, New York, pp 311–352

Paul D, Lade H (2014) Plant-growth-promoting rhizobacteria to improve crop growth in saline soils: a review. Agron Sustain Dev 34:737–752

Paul D, Dineshkumar N, Nair S (2006) Proteomics of a plant growth promoting rhizobacterium, *Pseudomonas fluorescens* MSP-393, subjected to salt shock. World J Microbiol Biotechnol 22:369–374

Peng G, Wang H, Zhang G et al (2006) *Azospirillum melinis* sp. nov., a group of diazotrophs isolated from tropical molasses grass. Int J Syst Evol Microbiol 56:1263–1271

Pérez-Montaño F, del Cerro P, Jiménez-Guerrero I, López-Baena FJ, Cubo MT, Hungria M, Megías M, Ollero FJ (2016) RNA-seq analysis of the *Rhizobium tropici* CIAT 899 transcriptome shows similarities in the activation patterns of symbiotic genes in the presence of apigenin and salt. BMC Genomics 17:198

Pikovskaya RI (1948) Mobilization of phosphorous in soil in connection with the vital activity of some microbial species. Mikrobiologiya 17:362–370

Qu LQ, Huang YY, Zhu CM, Zeng HQ, Shen CJ, Liu C, Zhao Y, Pi EX (2016) Rhizobia-inoculation enhances the soybean's tolerance to salt stress. Plant Soil 400:209–222

Ramadoss D, Lakkineni VK, Bose P, Ali S, Annapurna K (2013) Mitigation of salt stress in wheat seedlings by halotolerant bacteria isolated from saline habitats. Springerplus 2:1–7

Reinhold B, Hurek T, Niemann EG, Pendrik I (1986) Close association of *Azospirillum* and diazotrophic rods with different root zones of Kallar grass. Appl Environ Microbiol 52:520–526

Reinhold B, Hurek T, Fendrik I, Pot B, Gillis M, Kersters K, Thielemans S, De Ley J (1987) *Azospirillum halopraeferens* sp. nov., a nitrogen-fixing organism associated with roots of Kallar Grass (*Leptochloa fusca* (L.) Kunth). Int J Syst Bacteriol 37:43–51

Reinhold B, Hurek T, Baldani J, Dobereiner J (1988) Temperature and salt tolerance of Azospirillum spp. from salt affected soils in Brazil. In: Klingmueller W (ed) Azospirillum IV genetics, physiology, ecology. Springer, Berlin, pp 234–241

Reis VM, Olivares FL, Döbereiner J (1994) Improved methodology for isolation of *Acetobacter diazotrophicus* and confirmation of its endophytic habitat. World J Microbiol Biotechnol 10:401–405

Richardson AE (2007) Making microorganisms mobilize soil phosphorus. In: Velázquez E, Rodríguez-Barrueco C (eds) First international meeting on microbial phosphate solubilization

Rodriguez R, Redman R (2008) More than 400 million years of evolution and some plants still can't make it on their own: plant stress tolerance via fungal symbiosis. J Exp Bot 59:1109–1114

Rosenberg A (1983) *Pseudomonas halodurans* sp. nov., a halotolerant bacterium. Arch Microbiol 136:117–123

Sandhya V, ASK Z, Grover M, Reddy G, Venkateswarlu B (2009) Alleviation of drought stress effects in sunflower seedlings by the exopolysaccharides producing *Pseudomonas putida* strain GAPP45. Biol Fertil Soils 46:17–26

Santos H, da Costa MS (2002) Compatible solutes of organisms that live in hot saline environments. Environ Microbiol 4:501–509

Sauvage D, Hamelin J, Larher F (1983) Glycine betaine and other structurally related compounds improve the salt tolerance of *Rhizobium meliloti*. Plant Sci Lett 31:291–302

Seeman JR, Critchley C (1985) Effects of salt stress on the growth, ion content, stomatal behaviour and photosynthetic capacity of a salt-sensitive species *Phaseolus vulgaris* L. Plania 144:151–162

Shamseldin A, Werner D (2005) High salt and high pH tolerance of new isolated *Rhizobium* etli strains from Egyptian soils. Curr Microbiol 50:11–16

Sharma S, Kulkarni J, Jha B (2016) Halotolerant rhizobacteria promote growth and enhance salinity tolerance in peanut. Front Microbiol Front Microbiol 7:1600

Shukla PS, Agarwal PK, Jha B (2012) Improved salinity tolerance of *Arachis hypogaea* (L.) by the interaction of halotolerant plant-growth-promoting rhizobacteria. J Plant Growth Regul 31:195–206

Singleton PW, Bohlool BB (1984) Effect of salinity on nodule formation by soybean. Plant Physiol 74:72–76

Singleton PW, EL-Swaify SA, Bohlool BB (1982) Effect of salinity on *Rhizobium* growth and survival. Appl Environ Microbiol 44:884–890

Son HJ, Park GT, Cha MS, Heo MS (2006) Solubilization of insoluble inorganic phosphates by a novel salt and pH tolerant *Pantoea agglomerans* R-42 isolated from Soybean rhizophere. Bioresour Technol 97:204–210

Sorokin DY, Lysenko AM, Mityushina LL, Tourova TP, Jones BE, Rainey FA, Robertson LA, Kuenen JG (2001) *Thioalkalimicrobium aerophilum* gen. nov., sp. nov. and *Thioalkalimicrobium sibiricum*. sp. nov., and *Thioalkalivibrio versutus* gen. nov., sp. nov., *Thioalkalivibrio nitratis* sp. nov. and *Thioalkalivibrio denitrificans* sp. nov., novel obligately alkaliphilic and obligately chemolithoautotrophic sulfur-oxidizing bacteria from soda lakes. Int J Syst Evol Microbiol 51:565–580

Steil L, Hoffmann T, Budde I, Volker U, Bremer E (2003) Genome-wide transcriptional profiling analysis of adaptation of *Bacillus subtilis* to high salinity. J Bacteriol 185:6358–6370

Ventorino V, Caputo R, De Pascale S, Fagnano M, Pepe O, Moschetti G (2012) Response to salinity stress of *Rhizobium leguminosarum* bv. viciae strains in the presence of different legume host plants. Ann Microbiol 62:811–823

Videira SS, Araújo JLS, Rodrigues LS, Baldani VLD, Baldani JI (2009) Occurrence and diversity of nitrogen-fixing *Sphingomonas* bacteria associated with rice plants grown in Brazil. FEMS Microbiol Lett 293:11–19

Weber A, Jung K (2002) Profiling early osmostress-dependant gene expression in *Escherichia coli* using DNA macroarrays. J Bacteriol 184:5502–5507

Wei GH, Yang XY, Zhang ZX, Yang YZ, Lindstrom K (2008) Strain Mesorhizobium sp. CCNWGX035; A stress tolerant isolate from *Glycyrrhiza glabra* displaying a wide host range of nodulation. Pedosphere 18:102–112

Whatmore AM, Chudek JA, Reed RH (1990) The effects of osmotic upshock on the intracellular solute pools of *Bacillus subtilis*. J Gen Microbiol 136:2527–2535

Whitelaw MA (2000) Growth promotion of plants inoculated with phosphate solubilizing fungi. Adv Agron 69:99–151

Winogradsky S (1949) Microbiologie du sol. Masson et Cie, Paris, 861 p

Xiang W, Liang H, Liu S, Luo F, Tang J, Li M-Y, Che Z-M (2011) Isolation and performance evaluation of halotolerant phosphate solubilizing bacteria from the rhizospheric soils of historic Dagong Brine Well in China. World J Microbiol Biotechnol 27:2629–2637

Yelton MM, Yang SS, Edie SA, Lim ST (1983) Characterization of an effective salt tolerant fast growing strain of *Rhizobium japonicum*. J Gen Microbiol 129:1537–1547

Zarea MJ, Hajinia S, Karimi N, Mohammadi Goltapeh E, Rejali F, Varma A (2012) Effect of *Piriformospora indica* and *Azospirillum* strains from saline or non-saline soil on mitigation of the effects of NaCl. Soil Biol Biochem 45:139–146

Zarea MJ, Chordia P, Varma A (2013a) *Piriformospora indica* versus salt stress. In: Varma A, Kost G, Oelmüller R (eds) *Piriformospora indica*. Springer, Berlin, pp 263–281

Zarea MJ, Mohammadi Goltapeh E, Karimi N, Varma A (2013b) Sustainable agriculture in saline-arid and semiarid by use potential of AM fungi on mitigates NaCl effects. In: Mohammadi Goltapeh EM, Rezaidanesh Y, Varama A (eds) Fungi as bioremediators, Soil biology, vol 32. Springer, Berlin, pp 340–370

Zarea MJ, Miransari M, Karimi N (2014) Plant physiological mechanisms of salt tolerance induced by mycorrhizal fungi and *Piriformospora indica*. In: Miransari M (ed) Use of microbes for the alleviation of soil stresses. Springer, New York

Zhao XR, Lin QM (2001) A review of phosphate—dissolving microorganisms. Soil Fertil 3:7–11

Zhou S, Han L, Wang Y, Yang G, Zhuang L, Hu P (2012) *Azospirillum humicireducens* sp. nov., a nitrogen-fixing bacterium isolated from a microbial fuel cell. Int J Syst Evol Microbiol 63:2618–2624

Zhu F, Qu L, Hong X, Sun X (2011) Isolation and characterization of a phosphate-solubilizing halophilic bacterium *Kushneria* sp. YCWA18 from Daqiao Saltern on the coast of yellow sea of China. J Evid Based Complement Altern 2011:1–6

Chapter 13
The Mechanisms Involved in Improving the Tolerance of Plants to Salt Stress Using Arbuscular Mycorrhizal Fungi

Magdi T. Abdelhamid, Raafat R. El-Masry, Darwish S. Darwish, Mazhar M. F. Abdalla, Shinya Oba, and Ragab Ragab

Abstract Salinity is considered as one of the most harming stresses faced by the plant in regard to its survival and productivity. A new biological approach "plant-microbe interaction" such as arbuscular mycorrhizal (AM) fungi to address salinity problem has recently gained momentum. Therefore, this chapter aims to provide a general overview about salinity and its effects on plant and soil, and the use of AM fungal inoculant applied to plants to alleviate salinity effects, and the mechanism of AM fungi to increase the tolerance of plants to salt stress with a crucial discussion of major accomplishments reported in this area. The results show that some mechanism of how AM fungi can increase the plant salt tolerance might work well in this regard. AM fungi maintain a superior K^+:Na^+ ratio that is considered as one of AM fungi's strategies to improve tolerance to salt stress and boost absorption of P, K^+, and Ca^{2+} over harmful Na^+, thus sustaining lesser Na^+:K^+ ratio under salt stress. The improvement in chlorophyll as a result of AM fungi is owing in particular to the increased uptake of magnesium. AM fungi inoculation increases the activity of antioxidant enzymes in plants such as peroxidase (POD), superoxide dismutase (SOD), and catalase (CAT), which scavenge reactive oxygen species (ROS) and relieve salt stress. An additional salt tolerance mechanism by AM fungi increases the non-antioxidant of plants by accumulating osmolytes such as proline, which maintain the osmotic adjustment of plants under salinity stress. Thus, application of AM fungi inoculant is more probably due to its economic benefits under salt stress.

M. T. Abdelhamid (✉) · R. R. El-Masry
Botany Department, National Research Centre, Giza, Egypt

D. S. Darwish · M. M. F. Abdalla
Faculty of Agriculture, Agronomy Department, Cairo University, Giza, Egypt

S. Oba
Faculty of Applied Biological Sciences, Plant Production Control Lab, Gifu University, Gifu, Japan

R. Ragab
Centre for Ecology and Hydrology (CEH), Wallingford, UK

B. Giri, A. Varma (eds.), *Microorganisms in Saline Environments: Strategies and Functions*, Soil Biology 56, https://doi.org/10.1007/978-3-030-18975-4_13

Keywords Antioxidants · Arbuscular mycorrhizal (AM) fungi · Ionic balance · Oxidative stress · Salinity · Salt stress

13.1 Introduction

Soil degradation that is induced by salt is widespread in arid and semiarid regions to which irrigation is implemented without drainage systems. Insufficiency of drainage management in irrigated lands causes salt accumulation in the root zone and negatively affects soil properties and consequently plant productivity (Qadir et al. 2014). Salinization generally can refer to the accumulation of various salts, including potassium (K), magnesium (Mg), calcium (Ca), and sodium carbonate, chloride-bicarbonate, and sulfate (Bockheim and Gennadiyev 2000). Excessive use of chemical fertilizers, groundwater irrigation, flood irrigation, and no practice of crop rotation are also the major causes for increasing the salinization of farming land. Accordingly, a wide range of physiochemical properties result from this diverse ionic composition.

As stated by FAO Soils Portal, the soil affected by salt can be divided into saline, saline-sodic, and sodic, based on the amount of salt, type of salt, sodium present, and soil alkalinity. Saline soils have an electrical conductivity (EC) of more than 4 dS m^{-1} at 25 °C (approx. 40 mM NaCl), which produces an osmotic pressure of approximately 0.2 Mpa, with an osmotic pressure of approximately 0.2 Mpa (Grieve et al. 2008; Munns and Tester 2008). Figure 13.1 shows a general view of surface layers of salt-affected soil at Al Qantarah Sharq, Egypt, before planting. A soil's

Fig. 13.1 Salt-affected soil at Al Qantarah Sharq, Egypt, before planting

sodicity is a term more limited to the amount of Na^+ in the soil. High sodicity (more than 5% of Na^+'s total cation content) causes the clay to swell excessively when wet, thereby roughly restricting air and water movement and leading to poor drainage (Munns 2005; Hanin et al. 2016).

If a fertile soil is altered to saline soil, the agricultural productivity will be affected as salt concentration increased, therefore the soil fertility may decrease crop productivity by up to 20% (Porcel et al. 2012; Munns and Gilliham 2015). It is reported that salt-affected soils occur in more than half of irrigated land in some countries (Metternicht and Zinck 2003). However, widely reported data indicate that about 33% of the global irrigated lands have lesser productivity ascribable to poorly managed irrigation provoking waterlogging and salinity (Khan and Hanjra 2008). For example, multi-location studies carried out on saline-sodic fields in Pakistan reported that losses in the yield of wheat grain from salt-affected soils ranged from 20 to 43%, with a total average loss of 32%. High accumulation of salt in the soil reduces soil porosity, soil aeration, and water conductivity, which leads to drought and physiological water stress (Mahajan and Tuteja 2005). Therefore, salinity management strategies must be desirable for sustainable land management. Plant microbes are important for agriculture because they can improve plant growth and plant nutrition through biological fixation of nitrogen and other functions (Yadav et al. 2017b). Growth stimulation by microbes may result from biological nitrogen fixation (Pankievicz et al. 2015); biocontrol of phytopathogens by producing antifungal or antibacterial agents (Errakhi et al. 2016); the production of phytohormones, for example, cytokinins and IAA (Verma et al. 2015); substance induction achieved (Bach et al. 2016); nutritional competition (Bach et al. 2016); production of siderophores (Ellis 2017); induced host resistance (Pal and Gardener 2006); and improved availability of soil minerals (Haas and Défago 2005). Sustainable agriculture necessitates planning to enhance or sustain the current food production rate while reducing environmental and human health damage (Yadav et al. 2017a). An alternative to conventional agricultural technologies is the use of microbial plant growth promoters (Kour et al. 2017; Yadav et al. 2017c).

There are two main strategies for the use of salt-affected land: first, the use of reclamatory and preventive measures to ensure that salt-affected soils are suitable for agriculture and second, the use of salt-affected soils by cultivating halophytes or crops/cultivars tolerant to salt. The latter approach has been referred to as a biological approach (Ashraf 1994; Epstein 1977; Kingsbury and Epstein 1984) and has considerable potential to mitigate the problem of soil salinity around the world. A new biological approach "plant-microbe interaction" has recently gained momentum, such as arbuscular mycorrhizal (AM) fungi, to manage the salinity issue. Development of salt-resistant crops is not an economical approach to sustainable agriculture, while microbial inoculation to reduce salt stress is an improved alternative, as it minimizes production costs and environmental risks. This chapter aims to provide an overview on salinity and its effects on plant and soil, and the use of AM fungi inoculant to plants to alleviate salinity effects, and the AM fungi mechanism to increase plant tolerance to salt stress with critical discussion on the key accomplishments reported.

13.2 Salt Stress Effects on Plant Growth

The most recent reviews (e.g., Munns and Gilliham 2015; Hanin et al. 2016; Negrão et al. 2017) have discussed the influences of soil salinity on plant. The plant response to salinity can be elucidated in two major stages, reflecting the existing time frame of biological processes: shoot ion-independent responses (early) and ion-dependent responses (late) (Negrão et al. 2017). Negrão et al. (2017) stated that the plant's response to salinity can be explained in two main stages reflecting the existing biological process time frame: early (ion-independent reactions) and late (ion-dependent reactions). The initial reaction (a few minutes or a day after imposing stress), where the osmotic effect of salt is felt in the soil, is characterized as a rapid response by plants. The initial reaction is also called the osmotic phase (Roy et al. 2014). The reduced potential of soil water leads to a reduction in the water absorption of the plant, which causes stomatal closure, reduces photosynthesis, and inhibits the expansion of the leaf, which leads to a reduction in the growth rate of the shoot (Munns and Tester 2008; Das et al. 2015; Hanin et al. 2016; Nongpiur et al. 2016). There is an ionic phase when salts are toxic due to plant accumulation (Munns 2005). After several days to weeks after toxic Na^+ accumulation in photosynthetic tissues, the second phase is a slower response (Roy et al. 2014). Due to the accumulation of salt over time, growth in older leaves is slower, leading to senescence (Munns and Tester 2008). Unlike young leaves, the salt that enters their cells, which leads to their death, cannot be diluted. If the death rate is higher than the production rate, the photosynthetic capacity is affected, and the growth rate is reduced (Munns and Tester 2008; Nongpiur et al. 2016). Figure 13.2 shows response of wheat entries and cultivars grown under salt-affected soil and salt-free soil at Tag Elez Agricultural Research Station of the Agricultural Research Center, Mansoura, Dakahlia Governorate, Egypt.

The mechanisms of plant salt tolerance can be classified in three different classes according to Negrão et al. (2017), i.e., exclusion of ions, which refers to the net exclusion of toxic ions from shoot; tissue tolerance, which refers to the compartmentation of toxic ions, in particular tissues, cells, and subcellular organelles; and independent tolerance of shoot ions, which centers on the maintenance of growth and water uptake independent of the amount of accumulation of Na^+ in the shoot. The major cellular, metabolic, and physiological responses triggered in everyone are explained in detail by Munns and Tester (2008) and Roy et al. (2014). However, the basic element determining salt tolerance is the transport of ions via cell membranes (Ismail and Horie 2017). Salinity also causes oxidative stress in plants due to the production of reactive oxygen species (ROS), such as superoxide radicals (O^{2-}), hydrogen peroxide (H_2O_2), and hydroxyl radicals (OH^-), which can cause damage to DNA and cells (El-Mashad and Mohamed 2012). In addition, enzymatic antioxidant mechanisms such as superoxide dismutase (SOD), peroxidase (POX), and catalase (CAT), among others, activate antioxidant stress protection (Gharsallah et al. 2016). Incidental accumulation of compatible solutes under salt stress occurs when solutes accumulate in cytosol (Munns and Tester 2008). Low-weight solutes

Fig. 13.2 Wheat entries and cultivars were sown in salt-affected soil and salt-free soil at Tag Elez Agricultural Research Station of the Agricultural Research Center, Mansoura, Dakahlia Governorate, Egypt

such as proline, polyols, amino acids, proteins, and betaine were generally considered to be compatible solutes that function under high salt conditions in osmoprotection and osmotic adjustment (El-Mashad and Mohamed 2012; Gharsallah et al. 2016). Proline with overproduction in plants definitely represents the stress tolerance through maintaining cell turgor or osmotic balance, stabilizing membranes, preventing electrolyte leakage, and leading to ROS concentrations in normal ranges, thus stopping oxidative cracks in plants (Hayat et al. 2012).

The shaping of salinity tolerance degrees is depending upon its plant morphology, physiology, and biochemistry (Nongpiur et al. 2016). In the environment, there is a slope for the tolerance to NaCl ranging from very tolerant to very susceptible plant species. For example, the species of *Tecticornia* and *Atriplex* are salt-tolerant and mostly referred to as halophytes (Munns and Tester 2008; Flowers and Colmer 2015), while *Oryza sativa* is one of the most susceptible salt-tolerant species (Das et al. 2015). In field conditions with salinity up to 10 dS m^{-1}, approx. 100 mM NaCl can occur, and rice passes before maturity, while *Triticum aestivum* can produce less yield (Munns et al. 2006; El-Metwally et al. 2015). Therefore, wheat is considered moderately salt-tolerant. In contrast, *Hordeum vulgare* is rated as one of the most salt-tolerant cereals, although it fades when salt levels exceed 250 mM NaCl (approx. 50% seawater) which last for a period of time (Munns et al. 2006). Through these examples and in order to ensure our food security, it is necessary to understand the mechanism by which crop productivity can be sustained in the saline soil (Landi et al. 2017). Damage due to salinity could be in terms of reduced photosynthetic rate, reduced activity of antioxidants, reduced stomatal behavior, induced ion deficiencies, decreased membrane stability, and reduced plant water content (Renault 2012; Rady et al. 2015; Abouelsaad and Renault 2018). Furthermore, other reasons such as physiological drought, ion toxicity and imbalance, and soil degradation can reduce plant growth (Abdelhamid et al. 2010). Salt stress adversely affects the PSII and electron transport in the photosynthetic apparatus, thus decreasing net photosynthetic rate (Sheng et al. 2008; Talaat and Shawky 2012). Soil salinity also affects plant growth with toxic effects of Na^+ and Cl^-, which causes enzyme denaturation structures, damage to cell organelles, reduced photosynthesis, respiration, and osmotic imbalances, which will lead to physiological drought, plus restricting nutrients uptake. Such effects result in reduced plant growth as a result of salt stress and lower agricultural productivity (Adiku et al. 2001; Evelin et al. 2009).

Plant cells generally absorb large amounts of dissolved salts in saline soil, leading to plasmolysis of plant cells, and these cells begin to collapse, affecting the morphological characteristics of plants, such as leaf expansion and reduction of fresh and dry matter (Hernandez et al. 1995). In *Trifolium alexandrinum*, the length of shoot and root and total dry matter were decreased under salinity stress (Shokri and Maadi 2009). Plants accumulate different types of proteins and osmolytes such as proline, glycine betaine, and malondialdehyde to protect itself against osmotic shock in response to salt stress (Orabi and Abdelhamid 2016; Bargaz et al. 2016). Moreover, salinity is also responsible for the damage caused to the plant by reactive oxygen species (ROS) (Ahmad et al. 2010). After exposure to salt stress, these biochemical constituents accumulate in higher concentrations. Some authors reported no effect on growth of legumes under low salinity stress, although decreasing plant fresh and dry weights were found (Parida and Das 2005). There are many studies that have been conducted on different plant species to examine the adverse effects of osmotic stress caused by salinity, e.g., maize (Awad et al. 2012), faba bean (Bekheta et al. 2009; Hellal et al. 2012; Abdelhamid et al. 2013b; Dawood et al. 2014a, b; Sadak et al. 2015; Orabi and Abdelhamid 2016), common bean (Abdelhamid et al. 2013a; Dawood et al. 2014a, b), lupin (Dawood et al. 2016), soybean (Rady et al. 2015), and wheat (El-Lethy et al. 2013; Ouda et al. 2015; Rady et al. 2016a, b). In addition, several studies were conducted to explore the differences on plant varieties and genotypes, for example, faba bean varieties (Orabi and Abdelhamid 2016) and common bean genotypes (Bargaz et al. 2016; Rady et al. 2016a, b).

Salinity can have a significant impact on many plant morphophysiological functions by disrupting osmotic gradients and intracellular ion concentrations, which inhibit several key physiological traits, e.g., reduced photosynthesis (Renault 2012), reduced protein synthesis and enzymatic activity (Sadak and Abdelhamid 2015), nutritional imbalance (lowering N, P, K^+, Ca^{2+}) (Abdelhamid et al. 2010), decreased membrane stability index (Talaat et al. 2015), decreased relative water content (Rady et al. 2015), and decreased stem anatomy parameters (Dawood et al. 2014b; Semida et al. 2014). The induced alterations are mostly coincided with mineral toxicity and consequently reduced nutritional balance.

13.3 Arbuscular Mycorrhizal Fungi

Arbuscular mycorrhizal (AM) fungi are the most common type of mycorrhizal fungi in the root cortical cells that are microscopic and ramified forms (Manchanda and Garg 2007). These AM fungi are found in 80% of vascular plants and are involved in various functions from accelerating the absorption of nutrients in order to improve the development of plants and improve soil properties (Basu et al. 2018). AM fungus is a special type of symbiotic fungus that is associated with the terrestrial plant root in a symbiotic, nonpathogenic way. In saline land, AM fungi are widely spread (Yamato et al. 2008). A wide range of plants, including angiosperm, gymnosperm, pteridophytes, mosses, etc., can find this type of symbiotic partnership between plant fungi (Bago et al. 2000). Such symbiotic relationship enhances water and nutrient absorption in the host plant and protects the plant from various abiotic and biotic stresses (Zuccarini and Okurowska 2008; Gupta et al. 2000). AM fungi are soil fungi belonging to at least 460 million years ago, a newly approved phylum, *Glomeromycota*. AM fungi are obligatory symbionts whose life cycle requires plant hosts. AM fungi spores germinate in the soil in appropriate conditions and identify well-matched host roots to trigger their symbiotic association (Schubler et al. 2001; Prasad et al. 2017).

Several researches have shown the importance of AM fungi in relieving salt stress and improving plant growth and agricultural production. AM fungi have the capacity to quench salt stress due to their ability to reshape root morphology and extend the area rummaged by plants for nutrients by an extensive hyphal netting. AM fungi can also increase the host plant's intake of water and nutrients, relieving abiotic and biotic stresses. Glycoprotein glomalin is also produced by AM fungi, which improves soil structure and properties. This fungus also reshapes plant physiology and increases plant growth and nutrition, allowing reduced use of mineral fertilizers (Gianinazzi et al. 2010; Estrada et al. 2013). AM fungi are able to promote salinity tolerance of host plants by increasing the absorption of water and nutrients from the soil through extensive hyphae tissue (Gopal et al. 2012; Bharti et al. 2014), reshaping morphology and plant physiology, allowing host plants to cope with stress faster (Miransari et al. 2008), manufacturing plant hormones, and involving other soil microbes such as rhizobia and plant growth-promoting rhizobia (PGPR) (Miransari 2010; Mardukhi et al. 2015). Saxena et al. (2017) showed another approach by which AM plants could alleviate salt stress (Fig. 13.3). Under the condition of salt stress, the mycorrhizal plant (right) improves plant growth and yield by increasing the absorption of water and nutrients, reducing the harmful effects of ROS, managing lipid peroxidation and destruction of the cell membrane,

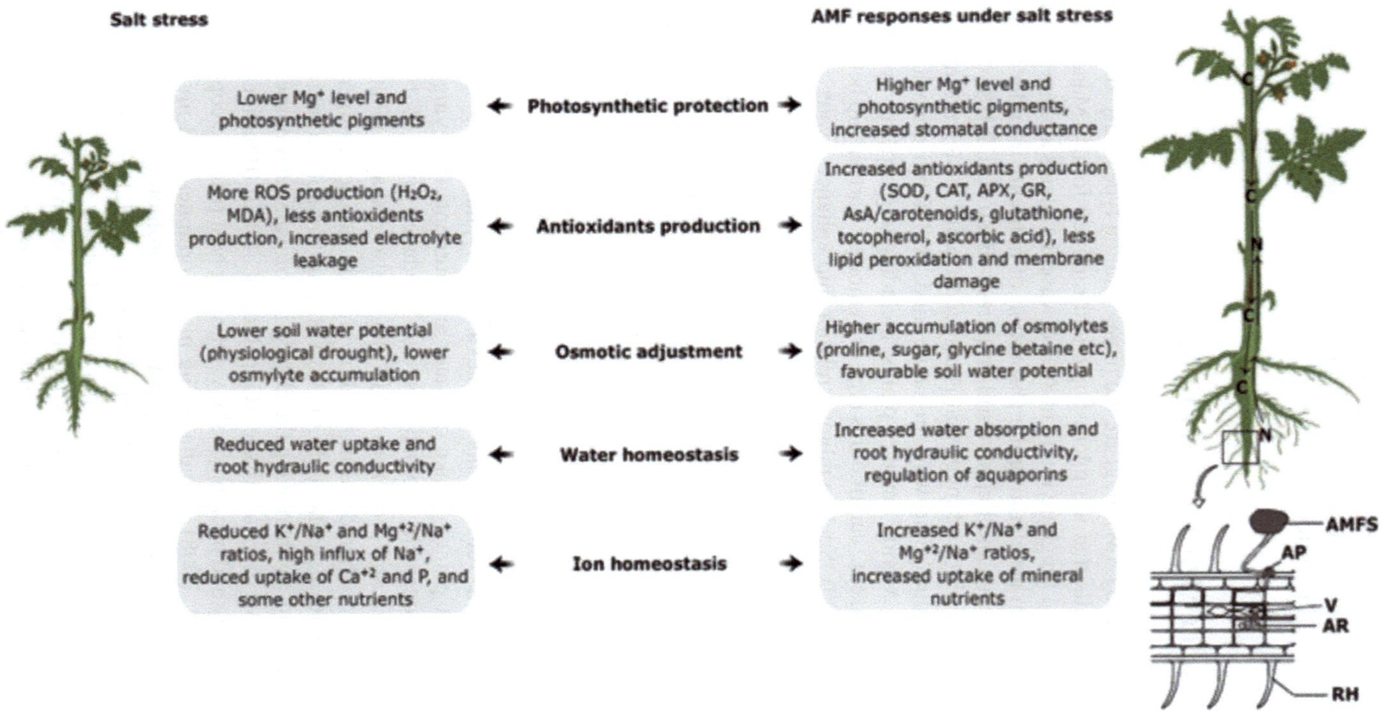

Fig. 13.3 Mycorrhizal (*right*) and non-mycorrhizal (*left*) plants response under salt stress (Reproduced with permission from Saxena et al. 2017, Springer Nature). *AMFS* arbuscular mycorrhizal fungi spores, *AP* appressorium, *V* vesicles, *AR* arbuscules, *RH* root hairs

collecting osmolytes, and maintaining a pleasant water potential in the soil and homeostasis of water and ions. In contrast, non-mycorrhizal plants (left) benefit from basic water and nutrient requirements, face the problem of reduced water potential in the soil, collect more ROS, and produce lower antioxidants, resulting in poor growth and productivity under salt stress (Saxena et al. 2017). AM fungi are worthwhile to plant performance and counter various salinity stress conditions. Inoculation with AM fungi in garlic plant under 100 mM NaCl salt stress enhanced morphological traits such as leaf area and plant biomass (Borde et al. 2010). Inoculation of tomato with *Glomus mosseae* under moderate salinity condition resulted in increase in the biomass (Al-Karaki and Hammad 2001).

The ability of AM fungi-inoculated plants to adapt to salt stress was examined in several researches. The favorable accumulative effect of nutrient intake, photosynthetic ability, and biochemical and physiological traits due to AM fungi inoculation improves plant growth and, thus, reduces salt stress (Borde et al. 2011). Many authors have reported that the AM fungi inoculation increased plant salinity stress tolerance capacity compared to the non-AM fungi-inoculated plant (Sannazzaro et al. 2007; Zuccarini and Okurowska 2008; Borde et al. 2011). AM fungi enhanced plant biomass and tomato fruit yield under salinity (Al-Karaki and Hammad 2001). Lots of reports on various plant species have shown that AM fungi improve plant salinity tolerance by improving photosynthetic ability, enhancing nutrient absorption, increasing antioxidant defense, and promoting osmolyte accumulation, which leads to improved plant growth performance and improved plant salinity tolerance (Table 13.1).

13.4 The Role of AM Fungi in Enhancing Plant Salt Stress Tolerance

13.4.1 Induced Salt Tolerance by AM Fungi Through Ionic Balance

It has been shown that AM fungi support plant growth, improve nutrient absorption from the soil, improve soil structure, and also increase plant tolerance under various abiotic and biotic stresses, such as salinity (Hameed et al. 2014). Furthermore, AM fungi play an important role in regulating the transport of ion and membrane proteins, which control host plants' homeostasis (Ramos et al. 2011). A good ratio of K^+:Na^+ is as well considered beneficial for the maintenance of ion balance in plant leaves and index of plant salt tolerance (Abdelhamid et al. 2010).

Higher levels of Na^+ with >40 mM have injuring influence on plant growth because of hyperosmotic stress, overabsorption, and ion imbalance (Munns and Tester 2008). K^+ and Na^+ ions participate on vital function in ion strength and osmotic pressure in salinity stress (Serrano and Rodriguez-Navarro 2001). The increase in salinity in irrigated land reduces the content of K^+, Ca^{2+}, and NO_3 while increases inorganic phosphate (PO_4^{3-}) and Na^+ and Cl^- ions to a degree

Table 13.1 Influence of AM fungi on plant under salinity stress conditions

Plant species	Effect of AM fungi	References
Cucumis sativus	Enhanced nutrient uptake, stand establishment rate, and photosynthetic rate	Sallakua et al. (2019)
	Elevated levels of antioxidant enzymes (SOD, POD, CAT) and lower uptake of Na^+	Ahmad et al. (2018)
	Increased fruit fresh and dry weights, proline and electrolyte leakage, root and shoot dry weights, photosynthesis and stomata conductance	Haghighi et al. (2017)
Trigonella foenum-graecum	Enhanced growth, proline, and antioxidant enzymes and phosphatase	Metwally and Abdelhameed (2018)
Triticum aestivum	Increased leaf area, stomatal density, stomatal conductance (*gs*) and maximum rate of *gs* to water vapor (*gsmax*), and leaf water potential	Zhu et al. (2018)
	Increased total phenol, proline content, leaf area, shoot and root dry weights, and lengths	Tofighi et al. (2017)
	Increased plant dry weight; stomatal conductance, density, size, and aperture; higher soluble sugar concentration; and K^+:Na^+ ratio, C assimilation, and N uptake	Zhu et al. (2016)
Triticum durum	Increased aboveground and root biomass, N uptake and aboveground N, and stability of plasma membranes	Fileccia et al. (2017)
Solanum lycopersicum	Increase plant metabolic plasticity	Rivero et al. (2018)
Medicago sativa	Increased photosynthesis-related performance, such as net photosynthesis, stomatal conductance, the maximum quantum yield (Fv/Fm), and effective quantum yield (ΦPSII) of PSII photochemistry	Fan et al. (2018)
	Increased K^+, K^+:Na^+, and Ca^{+2}:Na^+	Namdari et al. (2018)
Arundo donax	Increase in proline accumulation and a trend toward higher isoprene emission and higher H_2O_2	Pollastri et al. (2018)
Zea mays	Increased maize salt tolerance by decreasing salt concentrations, increasing soil macroaggregate proportions, soil bacterial diversity, maize mineral uptake, and photosynthesis. Enhanced K selected transportation from the root to the shoot and reduced the shoot malondialdehyde content to promote maize photosynthesis and transpiration	Zhang et al. (2018)
	Increased osmolyte accumulation, water potential, and photosynthetic activity	Sheng et al. (2011)
Chrysanthemum morifolium	Increased root length, shoot and root dry weight, total dry weight, and root N concentration	Wang et al. (2018)
Ricinus communis	Increased the net photosynthetic rate, stomatal conductance, transpiration rate, soluble protein and proline, and decreased intercellular CO_2 concentration and malondialdehyde (MDA)	Zhang et al. (2018)

(continued)

Table 13.1 (continued)

Plant species	Effect of AM fungi	References
Ocimum basilicum	Increased chlorophyll content and water use efficiency K, P and Ca, K/Na, and Ca/Na	Elhindi et al. (2017)
Jatropha curcas	Growth rate and photosynthetic activity	De Oliveira et al. (2017)
Capsicum annuum	Increased total fresh and dry weight, chlorophyll a and b concentrations, antioxidant content expressed as total soluble phenols and proline concentrations, relative water content, and decreased membrane permeability	Hegazi et al. (2017)
Cajanus cajan	Improved plant biomass and reduced oxidative burst by strengthening antioxidant enzymatic activities	Pandey and Garg (2017)
Leymus chinensis	Increased plant biomass, photosynthetic parameters, and contents of photosynthetic pigments	Lin et al. (2017)
Robinia pseudoacacia	Improved the net photosynthetic rate, quantum efficiency of photosystem II photochemistry, and K^+ content and relative water content, but Na^+ content and shoot/root ratio of Na^+	Chen et al. (2017)
	Increased chlorophyll content, net photosynthetic rate, intercellular CO_2 concentration, stomatal conductance, and transpiration rate, carbon content, calorific value, carbon, and energy accumulations	Zhu et al. (2017)
Solanum lycopersicum	Increased K^+ concentration, K^+:Na^+ ratio cytokinins, indoleacetic acid and abscisic acid concentrations	Khalloufia et al. (2017)
	Enhanced chlorophyll content	Hajiboland et al. (2010)
	Increased antioxidant activity	He et al. (2007)
Poncirus trifoliata	Increased plant height, stem diameter, leaf number, shoot biomass, and root biomass, length, surface area, and volume	Zhang et al. (2017)
Citrus aurantium	Increased shoot and root dry matter, root infections, spore production, and concentrations of N, P, K, Zn, Mn, and Cu in plant tissues	Satir et al. (2016)
Cicer arietinum	Enhanced nodulation and nitrogen fixation	Garg and Singla (2016)
	Improved growth and productivity, Si uptake	Garg and Bhandari (2016)
Oryza sativa	Decreased of Na^+ root-to-shoot distribution and an increase of Na^+ accumulation in rice roots	Porcel et al. (2016)
Vitis vinifera	Increased root growth, root and shoot biomass, and leaf area and improved leaf Ψw, $\Psi \pi$, P_n, and chlorophyll content, K^+:Na^+ ratio, and P_n	Upreti et al. (2016)
Acacia gerrardii	Increased osmoprotectants such as glycine, betaine, and proline and reduced lipid peroxidation	Hashem et al. (2016a, b)
Gossypium hirsutum	Promoted leaf proline and K^+:Na^+	Liu et al. (2016)
Lupinus termis	Increased proline, glycine betaine, and sugars and increased the activities of key antioxidant enzymes like	Hashem et al. (2016a, b)

(continued)

Table 13.1 (continued)

Plant species	Effect of AM fungi	References
	superoxide dismutase (SOD), catalase (CAT), ascorbate peroxidase (APX), and peroxidase (POD)	
Ocimum basilicum	Enhanced the biomass production and accumulation of nutrients, reducing sugars, total soluble carbohydrates, photosynthetic pigments, proline, and protein by reducing Na	Elhindi et al. (2016)
Vigna radiata	Improved plant growth	Rabie (2005)
Cucurbita pepo	Enhanced relative water content	Colla et al. (2008)
Sesbania grandiflora	Decreased potassium	Wu et al. (2010)
Sesbania aegyptiaca	Increased chlorophyll content and potassium	Giri et al. (2007)
Acacia nilotica	Increased potassium	Giri and Mukerji (2004)
Trifolium alexandrinum	Increased plant biomass	Shokri and Maadi (2009)
Glycine max	Increased CAT activity	Ghorbanli et al. (2004)
Cajanus cajan	Higher osmolyte accumulation	Garg and Manchanda (2009)
Phragmites australis	Enhanced osmolyte accumulation	Al-Garni (2006)
Oscimum basilium	Increased in proline accumulation	Shekoofeh et al. (2012)
Piper nigrum	Enhanced chlorophyll content	Kohler et al. (2009)
Citrus sinensis	Increased of potassium and calcium uptake	Wu et al. (2010)
Lactuca sativa	Increased growth, minerals content	Kohler et al. (2009)

that increases the concentration of ionic injuries and imbalances in osmotic and nutrients (Bothe 2012).

Hindered uptake of K^+ and P under salinity is reported by Wu et al. (2010). For several important metabolic processes, optimal K^+ in plant cells is essential (Tomar and Agarwal 2013). The inoculation of AM fungi in salt-stressed citrus plants significantly reduced the harmful effect on essential element uptake such as P, K^+, and Ca^{2+} (Wu et al. 2010). Inoculation of lettuce with AM fungi under salinity stress significantly improved plant growth by promoting the absorption of essential mineral elements compared to the lettuce plant inoculated with non-AM fungi (Kohler et al. 2009). AM fungi inoculation maintains a greater ratio of K^+:Na^+, which work well as index of salt tolerance (Wu et al. 2010; Tomar and Agarwal (2013), and increases uptake of Ca^{2+}, P, and K^+ greater than injurious Na^+, thus sustaining lesser ratio of Na^+:K^+ under salinity stress (Ahmad et al. 2014).

Under salinity stress, there is delaying in uptake of essential mineral elements in stressed plants. The high concentration of Na^+ in the root zone immediately affects the uptake of several essential elements as K^+ and Na^+, which express opposed relation with K^+ (Kohler et al. 2009). The inoculation of AM fungi under salinity stress reduces the absorption of Na^+ and increases the absorption of K^+ compared to non-AM fungi plants (Hammer et al. 2011). Excessive amounts of Na^+ and Cl^- ions in saline condition interrupt the ionic equilibrium of soil solution, thus affecting the absorption, transport, and use of essential nutrients (Roberts et al. 1984). AM fungi can selectively help plants to gain more nutrients such as K^+ and Ca^+, which play a role in osmotic adjustment, while excluding Na^+ absorption and thus reduce concentration of Na^+ in salt-stressed plants (Hammer et al. 2011). This could let AM fungi beneficial for salinity stress relieving for their host plant.

13.4.2 Induced Salt Tolerance by Microbes Through Plant Physiological Response

It has been reported that salinity stress reduces photosynthetic ability and induces physiological plant drought stress, leading to reduced plant productivity (Rady et al. 2016a, b; Bargaz et al. 2016). AM fungi sustain stressed soil and enhance plant growth and development, therefore improving plant biotic and abiotic stress tolerance (Abd-Alla et al. 2000) through controlling plant physiological and biochemical systems (Evelin et al. 2009). Lately, numerous authors proved that AM fungi improve the plant ability to survive under salinity stress (Yano-Melo et al. 2003; Rabie 2005; Jahromi et al. 2008) by enhancing water intake (Ruiz-Lozano and Azcon 1995), improving the absorption of nutrients (Cantrell and Linderman 2001; Asghari et al. 2005) and ionic balance (Giri et al. 2007), and securing activity of enzymes (Rabie and Almadini 2005; Giri and Mukerji 2004). Biological remediation, such as the use of AM fungi as bioinoculants in saline soils, could reduce the stress of salt in plants (Evelin et al. 2009). Therefore, AM fungi can boost more absorption of minerals (Evelin et al. 2012), ionic balance (Giri et al. 2007), protection of enzyme activities (Patel and Saraf 2013), and the promotion of water absorption (Aroca et al. 2007).

When plants are affected by salinity stress, the amount of photosynthetic pigments decreases (Bargaz et al. 2016). This reduced chlorophyll pigment content was confirmed as a result of salinity stress by Awad et al. (2012), El-Lethy et al. (2013), Dawood et al. (2014a, b), and Bargaz et al. (2016) in *Zea mays*, *Triticum aestivum*, *Vicia faba*, and *Phaseolus vulgaris*, respectively. Dawood et al. (2014a, b) and Bargaz et al. (2016) have shown a significant reduction in chlorophyll content as salinity levels increase. The reduced concentration of chlorophyllase under salinity stress is due to increased chlorophyllase activity, which causes pigment breakdown and thus reduces photosynthesis and affects growth. Compared to stressed plants, AM fungi-inoculated plants had a higher chlorophyll pigment content. The findings of Hajiboland et al. (2010) and Aroca et al. (2013) for *Solanum lycopersicum* and *Lactuca sativa* confirm the fact that AM fungi inoculation significantly improved chlorophyll content in many plants. Inoculation of AM fungi in normal and salt-stressed pepper plants increased chlorophyll content (Kaya et al. 2009). The improvement of chlorophyll content because of AM fungi might be because of increased mineral absorption, particularly magnesium, a key constituent of the chlorophyll molecule (Sheng et al. 2008). Briefly, a greater chlorophyll content in AM fungi-inoculated plants contributes to higher photosynthetic activity, which leads to improved growth.

13.4.3 Induced Salt Tolerance by Microbes Through Antioxidant Defense Response

Salinity inhibits plant growth and productivity, and therefore it is important to recognize the different mechanisms that permit plants to overcome the salt stress. Plant manages salt stress by increasing accumulation of certain osmolytes and antioxidants that protect the plant cell from oxidative injury (Rai et al. 2011). Plants developed certain protective systems which involve antioxidant molecules and enzymes to prevent or alleviate the oxidative stress (Jiang and Zhang 2002; Nunez et al. 2003). Once a pathogen attacked the plant, the plant reacts to the attack by activating its defense mechanisms such as catalase (CAT) and peroxidase (POD) which support the cell wall or work as an antioxidant (Mehdy 1994). Superoxide dismutase (SOD) is a basic enzyme that converts superoxide into oxygen and hydrogen peroxide (H_2O_2) (Alscher et al. 2002). CAT is also included in the scavenging of H_2O_2 by decay and transformation into water and oxygen in addition to SOD. Therefore, CAT and SOD are considered the main components that respond to antioxidant activities and adjust them by controlling concentration of O_2 and H_2O_2 induced during stress in the plant (Van Breusegem et al. 2001).

Salt stress in plants induces a number of oxidative types of stress (Hajiboland et al. 2009). Antioxidant enzymes are the best protection system to prevent stress-induced oxidative injuries. The production of ROS is considered to be a major factor in harming the plant during various environmental stresses, including salinity stress (Hernandez et al. 1995). The accumulation of Reactive oxygen species (ROS) has a significant impact on plants, in particular the disruption caused by oxidative damage to the plant's metabolism (Jiang and Zhang 2002). Plants have their own preservative system to reduce the injury caused by ROS, which reduces the incidence of oxidative destruction (Abdel Latef 2010).

Enzymes comprising POD, SOD, glutathione reductase (GR), and CAT play a role on detoxification of the ROS (Ahmad et al. 2008, 2010; Ahanger et al. 2014). The plant itself induces some antioxidant enzymes to reduce the adverse effect of stress, but mycorrhizal association in plants helps to promote the host plant's antioxidant protection system. In mitochondria, peroxisomes, and plant cytoplasm, these enzymes are continuously triggered. The antioxidant defense system is positively linked to mycorrhizal colonized plants' response to salt stress. In AM fungi-inoculated plants under salt stress, higher antioxidant activity has been associated with improved plant growth performance (Alguacil et al. 2003; ZhongQun et al. 2007). Moreover, in tomatoes AM fungi-inoculated salt-stressed plants, the activities of SOD, CAT, and POD were higher than in control salt-stressed seedlings (ZhongQun et al. 2007). Stomatal closure increases NADPH consumption under the condition of salinity stress. In saline conditions, the plant initiates an antioxidant defense system against the harmful effect of ROS.

Among the functions of osmolytes in plant cell under salt stress are detoxification of ROS, osmotic adjustment, protection of membrane integrity, and enzyme stability (Ashraf and Foolad 2007). Compatible solutes usually occur in extremely little amounts when a plant rises under standard conditions, while its concentration increases rapidly following exposure to salt stress. For host plants and AM fungi, the accumulation of these osmolytes under salt stress is different (Rabie and Almadini 2005). Positive physiological indicator under salt stress is the accumulation of compatible solutes in the host plant. These osmolytes support the maintenance of the osmotic balance and the unity of the membrane, and this is also vital energy and nitrogen pools for utilizing by plants under salt stress (Ashraf and Foolad 2007).

13.4.4 Induced Salt Tolerance by Microbes Through Proline Accumulation

Proline is one of the organic solutes, which is the most common osmolyte in plants that plays a significant function for increasing adaptability of plant under salinity stress. There are two enzymes that synthesize proline, namely, pyronine-5 carboxylate synthase (P-5 Cs) and pyronine-5 carboxylate reductase (P-5 Cr) (Kishor et al. 1995). Proline is a major organic compound that contributes to abiotic stress osmotic adjustment (Kishor et al. 1995). Osmolytes are compatible solutes such as glycine betaine and proline that contribute to osmoregulation by maintaining cell water content (Ahmad et al. 2008; Ahanger et al. 2014). Increased concentration of osmolytes in plants inoculated with AM fungi under salt stress has been documented by several researchers (Feng et al. 2002; Al-Garni 2006; Sharifi et al. 2007; Garg and Manchanda 2009) which has been confirmed by the higher level of osmolytes in AM fungi-inoculated plants compared to non-AM fungi-inoculated plants under salt stress in *Phragmites australis*, *Lotus glaber*, *Cajanus cajan*, and *Zea mays* (Al-Garni 2006; Sannazzaro et al. 2007; Garg and Manchanda 2009; Sheng et al. 2011).

In AM fungi-inoculated garlic shoots, the accumulation of proline is significantly higher than in non-AM fungi inoculated under low to moderate saline soil, while in AM fungi-inoculated plants, the accumulation of root proline was higher than in non-AM fungi-inoculated plants, and in general, more accumulation of proline is found in roots than in shoots, since the roots are the primary salinity stress sites

(Borde et al. 2011). The changes in accumulation of proline under AM fungi may be attributed to an increase in the activity of enzymes which are synthesizing proline and a reduction in enzymes catabolizing or their limited use throughout synthesis of protein (Ahmad et al. 2010). The finding of Shekoofeh et al. (2012) on basil supports increased proline accumulations due to AM fungi inoculation. Proline is one of the typically osmolytes that accumulates within species with low salinity tolerance, which regulates salt stress by osmotic adjustment (Yoshiba et al. 1995), protects macromolecules throughout dehydration (Sanchez et al. 1998), and acts as a new radical hydroxyl scavenger (Alia et al. 1995). Sannazzaro et al. (2007) demonstrated that inoculation of *Glomus intraradices* in two *Lotus glaber* genotypes affects the accumulation of polyamine and proline under salinity stress. In mycorrhiza-inoculated *Glycine max* and *Cyamopsis tetragonoloba* legumes, Datta and Kulkarni (2014) showed an increase in proline content under salinity stress. Therefore, AM fungi in saline soils are considered to act as a bio-reducer (Singh et al. 1997).

13.5 Concluding Remarks and Future Prospects

Salinity is considered to be one of the most harmful stresses on the survival and productivity of the plant. Salinity damages could be reduced photosynthetic rate, reduced the activity of antioxidant enzymes, reduced stomatal behavior, induced ion insufficiency, decreased membrane stability, and reduced the plant water content. Shaping of salinity tolerance degrees is depending upon its plant morphology, physiology, and biochemistry. Plant microbes are important for agriculture, because they can increase plant growth and nutrient uptake through enhancing biological nitrogen fixation and other systems. Therefore, a new biological approach "plant-microbe interaction" such as arbuscular mycorrhizal (AM) fungi to address salinity problem has recently gained momentum. AM fungi participate in regulating the transport of ion and membrane proteins that control host plants' ion homeostasis. AM fungi maintain a higher Na^+:K^+ ratio that is considered as one of AM fungi's strategies to reduce salt stress and increase P, K^+, and Ca^{2+} uptake greater than injurious Na^+, thus maintaining a lesser ratio of Na^+:K^+ under salinity stress. The improvement in chlorophyll pigments caused by AM fungi is due to the increased absorption of minerals, especially magnesium, an important component of the chlorophyll molecule. One of the salt tolerance systems caused by AM fungi inoculation is increasing the activity of antioxidant enzymes in plants such as POD, SOD, and CAT, which are scavenging ROS and relieving salt stress. Another mechanism for AM salinity tolerance could be increased accumulation of non-antioxidants in mycorrhizal plant, especially by accumulating osmolytes such as proline, which supports the osmotic adjustment of plants and, consequently, improves the plant performance and its salinity tolerance. Future studies need to be concentrated in order to understand the key elements that can be used to overcome salinity stress and improve salt tolerance for many plant species by inoculating plants with AM fungi with reference to AM fungi species, which produce different antioxidants, accumulation of proline, increasing absorption of P, K^+ and Ca^{+2}, exclusion of Na^+ and Cl^-, and increasing the growth of plants by changing the physiological status of salinity-stressed plants.

Acknowledgments The first author Dr. Magdi Abdelhamid would like to express his gratitude to his colleagues at the Department of Botany, National Research Center, Egypt, for their valuable contribution in some phases of this study.

References

Abd-Alla MH, Omar SA, Karanxha S (2000) The impact of pesticides on arbuscular mycorrhizal and nitrogen-fixing symbioses in legumes. Appl Soil Ecol 14:191–200

Abdel Latef AA (2010) Changes of antioxidative enzymes in salinity tolerance among different wheat cultivars. Cereal Res Commun 38:43–55

Abdelhamid MT, Shokr M, Bekheta MA (2010) Growth, root characteristics, and leaf nutrients accumulation of four faba bean (*Vicia faba* L.) cultivars differing in their broomrape tolerance and the soil properties in relation to salinity. Commun Soil Sci Plant Anal 41:2713–2728

Abdelhamid MT, Rady M, Osman A, Abdalla M (2013a) Exogenous application of proline alleviates salt-induced oxidative stress in *Phaseolus vulgaris* L. plants. J Hortic Sci Biotech 88:439–446

Abdelhamid MT, Sadak MSH, Schmidhalter U, El-Saady A (2013b) Interactive effects of salinity stress and nicotinamide on physiological and biochemical parameters of faba bean plant. Acta Biol Colomb 18:499–510

Abouelsaad I, Renault S (2018) Enhanced oxidative stress in the jasmonic acid-deficient tomato mutant def-1 exposed to NaCl stress. J Plant Physiol 226:136–144

Adiku G, Renger M, Wessolek G, Facklam M, Hech-Bischoltz C (2001) Simulation of dry matter production and seed yield of common beans under varying soil water and salinity conditions. Agric Water Manag 47:55–68

Ahanger MA, Tyagi SR, Wani MR, Ahmad P (2014) Drought tolerance: role of organic osmolytes, growth regulators, and mineral nutrients. In: Ahmad P, Wani MR (eds) Physiological mechanisms and adaptation strategies in plants under changing environment, vol 1. Springer, New York, pp 25–55

Ahmad P, Gadgil K, Sharma S (2008) Effect of cadmium and lead on growth, biochemical parameters and uptake in *Lemna polyrrhiza* L. Plant Soil Environ 54:262–270

Ahmad P, Jaleel CA, Sharma S (2010) Antioxidative defense system, lipid peroxidation, proline metabolizing enzymes and biochemical activity in two *Morus alba* genotypes subjected to NaCl stress. Russ J Plant Physiol 57:509–517

Ahmad P, Ozturk M, Sharma S, Gucel S (2014) Effect of sodium carbonate-induced salinityalkalinity on some key osmoprotectants, protein profile, antioxidant enzymes, and lipid peroxidation in two mulberry (*Morus alba* L.) cultivars. J Plant Interact 9:460–467

Ahmad H, Hayat S, Ali M, Liu T, Cheng Z (2018) The combination of arbuscular mycorrhizal fungi inoculation (*Glomus versiforme*) and 28-homobrassinolide spraying intervals improves growth by enhancing photosynthesis, nutrient absorption, and antioxidant system in cucumber (*Cucumis sativus* L.) under salinity. Ecol Evol 8:5724–5740

Al-Garni SMS (2006) Increasing NaCl-salt tolerance of a halophytic plant *Phragmites australis* by mycorrhizal symbiosis. Am Eurasian J Agric Environ Sci 1:119–126

Alguacil MM, Hernández JA, Caravaca F, Portillo B, Roldán A (2003) Antioxidant enzyme activities in shoots from three mycorrhizal shrub species afforested in a degraded semi-arid soil. Plant Physiol 118:562–570

Alia, Prasad KVSK, Saradhi PP (1995) Effect of zinc on free radicals and proline in Brassica and Cajanus. Phytochemistry 39:45–47

Al-Karaki G, Hammad R (2001) Mycorrhizal influence on fruit yield and mineral content of tomato grown under salt stress. J Plant Nutr 24:1311–1323

Alscher RG, Erturk N, Heath LS (2002) Role of superoxide dismutases (SODs) in controlling oxidative stress in plants. J Exp Bot 53:1331–1341

Aroca R, Porcel R, Ruiz-Lozano JM (2007) How does Arbuscular mycorrhizal symbiosis regulate root hydraulic properties and plasma membrane aquaporins in *Phaseolus vulgaris* under drought, cold or salinity stresses? New Phytol 173:808–816

Aroca R, Ruiz-Lozano JM, Zamarreno AM, Paz JA, García-Mina JM, Pozo MJ, López-Ráez JA (2013) Arbuscular mycorrhizal symbiosis influences strigolactone production under salinity and alleviates salt stress in lettuce plants. J Plant Physiol 170:47–55

Asghari HR, Marschner P, Smith SE, Smith FA (2005) Growth response of *Atriplex nummularia* to inoculation with arbuscular mycorrhizal fungi at different salinity levels. Plant Soil 273:245–256

Ashraf M (1994) Breeding for salinity tolerance in plants. Crit Rev Plant Sci 13:17–42

Ashraf M, Foolad MR (2007) Roles of glycine betaine and proline in improving plant abiotic stress resistance. Environ Exp Bot 59:207–216

Awad N, Turky A, Abdelhamid M, Attia M (2012) Ameliorate of environmental salt stress on the growth of *Zea mays* L plants by exopolysaccharides producing bacteria. J Appl Sci Res 8:2033–2044

Bach E, dos Santos Seger GD, de Carvalho Fernandes G, Lisboa BB, Passaglia LMP (2016) Evaluation of biological control and rhizosphere competence of plant growth promoting bacteria. Appl Soil Ecol 99:141–149

Bago B, Pfeffer PE, Shachar-Hill Y (2000) Carbon metabolism and transport in arbuscular mycorrhiza. Plant Physiol 124:949–958

Bargaz A, Nassar RMA, Rady MM, Gaballah MS, Thompson SM, Brestic M, Schmidhalter U, Abdelhamid MT (2016) Improved salinity tolerance by phosphorus fertilizer in two *Phaseolus vulgaris* recombinant inbred lines contrasting in their phosphorus deficiency sensitivity. J Agron Crop Sci 202:497–507

Basu S, Rabara RC, Negi S (2018) AMF: the future prospect for sustainable agriculture. Physiol Mol Plant Path 102:36–45

Bekheta MA, Abdelhamid MT, El-Morsi AA (2009) Physiological response of *Vicia faba* to prohexadione–calcium under saline conditions. Planta Daninha 27:769–779

Bharti N, Barnawal D, Awasthi A, Yadav A, Kalra A (2014) Plant growth promoting rhizobacteria alleviate salinity induced negative effects on growth, oil content and physiological status in *Mentha arvensis*. Acta Physiol Plant 36:45–60

Bockheim JG, Gennadiyev AN (2000) The role of soil-forming processes in the definition of taxa in soil taxonomy and the world soil reference base. Geoderma 95:53–72

Borde MY, Dudhane MP, Jite PK (2010) AM fungi influences the photosynthetic activity, growth and antioxidant enzymes in *Allium sativum* L. under salinity condition. Not Sci Biol 2:64–71

Borde MY, Dudhane MP, Jite PK (2011) Growth photosynthetic activity and antioxidant responses of mycorrhizal and non-mycorrhizal bajra (*Pennisetum glaucum*) crop under salinity stress condition. Crop Prot 30:265–271

Bothe H (2012) Arbuscular mycorrhiza and salt tolerance of plants. Symbiosis 58:7–16

Cantrell IC, Linderman RG (2001) Preinoculation of lettuce and onion with VA mycorrhizal fungi reduces deleterious effects of soil salinity. Plant Soil 233:269–281

Chen J, Zhang H, Zhang X, Tang M (2017) Arbuscular mycorrhizal symbiosis alleviates salt stress in black locust through improved photosynthesis, water status, and KC/NaC homeostasis. Front Plant Sci 8:1739. https://doi.org/10.3389/fpls.2017.01739

Colla G, Rouphael Y, Cardarelli M, Tullio M, Rivera CM, Rea E (2008) Alleviation of salt stress by arbuscular mycorrhizal in zucchini plants grown at low and high phosphorus concentration. Biol Fertil Soils 44:501–509

Das P, Nutan KK, Singla-Pareek SL, Pareek A (2015) Understanding salinity responses and adopting "omics-based" approaches to generate salinity tolerant cultivars of rice. Front Plant Sci 6:712. https://doi.org/10.3389/fpls.2015.00712
Datta P, Kulkarni M (2014) Arbuscular mycorrhizal colonization enhances biochemical status in and mitigates adverse salt effect on two legumes. Not Sci Biol 6:381–393
Dawood MG, Taie HAA, Nassar RMA, Abdelhamid MT, Schmidhalter U (2014a) The changes induced in the physiological, biochemical and anatomical structure of *Vicia faba* by the exogenous application of proline under seawater stress. S Afr J Bot 93:54–63
Dawood MG, Abdelhamid MT, Schmidhalter U (2014b) Potassium fertiliser enhances the salt-tolerance of common bean (*Phaseolus vulgaris* L.). J Hortic Sci Biotech 89:185–192
Dawood MG, El-Metwally IM, Abdelhamid MT (2016) Physiological response of lupine and associated weeds grown at salt-affected soil to α-tocopherol and hoeing treatments. Gesunde Pflanzen 68:117–127
De Oliveira DFB, Endres L, Silva JV, Clemente PRA (2017) Pre-colonized seedlings with arbuscular mycorrhizal fungi: an alternative for the cultivation of *Jatropha curcas* L. in salinized soils. Theor Exp Plant Physiol 29:129–142
Elhindi K, Sharaf El Din A, Abdel-Salam E, Elgorban A (2016) Amelioration of salinity stress in different basil (*Ocimum basilicum* L.) varieties by vesicular arbuscular mycorrhizal fungi. Acta Agric Scand B Soil Plant Sci 66:583–592
Elhindi KM, Sharaf El-Din A, Elgorban AM (2017) The impact of arbuscular mycorrhizal fungi in mitigating salt-induced adverse effects in sweet basil (*Ocimum basilicum* L.). Saudi J Biol Sci 24:170–179
El-Lethy SR, Abdelhamid MT, Reda F (2013) Effect of potassium application on wheat (*Triticum aestivum* L.) cultivars grown under salinity stress. World Appl Sci J 26:840–850
Ellis J (2017) Can plant microbiome studies lead to effective biocontrol of plant diseases? Mol Plant Microbe. https://doi.org/10.1094/MPMI-12-16-0252-CR
El-Mashad AAA, Mohamed HI (2012) Brassinolide alleviates salt stress and increases antioxidant activity of cowpea plants (*Vigna sinensis*). Protoplasma 249:625–635
El-Metwally IM, Ali OAM, Abdelhamid MT (2015) Response of wheat (*Triticum aestivum* L.) and associated grassy weeds grown in salt-affected soil to effects of graminicides and indole acetic acid. Agriculture 61:1–11
Epstein E (1977) Genetic potentials for solving problems of soil mineral stress: adaptation of crops to salinity. In: Wright MJ (ed) Plant adaptation to mineral stress in problem soils. Cornell University, Ithaca, pp 73–123
Errakhi R, Bouteau F, Barakate M, Lebrihi A (2016) Isolation and characterization of antibiotics produced by Streptomyces J-2 and their role in biocontrol of plant diseases, especially grey mould. In: Compant S, Mathieu F (eds) Biocontrol of major grapevine diseases: leading research. CAB International, Wallingford, pp 76–83
Estrada B, Aroca R, Maathuis FJ, Barea JM, Ruiz-Lozano JM (2013) Arbuscular mycorrhizal fungi native from a Mediterranean saline area enhance maize tolerance to salinity through improved ion homeostasis. Plant Cell Environ 36:1771–1782
Evelin H, Kapoor R, Giri B (2009) Arbuscular mycorrhizal fungi in alleviation of salt stress: a review. Ann Bot 104:1263–1280
Evelin H, Giri B, Kapoor R (2012) Contribution of *Glomus intraradices* inoculation to nutrient acquisition and mitigation of ionic imbalance in NaCl-stressed *Trigonella foenum-graecum*. Mycorrhiza 22:203–217
Fan X, Chang W, Fenga F, Song F (2018) Responses of photosynthesis-related parameters and chloroplast ultrastructure to atrazine in alfalfa (*Medicago sativa* L.) inoculated with arbuscular mycorrhizal fungi. Ecotoxicol Environ Saf 166:102–108
Feng G, Zhang FS, Li XL, Tian CY, Tang C, Rengel Z (2002) Improved tolerance of maize plants to salt stress by arbuscular mycorrhiza is related to higher accumulation of soluble sugars in roots. Mycorrhiza 12:185–190

Fileccia V, Ruisi P, Ingraffia R, Giambalvo D, Frenda AS, Martinelli F (2017) Arbuscular mycorrhizal symbiosis mitigates the negative effects of salinity on durum wheat. PLoS One 12(9):e0184158. https://doi.org/10.1371/journal.pone.0184158

Flowers TJ, Colmer TD (2015) Plant salt tolerance: adaptations in halophytes. Ann Bot 115:327–331

Garg N, Bhandari P (2016) Silicon nutrition and mycorrhizal inoculations improve growth, nutrient status, K^+/Na^+ ratio and yield of *Cicer arietinum* L. genotypes under salinity stress. Plant Growth Regul 78:371–387

Garg N, Manchanda G (2009) Role of arbuscular mycorrhizae in the alleviation of ionic, osmotic and oxidative stresses induced by Mycorrhiza salinity in *Cajanus cajan* (L.) Millsp. (pigeonpea). J Agron Crop Sci 195:110–123

Garg N, Singla P (2016) Stimulation of nitrogen fixation and trehalose biosynthesis by naringenin (Nar) and arbuscular mycorrhiza (AM) in chickpea under salinity stress. Plant Growth Regul 80:5–22

Gharsallah C, Fakhfakh H, Grubb D, Gorsane F (2016) Effect of salt stress on ion concentration, proline content, antioxidant enzyme activities and gene expression in tomato cultivars. AoB Plants. https://doi.org/10.1093/aobpla/plw055

Ghorbanli H, Ebrahimzadeh M, Sharifi M (2004) Effects of NaCl and mycorrhizal fungi on antioxidative enzymes in soybean. Biol Plant 48:575–581

Gianinazzi S, Gollotte A, Binet M, van Tuinen D, Redecker D, Wipf D (2010) Agroecology: the key role of arbuscular mycorrhizas in ecosystem services. Mycorrhiza 20:519–530

Giri B, Mukerji KG (2004) Mycorrhizal inoculants alleviate salt stress in *Sesbania aegyptiaca* and *Sesbania grandiflora* under field conditions: evidence for reduced sodium and improved magnesium uptake. Mycorrhiza 14:307–312

Giri B, Kapoor R, Mukerji KG (2007) Improved tolerance of *Acacia nilotica* to salt stress by arbuscular mycorrhiza, *Glomus fasciculatum* may be partly related to elevated K/Na ratios in root and shoot tissues. Microb Ecol 54:753–760

Gopal S, Chandrasekaran M, Shagol C, Kim K, Sa T (2012) Spore associated bacteria (SAB) of arbuscular mycorrhizal fungi (AMF) and plant growth promoting rhizobacteria (PGPR) increase nutrient uptake and plant growth under stress conditions. Korean J Soil Sci Fertil 45:582–592

Grieve CM, Grattan SR, Maas EV (2008) Plant salt tolerance. In: Wallender WW, Tanj KK (eds) Agricultural salinity assessment and management. American Society of Civil Engineers, Reston, pp 405–459

Gupta ML, Khaliq A, Pandey R, Shukla RS, Singh HK, Kumar S (2000) Vesicular–arbuscular mycorrhizal fungi associated with *Ocimum* spp. J Herbs Spices Med Plants 7:57–63

Haas D, Défago G (2005) Biological control of soil-borne pathogens by fluorescent pseudomonads. Nat Rev Microbiol 3:307–319

Haghighi M, Mohammadnia S, Attai Z, Pessarakli M (2017) Effects of mycorrhiza inoculation on cucumber growth irrigated with saline water. J Plant Nutr 40:128–137

Hajiboland R, Joudmand A, Fotouhi K (2009) The K/Na replacement and function of antioxidant defense system in sugar beet (*Beta vulgaris* L.) cultivars. Acta Agric Scand B Soil Plant Sci 59:246–259

Hajiboland R, Aliasgharzadeh S, Laiegh F, Poschenrieder C (2010) Colonization with arbuscular mycorrhizal fungi improves salinity tolerance of tomato (*Solanum lycopersicum* L.) plants. Plant Soil 331:313–327

Hameed A, Egamberdieva D, Abd Allah EF, Hashem A, Kumar A, Ahmad P (2014) Salinity stress and arbuscular mycorrhizal symbiosis in plants. In: Miransari M (ed) Use of microbes for the alleviation of soil stresses, vol 1. Springer, New York, pp 139–159

Hammer EC, Nasr H, Pallon J, Olsson PA, Wallander H (2011) Elemental composition of arbuscular mycorrhizal fungi at high salinity. Mycorrhiza 21:117–129

Hanin M, Ebel C, Ngom M, Laplaze L, Masmoudi K (2016) New insights on plant salt tolerance mechanisms and their potential use for breeding. Front Plant Sci 7:1787. https://doi.org/10.3389/fpls.2016.01787

Hashem A, Abd_Allah EF, Alqarawi AA, Wirth S, Egamberdieva D (2016a) Arbuscular mycorrhizal fungi alleviate salt stress in lupine (*Lupinus termis* Forsik) through modulation of antioxidant defense systems and physiological traits. Legume Res 39:198–207

Hashem AEF, Abd_Allah EF, Alqarawi AA, Al-Huqail AA, Shah MA (2016b) Induction of osmoregulation and modulation of salt stress in *Acacia gerrardii* Benth. by arbuscular mycorrhizal fungi and *Bacillus subtilis* (BERA 71). BioMed Res Int 2016, 6294098, 11 p. https://doi.org/10.1155/2016/6294098

Hayat S, Hayat Q, Alyemeni MN, Wani AS, Pichtel J, Ahmad A (2012) Role of proline under changing environments: a review. Plant Signal Behav 7:1456–1466

He Z, He C, Zhang Z, Zou Z, Wang H (2007) Changes of antioxidative enzymes and cell membrane osmosis in tomato colonized by arbuscular mycorrhizae under NaCl stress. Colloids Surf B Biointerfaces 59:128–133

Hegazi AM, El-Shraiy AM, Ghoname AA (2017) Mitigation of salt stress negative effects on sweet pepper using arbuscular mycorrhizal fungi (AMF), *Bacillus megaterium* and Brassinosteroids (BRs). Gesunde Pflanzen 69:91–102

Hellal FA, Abdelhameid MT, Abo-Basha DM, Zewainy RM (2012) Alleviation of the adverse effects of soil salinity stress by foliar application of silicon on faba bean (*Vicia faba* L.). J Appl Sci Res 8:4428–4433

Hernandez JA, Olmos E, Corpas FJ, Sevilla F, Delrio LA (1995) Salt induced oxidative stress in chloroplast of pea plants. Plant Sci 105:151–167

Ismail AM, Horie T (2017) Genomics, physiology, and molecular breeding approaches for improving salt tolerance. Annu Rev Plant Biol 68:405–434. https://doi.org/10.1146/annurev-arplant-042916-040936

Jahromi F, Aroca R, Porcel R, Ruiz-Lozano JM (2008) Influence of salinity on the in vitro development of *Glomus intraradices* and on the in vivo physiological and molecular responses of mycorrhizal lettuce plants. Microb Ecol 55:45–53

Jiang M, Zhang J (2002) Water stress-induced abscisic acid accumulation triggers the increased generation of reactive oxygen species and up-regulates the activities of antioxidant enzymes in maize-leaves. J Exp Bot 53:2401–2410

Kaya C, Ashraf M, Sonmez O, Aydemir S, Tuna AL, Cullu MA (2009) The influence of arbuscular mycorrhizal colonization on key growth parameters and fruit yield of pepper plants grown at high salinity. Sci Hortic 121:1–6

Khalloufia M, Martínez-Andújara C, Lachaâlb M, Karray-Bouraouib N, Pérez-Alfoceaa F, Albacete A (2017) The interaction between foliar GA3 application and arbuscular mycorrhizal fungi inoculation improves growth in salinized tomato (*Solanum lycopersicum* L.) plants by modifying the hormonal balance. J Plant Physiol 214:134–144

Khan S, Hanjra MA (2008) Sustainable land and water management policies and practices: a pathway to environmental sustainability in large irrigation systems. Land Degrad Dev 19:469–487

Kingsbury RW, Epstein E (1984) Selection for salt resistant spring wheat. Crop Sci 24:310–315

Kishor PBK, Hong Z, Miao GH, Hu CAA, Verma DPS (1995) Over expression of Δ1-pyrroline-5-carboxylate synthetase increase proline production and confers osmotolerance in transgenic plants. Plant Physiol 108:1387–1394

Kohler J, Hernández JA, Caravaca F, Roldan A (2009) Induction of antioxidant enzymes is involved in the greater effectiveness of a PGPR versus AM fungi with respect to increasing the tolerance of lettuce to severe salt stress. Environ Exp Bot 65:245–252

Kour D, Rana KL, Verma P, Yadav AN, Kumar V, Dhaliwal HS (2017) Biofertilizers: eco-friendly technologies and bioresources for sustainable agriculture. In: Proceeding of international conference on innovative research in engineering science and technology, IREST/PP/014

Landi S, Hausman J-F, Guerriero G, Esposito S (2017) Poaceae vs. abiotic stress: focus on drought and salt stress, recent insights and perspectives. Front Plant Sci 8:1214. https://doi.org/10.3389/fpls.2017.01214

Lin J, Wang Y, Sun S, Mu C, Yan X (2017) Effects of arbuscular mycorrhizal fungi on the growth, photosynthesis and photosynthetic pigments of *Leymus chinensis* seedlings under salt-alkali stress and nitrogen deposition. Sci Total Environ 576:234–241

Liu S, Guo X, Feng G, Maimaitiaili B, Fan J, He X (2016) Indigenous arbuscular mycorrhizal fungi can alleviate salt stress and promote growth of cotton and maize in saline fields. Plant Soil 398:195–206

Mahajan S, Tuteja N (2005) Cold, salinity and drought stresses: an overview. Arch Biochem Biophys 444:139–158

Manchanda G, Garg N (2007) Endomycorrhizal and rhizobial symbiosis: how much do they share? J Plant Interact 2:79–88

Mardukhi B, Rejali F, Daei G, Ardakani M, Malakouti MJ, Miransari M (2015) Mineral uptake of mycorrhizal wheat (*Triticum aestivum* L.) under salinity stress. Commun Soil Sci Plant Anal 46:343–357

Mehdy MC (1994) Active oxygen species in plant defense against pathogens. Plant Physiol 105:467–472

Metternicht GI, Zinck JA (2003) Remote sensing of soil salinity: potentials and constraints. Remote Sens Environ 85:1–20

Metwally RA, Abdelhameed RE (2018) Synergistic effect of arbuscular mycorrhizal fungi on growth and physiology of salt-stressed *Trigonella foenum*-graecum plants. Biocatal Agric Biotechnol 16:538–544

Miransari M (2010) Contribution of arbuscular mycorrhizal symbiosis to plant growth under different types of soil stress. Plant Biol 12:563–569

Miransari M, Bahrami H, Rejali F, Malakouti M (2008) Using arbuscular mycorrhiza to alleviate the stress of soil compaction on wheat (*Triticum aestivum* L.) growth. Soil Biol Biochem 40:1197–1206

Munns R (2005) Genes and salt tolerance: bringing them together. New Phytol 167:645–663

Munns R, Gilliham M (2015) Salinity tolerance of crops—what is the cost? New Phytol 208:668–673

Munns R, Tester M (2008) Mechanisms of salinity tolerance. Annu Rev Plant Biol 59:651–681

Munns R, James RA, Läuchli A (2006) Approaches to increasing the salt tolerance of wheat and other cereals. J Exp Bot 57:1025–1043

Namdari A, Arani AB, Moradi A (2018) Arbuscular mycorrhizal (*Funneliformis mosseae*) improves alfalfa (*Medicago sativa* L.) re-growth ability in saline soil through enhanced nitrogen remobilization and improved nutritional balance. J Cent Eur Agri 19:166–183

Negrão S, Schmöckel SM, Tester M (2017) Evaluating physiological responses of plants to salinity stress. Ann Bot 119:1–11

Nongpiur RC, Singla-Pareek SL, Pareek A (2016) Genomics approaches for improving salinity stress tolerance in crop plants. Curr Genomics 17:343–357

Nunez M, Mazzafera P, Mazorra LM, Siqueira WJ, Zullo MAT (2003) Influence of brassinosteroid analogue on antioxidant enzymes in rice grown in culture medium with NaCl. Biol Plant 47:67–70

Orabi SA, Abdelhamid MT (2016) Protective role of α-tocopherol on two *Vicia faba* cultivars against seawater-induced lipid peroxidation by enhancing capacity of anti-oxidative system. J Saudi Soc Agric Sci 15:145–154

Ouda S, Noreldin T, Mounzer O, Abdelhamid MT (2015) CropSyst model for wheat irrigation water management with fresh and poor quality water. J Water Land Devel 27:41–50

Pal KK, Gardener BM (2006) Biological control of plant pathogens. Plant Health Instructor 2:1117–1142

Pandey R, Garg N (2017) High effectiveness of *Rhizophagus irregularis* is linked to superior modulation of antioxidant defence mechanisms in *Cajanus cajan* (L.) Millsp. genotypes grown under salinity stress. Mycorrhiza 27:669–682

Pankievicz V, Amaral FP, Santos KF, Agtuca B, Xu Y, Schueller MJ, Arisi ACM, Steffens M, Souza EM, Pedrosa FO (2015) Robust biological nitrogen fixation in a model grass–bacterial association. Plant J 81:907–919

Parida SK, Das AB (2005) Salt tolerance and salinity effects on plants. Ecotoxicol Environ Saf 60:324–349

Patel D, Saraf M (2013) Influence of soil ameliorants and microflora on induction of antioxidant enzymes and growth promotion of *Jatropha curcas* L. under saline condition. Eur J Soil Biol 55:47–54

Pollastri S, Savvides A, Pesando M, Lumini E, Volpe MG, Ozudogru EA, Faccio A, De Cunzo F, Michelozzi M, Lambardi M, Fotopoulos V, Loreto F, Centritto M, Balestrini R (2018) Impact of two arbuscular mycorrhizal fungi on *Arundo donax* L. response to salt stress. Planta 247:573–585

Porcel R, Aroca R, Ruiz-Lozano JM (2012) Salinity stress alleviation using arbuscular mycorrhizal fungi. A review. Agron Sustain Dev 32:181–200

Porcel R, Aroca R, Azcon R, Ruiz-Lozano JM (2016) Regulation of cation transporter genes by the arbuscular mycorrhizal symbiosis in rice plants subjected to salinity suggests improved salt tolerance due to reduced Na^+ root-to-shoot distribution. Mycorrhiza 26:673–684

Prasad R, Bhola D, Akdi K, Cruz C, Sairam KVSS, Tuteja N, Varma A (2017) Introduction to mycorrhiza: historical development. In: Varma A, Prasad R, Tuteja N (eds) Mycorrhiza. Springer, Cham, pp 1–7

Qadir M, Quillérou E, Nangia V, Murtaza G, Singh M, Thomas RJ, Drechsel P, Noble AD (2014) Economics of salt-induced land degradation and restoration. Nat Resour Forum 38:282–295

Rabie GH (2005) Influence of arbuscular mycorrhizal fungi and kinetin on the response of mungbean plants to irrigation with seawater. Mycorrhiza 15:225–230

Rabie GH, Almadini AM (2005) Role of bioinoculants in development of salt-tolerance of *Vicia faba* plants under salinity stress. Afr J Biotechnol 4:210–220

Rady MM, Sadak MS, El-Lethy SR, Abdelhamid EM, Abdelhamid MT (2015) Exogenous α-tocopherol has a beneficial effect on *Glycine max* (L.) plants irrigated with diluted sea water. J Hortic Sci Biotech 90:195–202

Rady MM, Semida WM, Hemida KA, Abdelhamid MT (2016a) The effect of compost on growth and yield of *Phaseolus vulgaris* plants grown under saline soil. Int J Recycl Org Waste Agr 5:311–321

Rady MM, Mounzer OH, Alarcón JJ, Abdelhamid MT, Howladar SM (2016b) Growth, heavy metal status and yield of salt-stressed wheat (*Triticum aestivum* L.) plants as affected by the integrated application of bio-, organic and inorganic nitrogen-fertilizers. J Appl Bot Food Qual 89:21–28

Rai MK, Kalia RK, Singh R, Gangola MP, Dhawan AK (2011) Developing stress tolerant plants through in vitro selection – an overview of the recent progress. Environ Exp Bot 71:89–98

Ramos AC, Facanha AR, Palma LM, Okorokov LA, Cruz ZMA, Silva AG (2011) An outlook on ion signaling and ionome of mycorrhizal symbiosis. Braz J Plant Physiol 23:79–89

Renault S (2012) Salinity tolerance of *Cornus sericea* seedlings from three provenances. Acta Physiol Plant 34:1735–1746

Rivero J, Alvarez D, Flors V, Azcón-Aguilar C, Pozo MJ (2018) Root metabolic plasticity underlies functional diversity in mycorrhiza-enhanced stress tolerance in tomato. New Phytol 220:1322–1336. https://doi.org/10.1111/nph.15295

Roberts JKM, Linker CS, Benoit AG, Jardetzky O, Nieman RH (1984) Salt Stimulation of Phosphate Uptake in Maize Root Tips Studied by 31P Nuclear Magnetic Resonance. Plant Physiol 75:947–950

Roy SJ, Negrão S, Tester M et al (2014) Salt resistant crop plants. Curr Opin Biotechnol 26:115–124. https://doi.org/10.1016/j.copbio.2013.12.004

Ruiz-Lozano JM, Azcon R (1995) Hyphal contribution to water uptake in mycorrhizal plants as affected by the fungal species and water status. Physiol Plant 95:472–478

Sadak MS, Abdelhamid MT (2015) Influence of amino acids mixture application on some biochemical aspects, antioxidant enzymes and endogenous polyamines of *Vicia faba* plant grown under seawater salinity stress. Gesunde Pflanzen 67:119–129

Sadak MSH, Abdelhamid MT, Schmidhalter U (2015) Effect of foliar application of amino acids on plant yield and physiological parameters in bean plants irrigated with seawater. Acta Biolo Colomb 20:141–152

Sallakua G, Sandén H, Babaj I, Kaciu S, Balliu A, Rewald B (2019) Specific nutrient absorption rates of transplanted cucumber seedlings are highly related to RGR and influenced by grafting method, AMF inoculation and salinity. Sci Hortic 243:177–188

Sanchez FJ, Manzanares M, De Andres EF, Tenorio JL, Ayerbe L (1998) Turgor maintenance, osmotic adjustment and soluble sugar and proline accumulation in 49 pea cultivars in response to water stress. Field Crop Res 59:225–235

Sannazzaro AI, Echeverria M, Alberto EO, Ruiz OA, Menendez AB (2007) Modulation of polyamine balance in *Lotus glaber* by salinity and arbuscular mycorrhiza. Plant Physiol Biochem 45:39–46

Satir NY, Ortas I, Satir O (2016) The influence of mycorrhizal species on sour orange (*Citrus aurantium* L.) growth under saline soil conditions. Pak J Agri Sci 53:399–406

Saxena B, Shukla K, Giri B (2017) Arbuscular mycorrhizal fungi and tolerance of salt stress in plants. In: Wu Q-S (ed) Arbuscular mycorrhizas and stress tolerance of plants. Springer Nature, Singapore, pp 67–97

Schubler A, Schwarzott D, Walker C (2001) A new fungal phylum, the Glomeromycota: phylogeny and evolution. Mycol Res 105:1413–1421

Semida WM, Taha RS, Abdelhamid MT, Rady MM (2014) Foliar-applied α-tocopherol enhances salt-tolerance in *Vicia faba* L. plants grown under saline conditions. S Afr J Bot 95:24–31

Serrano R, Rodriguez-Navarro A (2001) Ion homeostasis during salt stress in plants. Curr Opin Cell Biol 13:399–404

Sharifi M, Ghorbanli M, Ebrahimzadeh H (2007) Improved growth of salinity-stressed soybean after inoculation with pre-treated mycorrhizal fungi. J Plant Physiol 164:1144–1151

Shekoofeh E, Sepideh H, Roya R (2012) Role of mycorrhizal fungi and salicylic acid in salinity tolerance of *Ocimum basilicum* resistance to salinity. Afr J Biotechnol 11:2223–2235

Sheng M, Tang M, Chan H, Yang B, Zhang F, Huang Y (2008) Influence of arbuscular mycorrhizae on photosynthesis and water status of maize plants under salt stress. Mycorrhiza 18:287–296

Sheng M, Tang M, Zhang F, Huang Y (2011) Influence of arbuscular mycorrhiza on organic solutes in maize leaves under salt stress. Mycorrhiza 21:423–430

Shokri S, Maadi B (2009) Effects of arbuscular mycorrhizal fungus on the mineral nutrition and yield of *Trifolium alexandrinum* plants under salinity stress. J Agron 8:79–83

Singh RP, Choudhary A, Gulati A, Dahiya HC, Jaiwal PK, Sengar RS (1997) Response of plants to salinity in interaction with other abiotic and biotic factors. In: Jaiwal PK, Singh RP, Gulati A (eds) Strategies for improving salt tolerance in higher plants. Science, Enfield, pp 25–39

Talaat NB, Shawky BT (2012) 24-Epibrassinolide ameliorates the saline stress and improves the productivity of wheat (*Triticum aestivum* L.). Environ Exp Bot 82:80–88

Talaat NB, Ghoniem AE, Abdelhamid MT, Shawky BT (2015) Effective microorganisms improve growth performance, alter nutrients acquisition and induce compatible solutes accumulation in common bean (*Phaseolus vulgaris* L.) plants subjected to salinity stress. Plant Growth Regul 75:281–295

Tofighi C, Khavari-Nejad RA, Najafi F, Razavi K, Rejali F (2017) Brassinosteroid (BR) and arbuscular mycorrhizal (AM) fungi alleviate salinity in wheat. J Plant Nutr 40:1091–1098

Tomar NS, Agarwal RM (2013) Influence of treatment of *Jatropha curcas* L. leachates and potassium on growth and phytochemical constituents of wheat (*Triticum aestivum* L.). Am J Plant Sci 4:1134–1150

Upreti KK, Bhatt RM, Panneerselvam P, Varalakshmi LR (2016) Morpho-physiological responses of grape rootstock ‘Dogridge’ to arbuscular mycorrhizal fungi inoculation under salinity stress. Int J Fruit Sci 16:191–209

Van Breusegem F, Vranova E, Dat JF, Inze (2001) The role of active oxygen species in plant signal transduction. Plant Sci 161:405–414

Verma P, Yadav A, Khannam K, Panjiar N, Kumar S, Saxena A, Suman A (2015) Assessment of genetic diversity and plant growth promoting attributes of psychrotolerant bacteria allied with wheat (*Triticum aestivum*) from the northern hills zone of India. Ann Microbiol 65:1885–1899

Wang Y, Wang M, Li Y, Wu A, Huang J (2018) Effects of arbuscular mycorrhizal fungi on growth and nitrogen uptake of *Chrysanthemum morifolium* under salt stress. PLoS One 13(4): e0196408. https://doi.org/10.1371/journal.pone.0196408

Wu QS, Zou YN, He XH (2010) Contributions of arbuscular mycorrhizal fungi to growth, photosynthesis, root morphology and ionic balance of citrus seedlings under salt stress. Acta Physiol Plant 32:297–304

Yadav AN, Kumar R, Kumar S, Kumar V, Sugitha TCK, Singh B, Chauhan VS, Dhaliwal HS, Saxena AK (2017a) Beneficial microbiomes: biodiversity and potential biotechnological applications for sustainable agriculture and human health. J Appl Biol Biotechnol 5:45–57

Yadav AN, Verma P, Kumar S, Kumar V, Kumar M, Singh BP, Saxena AK, Dhaliwal HS (2017b) Actinobacteria from rhizosphere: molecular diversity, distributions and potential biotechnological applications. In: Singh BP, Gupta VK, Passari AK (eds) New and future developments in microbial biotechnology and bioengineering actinobacteria: diversity and biotechnological applications. Elsevier, Atlanta, pp 13–41

Yadav AN, Verma P, Kour D, Rana KL, Kumar V, Singh B, Chauhan VS, Sugitha TCK, Saxena AK, Dhaliwal HS (2017c) Plant microbiomes and its beneficial multifunctional plant growth promoting attributes. Int J Environ Sci Nat Resour 3:1–8

Yamato M, Ikeda S, Iwase K (2008) Community of arbuscular mycorrhizal fungi in coastal vegetation on Okinawa Island and effect of the isolated fungi on growth of sorghum under salt-treated conditions. Mycorrhiza 18:241–249

Yano-Melo AM, Saggin OJ, Maia LC (2003) Tolerance of mycorrhized banana (Musa sp. cv. Pacovan) plantlets to saline stress. Agric Ecosyst Environ 95:343–348

Yoshiba Y, Kiyosue T, Katagiri T, Ueda H, Mizoguchi T, Yamaguchishinozaki K, Wada K, Harada Y, Shinozaki K (1995) Correlation between the induction of a gene for delta (1)-pyrroline-5-carboxylate synthetase and the accumulation of proline in *Arabidopsis thaliana* under osmotic stress. Plant J 7:751–760

Zhang Y, Wang P, Wu Q-H, Zou Y-N, Bao Q, Wu Q-S (2017) Arbuscular mycorrhizas improve plant growth and soil structure in trifoliate orange under salt stress. Arch Agron Soil Sci 63:491–500

Zhang W, Wang C, Lu T, Zheng Y (2018) Cooperation between arbuscular mycorrhizal fungi and earthworms promotes the physiological adaptation of maize under a high salt stress. Plant Soil 423:125–140

ZhongQun H, ChaoXing H, Zhang ZB, Zou ZR, Wang HS (2007) Changes of antioxidative enzymes and cell membrane osmosis in tomato colonized by arbuscular mycorrhizae under NaCl stress. Colloids Surf B Biointerfaces 59(2):128–133

Zhu X, Song F, Liu S, Liu F (2016) Role of arbuscular mycorrhiza in alleviating salinity stress in wheat (*Triticum aestivum* L.) grown under ambient and elevated CO_2. J Agro Crop Sci 202:486–496

Zhu XQ, Tang M, Zhang HQ (2017) Arbuscular mycorrhizal fungi enhanced the growth, photosynthesis, and calorific value of black locust under salt stress. Photosynthetica 55:378–385

Zhu X, Cao Q, Sun L, Yang X, Yang W, Zhang H (2018) Stomatal conductance and morphology of arbuscular mycorrhizal wheat plants response to elevated $CO_2$and NaCl stress. Front Plant Sci 9:1363. https://doi.org/10.3389/fpls.2018.01363

Zuccarini P, Okurowska P (2008) Effects of mycorrhizal colonization and fertilization on growth and photosynthesis of sweet basil under salt stress. J Plant Nutr 31:497–513

Chapter 14
Plant Growth-Promoting Bacteria for Improving Crops Under Saline Conditions

Leila Bensidhoum and El-hafid Nabti

Abstract Elevated concentration of salinity hinders plant life from germination to final step. So, numerous mechanisms are inhibited, such as phytohormone synthesis, nutrient uptake, root and shoot regulation, photosynthesis, and DNA replication. It is well documented that the plant growth-promoting bacteria/rhizobacteria (PGPB/R) colonize plants and stimulate plant growth by different processes. In this way, halotolerant or halophilic PGPB become important; therefore, their capability resides in the rise of resistance toward high salt levels by eminent osmoregulatory mechanism to preserve regular cell functions. This kind of bacteria provides plants with many activities to challenge osmotic stress such as phytohormone synthesis [auxins, gibberellic acid (GA), cytokinins (CK), abscisic acid (ABA)], phosphate solubilization, and siderophore production. Adding to that, ACC deaminase is an important enzyme implicated in ethylene reduction. Moreover, PGPB are also capable of colonizing plants endophytically and synthetizing different antimicrobial metabolites like antifungal and antibacterial substances and plant systemic resistance. PGPB are similar in restoration and bioremediation of soils.

Keywords Bacteria–rhizobacteria · Plant growth promotion · Osmoregulation · Salinity

14.1 Introduction

Salinity is a major edaphic stress that affects negatively agricultural land; it represents a serious problem since salinization can suppress plant development, disturb soil fertility, and reduce their microbial biomass. Salts in the soil occur as ions; they are released either from weathering of minerals in the soil or applied through

L. Bensidhoum · El.-h. Nabti (✉)
Faculté des Sciences de la Nature et de la Vie, Laboratoire de Maitrise des Energies Renouvelables, Université de Bejaia, Bejaia, Algeria

B. Giri, A. Varma (eds.), *Microorganisms in Saline Environments: Strategies and Functions*, Soil Biology 56, https://doi.org/10.1007/978-3-030-18975-4_14

irrigation water or as fertilizers or sometimes migrate upward in the soil from shallow groundwater (Shrivastava and Kumar 2015). When precipitation is insufficient to leach ions from the soil profile, salts accumulate in the soil, resulting in soil salinity (Blaylock 1994). The presence of salt ions in the soil, which are soluble in water like sodium (Na^+), bicarbonate (HCO_3^-), magnesium (Mg^{+2}), sulfate (SO_4^{-2}), potassium (K^+), chloride (Cl^-), calcium (Ca^+), and carbonate (CO_3^{-2}), is called salinization. In the majority of the salts, sodium chloride is the major constituent, and their chloride ions are considered toxic for plants (Numan et al. 2018). Irrigated agriculture is a major human activity, which often leads to secondary salinization of land and water resources in arid and semiarid conditions (Shrivastava and Kumar 2015). Globally, more than 770,000 km^2 of land is salt affected by secondary salinization: 20% of irrigated land and about 2% of dry agricultural land (FAO 2000), with a resultant monetary loss of US$12 billion in agricultural production (Shabala 2013). In 2004, the United Nations estimated that at the end of 2050, the human population would cross 10 billion. The loss in agricultural production due to soil salinity is directly in conflict with nutrient demand of this growing population and the challenge of maintaining the world food supplies (Joseph and Jini 2010).

Salinity was categorized as primary and secondary based on their source of cause. Primary salinity results from weathering of rocks, capillary rise from shallow brackish groundwater, intrusion of seawater along the coast, salt-laden sand blown by sea winds, and impeded drainage. Secondary salinization is due to human activities; it may occur when there is inadequate drainage, industrial effluents, overuse of fertilizers, removal of natural plant cover, flooding with salt-rich waters, high water table, and the use of poor quality groundwater for irrigation (Wang et al. 2003). It increases extremely in protected cultivation (Wang et al. 2003). Moreover, it affects the irrigated lands with water scarcity, high temperature, and evapotranspiration, using poor quality irrigation water and adopting poor irrigation management practices (Cano et al. 1998; Abed Alrahman et al. 2005).

Salinity causes inhibition and deterioration of several stages of plant life, such as seed germination, plant growth, reproduction development, biosynthesis of phytohormones and other plant growth-regulating factors, nutrient imbalance, enzyme activities, and photosynthesis by increasing photorespiration and alerting the normal homoeostasis of cells which generates an increased generation of reactive oxygen species (ROS) (Drew et al. 1990; Almansouri et al. 1999; Sudhir and Murthy 2004; Raafat et al. 2011; Nabti et al. 2015; Vaishnav et al. 2015). The most common effects of salinity on glycophytes are loss of turgor; reduction in growth, resulting in smaller leaves; shorter stature; early senescence; decreased photosynthesis; respiratory changes; loss of cellular integrity; tissue necrosis; and even plant death (Cheeseman 1988). Saline stress is also known by affecting many physiological activities related to the accumulation of ions and osmolytes such as proline (Lee et al. 2008). Osmotic stress generated by salt stress limits absorption of water from soil even though the soil appears quite moist (Selvakumar et al. 2012; Turan et al. 2012). This abiotic factor affects furthermore plant development either through perturbation of water balance or imbalance in ratio of ion homeostasis.

Numerous reports showed that salinity increased Na^+ accumulation and consequently decrease Ca^+ and K^+ uptake (Yildirim et al. 2006; Zhang et al. 2008; Tank and Saraf 2010; Shukla et al. 2011; Rojas-Tapias et al. 2012; Amir et al. 2013; Jha and Subramanian 2013; Younesi et al. 2013; Sharma et al. 2016). Accumulation of toxic ions (Na^+ and Cl^-) perturbs water balance and ion homeostasis, which eventually disturbs metabolic processes and hormonal status and affects enzymes, photosynthetic function, transpiration, and translocation of nutrients (Zhu 2001; Munns 2002; Tester and Davenport 2003). Na^+ and Cl^- toxicity occurs when these two ions reach the plant's system through water and then affect plant's physiological processes often resulting in reduced growth, leaf burn, and even plant death (Drew et al. 1990; Marschner 1995; Sairam and Tyagi 2004; Arora et al. 2012). High amounts of salts reduce the accessibility and the uptake of essential mineral nutrients like nitrogen, phosphorus, potassium, and magnesium, which are extremely essential for plant development (Günes et al. 1996; Abdel-Ghaffar et al. 1998; Munns 2002; Tester and Davenport 2003; Bianco and Defez 2011; Jouyban 2012). Salt stress causes competitive interactions between toxic ions and nutrient ions for binding sites and transport proteins in root cells and thereafter for translocation, deposition, and partitioning within the plant (Tester and Davenport 2003). Consequently, to this interaction, the concentration of nutrient elements will be reduced in whole plant. Salt stress is also related to an oxidative stress because of generation of ROS such as superoxide ion ($O2^-$), hydrogen peroxide (H_2O_2), and hydroxyl radicals (OH^-), which are harmful to plant survival under stress conditions (Turan et al. 2012). ROS can act on unsaturated fatty acids and damage plant membranes and finally affect the DNA (Selvakumar et al. 2012); likewise they cause oxidative damage to biomolecules such as proteins and lipids and eventually lead to plant death (Del Rio et al. 2003). Under salt stress, plant is exposed to premature death originated from enhanced ethylene and abscisic acid (ABA) production (Kefu et al. 1991; Munns 2002; Mayak et al. 2004) that consequently accelerate petal and leaf dropping (abscission) and tissue aging (senescence) and retardation of plant growth (Kefu et al. 1991; Mayak et al. 2004). Successful establishment of the symbiotic interaction involves chemotaxis of the bacteria toward the roots, root colonization, root hair deformation, and rapid division of root cortex cells (Van Rhijn and Vanderleyden 1995).

Another deleterious effect of salinity is the disturbance of the symbiotic interaction between plant and beneficial microorganisms. For example, a disturbance in the interaction of plants with bacteria is caused by alterations of proteins involved in the initial attachment steps (adsorption and anchoring) of bacteria to plant roots in symbiotic interactions as well as by the inhibition of nodulation and nitrogen fixation activities (Nabti et al. 2015). Dardanelli et al. (2008) showed the negative effect of NaCl on nod gene expression by *Rhizobium tropici* and *Rhizobium etli* and on nodulation factors' pattern. Additionally, salt stress disturbs structure of exopolysaccharide (EPS) and lipopolysaccharide (LPS) of bacterial cell surface, generating cuts off the molecular signal exchange between bacteria and plant host, bacterial chemotaxis, and root colonization of surface-associated or endophytic PGPB (Jofré et al. 1998). Numerous studies have shown that soil salinity causes decrease in nodulation (Rabie and Almadini 2005) and reduces dramatically N_2 fixation and nitrogenase activity of nodulated legumes such as soybean (Tu 1981;

Singleton and Bohlool 1984), common bean, faba bean (Rabie and Almadini 2005), and pea (Zahran and Sprent 1986; Georgiev and Atkins 1993). Absorption and transport of inorganic phosphate appears to be sensitive to salinity (Ehsanpour and Amini 2003), which may negatively affect the acid phosphatase involved in inorganic phosphate uptake in plants (Turan et al. 2012).

14.2 Plant Responses to Salinity Stress

Plants are divided into halophytes and glycophytes due to their abilities to grow on different salt concentrations (Flowers et al. 1977). Halophytes are naturally salt-tolerant plant. Glycophytes, which include most crop plants, cannot tolerate high concentration of salt; their growth is inhibited by NaCl concentrations of 100–200 mM (Munns et al. 2002). Several parameters influence plant tolerance to salinity, e.g., plant physiological mechanisms, duration of exposure to saline conditions, concentration of salt around roots, local soil–water relations, and microclimate conditions (temperature, humidity, etc.) (Ilangumaran and Smith 2017). Plants react differently once under stress; their responses differ in terms of toxic ion absorption, ion compartmentation and/or exclusion, osmotic regulation, CO_2 assimilation, photosynthetic electron transport, chlorophyll content and fluorescence, reactive oxygen species (ROS) generation, and antioxidant defenses (Stepien and Johnson 2009; Tang et al. 2015). Saline stress tolerance must maintain or rapidly regulate osmotic and ionic homeostasis in cells; plants respond immediately and effectively by initiating a succession of stress-specific signals like phytohormones, e.g., abscisic acid, jasmonic acid, and salicylic acid (Ortíz-Castro et al. 2009). In addition, an accumulation of phenolic acids and flavonoids will take place; adding to that, the development of numerous antioxidants and the activation of transcription factors (TF) are initiated at the same time as the expression of specific genes-stress to support the defense system (Austin and Ballare 2014). As reported previously, glycophytes cannot grow at high salinity level contrary to halophytes, which are characterized by their potential salt resistance mechanisms. Plant responses to salt stress depend on several parameters, e.g., stage of plant growth, growing conditions, agronomic and irrigation management practices, and the degree of plant tolerance (Alarcón et al. 1999).

Salt tolerance in plants depends mainly on the capability of roots for restricted or controlled uptake of Na^+ and Cl^- and continued uptake of essential elements, particularly K^+ and NO^{3-} (Ashraf et al. 2004). Once under saline habitats, plants employed several ways to prevent Na^+ accumulation in cytoplasm that include minimizing Na^+ influx, intracellular compartmentation of Na^+, maximizing Na^+ efflux via Na^+/H^+ antiporters in the plasma membrane, as well as recirculation of Na^+ out of the shoot via the phloem (Apse et al. 1999; Pérez-Alfocea et al. 2000; Acosta-Motos et al. 2017; Wang et al. 2018). The efflux of Na^+ out of plant cells is achieved by sodium/proton antiporters SOS1 (salt overly sensitive), which is situated on the plasma membrane (Qiu et al. 2002). To increase the potassium uptake, plant activates high-affinity K^+ transporters (HKT); hence K^+ concentration relative to Na^+ in cytoplasm increases salinity tolerance (Rodriguez-Navarro and Rubio 2006). The accumulation of Cl^- is toxic to chloroplasts and mitochondria, as the sodium, Na^+, requires compartmentalization and exclusion from plant cells that is catalyzed through $Cl^-/2H^+$ symporter (Yamashita et al. 1994).

Vacuolar compartmentation of Na^+ is another mechanism employed by plant to reduce cell water potential (Apse et al. 1999; Hasegawa et al. 2000; Gaxiola et al. 2001); however, it reduces osmotic potential in the vacuole. To equilibrate osmotic potential in the vacuole with that of the cytoplasm, plants accumulate compatibles solutes (Xiong and Zhu 2002). Some halophyte species have evolved specialized cells called salt glands in shoots to excrete salt on its surface and then removed by water or wind (Ilangumaran and Smith 2017).

Nevertheless, all these means cannot eliminate the generation of reactive oxygen species (ROS) constantly produced by diverse biological process, e.g., photosynthesis, photorespiration, and CO_2 assimilation. They may act as damaging, protective, or signaling factor, depending on the delicate equilibrium between ROS production and scavenging (Ahmad et al. 2008a, b; Ahmad 2010). Plants have developed a complex nonenzymatic as well as enzymatic antioxidative system including carotenoids, ascorbate, glutathione, tocopherols, superoxide dismutase (SOD), catalase (CAT), ascorbate peroxidase (APX), glutathione reductase (GR), etc. (Del Rio et al. 2003; Ahmad et al. 2008a; Ashraf 2009; Sharma and Dietz 2009; Gill and Tuteja 2010). Normally, each cellular compartment contains more than one enzymatic activity that detoxifies a particular ROS.

To be adapted to the fluctuation in osmolality once in salt environment, plant accumulates low molecular mass compounds known as compatible solutes or osmolytes (explained below). These solutes, once accumulated in plant cell cytosol, maintain osmotic balance both inside and outside the cell (Ilangumaran and Smith 2017). Kant et al. (2006) have highlighted the role of these solutes in alleviating salt stress, in view of their results obtained on two plant models, salt-sensitive *Arabidopsis thaliana* and the halophytic *Thellungiella halophile*. They suggested that *Thellungiella halophile* may be more salt tolerant than *Arabidopsis* due to more efficient production of osmoprotectants and that differential gene expression between glycophytes and halophytes contributes to the salt tolerance of halophytes.

14.3 Salt Stress Mitigation in Agricultural Crops

As discussed previously, salinity is a prime factor limiting crop productivity worldwide. The development of efficient, economical, and easily adaptable methods for the management of abiotic stress is a major challenge. There are different methods of treating saline soils for crop production including physical, chemical, and biological methods. Physical methods include scraping (removal of the salt layer of soil surface by mechanical means, and the lower layer is used for cultivation), flushing (washing the soil surface with water), and leaching (continuous ponding, intermittent ponding); this method has resulted in limited success and/or involves high cost (Qadir et al. 2000). Concerning chemical methods, gypsum is considered as one of the most helpful soil amendments to leach out cations (Na^+) from the soil; sulfuric acid and farmyard manure amendments are also applied especially in sodic or saline sodic soils. The use of chemical amendment has negative impacts on the environment and affects beneficial microorganisms, making crops more susceptible to disease. Salt stress may be also alleviated by using naturally tolerant plants or genetically modified ones (Günes et al. 1996; Han and Lee 2005; Wong et al. 2008; Schubert et al. 2009; Miransari 2012; Shrivastava and Kumar 2015), by changing farm management practices, and by supporting agriculture through better irrigation practices (Miransari 2012; Shrivastava and Kumar 2015); however, these methods are time-consuming and not economical. Plant growth-promoting rhizobacteria (PGPB) have been proved for their efficient salt stress alleviation via numerous biological activities.

14.4 Plant Salt Tolerance Mediated by PGPB

Plant growth-promoting bacteria (PGPB) present a pronounced scientific and agronomic interest because they play an important role in the biological functioning of the rhizosphere and therefore improve growth, yield, and health of agricultural crops (De Salamon et al. 2005). The use of PGPB to improve plant growth offers many possibilities for an agronomy respectful of the environment. Indeed, a better understanding of the use of these bacterial populations could allow a reduction of chemical inputs and monetary cost in agricultural soils (Table 14.1).

This group of bacteria increases seed germination rate, root development, shoot and root weights, yield, leaf area, chlorophyll content, tolerance to abiotic and biotic stress, protein content, and nutrient uptake including phosphorus and nitrogen (Mahaffee and Kloepper 1994; Raaijmakers et al. 1997; Bashan et al. 2004; Mantelin and Touraine 2004; Bakker et al. 2007; Yang et al. 2009; Dardanelli et al. 2010). They colonize the rhizosphere by means of root exudates as nutrient substrates, but unlike other rhizospheric bacteria, they stimulate plant growth and health via a variety of mechanisms (Vacheron et al. 2013). PGPR may include free bacteria living in soil without implication of root exudates for their growth or symbiotic bacteria using directly organic compounds of roots (Barraquio et al. 2000). Communication in beneficial plant-microbe interaction is regulated by root exudates released by plant (Ilangumaran and Smith 2017). These substances constitute a nutrient source for the colonizing bacteria, and they act as chemical cues that allow plants to communicate with the surrounding microbial flora (Badri et al. 2009).

In recent decade, many bacterial species have been identified as PGPR because of the numerous studies covering a wide range of plant species. The diversity of PGPR varies widely depending on the type of plant, soil, and nutrient availability. Among the identified PGPR, *Pseudomonas* and *Bacillus* are widely distributed and they are the most studied. The strains belonging to the genera *Aeromonas*, *Azoarcus*,

Table 14.1 Salt stress alleviation by plant growth-promoting bacteria

PGPR	Crops species	Beneficial effects	References
Enterobacter sp. UPMR18 (ACC deaminase)	*Abelmoschus esculentus*	Increase antioxidant enzyme activities (SOD, APX, and CAT) Upregulation of ROS pathway genes (CAT, APX, GR, and DHAR)	Habib et al. (2016)
Pseudomonas chlororaphis TSAU13, *Pseudomonas extremorientalis* TSAU20	*Lycopersicon esculentum* cv. Bella	Increase N, P uptake Increase proline amount Increase antioxidant enzyme activities (SOD, APX, and CAT)	Egamberdieva et al. (2017)
Brevibacterium linens RS16 (ACC deaminase)	*Oryza sativa*	Enhance photosynthesis Decrease ACC accumulation and the ACC oxidase activity	Chatterjee et al. (2018)
Bacillus amyloliquefaciens HM6	*Hordeum vulgare*	Enhance plant growth (seedling length, fresh and dry masses, relative water content)	Kasim et al. (2016)
Bacillus sp.	*Capsicum annuum*	Enhance plant growth (fresh and dry weight, shoot and root length) Enhance proline content Increase antioxidant enzyme activities	Wang et al. (2018)
Enterobacter sp. (ACC deaminase)	*Oryza sativa* cv. Ratna	Promote rice seedling growth Reduce antioxidant enzymes and ethylene production	Sarkar et al. (2018)
Serratia marcescens CDP-13	*Triticum aestivum*	Enhance wheat growth Increase proline, malondialdehyde, total soluble sugar, total protein, and indoleacetic acid level in plant	Singh and Jha (2016)
Dietzia natronolimnaea STR1	*Triticum aestivum*	Promote salinity tolerance of plant Enhance gene expression of antioxidant enzymes and expression of TaST, a salt stress-induced gene Modulate ABA signaling cascade and SOS pathway-related genes, tissue-specific responses of ion transporters	Bharti et al. (2016)
Pseudomonas simiae AU	*Glycine max*	Upregulation of vegetative storage proteins, gamma-glutamyl hydrolase (GGH), RuBisCO large-chain proteins Enhance N, P, and K^+ level Decrease Na^+ accumulation and increase proline and chlorophyll content	Vaishnav et al. (2015)

(continued)

Table 14.1 (continued)

PGPR	Crops species	Beneficial effects	References
Azospirillum lipoferum	*Brassica napus*	Enhance canola growth Decrease level of antioxidant enzyme Increase microelement uptake (Zn, Fe, and Mn)	Baniaghil et al. (2013)
A. tumefaciens *Pseudomonas* sp.	*Arachis hypogaea*	Enhance peanut plant growth Ion homeostasis and reduce reactive oxygen species (ROS) level	Sharma et al. (2016)
Pseudomonas putida R4 *Pseudomonas chlororaphis* R5	*Gossypium hirsutum*	Improve seed germination and seedling growth	Egamberdieva et al. (2015)
B. japonicum *Bacillus thuringiensis* NEB17	*Glycine max*	Upregulation of PEP carboxylase, RuBisCO oxygenase large subunit, pyruvate kinase, and isocitrate lyase and antioxidant glutathione-*S*-transferase and other stress-related proteins	Subramanian et al. (2016)

Azospirillum, *Azotobacter*, *Arthrobacter*, *Clostridium*, *Enterobacter*, *Gluconacetobacter*, *Klebsiella*, and *Serratia* are also classified as PGPB (Fernando et al. 2005; Fuentes-Ramirez and Caballero-Mellado 2005; Tripathi et al. 2007).

Based on their activity, PGPR are classified as biofertilizers (increase the availability of plant nutrients), phytostimulators (rise plant growth, usually by producing plant hormones), rhizo-remediants (degrade organic pollutants), and biocontrol agents (control plant diseases, mainly by competition, the production of antibiotics, and antifungal metabolites) (Dong et al. 1992; Somers et al. 2004).

The potential of PGPB to increase crop production involves several direct and indirect mechanisms like solubilization of inorganic phosphate and ferric iron minerals, increased uptake of mineral nutrients (Martinez-Viveros et al. 2010), N_2 fixation (Lucy et al. 2004), production of plant hormones (Glick 1995), production of hydrolytic enzymes, control of plant pathogens, and stimulation of plant resistance system (induced systemic resistance) (García-Fraile et al. 2015). PGPR may also contribute to sustaining the resistance of plant to salt stress by their ability to produce exopolysaccharides used to protect them against desiccation and ionic toxicity. The exopolysaccharides chelate sodium ions from rhizosphere zone and make it more suitable for root proliferation (Arora et al. 2012). PGPB are also capable of protecting plants from harmful effects of several abiotic stresses such as drought (Sandhya et al. 2009; Timmusk and Wagner 1999), chilling injury (Barka et al. 2006), salinity (Han and Lee 2005; Nabti et al. 2010), metal toxicity (Dell'Amico et al. 2008), and elevated temperature stress (Ali et al. 2009, 2011).

In a context of reduction of salt stress effect, PGPB seem to have quite a lot of potentials to make to restore plant growth and reinforce their resistance system. With their potential of and through their diversity, PGPB may mitigate salt stress. PGPB diversity and their mechanisms of salt stress alleviation allow them to restore the growth of different crops, e.g., pepper, tomato, soybean, wheat, barley, canola, and lettuce. Egamberdieva and Kucharova (2009) revealed the ability of two strains belonging to the genus *Pseudomonas* to improve growth of various crops under salt stress. The depressive effect of salt on shoot and root length and dry weight of soybean crop has been alleviated by PGPR (Naz et al. 2009; Khan et al. 2017). Jha and Subramanian (2013) showed the improvement of seed germination under salt stress of paddy plant inoculated with *Pseudomonas pseudoalcaligenes* and *Bacillus pumilus*. *Bacillus megaterium* increased seed germination of tomato and furthermore increased seedling dry weight at NaCl up to 120 mM (Chookietwattana and Maneewan 2012). *Bacillus amyloliquefaciens* (SQR9) induced maize salt tolerance by enhancing its chlorophyll production level, by excluding sodium from its roots, and by stimulating sugars and antioxidant production (Chen et al. 2016). Manaf and Zayed (2015) indicated that the co-inoculation of *Pseudomonas* and *endomycorrhizae* on cowpea plant cultivated under salt stress increased the carotenoid concentration. *Pseudomonas putida* R4 and *Pseudomonas chlororaphis* R5, two PGPB strains, alleviate salt stress on cotton rhizosphere by the production of IAA phytohormone (Egamberdieva et al. 2015, 2017). According to Paulucci et al. (2015), a newly identified rhizobacteria *Ochrobactrum intermedium* which have the ability to modify its lipid components in response to temperature and salinity, producing indoleacetic acid, siderophores, and deaminase enzyme, improve the growth (root and shoot) of peanut plant under salt stress. *Curtobacterium flaccumfaciens*, a new species in PGPB groups, are recently found to protect barley plant from salt stress; it increased significantly seed germination; salt stress resistance of leaves, stems, and roots; and biomass accumulation of leaves, stems, and roots of barley (Cardinale et al. 2015). Singh and Jha (2016) have demonstrated that *Serratia marcescens* CDP-13 strain, isolated from *Capparis decidua* plant, enhanced the growth of wheat cultivated under salt conditions. The inoculation of plant by this strain has also reduced the inhibition growth, modulated the concentrations of different osmoprotectants (proline, malondialdehyde, total soluble sugar, total protein content, and indoleacetic acid) in plants, and reduced the disease severity caused by fungal infection; thereby, *Serratia marcescens* may induce systemic resistance of wheat to both biotic and abiotic stress. *Enterobacter* sp. P23 was shown to promote rice seedling growth under salt stress, and this effect was correlated with a decrease in antioxidant enzymes and stress-induced ethylene (Sarkar et al. 2018) (Fig. 14.1). The potential of PGPB to reduce the depressive effect of abiotic stress and to restore plant growth was reported by different authors; however, the mechanisms involved in plant improvement are intricate and still unclear.

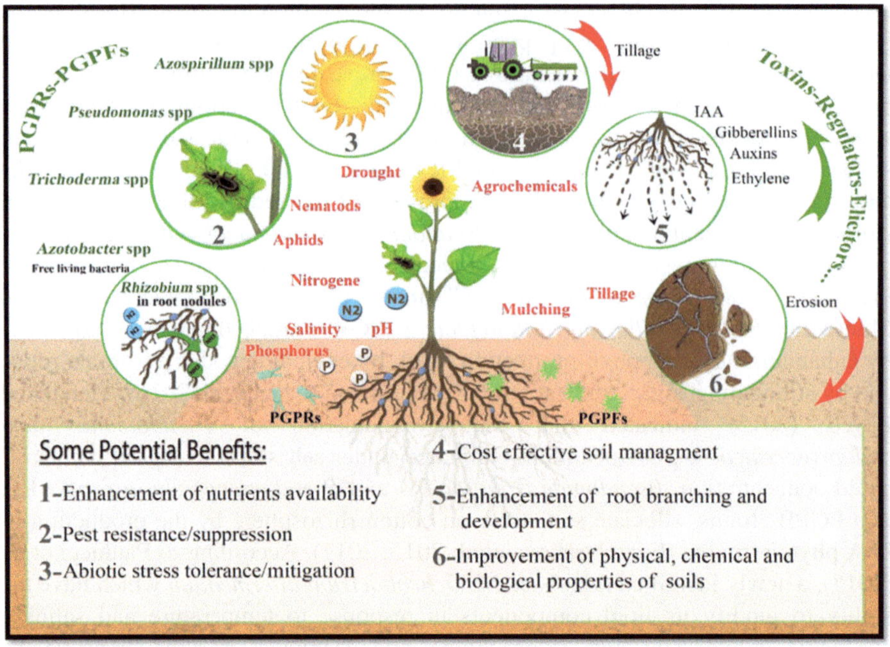

Fig. 14.1 Soil management microbiome and crop enhancement by PGPR

14.5 Induced Systemic Tolerance

Plants are sessile organisms; they are frequently exposed to several biotic and abiotic stress. Under stressed environment, plants developed various mechanisms that may lead them to adapt to any stress conditions. Pastor et al. (2013) suggested that plant possesses a memory, named <<memory>> defenses, allowing it to a rapid and efficient activation of their mechanisms of defense once under stress. The term "induced systemic tolerance" (IST) was proposed for PGPR-induced physical and chemical changes in plants that result in enhanced tolerance to abiotic stress (Mantelin and Touraine 2004). Many studies concerning the role of PGPR in eliciting plants to tolerate abiotic stresses, e.g., drought, salt, cold, and high temperature, were published. PGPR by their several traits (explained below) help plants directly or indirectly to overcome salt stress.

14.6 Phytohormone Biosynthesis of PGPB

It is well understood that the most important mechanism involved in PGPB in the stimulation of plant growth is the production of growth-regulating substances. The commonly recognized classes of phytohormones, regarded as the "classical five," are the auxins, gibberellins, cytokinins, abscisic acid, and ethylene (Baca and Elmerich 2007). Cytokinins and gibberellins are involved in plant morphology and in growth promotion of the aerial part (Van Loon 2007). The abscisic acid is involved in the regulation of seed germination; it acts as cellular signal and modulates various genes in response to water deficiency and salt stress (Numan et al. 2018). Auxins such as indole-3-butyric acid (IBA) and indole-3-acetic acid (IAA) or their precursors are produced by a wide range of bacteria as part of their metabolism (Martínez-Morales et al. 2003; Spaepen et al. 2007). Indole-3-acetic acid (IAA) is one of the most important phytohormones that regulate many aspects of plant growth and development, e.g., cell division, seed germination, differentiation of tissues, leaf expansion, root elongation, and resistance to stressful conditions (Egamberdieva 2008; Maleki et al. 2010; Martinez-Viveros et al. 2010). According to Zakharova et al. (1999), about 80% of rhizosphere bacteria produce IAA. The production of this hormone is influenced by culture conditions, growth stage, and substrate availability in the medium. Bacteria producing auxin stimulate seed germination, cell division, and tissue expansion and play a major role in root elongation (Egamberdieva 2008; Maleki et al. 2010; Martinez-Viveros et al. 2010).

Growth-regulating mechanisms are speculated to be long-distance signals of hormones and their precursors from roots to shoots. Bacteria release exogenous hormones, metabolites, and enzymes by which they increase salt tolerance of plant (Ilangumaran and Smith 2017). Asim et al. (2013) reported that the IAA level in soybean seedlings has decreased under stress; however, IAA level has been established after plant inoculation by PGPR, thus leading to plant growth parameter's promotion. According to Egamberdieva et al. (2015), *Pseudomonas putida* R4 and *Pseudomonas chlororaphis* R5, producing IAA, alleviate salt stress on cotton rhizosphere. Wu et al. (2014) have also indicated the implication of hormone production trait in the alleviation of salt stress effect; they showed that *Klebsiella oxytoca* (Rs-5) increase seed germination (15.40%), growth, and overall tolerance of cotton to salinity stress. In another study, an increase in ABA and IAA level in wheat leaves, inoculated by *Bacillus cereus* and *Pseudomonas* species, was observed (Numan et al. 2018). Khalid et al. (2013) revealed also the role of exogenous application of IAA in alleviating the deleterious effect of salt stress in maize plant. It was found that IAA overproducing PGPR strains are able to alleviate different stress conditions by increasing proline amount in *Medicago* plants (Saghafi et al. 2013). Under stress conditions, IAA level decreases in plant, and, consequently, a reduction in growth and germination speed occurs (Naqvi 1999). One of the mechanisms used by plant to counteract the stress situation is the decrease of IAA production (Yue et al. 2007), which requires extra metabolic energy. However, this extra energy consumption decreases plant growth. Plant inoculated by PGPR producing IAA may minimize the energy requirement of plant under salt stress (Singh and Jha 2016).

14.7 ACC Deaminase

Under salt stress, 1-aminocyclopropane-1-carboxylic acid (ACC) and ethylene level in plant increase (Arbona et al. 2005). Ethylene is a gaseous plant hormone acting as a secondary signaling molecule in the induction of plant defenses (Ecker 1995) and in the regulation of plant growth at moderate concentrations. Under stress conditions such as salinity, plant hormone ethylene endogenously regulates plant homoeostasis resulting in reduced root and shoot growth (Shrivastava and Kumar 2015; Vurukonda et al. 2016); it is also involved in the transduction of a signal for the recognition of salt stress (Selvakumar et al. 2012). Many authors exhibited that microorganisms producing deaminase enzyme (1-aminocyclopropane-1-carboxylate deaminase) may alleviate salt stress and help plants to survive under stress conditions. This enzyme degrades aminocyclopropane-1-carboxylate, which is a plant product acting as a precursor of ethylene (Honma and Shimomura 1978; Adams and Yang 1979; Dong et al. 1992; Glick 1995, 2014; Penrose et al. 2001; Glick et al. 2007; Numan et al. 2018). ACC deaminase can hydrolyze the endogenous levels of ACC, which is an immediate precursor of ethylene biosynthesis in plants, into ammonia and α-ketobutyrate resulting in reduced level of ethylene (Hontzeas et al. 2004; Glick et al. 2007; Nascimento et al. 2014; Santoyo et al. 2016; Khan et al. 2017). PGPR producing this enzyme protect plant from deleterious effect of salinity and drought by reducing salt stress-induced ethylene (Glick et al. 1998; Saleem et al. 2007; Zahir et al. 2009). ACC deaminase activity was expected to be the best indicator of the most effective PGPR under soil salinity (Cardinale et al. 2015). Both in laboratory experiments and in the field, bacteria containing ACC deaminase mitigated salt stress and improved growth of plant used in phytoremediation (Gerhardt et al. 2006). Ethylene production by tomato seedlings was reduced by the bacterium *Achromobacter piechaudii* as reported by Mayak et al. (2004). They revealed the significant increase of fresh and dry weights of tomato grown in NaCl (up to 172 mM). *Enterobacter* sp. P23 was shown to promote rice seedling growth under salt stress and reduce ethylene level, which is linked to ACC deaminase activity of this strain. The amount of ethylene produced in rice seedlings inoculated with the acdS mutant (unable to produce ACC deaminase) in the presence of 150 mM NaCl was similar to that of non-inoculated seedlings, while a significant reduction of stress-induced ethylene was obtained in plant inoculated with wild type (Sarkar et al. 2018). Nadeem et al. (2006) have also mentioned that inoculation of canola by ACC deaminase producing *Pseudomonas syringae* and *Pseudomonas* sp. improves yield and growth under salinity (12 dS m^{-1}). The role of PGPR containing ACC deaminase in improving nodulation in legumes has been demonstrated by Shaharoona et al. (2006) showing that inoculation with PGPR containing ACC deaminase promoted nodule formation by indigenous competitive *Bradyrhizobium.* Thereby, these bacteria could be a promising approach for successful nodulation in legumes (Zafar-ul-Hye et al. 2013). Similarly, Chaudhary and Sindhu (2017) have noted the ability of two strains, *Mesorhizobium* MBD26 and

Rhizobacteria RHD18 which are producers of ACC deaminase, to increase nodulation capacity and N_2 level of chickpea plant under salt stress.

14.8 Compatible Solutes

Compatible solutes are so named because, even at high concentration in the cytoplasm, they do not interfere with the vital cellular processes (Kempf and Bremer 1998) and they can be synthesized by halophile organisms or accumulated by uptake from surrounding medium (Mohammadipanah et al. 2015). These solutes, once accumulated in cell cytosol, maintain osmotic balance both inside and outside the cell (Ilangumaran and Smith 2017). They harbor different classes of molecules including soluble sugars such as sucrose, trehalose (Breedveld et al. 1993), maltose, cellobiose, turanose, gentiobiose, and palatinose (Gouffi et al. 1999) or solutes like amino acids, e.g., glutamate, proline, alanine, serine, threonine, and aspartic acid; quaternary amines, e.g., glycine betaine and carnitine; imino acids (pipecolate); K^+; and tetrahydropyrimidines (ectoines) (Mishra et al. 2018). The accumulation of osmoprotectant molecules is the major mechanism involved in both prokaryote and eukaryote organisms to alleviate salt stress. These solutes do not inhibit biochemical cellular processes, but rather protect them from inorganic ion damage. Compatible solutes can also act as free radical scavengers or chemical chaperones by directly stabilizing membranes and/or proteins (Mc Neil et al. 1999; Diamant et al. 2001; Rhodes et al. 2002). They are also involved in osmoprotection process by preventing membrane desiccation and by stabilizing dehydrated enzymes involved in osmoregulation (Ilangumaran and Smith 2017). Compatible solutes can be synthesized by bacteria or accumulated by uptake from surrounding medium (Mohammadipanah et al. 2015). Oren (1999) reported that uptake is preferable than synthesizing because the latter consumes more energy, and under stress environment, energy conservation is better for survival. The concept of osmoprotectant supposes an ecological cycle in which compatible solutes are transported from producers to consumers injured by osmotic stress (Arora et al. 2012). Inoculation of salt-tolerant PGPR (produce or accumulate compatible solutes) improves plant growth by alleviating the toxicity of Na^+ and ion imbalance through enhanced level of osmoprotectant synthesis in plants (Azooz et al. 2004; Ramezani et al. 2011). Proline accumulation is a very common response in plants exposed to abiotic stress; it contributes to osmotic adjustment by alleviating the ROS-induced oxidative damage (Gupta and Huang 2014). Saghafi et al. (2013) showed that PGPR strains increased proline contents in wheat cultivars. Similarly, Naz et al. (2009) reported that PGPR isolates increase significantly proline contents of shoots and roots of soybean plant under salt stress. Singh and Jha (2016) showed that wheat inoculated by *Serratia marcescens* CDP-13 under saline conditions modulated the plant concentration of proline and soluble sugars in comparison to uninoculated control plants. This strain has significantly increased the proline level in plant which can reduce the depressive effect of salt stress. Qurashi and Sabri (2011) found that *Lens esculenta* inoculated with *Oceanobacillus profundus* Ptm2 strain under salt conditions showed 98% increase in the accumulation of endogenous proline, which thereby reduces NaCl toxicity and improved plant growth. The inoculation of *Zea mays* by *Achromobacter chroococcum* C5 and C9 under salt stress reduces and increases Na^+ and K^+ concentration, respectively, and reduces salt toxicity (Rojas-Tapias et al. 2012). Synthesis of osmoprotective molecules in plant is a mechanism that requires a high amount of energy (Raven 1985); under such circumstances, biosynthesis of osmolytes by PGPR is most likely to be quicker than their associated host plants (Ilangumaran and Smith 2017).

14.9 Antioxidant Enzyme Activity

The major problem caused by salt stress is the generation of reactive oxygen species (ROS) including superoxide radical (O^{2-}), hydroxyl radical (OH^-), and hydrogen peroxide (H_2O_2) which causes oxidative damage to biomolecules such as lipids and proteins and eventually leads to plant death (Del Rio et al. 2003). Perturbation of the cellular redox state is adjusted by the activation of antioxidant system in plant (Gill and Tuteja 2010; Jha and Subramanian 2013), e.g., superoxide dismutase (SOD), peroxidase (POX), catalase (CAT), and nonenzymatic antioxidant such as ascorbate, glutathione, and a-tocopherol. PGPR induce antioxidant enzyme production in salt-stressed plant, which is an essential mechanism to detoxify reactive oxygen species (ROS). The ability of PGPB to enhance the activity of antioxidant enzymes was already described. Kumar et al. (2003) and Jha and Subramanian (2013) reported that the antioxidant activity is more significant when PGPR are present. A rhizobial strain *Mesorhizobium ciceri* enhances chickpea tolerance to salt stress by increasing peroxidase POD activity (Mhadhbi et al. 2004). To monitor the putative changes in the expression of genes encoding the ROS scavenging enzymes, Gururani et al. (2013) subjected the PGPR-inoculated and PGPR-non-inoculated potato plants to different stresses. Thy showed that the mRNA expression of SOD and APX, in bacteria-inoculated plants growing under stress conditions, increased considerably in comparison to non-inoculated stressed plants. Likewise, mRNA expression levels of genes, encoding for catalase and glutathione reductase, have increased in bacteria-inoculated plants. Bharti et al. (2016) noticed that shoot ascorbate peroxidase (APX), catalase (CAT), and peroxidase (POD) of PGPR-inoculated salt-stressed plants recorded a threefold increase in gene expression as compared to uninoculated non-stressed plants.

14.10 Ion Homeostasis

One of the most depressive effects of salt stress on plants consists of ion homeostasis perturbation, increased Na^+ accumulation in plant cell, and decreased K^+ uptake. As explained previously, perturbation in the ion distribution on both sides of cell membrane disturbs plant growth (metabolic processes and hormonal status, enzymes, photosynthetic function, transpiration and translocation of nutrients, etc.) (Munns 2002). Maintenance of low Na^+/K^+ ratio is one of the predominant mechanisms involved in PGPR to improve plant growth under high salinity (Munns and Tester 2008). They reduce ion toxicity and uptake by regulating the expression of ion transporter in plant; additionally, they can produce a physical barrier around roots using their exopolysaccharides (Vaishnav et al. 2016). Siddikee et al. (2011) reported that the reduction of Na^+ ion uptake by red pepper roots is due to the ability of *B. linens* RS16 with surface polysaccharides. Salt stress affects also gene expression involved in ion homeostasis. Pinedo et al. (2015) clarified the deleterious effect of salt stress on the expression of ion homeostasis-related gene (*Arabidopsis* K^+ transporter 1, high-affinity K^+ transporter 1, sodium hydrogen exchanger 2, and *Arabidopsis* salt overly sensitive 1). They displayed that PGPR strain *Burkholderia phytofirmans* PsJN regulates the expression of important ion homeostasis-related genes after short- and long-term exposure to salt stress. *Bacillus subtilis* GB03 upregulated PtSOS1 gene expression under salt stress and therefore enhanced *P. tenuiflora* tolerance and reduced Na^+ accumulation (Niu et al. 2016). According to Panwar et al. (2016), a correlation between salt tolerance enhancement and reduction of Na^+ uptake and electrolyte leakage by chickpea (*Cicer arietinum*) inoculated by *Pantoea dispersa* PSB3 was noted. Additionally, plant inoculated showed increase of relative leaf water content, chlorophyll content, and K^+ uptake. PGPB have also the ability to provide plants with nutrient elements that compete with toxic ions; otherwise, providing nutrient to plant under stress alleviates toxic ion accumulation in plant cells. It is well known that PGPB provide nutrients to plant in stressed environment by siderophore production (iron acquisition), phosphate solubilization, and nitrogen fixation. The results obtained by Sapsirisopa et al. (2009) revealed that treatments of rice with *Bacillus megaterium* A12ag significantly increased N and P content in shoots. Jha and Subramanian (2013) showed that PGPR strains increased N, P, and K concentration and reduced Na^+ and Ca^+ accumulation. Inoculation of maize seeds with rhizobacterial strains augmented chlorophyll a and b and carotenoid contents of fresh leaves (Nadeem et al. 2006). The enzymes produced by PGPB in soil improve the bioavailability of nutrient elements essential for plant growth and health. These biological enzymes convert the insoluble nutrients present in soil into available form by plant (sugars, ferrous iron (Fe^{2+}), amino acids, NH^{4+}, PO_4^{-3}) (Burns and Wallenstein 2010). The inoculation of plant seeds by PGPR producing siderophores increases plant growth and chlorophyll levels (Sharma et al. 2003). Nitrogen fixation plays a potential role in improving the fertility and productivity of low-nitrogen soils. Many authors reported the improvement of nitrogen fixation level and nodulation in root plants.

14.11 Conclusion

The positive effect of halotolerant plant growth-promoting bacteria/rhizobacteria on plant growth under high salinity could be a sustainable approach for successful formulation to improve crop growth and yield in saline soils.

References

Abdel-Ghaffar BA, El-Shourbagy MN, Basha EM (1998) Responses of NaCl stressed wheat to IAA. In: Proceedings of the 6th Egypt botany conference, Cairo University, Giza, vol 6, pp 79–88

Abed Alrahman NM, Shibli RA, Ereifej KI, Hindiyeh MY (2005) Influence of salinity on growth and physiology of in vitro grown cucumber (*Cucumis sativus* L.). Jordan J Agric Sci 1:93–106

Acosta-Motos JR, Ortuño MF, Bernal-Vicente A, Diaz-Vivancos P, Sanchez-Blanco MJ, Hernandez JA (2017) Plant responses to salt stress: adaptive mechanisms. Agronomy 7:18. https://doi.org/10.3390/agronomy7010018

Adams DO, Yang SF (1979) Ethylene biosynthesis: identification of 1-aminocyclopropane-1-carboxylic acid as an intermediate in the conversion of methionine to ethylene. Proc Natl Acad Sci USA 76:170–174

Ahmad P (2010) Growth and antioxidant responses in mustard (*Brassica juncea* L.) plants subjected to combined effect of gibberellic acid and salinity. Arch Agron Soil Sci 56 (5):575–588

Ahmad P, Jhon R, Sarwat M, Umar S (2008a) Responses of proline, lipid peroxidation and antioxidative enzymes in two varieties of *Pisum sativum* L. under salt stress. Int J Plant Produc 2:353–366

Ahmad P, Sarwat M, Sharma S (2008b) Reactive oxygen species, antioxidants and signaling in plants. J Plant Biol 51:167–173

Alarcón JJ, Morales MA, Torrecillas A, Sánchez-Blanco MJ (1999) Growth, water relations and accumulation of organic and inorganic solutes in the halophyte *Limonium latifolium* cv. *Avignon* and its interspecific hybrid *Limonium caspia Limonium latifolium* cv. *Beltlaard* during salt stress. J Plant Physiol 154:795–801

Ali SZ, Sandhya V, Grover M, Rao LV, Kishore VN, Venkateswarlu B (2009) Pseudomonas sp. strain AKM-P6 enhances tolerance of sorghum seedlings to elevated temperatures. Biol Fertil Soils 46:45–55

Ali SKZ, Sandhya V, Grover M, Lingab VR, Bandi V (2011) Effect of inoculation with a thermotolerant plant growth promoting *Pseudomonas putida* strain AKMP7 on growth of wheat (*Triticum* spp.) under heat stress. J Plant Interact 6:239–246

Almansouri M, Kinet J, Lutts S (1999) Compared effects of sudden and progressive impositions of salt stress in three-durum wheat (*Triticum durum* Desf.) Cultivars. J Plant Physiol 154:743–752

Amir M, Aslem A, Khan MY, Jamshaid MU, Ahmed M, Asghar HN, Zahir ZA (2013) Co-inoculation with *Rhizobium* and plant growth promoting rhizobacteria (PGPR) for inducing salinity tolerance in mung bean under field condition of semi-arid climate. Asian J Biol 1:17–22

Apse MP, Aharon GS, Snedden WA, Blumwald E (1999) Salt tolerance conferred by overexpression of a vacuolar Na+/H+ antiporter in Arabidopsis. Science 285:1256–1258

Arbona V, Macro AJ, Iglessias DJ, Lopez-Climent MF, Talon M, Gomez-Cadnas A (2005) Carbohydrate depletion in roots and leaves of salt-affected potted *Citrus clementina* L. Plant Growth Regul 46:153–160

Arora NK, Tewari S, Singh S, Lal N, Maheshwari DK (2012) PGPR for protection of plant health under saline conditions. In: Maheshwari DK (ed) Bacteria in agrobiology-stress management. Springer, Berlin, pp 239–258

Ashraf M (2009) Biotechnological approach of improving plant salt tolerance using antioxidants as markers. Biotechnol Adv 27:84–93

Ashraf M, Hasnain S, Berge O, Mahmood T (2004) Inoculating wheat seedling with exopolysaccharide-producing bacteria restricts sodium uptake and stimulates plant growth under salt stress. Biol Fertil Soils 40:152–157

Asim M, Aslam M, Bano A, Munir M, Majeed A, Abbas SH (2013) Role of phytohormones in root nodulation and yield of soybean under salt stress. Am J Res Commun 1:191–208

Austin AT, Ballare CL (2014) Plant interactions with other organisms: molecules, ecology and evolution. New Phytol 204:257–260

Azooz MM, Shaddad MA, Abdel-Latef AA (2004) Leaf growth and K+/Na+ ratio as an indication of the salt tolerance of three sorghum cultivars grown under salinity stress and IAA treatment. Acta Agron Hung 52:287–296

Baca BE, Elmerich C (2007) Microbial production of plant hormones. In: Elmerich C, Newton WE (eds) Associative and endophytic nitrogen-fixing bacteria and cyanobacterial associations. Springer, Cham, pp 113–143

Badri DV, Weir TL, Van der Lelie D, Vivanco JM (2009) Rhizosphere chemical dialogues: plant–microbe interactions. Curr Opin Biotechnol 20:642–650. https://doi.org/10.1016/j.copbio.2009.09.014

Bakker PAHM, Pieterse CMJ, Van Loon LC (2007) Induced systemic resistance by fluorescent *Pseudomonas* spp. Phytopathology 97:239–243

Barka EA, Nowak J, Clement C (2006) Enhancement of chilling resistance of inoculated grapevine plantlets with a plant growth-promoting rhizobacterium Burkholderia phytofirmans strain PsJN. Appl Environ Microbiol 72:7246–7252

Barraquio WL, Segurbre EM, Gonzalez MS, Verma SC, James EK, Ladha JK, Tripathi AK (2000) Diazotrophic enterobacteria: what is their role in Rhizosphere of rice. In: Ladha JK, Reddy PM (eds) The Quest for nitrogen fixation in rice. IRRI, Los Banos, pp 93–118

Bashan Y, Holguin G, De Bashan LE (2004) *Azospirillum*–plant relationships: physiological, molecular, agricultural, and environmental advances (1997–2003). Can J Microbiol 50:521–577

Bharti N, Pandey SS, Barnawal D, Patel VK, Kalra A (2016) Plant growth promoting rhizobacteria *Dietzia natronolimnaea* modulates the expression of stress responsive genes providing protection of wheat from salinity stress. Sci Rep 6:1–16. https://doi.org/10.1038/srep34768

Bianco C, Defez R (2011) Soil bacteria support and protect plants against abiotic stress. In: Shanker A (ed) Abiotic stress in plants – mechanisms and adaptation. InTech Europe, Rijeka, pp 143–169

Blaylock AD (1994) Soil salinity, salt tolerance and growth potential of horticultural and landscape plants. Co-operative Extension Service, University of Wyoming, Department of Plant, Soil and Insect Science, College of Agriculture, Laramie, Wyoming

Breedveld MW, Dijkema C, Zevenhuizen L, Zehnder AJB (1993) Response of intracellular carbohydrates to a NaCl shock in *Rhizobium leguminosarum biovar trifolii* TA1 and *Rhizobium meliloti* SU47. J Gen Microbiol 139:3157–3163

Burns RG, Wallenstein MD (2010) Microbial extracellular enzymes and natural and synthetic polymer degradation in soil: current research and future prospects. In Gilkes R, Prakongkep N (eds) Proceedings of the 19th World Congress of soil science: soil solutions for a changing world, Brisbane, pp 67–69

Cano EA, Perez-Alfocea F, Moreno V, Caro M, Bolarin MC (1998) Evaluation of salt tolerance in cultivated and wild tomato species through in vitro shoot apex culture. Plant Cell Tissue Organ Cult 53:19–26

Cardinale M, Ratering S, Suarez C, Montoya AMZ, Geissler-Plaum R, Schnell S (2015) Paradox of plant growth promotion potential of rhizobacteria and their actual promotion effect on growth of barley (*Hordeum vulgare* L.) under salt stress. Microbiol Res 181:22–32

Chaudhary D, Sindhu SS (2017) Amelioration of salt stress in chickpea (*Cicer arietinum* L.) by coinculation of ACC deaminase-containing rhizospheric bacteria with Mesorhizobium strains. Legume Res 40:80–86

Cheeseman JM (1988) Mechanisms of salinity tolerance in plants. Plant Physiol 87:547–550

Chen L, Liu Y, Wu G, Njeri KV, Shen Q, Zhang N, Zhang R (2016) Induced maize salt tolerance by rhizosphere inoculation of *Bacillus amyloliquefaciens* SQR9. Physiol Plant 158:34–44. https://doi.org/10.1111/ppl.12441

Chookietwattana K, Maneewan K (2012) Screening of efficient halotolerant solubilizing bacterium and its effect on promoting plant growth under saline conditions. World Appl Sci J 16:1110–1117

Dardanelli MS, de Cordoba FJF, Espuny MR, Carvajal MAR, Diaz MES, Serrano AMG (2008) Effect of *Azospirillum brasilense* coinoculated with rhizobium on *Phaseolus vulgaris* flavonoids and Nod factor production under salt stress. Soil Biol Biochem 40:2713–2721

Dardanelli M, Manyani H, González-Barroso S, Rodríguez-Carvajal M, Gil-Serrano A, Espuny M, López-Baena F, Bellogín R, Megías M, Ollero F (2010) Effect of the presence of the plant growth promoting rhizobacterium (PGPR) *Chryseobacterium balustinum* Aur9 and salt stress in the pattern of flavonoids exuded by soybean roots. Plant Soil 328:483–493

De Salamon DIG, Hynes RK, Nelson LM (2005) Role of Cytokinins in plant growth promotion by rhizosphère bacteria. In: Siddiqui AA (ed) PGPR: biocontrol and biofertilization. Springer, Amsterdam, pp 173–195

Dell'Amico E, Cavalca L, Andreoni V (2008) Improvement of *Brassica napus* growth under cadmium stress by cadmium resistant rhizobacteria. Soil Biol Biochem 40:74–84

Del Rio LA, Corpas FJ, Sandalio LM, Palma JM, Barroso JB (2003) Plant peroxisomes, reactive oxygen metabolism and nitric oxide. IUBMB Life 55:71–81

Diamant S, Eliahu N, Rosenthal D, Goloubino P (2001) Chemical chaperones regulate molecular chaperones in vitro and in cells under combined salt and heat stresses. J Biol Chem 276:39586–39591

Dong JG, Fernandez-Maculet JC, Yang SF (1992) Purification and characterization of 1-aminocyclopropane-1-carboxylate oxidase from apple fruit. Proc Natl Acad Sci USA 89:9789–9793

Drew MC, Hole PC, Picchioni GA (1990) Inhibition by NaCl of net CO_2 fixation and yield of cucumber. J Am Soc Hortic Sci 115:472–477

Ecker JR (1995) The ethylene signal transduction pathway in plants. Science 268:667–675

Egamberdieva D (2008) Plant growth promoting properties of rhizobacteria isolated from wheat and pea grown in loamy sand soil. Turk J Biol 32:9–15

Egamberdieva D, Kucharova Z (2009) Selection for root colonizing bacteria stimulating wheat growth in saline soils. Biol Fertil Soils 45:563–571

Egamberdieva D, Jabborova D, Hashem A (2015) *Pseudomonas* induces salinity tolerance in cotton (*Gossypium hirsutum*) and resistance to *Fusarium* root rot through the modulation of indole-3-acetic acid. Saudi J Biol Sci 22:773–779

Egamberdieva D, Davranov K, Wirth S, Hashem A, Abd-Allah EF (2017) Impact of soil salinity on the plant-growth – promoting and biological control abilities of root associated bacteria. Saudi J Biol Sci 24:1601–1608

Ehsanpour AA, Amini F (2003) Effect of salt and drought stress on acid phosphatase activities in alfalfa (*Medicago sativa* L.) explants under in vitro culture. Afr J Biotechnol 2:133–135

FAO (2000) Global network on integrated soil management for sustainable use of salt effected soils. http://www.fao.org/ag/AGL/agll/spush/intro.htm

Fernando WGD, Nakkeeran S, Zhang Y (2005) Biosynthesis of antibiotics by PGPR and its relation in biocontrol of plant diseases. In: Siddiqui ZA (ed) PGPR: biocontrol and biofertilization. Springer, Cham, pp 67–109

Flowers TJ, Troke PF, Yeo AR (1977) The mechanism of salt tolerance in halophytes. Annu Rev Plant Physiol 28:89–121

Fuentes-Ramirez LE, Caballero-Mellado J (2005) Bacterial biofertilizers. In: Siddiqui ZA (ed) PGPR: biocontrol and biofertilization. Springer, Cham, pp 143–172
García-Fraile P, Menéndez E, Rivas R (2015) Role of bacterial biofertilizers in agriculture and forestry. AIMS Bioeng 2:183–205
Gaxiola R, Li J, Undurraga S, Dang LM, Allen GJ, Alper SL, Fink GR (2001) Drought- and salt-tolerant plants result from overexpression of the AVP1 Hpump. Proc Natl Acad Sci USA 99:11444–11449
Georgiev GI, Atkins CA (1993) Effects of salinity on N_2 fixation, nitrogen metabolism and export and diffusive conductance of cowpea nodules. Symbiosis 15:239–255
Gerhardt KE, Greenberg BM, Glick BR (2006) The role of ACC deaminase in facilitating the phytoremediation of organics, metals and salt. Curr Trends Microbiol 2:61–73
Gill SS, Tuteja N (2010) Reactive oxygen species and antioxidant machinery in abiotic stress. Plant Physiol Biochem 48:909–930
Glick BR (1995) The enhancement of plant growth by free-living bacteria. Can J Microbiol 41:109–117
Glick BR (2014) Bacteria with ACC deaminase can promote plant growth and help tofeed the world. Microbiol Res 169:30–39
Glick BR, Penrose DM, Li J (1998) A model for the lowering of plant ethylene concentrations by plant growth promoting bacteria. J Theor Biol 190:63–68
Glick BR, Chenz Z, Czarny J, Duan J (2007) Promotion of plant growth by ACC-deaminase producing soil bacteria. Eur J Plant Pathol 119:329–339
Gouffi K, Pica N, Pichereau V, Blanco B (1999) Disaccharides as a new class of non-accumulated osmoprotectants for *Sinorhizobium meliloti*. Appl Environ Microbiol 65:1491–1500
Günes A, Inal A, Alpaslan M (1996) Effect of salinity on stomatal resistance, proline and mineral composition of pepper. J Plant Nutr 19:389–396
Gupta B, Huang B (2014) Mechanism of salinity tolerance in plants: physiological, biochemical, and molecular characterization. Int J Genom 70:15–96
Gururani MA, Upadhyaya CP, Baskar V, Venkatesh J, Nookaraju A (2013) Plant growth-promoting rhizobacteria enhance abiotic stress tolerance in *Solanum tuberosum* through inducing changes in the expression of ROS-scavenging enzymes and improved photosynthetic performance. J Plant Growth Regul 32:245–258
Han HS, Lee KD (2005) Plant growth promoting rhizobacteria effect on antioxidant status, photosynthesis, mineral uptake and growth of lettuce under soil salinity. Res J Agric Biol Sci 1:210–215
Hasegawa PM, Bressan RA, Zhu J-K, Bonhert HJ (2000) Plant cellular and molecular responses to high salinity. Annu Rev Plant Physiol Plant Mol Biol 51:463–499
Honma M, Shimomura T (1978) Metabolism of 1-aminocyclopropane-1-carboxylic acid. Agric Biol Chem 42:1825–1831
Hontzeas N, Saleh SS, Glick BR (2004) Changes in gene expression in canola roots by ACC deaminase-containing plant growth-promoting bacteria. Mol Plant Microbe Interact 17:865–871
Ilangumaran G, Smith DL (2017) Plant growth promoting rhizobacteria in amelioration of salinity stress: a systems biology perspective. Front Plant Sci 8:1768. https://doi.org/10.3389/fpls.2017.01768
Jha Y, Subramanian RB (2013) Paddy plants inoculated with PGPR show better growth physiology and nutrient content under saline conditions. Chil J Agric Res 73:213–219
Jofré E, Fischer S, Rivarola V, Balegno H, Mori G (1998) Saline stress affects the attachment of *Azospirillum brasilense* Cd to maize and wheat roots. Can J Microbiol 44:416–422
Joseph B, Jini D (2010) Salinity induced programmed cell death in plants: challenges and opportunities for salt-tolerant plants. J Plant Sci 5:376–390
Jouyban Z (2012) Ethylene biosynthesis. Tech J Eng Appl Sci 1:107–100
Kant S, Kant P, Raveh E, Barak S (2006) Evidence that differential gene expression between the halophyte, *Thellungiella halophila*, and *Arabidopsis thaliana* is responsible for higher levels of

the compatible osmolyte proline and tight control of Na+ uptake in *T. halophile*. Plant Cell Environ 29:1220–1234

Kefu Z, Munns R, King RW (1991) Abscisic acid levels in NaCl-treated barley, cotton, and saltbush. Aust J Plant Physiol 18:17–24

Kempf B, Bremer E (1998) Uptake and synthesis of compatible solutes as microbial stress responses to high-osmolality environments. Arch Microbiol 170:319–330

Khan MY, Zahir ZA, Asghar HN, Waraich EA (2017) Preliminary investigations on selection of synergistic halotolerant plant growth promoting rhizobacteria for inducing salinity tolerance in wheat. Pak J Bot 49:1541–1551

Khalid S, Parvaiz M, Nawaz K, Hussain H, Arshad A, Shawaka S, Sarfaraz ZN, Waheed T (2013) Effect of Indole Acetic Acid (IAA) on morphological, biochemical and chemical attributes of Two varieties of maize (*Zea mays* L.) under salt stress. World Appl Sci J 26:1150–1159

Kumar SG, Reddy AM, Sudhakar C (2003) NaCl effects on proline metabolism in two high yielding genotypes of mulberry (*Morus alba* L.) with contrasting salt tolerance. Plant Sci 165:1245–1251

Lee G, Carrow RN, Duncan RR, Eiteman MA, Rieger MW (2008) Synthesis of organic osmolytes and salt tolerance mechanisms in *Paspalum vaginatum*. Environ Exp Bot 63:19–27

Lucy M, Reed E, Glick BR (2004) Applications of free living plant growth-promoting rhizobacteria. Antonie Van Leeu Wen Hook 86:1–25

Mahaffee WF, Kloepper JW (1994) Applications of plant growth promoting rhizobacteria in sustainable agriculture. In: Pankhurst CE, Doube BM, Gupta VVSR, Grace PR (eds) Soil biota: management in sustainable farming systems. CSIRO, Melbourne, pp 23–31

Maleki M, Mostafaee S, Mokhtarnejad L, Farzaneh M (2010) Characterization of *Pseudomonas fluorescens* strain CV6 isolated from cucumber rhizosphere in Varamin as a potential biocontrol agent. AJCS 4:676–683

Manaf HH, Zayed MS (2015) Productivity of cowpea as affected by salt stress in presence of endomycorrhizae and Pseudomonas fluorescens. Ann Agric Sci 60:219–226

Mantelin S, Touraine B (2004) Plant growth-promoting bacteria and nitrate availability: impacts on root development and nitrate uptake. J Exp Bot 55:27–34

Marschner H (1995) Functions of mineral nutrients. In: Marschner H (ed) Mineral nutrition of higher plants, vol 2. Academic, London, pp 229–312

Martínez-Morales LJ, Soto-Urzúa L, Baca BE, Sánchez-Ahédo JA (2003) Indole-3- butyric acid (IBA) production in culture medium by wild strain *Azospirillum brasilense*. FEMS Microbiol Lett 228:167–173

Martinez-Viveros O, Jorquera MA, Crowley DE, Gajardo G, Mora ML (2010) Mechanisms and practical considerations involved in plant growth promotion by rhizobacteria. J Soil Sci Plant Nutr 10:293–319

Mayak S, Tirosh T, Glick BR (2004) Plant growth-promoting bacteria that confer resistance in tomato and pepper to salt stress. Plant Physiol Biochem 42:565–572

Mc Neil SD, Nuccio ML, Hanson AD (1999) Betaines and related osmoprotectants: targets for metabolic engineering of stress resistance. Plant Physiol 120:945–949

Mhadhbi H, Jebara M, Limam F, Aouani ME (2004) Rhizobial strain involvement in plant growth, nodule protein composition and antioxidant enzyme activities of chickpea-rhizobia symbioses: modulation by salt stress. Plant Physiol Biochem 42(9):717–722

Miransari M (2012) Microbial products and soil stresses. In: Maheshwari DK (ed) Bacteria in agrobiology: stress management. Springer, Berlin, pp 65–75

Mishra J, Fatima T, Arora NK (2018) Role of secondary metabolites from plant growth-promoting rhizobacteria in combating salinity stress. In: Egamberdieva D, Ahmad P (eds) Plant microbiome: stress response, microorganisms for sustainability. Springer, Singapore, pp 127–163

Mohammadipanah F, Hamedi J, Dehhaghi M (2015) Halophilic bacteria: potentials and applications in biotechnology. In: Maheshwari DK, Saraf M (eds) Halophile: biodiversity and sustainable exploitation. Springer, Cham, pp 277–321

Munns R (2002) Comparative physiology of salt and water stress. Plant Cell Environ 25:239–250

Munns R, Tester M (2008) Mechanisms of salinity tolerance. Annu Rev Plant Biol 59:651–681

Munns R, Husain S, Rivelli AR, James RA, Condon AG, Lindsay MP, Lagudah ES, Schachtman DP, Hare RA (2002) Avenues for increasing salt tolerance of crops, and the role of physiologically based selection traits. Plant Soil 247(1):93–105

Nabti E, Sahnoune M, Ghoul M, Fischer D, Hofmann A, Rothballer M, Schmid M, Hartmann A (2010) Restoration of growth of durum wheat (*Triticum durum* var. Waha) under saline conditions due to inoculation with the rhizosphere bacterium *Azospirillum brasilense* NH and extracts of the marine alga *Ulva lactuca*. J Plant Growth Regul 29:6–22

Nabti E, Schmid M, Hartmann A (2015) Application of halotolerant bacteria to restore plant growth under salt stress. In: Maheshwari DK, Saraf M (eds) Halophile: biodiversity and sustainable exploitation. Springer, Cham, pp 235–359

Nadeem SM, Hussain I, Naveed M, Asghar HN, Zahir ZA, Arshad M (2006) Performance of plant growth promoting rhizobacteria containing ACC-deaminase activity for improving growth of maize under salt-stressed conditions. Pak J Sci 43:114–120

Naqvi SSM (1999) Plant hormones and stress phenomena. In: Pessarakli M (ed) Handbook of plant and crop stress. Marcel Dekker, New York, pp 709–730

Nascimento FX, Rossi MJ, Soares CRFS, McConkey BJ, Glick BR (2014) New insights into 1-aminocyclopropane-1-carboxylate (ACC) deaminase phylogeny, evolution and ecological significance. PLoS One 9. https://doi.org/10.1371/journal.pone.0099168

Naz I, Bano A, Ul-Hassan T (2009) Isolation of phytohormones producing plant growth promoting rhizobacteria from weeds growing in Khewra salt range, Pakistan, and their implication in providing salt tolerance to *Glycine max* (L.). Afr J Biotechnol 8:5762–5766

Niu SQ, Li HR, Paré PW, Aziz M, Wang SM, Shi H, Li J, Han QQ, Guo SQ, Li J, Guo Q, Ma Q, Zhang JL (2016) Induced growth promotion and higher salt tolerance in the halophyte grass *Puccinellia tenuiflora* by beneficial rhizobacteria. Plant Soil 407:217–230. https://doi.org/10.1007/s11104-015-2767-z

Numan M, Bashira S, Khana Y, Mumtaza R, KShinwaric Z, Khanb AL, Khanb A, Harrasib AAL (2018) Plant growth promoting bacteria as an alternative strategy for salt tolerance in plants: a review. Microbiol Res 209:21–32

Oren A (1999) Bioenergetic aspects of halophilism. Microbiol Mol Biol Rev 63:334–340

Ortíz-Castro R, Contreras-Cornejo HA, Macías-Rodríguez L, López-Bucio J (2009) The role of microbial signals in plant growth and development. Plant Signal Behav 4(8):701–712

Panwar M, Tewari R, Gulati A, Nayyar H (2016) Indigenous salt-tolerant rhizobacterium *Pantoea dispersa* (PSB3) reduces sodium uptake and mitigates the effects of salt stress on growth and yield of chickpea. Acta Physiol Plant 38:278. https://doi.org/10.1007/s11738-016-2284-6

Pastor V, Luna E, Mauch-Mani B, Ton J, Flors V (2013) Primed plants do not forget. Environ Exp Bot 94:46–56

Paulucci NS, Gallarato LA, Reguera YB, Vicario JC, Cesari AB, García de Lema MB, Dardanelli MS (2015) *Arachis hypogaea* PGPR isolated from Argentine soil modifies its lipids components in response to temperature and salinity. Microbiol Res 173:1–9

Penrose DM, Moffatt BA, Glick BR (2001) Determination of 1-aminocyclopropane-l-carboxylic acid (ACC) to assess the effects of ACC deaminase-containing bacteria on roots of canola seedlings. Can J Microbiol 47:77–80

Pérez-Alfocea F, Balibrea EM, Alarcón JJ, Bolarín CM (2000) Composition of Xylem and Phloem exudates in relation to the salt-tolerance of domestic and wild tomato species. J Plant Physiol 156(3):367–374

Pinedo I, Ledger T, Greve M, Poupin MJ (2015) *Burkholderia phytofirmans* PsJN induces long-term metabolic and transcriptional changes involved in *Arabidopsis thaliana* salt tolerance. Front Plant Sci 6:466. https://doi.org/10.3389/fpls.2015.00466

Qadir M, Ghafoor A, Murtaza G (2000) Amelioration strategies for saline soils: a review. Land Degrad Dev 11:501–521

Qiu QS, Guo Y, Dietrich MA, Schumaker KS, Zhu JK (2002) Regulation of SOS1, a plasma membrane NaC/HC exchanger in *Arabidopsis thaliana*, by SOS2 and SOS3. Proc Natl Acad Sci USA 99:8436–8441

Qurashi AW, Sabri SN (2011) Osmoadaptation and plant growth promotion by salt tolerant bacteria under salt stress. Afr J Microbiol Res 5(2):3546–3554

Raafat N, Tharwat E, Radwan E (2011) Improving wheat grain yield and its quality under salinity conditions at a newly reclaimed soil by using different organic sources as soil or foliar applications. J Appl Sci Res 7(1):42–55

Raaijmakers JM, Weller DM, Thomashow LS (1997) Frequency of antibiotic producing *Pseudomonas* spp. in natural environments. Appl Environ Microbiol 63:881–887

Rabie GH, Almadini AM (2005) Role of bioinoculants in development of salt-tolerance of *Vicia faba* plants under salinity stress. Afr J Biotechnol 4:210–222

Ramezani E, Sepanlou MG, Badi HAN (2011) The effect of salinity on the growth, morphology and physiology of *Echium amoenum* Fisch & Mey. Afr J Biotechnol 10:8765–8773

Raven JA (1985) Regulation of Ph and generation of osmolarity in vascular plants – a cost-benefit analysis in relation to efficiency of use of energy, nitrogen and water. New Phytol 101:25–77. https://doi.org/10.1111/j.1469-8137.1985.tb02816.x

Rhodes D, Nadolska-Orczyk A, Rich PJ (2002) Salinity, osmolytes and compatible solutes. In: Läuchli A, Lüttge U (eds) Salinity: environment – plants – molecules. Springer, Dordrecht, pp 181–204

Rodriguez-Navarro A, Rubio F (2006) High-affinity potassium and sodium transport systems in plants. J Exp Bot 57:1149–1160

Rojas-Tapias D, Moreno-Galvan A, Pardo-Diaz S, Obando M, Rivera D, Bonilla R (2012) Effect of inoculation with plant growth promoting bacteria (PGPB) on amelioration of saline stress in maize (*Zea mays*). Appl Soil Ecol 61:264–272

Saghafi K, Ahmadi J, Asgharzadeh A, Bakhtiari S (2013) The effect of microbial inoculants on physiological responses of two wheat cultivars under salt stress. Int J Adv Biol Biomed Res 4:421–431

Sairam RK, Tyagi A (2004) Physiology and molecular biology of salinity stress tolerance in plants. Curr Sci 86:407–421

Saleem M, Arshad M, Hussain S, Bhatti AS (2007) Perspective of plant growth promoting rhizobacteria (PGPR) containing ACC deaminase in stress agriculture. J Ind Microbiol Biotechnol 34:635–648

Sandhya V, Ali S, Grover M, Reddy G, Venkateswarlu B (2009) Alleviation of drought stress effects in sunflower seedlings by the exopolysaccharides producing *Pseudomonas putida* strain GAP-P45. Biol Fertil Soils 46:17–26

Santoyo G, Moreno-Hagelsieb G, Orozco-Mosqueda MC, Glick BR (2016) Plant growth-promoting bacterial endophytes. Microbiol Res 183:92–99

Sapsirisopa S, Chookietwattana K, Maneewan K, khaengkhan P (2009) Effect of salt-tolerant *Bacillus* inoculum on rice KDML 105 cultivated in saline soil. Asian J Food Agro-Ind S69–S74

Sarkar A, Ghosh PK, Pramanik K, Mitra S, Soren T, Pandey S, Mondal MH, Maiti TK (2018) A halotolerant Enterobacter sp. displaying ACC deaminase activity promotes rice seedling growth under salt stress. Res Microbiol 169:20–32

Schubert S, Neubert A, Schierholt A, Sümer A, Zörb C (2009) Development of salt-resistant maize hybrids: the combination of physiological strategies using conventional breeding methods. Plant Sci 177:196–202

Selvakumar G, Panneerselvam P, Ganeshamurthy AN (2012) Bacterial mediated alleviation of abiotic stress in crops. In: Maheshwari DK (ed) Bacteria in agrobiology: stress management. Springer, Berlin, pp 205–223

Shabala S (2013) Learning from halophytes: physiological basis and strategies to improve abiotic stress tolerance in crops. Ann Bot 112:1209–1221

Shaharoona B, Arshad M, Zahir ZA, Khalid A (2006) Performance of *Pseudomonas* spp. containing ACC-deaminase for improving growth and yield of maize (*Zea mays* L.) in the presence of nitrogenous fertilizer. Soil Biol Biochem 38:2971–2975

Sharma SS, Dietz K-J (2009) The relationship between metal toxicity and cellular redox imbalance. Trends Plant Sci 14:43–50

Sharma A, Johri BN, Sharma AK, Glick BR (2003) Plant growth-promoting bacterium *Pseudomonas* sp. strain GRP (3) influences iron acquisition in mung bean (*Vigna radiata* L. Wilczek). Soil Biol Biochem 35:887–894

Sharma S, Kulkarni J, Jha B (2016) Halotolerant rhizobacteria promote growth and enhance salinity tolerance in peanut. Front Microbiol 7:1600. https://doi.org/10.3389/fmicb.2016.01600

Shrivastava P, Kumar R (2015) Soil salinity: a serious environmental issue and plant growth promoting bacteria as one of the tools for its alleviation. Saudi J Biol Sci 22:123–131

Shukla PS, Agarwal PK, Jha B (2011) Improved salinity tolerance of *Arachis hypogaea* (L.) by the interaction of halotolerant plant-growth-promoting rhizobacteria. J Plant Growth Regul 31:195–206. https://doi.org/10.1007/s00344-011-9231-y

Siddikee MA, Glick BR, Chauhan PS, Sa T (2011) Enhancement of growth and salt tolerance of red pepper seedlings (*Capsicum annuum* L.) by regulating stress ethylene synthesis with halotolerant bacteria containing 1-aminocyclopropane-1-carboxylic acid deaminase activity. Plant Physiol Biochem 49:427–434

Singh RP, Jha PN (2016) The multifarious PGPR *Serratia marcescens* CDP-13 augments induced systemic resistance and enhanced salinity tolerance of wheat (*Triticum aestivum* L.). PLoS One 11(6):e0155026

Singleton PW, Bohlool BB (1984) Effect of salinity on nodule formation by soybean. Plant Physiol 74:72–76

Somers E, Vanderleyden J, Srinivasan M (2004) Rhizosphere bacterial signalling: a love parade beneath our feet. Crit Rev Microbiol 304:205–240

Spaepen S, Vanderleyden J, Remans R (2007) Indole-3-acetic acid in microbial and microorganism-plant signaling. FEMS Microbiol Rev 31:425–448

Stepien P, Johnson GN (2009) Contrasting responses of photosynthesis to salt stress in the glycophyte *Arabidopsis* and the halophyte *Thellungiella*: role of the plastid terminal oxidase as an alternative electron sink. Plant Physiol 149:1154–1165

Sudhir P, Murthy S (2004) Effects of salt stress on basic processes of photosynthesis. Photosynthetica 42:481–486

Tang X, Mu X, Shao H, Wang H, Brestic M (2015) Global plant-responding mechanisms to salt stress: physiological and molecular levels and implications in biotechnology. Crit Rev Biotechnol 35:425–437

Tank N, Saraf M (2010) Salinity-resistant plant growth promoting rhizobacteria ameliorates sodium chloride stress on tomato plants. J Plant Interact 5:51–58

Tester M, Davenport R (2003) Na^+ tolerance and Na^+ transport in higher plants. Ann Bot 91:503–527

Timmusk S, Wagner GH (1999) The plant-growth-promoting rhizobacterium *Paenibacillus polymyxa* induces changes in *Arabidopsis thaliana* gene expression: a possible connection between biotic and abiotic stress responses. Mol Plant Microbe Interact 12:951–959

Tripathi M, Munot HP, Van Loon LC (2007) Plant responses to plant-growth promoting rhizobacteria. Eur J Plant Pathol 119:243–254

Tu JC (1981) Effect of salinity on rhizobium-root-hair interaction, nodulation and growth of soybean. Can J Plant Sci 61:231–239

Turan M, Esitken A, Sahin F (2012) Plant growth promoting rhizobacteria as alleviators for soil degradation. In: Maheshwari DK (ed) Bacteria in agrobiology-stress management. Springer, Berlin, pp 41–63

Vacheron J, Desbrosses G, Bouffaud ML, Touraine B, Moënne-Loccoz Y, Muller D, Legendre L, Wisniewski-Dyé F, Prigent-Combaret C (2013) Plant growth promoting rhizobacteria and root system functioning. Front Plant Sci 4:1–19

Vaishnav A, Kumari S, Jain S, Varma A, Choudhary DK (2015) Putative bacterial volatile-mediated growth in soybean (*Glycine max* L. Merrill) and expression of induced proteins under salt stress. J Appl Microbiol 119:539–551. https://doi.org/10.1111/jam.12866

Vaishnav A, Varma A, Tuteja N, Choudhary DK (2016) PGPR-mediated amelioration of crops under salt stress. In: Choudhary DK et al (eds) Plant-microbe interaction: an approach to sustainable agriculture. Springer, Singapore, pp 205–226

Van Loon LC (2007) Plant responses to plant growth –promoting rhizobacteria. Eur J Plant Pathol 119:243–254

Van Rhijn P, Vanderleyden J (1995) The rhizobium-plant symbiosis. Microbiol Mol Biol Rev 59:124–142

Vurukonda SSKP, Vardharajula S, Shrivastava M, SkZ A (2016) Enhancement of drought stress tolerance in crops by plant growth promoting rhizobacteria. Microbiol Res 184:13–24

Wang Z, Pote J, Huang B (2003) Responses of cytokinins, antioxidant enzymes, and lipid peroxidation in shoots of creeping bentgrass to high root-zone temperatures. J Am Soc Hortic Sci 128:648–655

Wang W, Wu Z, He Y, Huang Y, Li X, Ye B-C (2018) Plant growth promotion and alleviation of salinity stress in *Capsicum annuum* L. by *Bacillus* isolated from saline soil in Xinjiang. Ecotox Environ Safe 164:520–529

Wong V, Dalal R, Greene R (2008) Salinity and sodicity effects on respiration and microbial biomass of soil. Biol Fertil Soils 44:943–953

Wu Z, Peng Y, Guo L, Li C (2014) Root colonization of encapsulated *Klebsiella oxytoca* Rs-5 on cotton plants and its promoting growth performance under salinity stress. Eur J Soil Biol 60:81–87

Xiong L, Zhu JK (2002) Salt-stress signal transduction. In: Scheel D, Wasternack C (eds) Plant signal transduction. frontiers in molecular biology series. Oxford University Press, Oxford, pp 165–197

Yamashita K, Kasai M, Yamamoto Y, Matsumoto H (1994) Stimulation of plasma-membrane HC-transport activity in barley roots by salt stress – possible role of increase in chloride permeability. Soil Sci Plant Nutr 40:555–563

Yang J, Kloepper JW, Ryu CM (2009) Rhizosphere bacteria help plants tolerate abiotic stress. Trends Plant Sci 14:1–4

Yildirim E, Taylor AG, Spittler TD (2006) Ameliorative effects of biological treatments on growth of squash plants under salt stress. Sci Hortic 111:1–6

Younesi O, Chaichi MR, Postini K (2013) Salt tolerance in alfalfa following inoculation with *Pseudomonas*. Middle-East J Sci Res 16:101–107

Yue H, Mo W, Li C, Zheng Y, Li H (2007) The salt stress relief and growth promotion effect of RS-5 on cotton. Plant Soil 297:139–145

Zafar-ul-Hye M, Ahmed M, Shahzad SM (2013) Synergistic effect of Rhizobia and plant growth promoting rhizobacteria on the growth and nodulation of lentil seedlings under axenic conditions. Soil Environ 32:79–86

Zahir ZA, Ghani U, Naveed M, Nadeem SM, Asghar HN (2009) Comparative effectiveness of *Pseudomonas* and *Serratia* sp. containing ACC-deaminase for improving growth and yield of wheat (*Triticum aestivum* L.) under salt-stressed conditions. Arch Microbiol 191:415–424

Zahran HH, Sprent JI (1986) Effects of sodium chloride and polyethylene glycol on root-hair infection and nodulation of *Vicia faba* L. plants by *Rhizobium leguminosarum*. Planta 167:303–309

Zakharova EA, Shcherbakov AA, Brudnik VV, Skripko NG, Bulkhin NS, Ignatov VV (1999) Biosynthesis of indole-3-acetic acid in *Azospirillum brasilense* insights from quantum chemistry. Eur J Biochem 259:572–576

Zhang H, Kim M-S, Dowd SE, Shi H, Pare PW (2008) Soil bacterial confer plant tolerance by tissue-specific regulation of the sodium transporter HKT1. Mol Plant Microbe Interact 21:737–744

Zhu J-K (2001) Plant salt tolerance. Trends Plant Sci 6:66–72

Chapter 15
Insights in the Physiological, Biochemical and Molecular Basis of Salt Stress Tolerance in Plants

Nisha Kumari, Kamla Malik, Babita Rani, Minakshi Jattan, Sushil, Ram Avtar, Sarita Devi, and Sunder Singh Arya

Abstract Salt stress is one of the most brutal abiotic stresses that arrests crop survival and productivity. It affects various physiological, biochemical and metabolic processes in plants, depending on severity and duration of the stress. Plant adaptation or tolerance to salinity stress involves complex physiological traits, metabolic pathways, and molecular or gene networks. In response to high salinity stress, various genes get upregulated, the products of which are involved either directly or indirectly in plant protection. Some of the genes encoding osmolytes, ion channels, receptors, components of calcium signaling and some other regulatory signaling factors or enzymes are able to confer salinity tolerance, when transferred to sensitive plants. Overall, the susceptibility or tolerance to high salinity stress in plants is a coordinated action of multiple stress-responsive genes. Plants that grow under different soil conditions are usually inhabited by microbes which are beneficial for the enhancement of their salinity tolerance mechanisms. Microbes, either independently or by interacting with plants, synthesize some organic osmolytes and other substances which offer an adaptive strategy to salinity stress. The partnership of plants with nitrogen-fixing bacteria and arbuscular mycorrhizal fungi can serve as a second possible and sustainable strategy to increase crop yields in saline soil. A comprehensive understanding on how plants respond to salinity stress at different

N. Kumari (✉) · B. Rani
Department of Biochemistry, CCS Haryana Agricultural University, Hisar, India

K. Malik
Department of Microbiology, CCS Haryana Agricultural University, Hisar, India

M. Jattan · R. Avtar
Department of Genetics and Plant Breeding, CCS Haryana Agricultural University, Hisar, India

Sushil
Department of Chemistry, CCS Haryana Agricultural University, Hisar, India

S. Devi
Department of Botany and Plant Physiology, CCS Haryana Agricultural University, Hisar, India

S. S. Arya
Department of Botany, Maharshi Dayanand University, Rohtak, India

B. Giri, A. Varma (eds.), *Microorganisms in Saline Environments: Strategies and Functions*, Soil Biology 56, https://doi.org/10.1007/978-3-030-18975-4_15

levels and an integrated approach of combining molecular tools with physiological and biochemical techniques are inevitable for the development of salt-tolerant varieties of plants in salt affected areas.

Keywords Salinity · Stress · Physiological · Biochemical · Microbe and tolerance

15.1 Introduction

Abiotic stress is one of the dreadful environmental stresses that affect the plant growth and crop yield even on irrigated areas throughout the world. Salinity stress is a major abiotic stress that limits crop production and agricultural sustainability in many areas of the world due to increasing use of low quality of water for irrigation. Salinization can also occur due to the close proximity to coastal areas or by anthropogenic activities like the use of poor irrigation practices that lead to the accumulation of high concentrations of salts. The reasons for increased salinity and sodicity in arid and semiarid regions are higher rate of evapotranspiration and unleached inorganic salts from the soil surface layers (Shannon et al. 1994). The worldwide salt stress affected area is more than 20% of cultivated land, and it is further increasing (Gupta and Huang 2014). The electrical conductivity of the soil saturation extract (EC) is the basis of classification for saline soils. The soils with ECs of 4 dS/m or more are accounted as saline soils, and soils with ECs exceeding 15 dS/m are considered strongly saline soils (Omami 2005). Saline soils contain high concentrations of salts such as $CaSO_4$ and Na_2CO_3, although sodium chloride (NaCl) is the most soluble, pervasive, and super abundant salt in soils all over the world. The key factors in the determination of salinity intensity are salt concentration in irrigation water, chemistry and physics of soils, type of plants, plant growth stages, and irrigation schedules (Vicente et al. 2004). Salt stress significantly affects the plant growth and development. Plants experience osmotic stress as the first stress when exposed to saline soil and it immediately affects plant growth (Horie et al. 2011). When salt level crosses the threshold limit, plants are unable to maintain ion homeostasis and growth, and it ultimately leads to ion toxicity (Munns and Tester 2008). Ion toxicity and osmotic stress cause oxidative stress and a series of secondary stresses. Plants have developed a large number of physiological and biochemical strategies to cope with primary and secondary effects of salt stress.

Microbial life which is the most fundamental and live system on the earth also plays an important role in salt stress mitigation in crop plants grown in saline soils. Microbes have become natural integral part of the crop production system. The intrinsic metabolic and genetic competencies make the microbes suitable organisms to combat extreme environmental conditions (Singh et al. 2014). Moreover, soil microbes also enhance water uptake through roots under both salinity and drought stresses (Dodd and Perez-Alfocea 2012). Microorganisms are important inhabitants of seeds and form symbiotic associations with the plants either at the surface or inside the roots, stems, or leaves through endophytic interactions. Approximately 80% of the terrestrial plant species make symbiotic associations with arbuscular mycorrhizal fungi (AMF) (Smith and Smith 2011). These micro symbionts play a critical role in plant nutrition and in return profited by the plant carbon. Therefore, the microbes interact with plant tissues and cells to different degrees, showing a positive impact on plant fitness and physiology in diverse ways by direct and indirect mechanisms (Kim et al. 2011).

Therefore, a comprehensive biological understanding and the knowledge of the effects of salt stress on plants are necessary to provide information on the plant responses to salinity stress which can be further utilized in the development of salt-tolerant crop varieties. Also, the plant-microbe interaction studies have crucial roles in ameliorating the impact of salinity on plants for agricultural sustainability.

15.2 Effects of Salinity on Plant Life

The saline conditions retard the plant growth however; the degree of reduction depends on salt level, plant type, and environmental conditions. Sometimes, a complete halt in plant growth may be reported as a result of lower osmotic potential that inhibits water and nutrients absorption by plant roots. High concentration of soluble salts enhanced their accumulation in the cells, which creates toxicity that attributes to chlorosis and necrosis of the leaf tissues (Wahome et al. 2001). Salinity stress may cause damage to plant physiological processes by overaccumulation of reactive oxygen species (ROS), ion toxicity, impairment of antioxidant defense systems, photosynthetic function, and imbalance of hormones (Kim et al. 2016). Plants grown under saline conditions may encounter four types of stresses, i.e., reduction in plant water uptake, severe ion toxicity, nutrient imbalances or deficiencies, and cell membrane damage or cell death due to reactive oxygen species generation. The morphological, physiological, and biochemical changes in plants grown under salt stress are presented in Fig. 15.1. The effects of salinity on plant life are broadly classified as physiological and biochemical effects.

15.2.1 Physiological Effects of Salt Stress

Salt stress affects all the major processes such as growth, nutrient imbalance, water relation, germination, photosynthetic pigments, photosynthesis, and yield (Hanin et al. 2016). The high osmotic potential and decrease in water availability could result in plant cell membrane dehydration and reduce the permeability of CO_2. Consequently, photosynthetic electron transport decreases via shrinkage of intercellular spaces. The smaller stomatal aperture and lower stomatal conductance due to disturbed water relations and sensitivity to abscisic acid (ABA) cause reduction in CO_2 supply so that the availability of CO_2 for carboxylation reactions

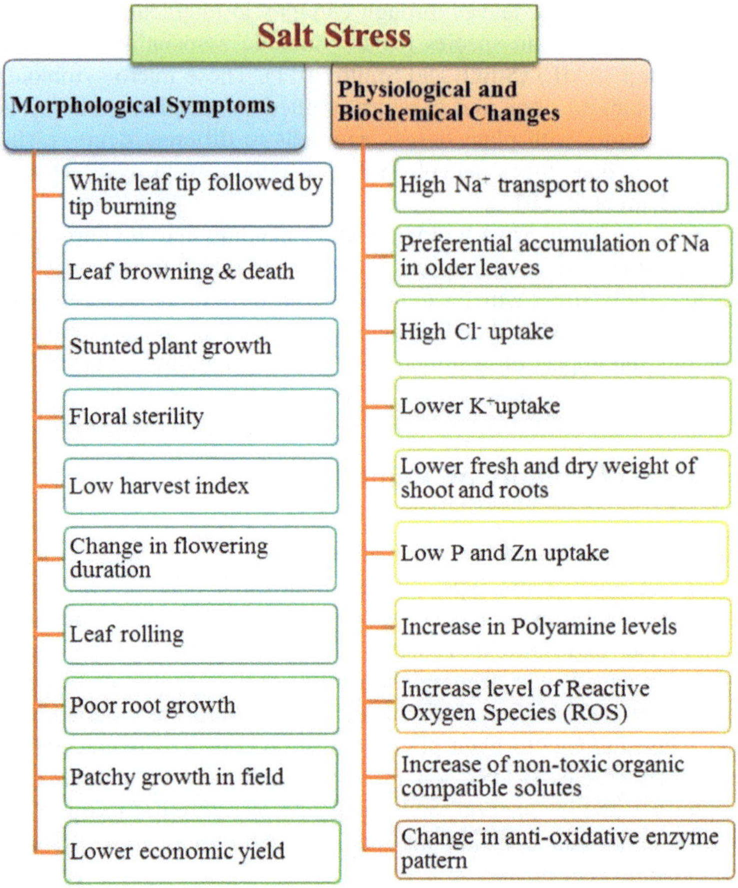

Fig. 15.1 An overview of morphological, physiological, and biochemical effects of salt stress in plants

is restricted (Zhao et al. 2017). The transpiration rate also decreases due to salt stress (Hnilikova et al. 2017). The chlorophyll a, chlorophyll b, and carotenoid concentration decreases under salt stress (Rahneshan et al. 2018; Hafsi et al. 2017). The chlorophyll fluorescence also decreases (Zhao et al. 2017). The leaf anatomy changes under salt stress, such as decrease in the thickness of the epidermis and mesophyll and decrease in intercellular spaces in the leaves (Parida et al. 2004) and a reduction in root length density (Vaughan et al. 2002). Salt stress results in nutrient imbalance with a decrease in the content of phosphorus, nitrogen, Ca^{2+}, Mg^{2+}, and K^+ (Gomes et al. 2011). The leaf relative water content decreases (Kapoor and Pande 2015; Roy and Chakraborty 2018) and membrane permeability increases (Dkhil and Denden 2012).

15.2.2 *Biochemical Effects of Salt Stress*

Besides direct impact of salinity on plants, a common consequence of salinity is induction of excessive accumulation of reactive oxygen species (ROS). Salt stress can lead to stomatal closure, which often reduces CO_2 availability in the leaves, inhibits carbon fixation, and exposes chloroplasts to excessive excitation energy, which in turn increase the generation of ROS such as superoxide ($O_2\cdot^-$), hydrogen peroxide (H_2O_2), hydroxyl radical (OH·), and singlet oxygen ($^1O_2$) (Parida and Das 2005; Ahmad and Sharma 2008). These ROS are highly reactive and may cause cellular damage through oxidation of lipids, proteins, and nucleic acids (Apel and Hirt 2004). The ROS-mediated membrane damage has been demonstrated to be a major cause of the cellular toxicity by salinity in different crop plants such as rice, tomato, citrus, pea, and mustard (Gueta-Dahan et al. 1997; Mittova et al. 2004; Ahmad et al. 2010). Proteins which serve as a reservoir of energy may also help in adjustment of osmotic potential in plants subjected to salinity. They may be synthesized de novo in response to salt stress or may be present constitutively at low concentration (Parvaiz and Satyawati 2008). It has been found that soluble protein content of leaves decreases in response to salinity (Wang and Nii 2000; Parida et al. 2002). Hence, osmotic potential in plant cells gets disturbed due to decrease in soluble proteins.

15.3 Strategies of Salt Stress Tolerance

There are various physiological, biochemical, and microbe-mediated mechanisms by which plants cope with the situations of salt stress.

15.3.1 *Physiological and Biochemical Basis of Salt Stress Tolerance*

Plants develop various physiological and biochemical strategies for the survival under high salt soils. Principal mechanisms include ion homeostasis and compartmentalization, ion transport and uptake, biosynthesis of osmoprotectants and compatible solutes, activation of antioxidant enzymes and synthesis of antioxidant compounds, synthesis of polyamines, generation of nitric oxide (NO), and modulation of hormones.

15.3.1.1 Ion Homeostasis

Glycophytes and halophytes although different in nature cannot tolerate the high salt concentration in their cytoplasm. In plant cells, high salt concentration causes ion toxicity via a large influx of Na^+ in the cells. The surplus salt is either transported to the vacuole or seize in older tissues which are ultimately sacrificed, thereby protecting the plant from salinity stress (Flowers et al. 1977). When plants absorbed Na^+, it can be discharged from cells or transported to the inactive metabolic regions. The vacuolar Na^+/H^+ antiporter is in charge of dividing Na^+ to the vacuole, in this way diminishing the Na^+ fixation in the cytoplasm. Two kinds of H^+ pumps are available in the vacuolar layer, for example, vacuolar type H^+-ATPase (V-ATPase) and the vacuolar pyrophosphatase (V-PPase) (De Lourdes Oliveira Otoch et al. 2001; Dietz et al. 2001). Among these, V-ATPase is the most prevailing H^+ pump present inside the plant cells. Amid nonstress condition it assumes a critical role in keeping up solute homeostasis, empowering secondary transport, and facilitating vesicle fusion. Under stressed conditions activity of V-ATPase determines the survivability of the plants (Dietz et al. 2001). The Na^+ is compartmentalized into vacuoles by Na^+/H^+ exchangers (NHXs). The Na^+ efflux from the plant roots is mediated by the activity of Na^+/H^+ antiporters catalyzed by the salt overly sensitive 1 (SOS1) protein (Keisham et al. 2018). The Na^+/H^+ antiporter and its effects on salt tolerance have been studied in many species. These studies indicate the role of a salt overly sensitive (SOS) stress signaling pathway in ion homeostasis, and three major proteins, SOS1, SOS2, and SOS3, are involved in salt tolerance (Hasegawa et al. 2000; Sanders 2000). Plant salt tolerance is not basically the tolerance against toxicity of Na^+ but also an adaptation to its secondary effects like water deficit, accessibility of K^+, etc. Plants maintain a high level of K^+ (ideally 100 mM) within the cytosol for cytoplasmic enzyme activities. Within the cell vacuole, K^+ concentration varied from 10 to 200 mM. Cell vacuole is the largest pool of K^+, and it has an important role in maintaining the turgor within the cell. It is transported against the concentration gradient via K^+ transporter and membrane channels. When the extracellular K^+ concentration is low, then high affinity K^+ uptake mechanism is active, while at high concentration of extracellular K^+, low affinity uptake mechanism is active. Therefore, the concentration of K^+ in the soil determines the uptake mechanism. In contrast, a very low concentration of Na^+ ion (about 1 mM or less) is maintained in the cytosol. Because of high concentration of Na^+ in the soil during salt stress, there is a competition between Na^+ and K^+ as they both offer a similar transport mechanism, in this way diminishing the uptake of K^+ (Munns and Tester 2008). Hence, keeping up cell Na^+/K^+ homeostasis has a critical role in plant survival under saline condition.

15.3.1.2 Osmotic Protection Through Compatible Solute Accumulation

Plant metabolism is adversely affected via osmotic stress under salinity. The response to osmotic stress predominantly includes osmotic changes. For keeping up cell turgor, osmotic modification is vital, which allows the running of plant metabolic movements and accordingly plant development and productivity (Sharp et al. 1990). Plants synthesize various compatible solutes known as osmolytes which are responsible for osmotic balance at the cellular level. These are a group of chemically diverse organic compounds that are uncharged, polar, and soluble in nature and do not interfere with the cellular metabolism even at high concentration. They mainly include proline, glycine betaine, sugar, and polyols. Organic osmolytes are synthesized and accumulated in varying amounts among different plant species. For example, glycine betaine accumulation is restricted among few members of Plumbaginaceae (Hanson et al. 1994), whereas accumulation of amino acid proline occurs in taxonomically diverse sets of plants (Saxena et al. 2013).

Proline is an important osmolyte with high water solubility, low molecular weight, and no net charge in the physiological pH range. When plants are exposed to salt stress, there is an increase in proline concentration. However, concentration of total free amino acids (i.e., cysteine, arginine, and methionine) decreases (El-Shintinawy and El-Shourbagy 2001). In plants, proline occurs in free state. Its biosynthesis is triggered to lessen osmotic stress under salinity, and it tends to accumulate in the plant cells (Zhang and Shi 2013). Proline accumulation is a well-known measure adopted for alleviation of salinity stress (Ben Ahmed et al. 2010; Saxena et al. 2013). Intracellular proline which is amassed in the saline conditions gives tolerance toward stress as well as acts as a natural nitrogen reserve throughout stress recovery. It functions as an O_2 quencher thereby revealing its antioxidant capability. Ben Ahmed et al. (2010) observed that proline supplements enhanced salt tolerance in olive (*Olea europaea*) by amelioration of some antioxidative enzyme activities, photosynthetic activity, and plant growth and the preservation of a suitable plant water status under salinity conditions. It has been reported that proline improves salt tolerance in *Nicotiana tabacum* by increasing the activity of enzymes involved in antioxidant defense system (Hoque et al. 2008).

Glycine betaine is an amphoteric quaternary ammonium compound usually found in microorganisms, higher plants, and animals and is electrically neutral over a wide range of pH. It is very much soluble in water but also has nonpolar moiety constituting 3-methyl groups. It can interact with both hydrophobic and hydrophilic regions of the macromolecules, such as enzymes and protein complexes because of its distinctive basic component. Glycine betaine raises the osmolarity of the cell under salinity; in this way, it plays an imperative role in stress mitigation. Likewise, it secures the cell by osmotic alteration (Gadallah 1999), protein stabilization (Makela et al. 2000), and reduction of ROS (Saxena et al. 2013). Studies have demonstrated that glycine betaine can hinder the transportation speed and amount of Na^+ and Cl^- from roots to the aerial parts and elevate K^+ transport to improve salt tolerance in barley (Dietz et al. 2001). Spraying glycine betaine to the maize leaves resulted in an augment in relative water content and decrease in the osmotic potential of leaves under salinity (De Lourdes Oliveira Otoch et al. 2001).

Polyols are compounds with various hydroxyl groups accessible for organic reactions. Sugar alcohols are a class of polyols working as companionable solutes, as low molecular weight chaperones, and as ROS-scavenging compounds (Saxena et al. 2013). They can be grouped into two noteworthy sorts, cyclic (e.g., pinitol) and noncyclic (e.g., mannitol). These solutes work as a defender or stabilizer of proteins or membrane structures that are susceptible to stress or harm provoked by ions. Therefore, these solutes assume an important role in mitigation of stress.

The important function of carbohydrates such as sugars in alleviation of stress includes osmoprotection, carbon storage, and ROS scavenging. Salinity enhances the level of reducing sugars (i.e., sucrose and fructans) in different plants (Kerepesi and Galiba 2000). Sucrose content was found to increase in tomato (*Lycopersicon lycopersicum*) under salinity due to increased activity of sucrose phosphate synthase (Gao et al. 1998). Similarly, sugar content was found to increase in some of the rice genotypes in a study conducted by Alamgir and Yousuf Ali (1999).

15.3.1.3 Role of Antioxidants in Reactive Oxygen Species Scavenging

Oxygen is one of the obligatory components in plants. It is associated with mito chondrial respiration, metabolism, and oxidative phosphorylation to generate energy. However, during metabolism oxygen is converted into reactive oxygen species. Reactive oxygen species (ROS), i.e., singlet oxygen, hydroxyl radical, superoxide radical, hydrogen peroxide, etc., increase in the plants during salinity stress, thereby causing cytoplasmic membrane damage, irreversible metabolic dysfunction, and cell death (Mittler 2002). When plants are subjected to salt stress, malondialdehyde (MDA) is the major artifact of membrane lipid peroxidation, and its level corresponds to degree of cell membrane damage. In this manner, MDA substance can be a sign of salt stress and tolerance.

Antioxidant enzymes and nonenzymatic compounds play a critical role in detoxifying ROS induced by salinity stress. Salinity tolerance is positively correlated with the activity of antioxidant enzymes like superoxide dismutase (SOD), catalase (CAT), glutathione peroxidase (GPX), ascorbate peroxidase (APX), glutathione reductase (GR), and with the accumulation of nonenzymatic antioxidant compounds (Asada 1999).

Superoxide dismutase (SOD) can evacuate the surplus superoxide anions in the cells and hence act as the first line of defense. It is classified into three groups in higher plants. Based on the metal ions present on auxiliary sites, they are iron SOD (Fe-SOD), copper/zinc SOD (Cu/Zn-SOD), and manganese SOD (Mn-SOD). The SOD can dismutate the superoxide anion into the O_2 and H_2O_2 and abolish the superoxide toxicity. In normal conditions, the ROS delivered in chloroplast are converted to H_2O_2 and O_2 by SOD, and H_2O_2 and its subordinates are quickly expelled or diminished by APX or GPX (Ma et al. 2012).

Ascorbate peroxidase (APX) is one of the important enzymes that eradicate H_2O_2. Hydrogen peroxide formed by Miller reaction in chloroplast mainly reduced by APX. There was an augmentation of both *S*-nitrosylated APX and APX activity in pea plants grown under saline condition (150 mM NaCl) (Begara-Morales et al. 2014). The adverse effects of salt stress in different plant species are mitigated by exogenous application of ascorbate, and it endorses plant recovery from the salt stress (Agarwal and Shaheen 2007; Munir and Aftab 2011).

Catalase (CAT) is another important enzyme that reduces H_2O_2 and it is present in peroxisomes. It differs from the APX because it has high affinity for hydrogen peroxide and it necessitates a reductive substrate. Hydrogen peroxide formed by light reaction is mainly removed by CAT. The effects of NaCl on the H_2O_2 content and CAT activity were studied in diverse groups of plants, such as unicellular alga, *Chlorella* sp., an aquatic macrophyte, *Najas graminea*, and a mangrove plant, *Suaeda maritima*, all showing high tolerance to NaCl. Significant accumulation of H_2O_2 was observed in all the tested plants upon their exposure to high levels of NaCl, and CAT activity increased significantly in response to the NaCl treatment (Liu et al. 2014).

Glutathione acts as scavenger of free radical as it can respond with hydrogen peroxide, hydroxyl radical, and superoxide radical. It can also take part in the renewal of ascorbate via ascorbate-glutathione cycle (Foyer et al. 1997). Glutathione assisted in preserving the permeability of plasma membrane and viability of cell during salt stress in *Allium cepa* when applied exogenously (Aly-Salama and Al-Mutawa 2009).

15.3.1.4 Role of Polyamines

Polyamines (PA) are low molecular weight, ubiquitous, small, and polycationic aliphatic molecules extensively distributed throughout the plant kingdom. Polyamines become involved in different roles such as development (i.e., somatic embryogenesis, dormancy breaking, differentiation, morphogenesis, and regulation of cell proliferation), seed germination, normal growth, etc. (Knott et al. 2007; Gupta et al. 2013). It has been observed that when the plants are exposed to salt stress, there is an increase in level of endogenous polyamines. The most common polyamines are diamine putrescine (PUT), triamine spermidine (SPD), and tetra-amine spermine (SPM) (Hussain et al. 2011; Shu et al. 2012).

The smallest polyamine is the diamine putrescine and is formed from either arginine or ornithine (Gupta et al. 2013; Hasanuzzaman et al. 2014). The diamine putrescine acts as a main substrate for the biosynthesis of higher polyamines (i.e., triamine spermidine and tetra-amine spermine). The significant consequences of polyamines have been connected with the regulation of gene expression, reduction in ROS production, for the synthesis of osmotically active solutes, maintenance of membrane integrity, and managing reserve of Na^+ and Cl^- ion in different organs (Afzal et al. 2009).

15.3.1.5 Role of Nitric Oxide

Nitric oxide (NO) is a small volatile gaseous molecule, which is involved in the regulation of various plant growth and developmental processes, such as root growth, respiration, stomata closure, flowering, seed germination, cell death and stress responses, as well as a stress signaling molecule (Lamattina et al. 2003; Crawford 2006; Besson-Bard et al. 2008; Zhao et al. 2009). Nitric oxide reacts with lipid radicals, thus preventing lipid oxidation and exerting a protective effect by scavenging superoxide radical and formation of peroxynitrite that can be neutralized by other cellular processes. It also helps in the activation of antioxidant enzymes (Bajguz 2014). Thus, positive effects of NO on salinity tolerance or stress mitigation have been attributed to antioxidant activities and modulation of ROS detoxification system (Mishra et al. 2011). Improved plant growth under salinity stress by exogenous application of NO was associated with increase in antioxidant enzymes such as SOD, CAT, GPX, APX, and GR (Zhao et al. 2004) and suppression of malondialdehyde (MDA) production or lipid peroxidation (Nalousi et al. 2012). Effects of nitric oxide on salinity tolerance are also related to its regulation of plasma membrane H^+-ATPase and Na^+/K^+ ratio (Crawford 2006).

15.3.1.6 Role of Hormone and Hormone-Like Substances

Abscisic acid (ABA) is an important phytohormone whose application to plant ameliorates the effect of stress conditions. It has long been recognized as a hormone which is upregulated due to soil water deficit around the root. Salinity stress causes osmotic stress and water deficit, increasing the production of ABA in shoots and roots (Cabot et al. 2009). The accumulation of ABA can mitigate the inhibitory effect of salinity on photosynthesis, growth, and translocation of assimilates (Popova et al. 1995; Jeschke et al. 1997). The positive relationship between ABA accumulation and salinity tolerance has been at least partially attributed to the accumulation of K^+, Ca^{2+}, and compatible solutes (proline and sugars) in vacuoles of roots, which counteract with the uptake of Na^+ and Cl^- (Gurmani et al. 2011).

Compounds such as salicylic acid (SA) and brassinosteroids (BR) have hormonal properties and participate in plant abiotic stress responses. Under salinity stress, endogenous level of SA increased along with the increase in the activity of salicylic acid biosynthetic enzyme as reported in rice seedlings (Sawada et al. 2006). The application of SA also promoted salinity tolerance in barley, as manifested by increase in the content of chlorophyll and carotenoid and maintaining membrane integrity (El-Tayeb 2005). However, according to Nazar et al. (2011), alleviation of decreased photosynthetic rate in mung bean cultivars under salt stress is due to enhanced nitrogen, sulfur assimilation, and antioxidant metabolism. The negative effects of salinity may also be mitigated by BR (Krishna 2003; Ashraf et al. 2010). Application of BR enhanced the activity of antioxidant enzymes (SOD, POX, APX, and GPX) and the accumulation of nonenzymatic antioxidant compounds (tocopherol, ascorbate, and reduced glutathione) (El-Mashad and Mohamed 2012). Hence, both BRs and SA are ubiquitous in the plant kingdom, affecting plant growth and development in many different ways and are known to improve plant salt stress tolerance.

15.4 Molecular Basis of Salt Tolerance

Salt tolerance is a multigenic feature. A large number of salt-responsive transcription factors and genes which are either upregulated or downregulated in different plant species in response to salinity stress have been identified and characterized using transcriptomic and genomic approaches.

The *SOS* gene family is among the stress-responsive genes, and it is believed to play a very intriguing role in ion homeostasis, thereby conferring salt tolerance (Sairam and Tyagi 2004). The *SOS1* gene encodes a protein which is characterized by a long cytosolic C-terminal tail, comprising a putative nucleotide binding motif and an autoinhibitory domain. This protein acts as a plasma membrane Na^+/H^+ antiporter, essential in regulating Na^+ efflux at cellular level. It also facilitates long-distance transport of Na^+ from root to shoot. Overexpression of this gene confers salt tolerance in plants (Shi et al. 2002). Another gene, *SOS2*, encodes a protein called as serine/threonine kinase. This protein consists of a well-developed N-terminal catalytic domain and a C-terminal regulatory domain (Liu et al. 2000). The C-terminal regulatory domain of SOS2 protein interacts with salt stress elicited Ca^{2+} through a FISL motif (also known as NAF domain). The third type of protein involved in the SOS stress signaling pathway is the SOS3 protein encoded by *SOS3* gene. It is a myristoylated Ca^{2+} binding protein and contains a myristoylation site at its N-terminus. This site plays an essential role in conferring salt tolerance (Ishitani et al. 2000). The Ca^{2+} along with the SOS proteins act as a modulator of intracellular Na^+ homeostasis. An increased Na^+ concentration stimulates a rapid increase in the intracellular Ca^{2+} level which in turn promotes SOS3 protein binding with it. The SOS3 protein then interacts with the SOS2 protein, and this protein complex in bound state activates the kinase (Guo et al. 2004). The activated kinase then phosphorylates SOS1 protein and hence increases the transport activity (Quintero et al. 2002). The phosphorylated SOS1 results in the increased Na^+ efflux, reducing Na^+ toxicity (Martinez-Atienza et al. 2007). Overexpression of the plasma membrane Na^+/H^+ antiporter gene *SOS1* has been shown to increase salt tolerance of *Arabidopsis* (Ishitani et al. 2000).

Proline is overproduced in salt-stressed plants. It is due to the upregulation of genes for two major enzymes, i.e., pyrroline carboxylic acid synthetase and pyrroline carboxylic acid reductase (Sairam and Tyagi 2004). Some ROS-scavenging and osmotic regulator genes are also upregulated by salinity in some plant species. In angiosperms, three *CAT* genes, *CAT1*, *CAT2*, and *CAT3*, are upregulated to overproduce CAT which removes H_2O_2 in an energy efficient way (Mhamdi et al. 2010).

There are some transcription factors which are regulated by different kinases and have been found significant players of plant adaptation to salinity stress. The transcription factor genes are capable of controlling the expression of a broad range of target genes by binding to the specific *cis*-acting element in the promoters of these genes. Schmidt et al. (2013) identified salt-responsive ERF1 (*SERF1*), a rice (*Oryza sativa*) transcription factor gene that showed a root-specific induction upon salt and H_2O_2 treatment. They also revealed that SERF1-dependent genes are H_2O_2 responsive and SERF1 binds to the promoters of *mitogen-activated protein kinase kinase kinase 6* (*MAP 3K6*), *MAPK5*, *dehydration-responsive element binding2A* (*DREB2A*), and *zinc finger protein179* (*ZFP179*) in vitro and in vivo. SERF1 also directly induces its own gene expression. Hence, the plants deficient for SERF1 are more sensitive to salt stress compared with the wild type, while constitutive overexpression of *SERF1* gene improves salinity tolerance. Overexpression of a NAC transcription factor in both rice and wheat confers salt tolerance, thereby predicting their role in stress mitigation (Nakashima et al. 2007).

Abscisic acid modulates the expression of a number of salt and water deficit-responsive genes. Fukuda and Tanaka (2006) demonstrated the effects of ABA on the expression of two genes, *HVP1* and *HVP10*, in *Hordeum vulgare* under salinity stress. ABA treatment in wheat induced the expression of *MAPK4*-like, *TIP1*, and *GLP1* genes under salinity stress (Keskin et al. 2010). Another factor which plays a chief role at molecular level is alternative splicing. It is prevalent in plants exposed to environmental stresses. A suite of Ser/Arg-rich proteins that are key regulators of alternative splicing undergoes alternative themselves in response to various abiotic stresses, such as salt stress (Tanabe et al. 2007; Staiger and Brown 2013).

Downregulated genes are also emerging as essential components of the response to salinity. For example, downregulation of carotene hydroxylase increases carotene and total carotenoids enhancing salt stress tolerance in transgenic cultured cells of sweet potato (Kim et al. 2012). Hence, mutual regulation mechanisms exist between different genes and proteins and signals underlying different processes of plant adaptation to salt stress.

15.5 Microbe-Mediated Salt Tolerance

Microorganisms live in diverse environmental conditions and have natural metabolic and genetic competencies to aggravate different abiotic stresses in the plants (Gopalakrishnan et al. 2015). Under adverse environmental conditions, microbes interact with the plants and modulate their local and systemic mechanisms to cope with these conditions. Microorganisms which are occupied inside the seed proliferate with the seed germination and growth to form symbiotic relations either at the surface or inside the tissues like roots, stems, or leaves. Bacteria and fungi behave as active partners with host plants by forming associations with the roots (Lebeis 2014). Therefore, the interactions between the roots and microbes inside the soil play a crucial role in normal growth and fighting against various types of stresses. The

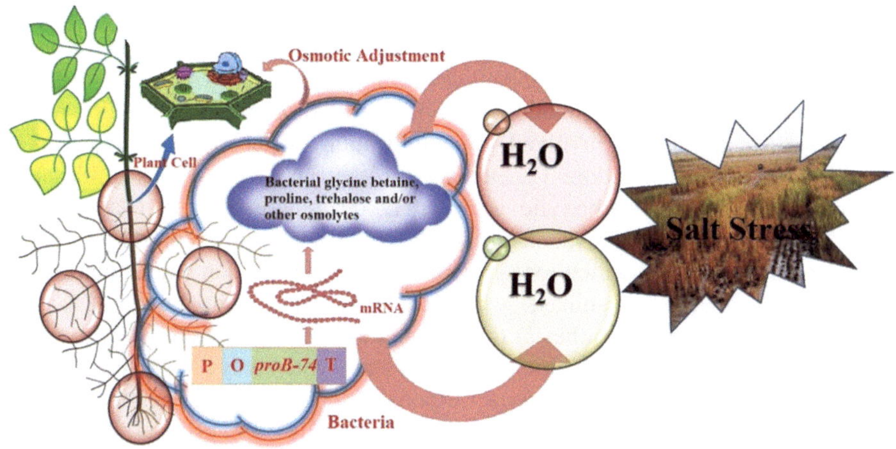

Fig. 15.2 Generalized mechanism depicting microbe-mediated salinity stress tolerance in plants

microbes interact with plant tissues and cells to different degrees (Montesinos et al. 2002), showing a positive impact on plant fitness and physiology in diverse ways by direct and indirect mechanisms (Kim et al. 2011). In direct mechanisms, bacteria release compounds which enhance uptake of essential macro- and micronutrients from the soil. Bacteria produce plant growth regulators and siderophores. They are also responsible for solubilization of phosphorus, potassium, and atmospheric nitrogen fixation. The indirect mechanisms involve production of HCN and antifungal compounds. Some antagonistic activities toward plant pathogens also occur in indirect mechanisms of action. Therefore, the microbes which are salt stress tolerant can be isolated by various selection and screening techniques and these can serve as a boon to stress-prone regions.

Under the salinity stress microorganisms trigger rapid fluxes of cell water along the osmotic gradient out of the cell and accumulate large amounts of organic osmolytes. The organic osmolytes such as trehalose, proline, glycine betaine, etc. offer an adaptive strategy to abiotic stresses, including high salinity. In the plant-microbe interaction, microbes modulate the transcriptional activator machinery of plants and enhance the salinity stress by several folds (Fig. 15.2).

Most of the root-associated microbes belong to the endophytic bacteria and plant growth-promoting rhizobacteria (PGPR). Both PGPR and endophytic bacteria behave in a similar way (Dimkpa et al. 2009), as most of the rhizobacteria also undergo colonization on the surface of root called rhizoplane and even penetrate the root cortex zone called endosphere. According to Gray and Smith (2005), there are two categories of PGPR, i.e., extracellular PGPR (ePGPR) and intracellular PGPR (iPGPR). The ePGPR is found on the rhizoplane and in the apoplastic space of the endosphere. The iPGPR is found inside the roots forming nodular structures (*Rhizobia* and *Frankia* species). Both rhizospheric and endophytic bacteria play a crucial role in improving salinity tolerance in plants. The PGPR is used worldwide as biofertilizers, phytostimulators, or biocontrol agents because of their remarkable positive impact on crop yield and tolerance to various biotic and abiotic stresses (Lugtenberg and Kamilova 2009; Upadhyay et al. 2012). Induced systemic tolerance (IST) or microbe-mediated induction of abiotic stress responses is caused by multiple bacterial determinants involved in solubilization of phosphate, production of volatiles, siderophores, indole-3-acetic acid (IAA), and exopolysaccharides and 1-aminocyclopropane-1-carboxylic acid (ACC)-deaminase activity (Farag et al. 2013; Kumari et al. 2015; Nadeem et al. 2016). Under saline conditions, plants produce ethylene due to which root growth is retarded (Mahajan and Tuteja 2005). PGPR produces ACC deaminase, which catalyzes the conversion of ACC (the precursor of ethylene biosynthesis) to ammonia and α-ketobutyrate. Hence, plants can overcome root growth inhibition under salt stress when they associate with these ACC deaminase-producing bacteria responsible for reduced ethylene levels (Hardoim et al. 2008; Nadeem et al. 2010; Ali et al. 2014; Deepti et al. 2014). The PGPR can enhance the uptake of macro- and micronutrients in plants. The rhizospheric bacteria belonging to the genera *Pseudomonas* (Sorty et al. 2016), *Azotobacter* (Sahoo et al. 2014), *Azospirillum* (Omar et al. 2009), *Rhizobium*, *Pantoea* (Sorty et al. 2016), and *Bacillus* (Sorty et al. 2016) play an important role in enhancing plant growth and mitigation of various abiotic stresses. When the tomato plants (*Solanum lycopersicum*) grown under saline conditions were inoculated with *Achromobacter piechaudii* (Betaproteobacteria), increased the uptake of phosphorous and potassium which improved water use efficiency (Mayak et al. 2004). Similarly, in wheat leaves, nitrogen, phosphorus, and potassium levels were increased significantly when plants grown under salt stress were inoculated with *Bacillus aquimaris* (Bacilli) (Upadhyay and Singh 2015). Hence, these studies favor the role of PGPR in P solubilization and nitrogen fixation asymbiotically. Also, PGPR has role in Na+ exclusion and K^+ uptake in plants to keep the K+/Na+ ratio high (Han et al. 2014). The PGPR secreted exopolysaccharides promote biofilm formation on root surfaces, thereby preventing influx of Na^+ in the roots (Dodd and Perez-Alfocea 2012).

Among the microbes, mycorrhiza fungi are also important plant growth promoters. These have two categories, i.e., arbuscular mycorrhizal fungi (AMF) and vesicular-arbuscular mycorrhizal (VAM) fungi. These form associations with the host plant externally (ectomycorrhizae) or endosymbiotically (VAM). The extensive networking by very fine hyphae of fungi increases overall nutrient uptake by the roots. *Piriformospora indica* (a root fungal endophyte) increased the antioxidant and hence improved the growth of barley plants under salt (Baltruschat et al. 2008; Franken 2012). The application of endophytic fungus, *Trichoderma harzianum*, improved the uptake of nutrients, enhanced antioxidant and osmolyte production, and decreased NaCl uptake in salt affected Indian mustard (*Brassica juncea*). The oil content was also reported to be enhanced (Ahmad et al. 2015). The plants associated with AMF maintain K^+/Na^+ ratio high by increasing absorption of K^+ and Ca^{2+} under salinity. However, the uptake or translocation of toxic Na^+ is avoided (Hammer et al. 2011; Evelin et al. 2013). Ruiz-Lozano and Aroca (2010) reported the upregulation of aquaporin genes by the AMF allowing an efficient water intake. The multifaceted actions of microbes advocate them as a strong and viable option for abiotic stress mitigation in crop plants.

15.6 Genetic Engineering for Salinity Stress Tolerance

Salt tolerance is generally controlled by large numbers of genes and by different physiological mechanisms. Consequently, examining the large number of salt related genes is obligatory. The analysis of genes related to salt gives a hypothetical premise for augmentation of the stress signal system and upgrading plant tolerance to stress (Liang et al. 2018). Genetic transformation technology enables scientists to achieve gene transfer in precise and predictable manner. Therefore, they focus on the transformation of plants for improving salinity tolerance by manipulating the osmoprotectant biosynthetic pathways for accumulating molecules that act by scavenging ROS, reducing lipid peroxidation, maintaining protein structure and functions, etc. (Abogadallah 2010; Saxena et al. 2013). Engineering plants for overexpression of these genes seems to be an effective method for generating salt-tolerant plants. Gene expression studies using constitutive promoters provide limited biological information compared with the use of inducible promoters or cell type-specific promoters. As strong abiotic stress-inducible promoters are required for transgene expression responsible for different abiotic stress tolerance at different growth stages, selection of a suitable promoter is critical. The broad spectrum of stress-inducible promoters is available that differ in their ability to regulate the expression patterns of transgenes for improved salinity stress tolerance in crop plants. The selection of these promoters can significantly affect the results from a transgenic manipulation.

In transgenic *A. thaliana*, salt tolerance is fundamentally enhanced by vacuolar Na^+/H^+ antiporter gene *AtNHX1* and plasma membrane Na^+/H^+ antiporter gene *SOS1* (He et al. 2005). A substantial number of genes, for example, HKT and NHX, encoding K^+ transporters and channels have been distinguished and cloned in different plant species. The class 1 HKT transporters, distinguished in Arabidopsis, shield the plant from the antagonistic impacts of salinity by anticipating overabundance aggregation of Na^+ in leaves (An et al. 2008). Similar results were examined in the experiment with rice where class 1 HKT transporter eliminates surplus Na^+ from xylem, therefore guarding the photosynthetic leaf tissues from the toxic cause of Na^+ (Schroeder et al. 2013). In transgenic wheat plant genes such as *Ta-UnP*, *TaZNF*, *TaSST*, *TaDUF1*, and *TaSP* can considerably develop the salt tolerance (Liang et al. 2014; Ma et al. 2016).

Studies conducted by Sun et al. (2010) revealed that an overexpressing salt-responsive gene, *ZFP179*, encoding a Cys2/His2 zinc finger protein enhanced salt tolerance in rice. There are other reports which suggest that the overexpression of genes for signaling proteins like kinases and phosphatases also have a role in salt tolerance. The mitogen-activated protein kinases (MAPKs) which are conserved signaling proteins play a pivotal role in various stresses faced by crop plants. Various transgenic plants engineered with MAPK cascade have been found tolerant to salt stress (Zhang et al. 2011).

In conclusion, for the effective use of genetic engineering methods for the production of salt-tolerant crops, an extensive number of salt-responsive genes, method of gene transfer, and their promoter sequences should be distinguished.

15.7 Future Perspectives

Salinity tolerance involves a complex of responses either at cellular or whole-plant levels through physiological, biochemical and molecular mechanisms. Extensive research through cellular, metabolic, and physiological analyses has elucidated that among various salinity responses, mechanisms controlling ion uptake, transport and balance, osmotic regulation, hormone metabolism, antioxidant metabolism, and stress signaling play critical roles in plant adaptation to salinity stress. Taking advantage of the recent techniques of genomic, transcriptomic, proteomic, and metabolomic, plant biologists are focusing on the development of a complete profile of genes, proteins, and metabolites responsible for different mechanisms of salinity tolerance in different plant species. However, there is lack of the integration of information from genomic, transcriptomic, proteomic, and metabolomics studies, and the combined approach is essential for the determination of the key pathways or processes controlling salinity tolerance. In spite of significant progress in the understanding of plant stress responses, more work is necessary to unravel the complexities of the innumerable signal transduction networks in plants. The salt overly sensitive signal transduction pathway is clearly important in salinity tolerance, although the mechanism of action at the whole-plant level remains to be established. To benefit more from the new genomics approaches, molecular studies with plants grown in physiologically realistic conditions are needed. Signaling pathways at the intracellular level have been well described, but long-distance signaling requires more attention. The future focus should be on the study of intercellular and intracellular molecular interactions involved in salinity stress response. Genetic engineering has been proved to be an efficient approach to the development of salinity-tolerant plants. However, this approach will prove more powerful only when large number of candidate genes associated with salinity tolerance will be identified and utilized.

References

Abogadallah GM (2010) Antioxidative defense under salt stress. Plant Signal Behav 5:369–374

Afzal I, Munir F, Ayub CM (2009) Changes in antioxidant enzymes, germination capacity and vigour of tomato seeds in response of priming with polyamines. Seed Sci Technol 37:765–770

Agarwal S, Shaheen R (2007) Stimulation of antioxidant system and lipid peroxidation by abiotic stresses in leaves of *Momordica charantia*. Braz J Plant Physiol 19:149–161

Ahmad P, Sharma S (2008) Salt stress and phyto-biochemical responses of plants-a review. Plant, Soil Environ 54:89–99

Ahmad P, Jaleel CA, Sharma S (2010) Antioxidant defense system, lipid peroxidation, proline-metabolizing enzymes, and biochemical activities in two *Morus alba* genotypes subjected to NaCl stress. Russ J Plant Physiol 57:509–517

Ahmad P, Hashem A, Abd-Allah EF (2015) Role of *Trichoderma harzianum* in mitigating NaCl stress in Indian mustard (*Brassica juncea* L) through antioxidative defense system. Front Plant Sci 6:868

Alamgir ANM, Yousuf Ali M (1999) Effect of salinity on leaf pigments, sugar and protein concentrations and chloroplast ATPase activity of rice (*Oryza sativa* L.). Bangladesh J Bot 28:145–149

Ali S, Charles TC, Glick BR (2014) Amelioration of high salinity stress damage by plant growth-promoting bacterial endophytes that contain ACC deaminase. Plant Physiol Biochem 80:160–167

Aly-Salama KH, Al-Mutawa MM (2009) Glutathione-triggered mitigation in salt-induced alterations in plasmalemma of onion epidermal cells. Int J Agric Biol 11:639–642

An BY, Luo Y, Li JR, Qiao WH, Zhang XS, Gao XQ (2008) Expression of a vacuolar Na+/H+ antiporter gene of alfalfa enhances salinity tolerance in transgenic Arabidopsis. Acta Agron Sinica 34:557–564

Apel K, Hirt H (2004) Reactive oxygen species: metabolism, oxidative stress, and signal transduction. Annu Rev Plant Biol 55:373–399

Asada K (1999) The water-water cycle in chloroplasts: scavenging of active oxygens and dissipation of excess photons. Annu Rev Plant Physiol Plant Mol Biol 50:601–639

Ashraf M, Akram NA, Arteca RN, Foolad MR (2010) The physiological, biochemical and molecular roles of brassinosteroids and salicylic acid in plant processes and salt tolerance. CRC Crit Rev Plant Sci 29:162–190

Bajguz A (2014) Nitric oxide: role in plants under abiotic stress. In: Ahmad P, Wani MR (eds) Physiological mechanisms and adaptation strategies in plants under changing environment. Springer, New York, pp 137–159

Baltruschat H, Fodor J, Harrach BD (2008) Salt tolerance of barley induced by the root endophyte *Piriformospora indica* is associated with a strong increase in antioxidants. New Phytol 180:501–510

Begara-Morales JC, Sanchez-Calvo B, Chaki M (2014) Dual regulation of cytosolic ascorbate peroxidase (APX) by tyrosine nitration and S-nitrosylation. J Exp Bot 65:527–538

Ben Ahmed C, Ben Rouina B, Sensoy S (2010) Exogenous proline effects on photosynthetic performance and antioxidant defense system of young olive tree. J Agric Food Chem 58:4216–4222

Besson-Bard A, Pugin A, Wendehenne D (2008) New insights into nitric oxide signaling in plants. Annu Rev Plant Biol 59:21–39

Cabot C, Sibole JV, Barcelo J, Poschenrieder C (2009) Abscisic acid decreases leaf Na^+ exclusion in salt-treated *Phaseolus vulgaris*. J Plant Growth Regul 28:187–192

Crawford NM (2006) Mechanisms for nitric oxide synthesis in plants. J Exp Bot 57:471–478

De Lourdes Oliveira Otoch M, Menezes Sobreira AC, Farias de Aragao ME (2001) Salt modulation of vacuolar H+-ATPase and H^+-Pyrophosphatase activities in *Vigna unguiculata*. J Plant Physiol 158:545–551

Deepti B, Nidhi B, Deepamala M, Chandan SC, Alok K (2014) ACC deaminase containing *Arthrobacter protophormiae* induces NaCl stress tolerance through reduced ACC oxidase activity and ethylene production resulting in improved nodulation and mycorrhization in *Pisum sativum*. J Plant Physiol 171:884–894

Dietz KJ, Tavakoli N, Kluge C (2001) Significance of the V-type ATPase for the adaptation to stressful growth conditions and its regulation on the molecular and biochemical level. J Exp Bot 52:1969–1980

Dimkpa C, Weinand T, Asch F (2009) Plant-rhizobacteria interactions alleviate abiotic stress conditions. Plant Cell Environ 32:1682–1694

Dkhil BB, Denden M (2012) Effect of salt stress on growth, anthocyanins, membrane permeability and chlorophyll fluorescence of Okra (*Abelmoschus esculentus* L.) seedlings. Am J Plant Physiol 7:174–183

Dodd IC, Perez-Alfocea F (2012) Microbial amelioration of crop salinity stress. J Exp Bot 63:3415–3428

El-Mashad AAA, Mohamed HI (2012) Brassinolide alleviates salt stress and increases antioxidant activity of cowpea plants (*Vigna sinensis*). Protoplasma 249:625–635

El-Shintinawy F, El-Shourbagy MN (2001) Alleviation of changes in protein metabolism in NaCl-stressed wheat seedlings by thiamine. Biol Plant 44:541–545

El-Tayeb MA (2005) Response of barley grains to the interactive effects of salinity and salicylic acid. Plant Growth Regul 45:215–224

Evelin H, Giri B, Kapoor R (2013) Ultrastructural evidence for AMF mediated salt stress mitigation in *Trigonella foenum-graecum*. Mycorrhiza 23:71–86

Farag MA, Zhang H, Ryu CM (2013) Dynamic chemical communication between plants and bacteria through airborne signals: induced resistance by bacterial volatiles. J Chem Ecol 39:1007–1018

Flowers TJ, Troke PF, Yeo AR (1977) The mechanism of salt tolerance in halophytes. Annu Rev Plant Physiol 28:89–121

Foyer CH, Lopez-Delgado H, Dat JF, Scott IM (1997) Hydrogen peroxide and glutathione associated mechanisms of acclimatory stress tolerance and signalling. Physiol Plant 100:241–254

Franken P (2012) The plant strengthening root endophyte *Piriformospora indica*: potential application and the biology behind. Appl Microbiol Biotechnol 96:1455–1464

Fukuda A, Tanaka Y (2006) Effects of ABA, auxin, and gibberellin on the expression of genes for vacuolar H^+-inorganic pyrophosphatase, H+-ATPase subunit A, and Na^+/H^+ antiporter in barley. Plant Physiol Biochem 44:351–358

Gadallah MAA (1999) Effects of proline and glycine betaine on *Vicia faba* responses to salt stress. Biol Plant 42:249–257

Gao Z, Sagi M, Lips S (1998) Carbohydrate metabolism in leaves and assimilate partitioning in fruits of tomato (*Lycopersicon esculentum* L.) as affected by salinity. Plant Sci 135:149–159

Gomes MADC, Suzuki MS, Cunha MD, Tullii CF (2011) Effect of salt stress on nutrient concentration, photosynthetic pigments, proline and foliar morphology of *Salvinia auriculata* Aubl. Acta Limnol Bras 23:164–176

Gopalakrishnan S, Sathya A, Vijayabharathi R (2015) Plant growth promoting rhizobia: challenges and opportunities. Biotech 5:355–377

Gray EJ, Smith DL (2005) Intracellular and extracellular PGPR: commonalities and distinctions in the plant-bacterium signaling processes. Soil Biol Biochem 37:395–412

Gueta-Dahan Y, Yaniv Z, Zilinskas BA, Ben-Hayyim G (1997) Salt and oxidative stress: similar and specific responses and their relation to salt tolerance in Citrus. Planta 203:460–469

Guo Y, Qiu QS, Quintero FJ (2004) Transgenic evaluation of activated mutant alleles of SOS2 reveals a critical requirement for its kinase activity and C-terminal regulatory domain for salt tolerance in *Arabidopsis thaliana*. Plant Cell Online 16:435–449

Gupta B, Huang B (2014) Mechanism of salinity tolerance in plants: physiological, biochemical, and molecular characterization. Int J Genomics 2014:1–18

Gupta K, Dey A, Gupta B (2013) Plant polyamines in abiotic stress responses. Acta Physiol Plant 35:2015–2036

Gurmani AR, Bano A, Khan SU, Din J, Zhang JL (2011) Alleviation of salt stress by seed treatment with abscisic acid (ABA), 6-benzylaminopurine (BA) and chlormequat chloride (CCC) optimizes ion and organic matter accumulation and increases yield of rice (*Oryza sativa* L.). Aust J Crop Sci 5:1278–1285

Hafsi C, Falleh H, Saada M (2017) Potassium deficiency alters growth, photosynthetic performance, secondary metabolites content, and related antioxidant capacity in *Sulla carnosa* grown under moderate salinity. Plant Physiol Biochem 118:609–617

Hammer EC, Nasr H, Pallon J (2011) Elemental composition of arbuscular mycorrhizal fungi at high salinity. Mycorrhiza 21:117–129

Han QQ, Lu XP, Bai JP, Qiao Y, Pare PW, Wang SM (2014) Beneficial soil bacterium *Bacillus subtilis* (GB03) augments salt tolerance of white clover. Front Plant Sci 5:525

Hanin M, Ebel C, Ngom M, Laplaze L, Masmoudi K (2016) New insights on plant salt tolerance mechanisms and their potential use for breeding. Front Plant Sci 7:1787

Hanson AD, Rathinasabapathi B, Rivoal J (1994) Osmoprotective compounds in the *Plumbaginaceae*: a natural experiment in metabolic engineering of stress tolerance. Proc Natl Acad Sci USA 91:306–310

Hardoim PR, Overbeek LSV, Elsas JDV (2008) Properties of bacterial endophytes and their proposed role in plant growth. Trends Microbiol 16:463–471

Hasanuzzaman M, Nahar K, Fujita M (2014) Regulatory role of polyamines in growth, development and abiotic stress tolerance in plants. In: Plant adaptation to environmental change: significance of amino acids and their derivatives. CAB ebooks. CABI, Wallingford, pp 157–193

Hasegawa PM, Bressan RA, Zhu JK, Bohnert HJ (2000) Plant cellular and molecular responses to high salinity. Annu Rev Plant Physiol Plant Mol Biol 51:463–499

He C, Yan J, Shen G (2005) Expression of an arabidopsis vacuolar sodium/proton antiporter gene in cotton improves photosynthetic performance under salt conditions and increases fiber yield in the field. Plant Cell Physiol 46:1848–1854

Hnilikova F, Martinkova J, Kraus K (2017) Effects of salt stress on water status, photosynthesis and chlorophyll fluorescence of rocket. Plant Soil Environ 63:362–367

Hoque MA, Banu MNA, Nakamura Y (2008) Proline and glycine betaine enhance antioxidant defense and methylglyoxal detoxification systems and reduce NaCl-induced damage in cultured tobacco cells. J Plant Physiol 165:813–824

Horie T, Kaneko T, Sugimoto G (2011) Mechanisms of water transport mediated by pip aquaporins and their regulation via phosphorylation events under salinity stress in barley roots. Plant Cell Physiol 52:663–675

Hussain SS, Ali M, Ahmad M, Siddique KHM (2011) Polyamines: natural and engineered abiotic and biotic stress tolerance in plants. Biotechnol Adv 29:300–311

Ishitani M, Liu J, Halfter U, Kim CS, Shi W, Zhu JK (2000) SOS3 function in plant salt tolerance requires N-myristoylation and calcium binding. Plant Cell 12:1667–1677

Jeschke WD, Peuke AD, Pate JS, Hartung W (1997) Transport, synthesis and catabolism of abscisic acid (ABA) in intact plants of castor bean (*Ricinus communis* L.) under phosphate deficiency and moderate salinity. J Exp Bot 48:1737–1747

Kapoor N, Pande V (2015) Effect of salt stress on growth parameters, moisture content, relative water content and photosynthetic pigments of fenugreek variety rmt-1. J Plant Sci 10:210–221

Keisham M, Mukherjee S, Bhatla SC (2018) Mechanisms of sodium transport in plants-progresses and challenges. Int J Mol Sci 19:647

Kerepesi I, Galiba G (2000) Osmotic and salt stress-induced alteration in soluble carbohydrate content in wheat seedlings. Crop Sci 40:482–487

Keskin BC, Sarikaya AT, Yuksel B, Memon AR (2010) Abscisic acid regulated gene expression in bread wheat (*Triticum aestivum* L.). Aust J Crop Sci 48:617–625

Kim YC, Leveau J, McSpadden Gardener BB (2011) The multifactorial basis for plant health promotion by plant-associated bacteria. Appl Environ Microbiol 77:1548–1555

Kim SH, Ahn YO, Ahn MJ (2012) Down-regulation of β-carotene hydroxylase increases β-carotene and total carotenoids enhancing salt stress tolerance in transgenic cultured cells of sweet potato. Phytochemistry 74:69–78

Kim J, Liu Y, Zhang X, Zhao B, Childs K (2016) Analysis of salt-induced physiological and proline changes in 46 switchgrass (*Panicum virgatum*) lines indicates multiple response modes. Plant Physiol Biochem 105:203–212

Knott JM, Romer P, Sumper M (2007) Putative spermine synthases from *Thalassiosira pseudonana* and *Arabidopsis thaliana* synthesize thermospermine rather than spermine. FEBS Lett 581:3081–3086

Krishna P (2003) Brassinosteroid-mediated stress responses. J Plant Growth Regul 22:289–297

Kumari S, Vaishnav A, Jain S (2015) Bacterial-mediated induction of systemic tolerance to salinity with expression of stress alleviating enzymes in soybean (*Glycine max* L. Merrill). J Plant Growth Regul 34:558–573

Lamattina L, Garcia-Mata C, Graziano M, Pagnussat G (2003) Nitric oxide: the versatility of an extensive signal molecule. Annu Rev Plant Biol 54:109–136

Lebeis SL (2014) The potential for give and take in planta microbiome relationships. Front Plant Sci 5:287

Liang W, Cui W, Ma X (2014) Function of wheat Ta-UnP gene in enhancing salt tolerance in transgenic Arabidopsis and rice. Biochem Biophys Res Commun 450:794–801

Liang W, Ma X, Wan P, Liu L (2018) Plant salt-tolerance mechanism: a review. Biochem Biophys Res Commun 495:286–291

Liu J, Ishitani M, Halfter U (2000) The *Arabidopsis thaliana* SOS2 gene encodes a protein kinase that is required for salt tolerance. Proc Natl Acad Sci USA 97:3730–3734

Liu C, Mao B, Ou S (2014) OsbZIP71, a bZIP transcription factor, confers salinity and drought tolerance in rice. Plant Mol Biol 84:19–36

Lugtenberg B, Kamilova F (2009) Plant growth promoting Rhizobacteria. Annu Rev Microbiol 63:541–556

Ma L, Zhang H, Sun L (2012) NADPH oxidase AtrbohD and AtrbohF function in ROS-dependent regulation of Na^+/K^+ homeostasis in Arabidopsis under salt stress. J Exp Bot 63:305–317

Ma X, Liang W, Gu P, Huang Z (2016) Salt tolerance function of the novel C2H2-type zinc finger protein TaZNF in wheat. Plant Physiol Biochem 106:129–140

Mahajan S, Tuteja N (2005) Cold, salinity and drought stresses: an overview. Arch Biochem Biophys 444:139–158

Makela P, Karkkainen J, Somersalo S (2000) Effect of glycine betaine on chloroplast ultrastructure, chlorophyll and protein content, and RuBPCO activities in tomato grown under drought or salinity. Biol Plant 43:471–475

Martinez-Atienza J, Jiang X, Garciadeblas B (2007) Conservation of the salt overly sensitive pathway in rice. Plant Physiol 143:1001–1012

Mayak S, Tirosh T, Glick BR (2004) Plant growth-promoting bacteria confer resistance in tomato plants to salt stress. Plant Physiol Biochem 42:565–572

Mhamdi A, Queval G, Chaouch S (2010) Catalase function in plants: a focus on Arabidopsis mutants as stress-mimic models. J Exp Bot 61:4197–4220

Mishra S, Jha AB, Dubey RS (2011) Arsenite treatment induces oxidative stress, upregulates antioxidant system, and causes phytochelatin synthesis in rice seedlings. Protoplasma 248:565–577

Mittler R (2002) Oxidative stress, antioxidants and stress tolerance. Trends Plant Sci 7:405–410

Mittova V, Guy M, Tal M, Volokita M (2004) Salinity up-regulates the antioxidative system in root mitochondria and peroxisomes of the wild salt-tolerant tomato species *Lycopersicon pennellii*. J Exp Bot 55:1105–1113

Montesinos E, Bonaterra A, Badosa E, Frances J, Alemany J, Llorente I, Moragrega C (2002) Plant-microbe interactions and the new biotechnological methods of plant disease control. Int Microbiol 5:169–175

Munir N, Aftab F (2011) Enhancement of salt tolerance in sugarcane by ascorbic acid pre-treatment. Afr J Biotechnol 10:18362–18370

Munns R, Tester M (2008) Mechanisms of salinity tolerance. Annu Rev Plant Biol 59:651–681

Nadeem SM, Zahir ZA, Naveed M, Ashraf M (2010) Microbial ACC-Deaminase: prospects and applications for inducing salt tolerance in plants. CRC Crit Rev Plant Sci 29:360–393
Nadeem SM, Ahmad M, Naveed M (2016) Relationship between in vitro characterization and comparative efficacy of plant growth-promoting rhizobacteria for improving cucumber salt tolerance. Arch Microbiol 198:379–387
Nakashima K, Tran LSP, Van Nguyen D (2007) Functional analysis of a NAC-type transcription factor OsNAC6 involved in abiotic and biotic stress-responsive gene expression in rice. Plant J 51:617–630
Nalousi AM, Ahmadiyan S, Hatamzadeh A, Ghasemnezhad M (2012) Protective role of exogenous nitric oxide against oxidative stress induced by salt stress in bell-pepper (*Capsicum annum* L.). Ame-Eura J Agri Env Sci 12:1085–1090
Nazar R, Iqbal N, Syeed S, Khan NA (2011) Salicylic acid alleviates decreases in photosynthesis under salt stress by enhancing nitrogen and sulfur assimilation and antioxidant metabolism differentially in two mungbean cultivars. J Plant Physiol 168:807–815
Omami NE (2005) Response of amaranth to salinity stress. PhD Thesis. University of Pretoria etd, Pretoria, 235 p
Omar MNA, Osman MEH, Kasim WA, Abd El-Daim IA (2009) Improvement of salt tolerance mechanisms of barley cultivated under salt stress using *Azospirillum brasiliense*. Tasks Veg Sci 44:133–147
Parida AK, Das AB (2005) Salt tolerance and salinity effects on plants: a review. Ecotoxicol Environ Saf 60:324–349
Parida A, Das AB, Das P (2002) NaCl stress causes changes in photosynthetic pigments, proteins, and other metabolic components in the leaves of a true mangrove, *Bruguiera parviflora*, in hydroponic cultures. J Plant Biol 45:28–36
Parida A, Das A, Mittra B (2004) Effects of salt on growth, ion accumulation, photosynthesis and leaf anatomy of the mangrove, *Bruguiera parviflora*. Trees 18:167–174
Parvaiz A, Satyawati S (2008) Salt stress and phyto-biochemical responses of plants – a review. Plant Soil Environ 54:89–99
Popova LP, Stoinova ZG, Maslenkova LT (1995) Involvement of abscisic acid in hotosynthetic process in *Hordeum vulgare* L. during salinity stress. J Plant Growth Regul 14:211–218
Quintero FJ, Ohta M, Shi H (2002) Reconstitution in yeast of the Arabidopsis SOS signaling pathway for Na^+ homeostasis. Proc Natl Acad Sci 99:9061–9066
Rahneshan Z, Nasibi F, Moghadam AA (2018) Effects of salinity stress on some growth, physiological, biochemical parameters and nutrients in two pistachio (*Pistacia vera* L.) rootstocks. J Plant Int 13:73–82
Roy S, Chakraborty U (2018) Role of sodium ion transporters and osmotic adjustments in stress alleviation of *Cynodon dactylon* under NaCl treatment: a parallel investigation with rice. Protoplasma 255:175–191
Ruiz-Lozano JM, Aroca R (2010) Modulation of aquaporin genes by the arbuscular mycorrhizal symbiosis in relation to osmotic stress tolerance. In: Seckbach J, Grube M (eds) Symbioses and stress: joint ventures in biology, cellular origin, life in extreme habitats and astrobiology. Springer, Berlin, pp 359–374
Sahoo RK, Ansari MW, Dangar TK (2014) Phenotypic and molecular characterisation of efficient nitrogen-fixing Azotobacter strains from rice fields for crop improvement. Protoplasma 251:511–523
Sairam RK, Tyagi A (2004) Physiology and molecular biology of salinity stress tolerance in plants. Curr Sci 86:407–421
Sanders D (2000) Plant biology: the salty tale of Arabidopsis. Curr Biol 10:486–488
Sawada H, Shim IS, Usui K (2006) Induction of benzoic acid 2-hydroxylase and salicylic acid biosynthesis-modulation by salt stress in rice seedlings. Plant Sci 71:263–270
Saxena SC, Kaur H, Verma P, Petla BP, Andugula VR, Majee M (2013) Osmoprotectants: potential for crop improvement under adverse conditions. In: Tuteja N, Gill SS (eds) Plant acclimation to environmental stress. Springer, New York, pp 197–232

Schmidt R, Mieulet D, Hubberten HM (2013) Salt-responsive ERF1 regulates reactive oxygen species-dependent signaling during the initial response to salt stress in rice. Plant Cell 25:2115–2131

Schroeder JI, Delhaize E, Frommer WB (2013) Using membrane transporters to improve crops for sustainable food production. Nature 497:60–66

Shannon MC, Grieve CM, Francois LE (1994) Whole-plant response to salinity. In: Wilkinson RE (ed) Plant-environment interactions. Marcel-Decker, New York, pp 199–244

Sharp RE, Hsiao TC, Silk WK (1990) Growth of the maize primary root at low water potentials: role of growth and deposition of hexose and potassium in osmotic adjustment. Plant Physiol 93:1337–1346

Shi H, Quintero FJ, Pardo JM, Zhu JK (2002) The putative plasma membrane Na^+/H^+antiporter SOS1 controls long-distance Na^+ transport in plants. Plant Cell 14:465–477

Shu S, Guo SR, Yuan LY (2012) A review: polyamines and photosynthesis. In: Najafpour MM (ed) Advances in photosynthesis-fundamental aspects. In Tech, Rijeka, pp 439–464

Singh R, Singh P, Sharma R (2014) Microorganism as a tool of bioremediation technology for cleaning environment: a review. Proc Int Acad Ecol Environ Sci 4:1–6

Smith SE, Smith FA (2011) Roles of arbuscular mycorrhizas in plant nutrition and growth: new paradigms from cellular to ecosystem scales. Annu Rev Plant Biol 62:227–250

Sorty AM, Meena KK, Choudhary K (2016) Effect of plant growth promoting bacteria associated with halophytic weed (*psoralea corylifolia* l.) on germination and seedling growth of wheat under saline conditions. Appl Biochem Biotechnol 180:872–882

Staiger D, Brown JWS (2013) Alternative splicing at the intersection of biological timing, development, and stress responses. Plant Cell 25:3640–3656

Sun SJ, Guo SQ, Yang X, Bao YM, Tang HJ, Sun H (2010) Functional analysis of a novel Cys2/His2 type zinc finger protein involved in salt tolerance in rice. J Exp Bot 61:2807–2818

Tanabe N, Yoshimura K, Kimura K, Yabuta Y, Shigeoka S (2007) Differential expression of alternatively spliced mRNAs of Arabidopsis SR protein homologs, atSR30 and atSR45a, in response to environmental stress. Plant Cell Physiol 48:1036–1049

Upadhyay SK, Singh DP (2015) Effect of salt-tolerant plant growth-promoting rhizobacteria on wheat plants and soil health in a saline environment. Plant Biol 17:288–293

Upadhyay SK, Singh JS, Saxena AK, Singh DP (2012) Impact of PGPR inoculation on growth and antioxidant status of wheat under saline conditions. Plant Biol 14:605–611

Vaughan LV, MacAdam JW, Smith SE, Dudley LM (2002) Root growth and yield of differing alfalfa rooting populations under increasing salinity and zero leaching. Crop Sci 42:2064

Vicente O, Boscaiu M, Naranjo MA, Estrelles E, Belles JM, Soriano P (2004) Responses to salt stress in the halophyte *Plantago crassifolia* (*Plantaginaceae*). J Arid Environ 58:463–481

Wahome P, Jesch H, Grittner I (2001) Mechanisms of salt stress tolerance in two rose rootstocks: *Rosa chinensis* major and *Rosa rubiginosa*. Sci Hortic 87:207–216

Wang Y, Nii N (2000) Changes in chlorophyll, ribulose bisphosphate carboxylase-oxygenase, glycine betaine content, photosynthesis and transpiration in *Amaranthus tricolor* leaves during salt stress. J Hortic Sci Biotechnol 75:623–627

Zhang JL, Shi H (2013) Physiological and molecular mechanisms of plant salt tolerance. Photosynth Res 115:1–22

Zhang L, Xi D, Li S, Gao Z, Zhao S, Shi J (2011) A cotton group CMAP kinase gene, GhMPK2, positively regulates salt and drought tolerance in tobacco. Plant Mol Biol 77:17–31

Zhao L, Zhang F, Guo J (2004) Nitric oxide functions as a signal in salt resistance in the calluses from two ecotypes of reed. Plant Physiol 134:849–857

Zhao MG, Chen L, Zhang LL, Zhang WH (2009) Nitric reductase-dependent nitric oxide production is involved in cold acclimation and freezing tolerance in Arabidopsis. Plant Physiol 151:755–767

Zhao CY, Si JH, Feng Q, Deo RC, Yu TF, Li PD (2017) Physiological response to salinity stress and tolerance mechanics of *Populus euphratica*. Environ Monit Assess 189:533

Chapter 16
Arbuscular Mycorrhizal Fungi Alleviate Soil Salinity Stress in Arid and Semiarid Areas

Karima Bencherif, Yolande Dalpé, and Anissa Lounès Hadj-Sahraoui

Abstract Soil salinization is an increasingly important problem in many parts of the world, particularly in the arid and semiarid areas. Soil salinity affects plant growth adversely, by disrupting their physiological mechanisms due to excessive Na^+ and Cl^- ions toxicity toward cells. The toxic effects include disruption of enzyme structure and other macromolecules, disruption of photosynthetic efficiency, gas exchange, membrane organization, and water status. Salinity may directly or indirectly inhibit plant cell division, development, and productivity. To overcome the detrimental effects and to improve plant tolerance to stresses, particularly salt stress, plants adopt a wide variety of strategies including symbiotic association with soil fungi such as the arbuscular mycorrhizal fungi (AMF). The use of arbuscular mycorrhizal fungi proved to be an interesting way for the management of native flora and restoration of natural habitats with minimal chemical inputs. Unfortunately, the knowledge about restoration of salt-affected ecosystems using AMF biofertilizer is limited. This chapter aims to review the impact of salinity stress on plants and on AMF life cycle and physiology and to describe the effect of AMF biofertilizers on plant development underlying physiological, biochemical, and molecular plant mechanisms within the context of salinity stress.

Keywords Mycorrhizal colonization · Salinity stress · Biofertilization · Steppic areas · Abiotic stress alleviation

K. Bencherif (✉)
Faculté des Sciences de la Nature et de la vie, Université de Djelfa, Djelfa, Algeria

Unité de Chimie Environnementale et Interactions sur le Vivant (UCEIV), Université du Littoral Côte d'Opale, Calais cedex, France

Y. Dalpé
Agriculture et agroalimentaire Canada, Centre de recherche et développement d'Ottawa, Ottawa, ON, Canada

A. Lounès Hadj-Sahraoui
Unité de Chimie Environnementale et Interactions sur le Vivant (UCEIV), Université du Littoral Côte d'Opale, Calais cedex, France
e-mail: lounes@univ-littoral.fr

B. Giri, A. Varma (eds.), *Microorganisms in Saline Environments: Strategies and Functions*, Soil Biology 56, https://doi.org/10.1007/978-3-030-18975-4_16

16.1 Introduction

The increase in harsh abiotic stresses such as drought, salinity, and abrupt changes in temperature are main consequences of climate change. These stresses have led to loss of soil organic matter and other forms of soil degradation that negatively affect agricultural productivity (Ahmad et al. 2015; Egamberdieva et al. 2017). Indeed, soil salinity is a serious problem throughout the world, particularly in arid and semiarid areas (Badda et al. 2014; Bencherif et al. 2015). Soil salinity affects seed germination, plant growth and vigor, and thus crop productivity (Giri et al. 2007; Mathur et al. 2007; Carillo et al. 2011; Plouznikoof et al. 2016). Furthermore, salinity stress has a direct impact on water depleting in the root zone, consequently affecting the soil water potential (Sabir et al. 2009). The high ratios of toxic ions like Na^+ and Cl^- damage the balance ions, reducing the plant ability to absorb other ions like K^+, Ca^+, and Mn^{+2} (ElHindi et al. 2017). This salt deposition in the soil results in hyper-ionic and hyperosmotic stresses in organisms (Evelin et al. 2009), which may limit the growth of other organisms than plants, such as fungi due to specific ion toxicity or osmotic stress (Estrada et al. 2013).

To counteract these problems, the use of nonhazardous biological method, such as symbiotic microorganisms like rhizobacteria and mycorrhizal fungi, to mitigate the negative effects of salinity stress combined with the cultivation of moderately salt-tolerant plants is an excellent environmental choice to make (Mayak et al. 2004; Badda et al. 2014; ElHindi et al. 2017). Among the microorganisms, AMF are of particular interest. An increasing number of scientists and farmers alike think their exploitation and valorization represent the next revolution in agriculture (Plouznikoof et al. 2016). Indeed, AMF potential have been repeatedly studied and tested under saline environments (Allen and Cunningham 1983; Pond et al. 1984; Rozema et al. 1986; Juniper and Abbott 1993; Bencherif et al. 2015). It has been repeatedly demonstrated that AMF can promote plant salinity tolerance through a diversity of mechanisms, such as enhancing nutrient uptake (Badda et al. 2014), stimulating plant growth hormones, improving rhizosphere microbial diversity and consequently soil quality (Evelin et al. 2009), while reducing the uptake of sodium and chloride ions and impairing their movement to plant leaves (ElHindi et al. 2017). As such, soil biofertilization with AMF is recognized to globally enhance plant and crop yields and to protect plants against biotic and abiotic stress (Garg and Pandey 2016; Plouznikoof et al. 2016). In this chapter, we discuss the current knowledge on impact of AMF biofertilization in alleviating salt salinity stress.

16.2 Impact of Soil Salinity on Plants Development

Soil salinity is one of the major problems in arid and semiarid areas. More than 800 million ha of the total land surface suffers from soil salinization, significantly impacting on growth and productivity of agricultural crops (Yun et al. 2018). A saline soil is generally defined as one in which the electrical conductivity (EC) of the saturation extract (ECe) in the root zone exceeds 4 dS m^{-1} (approximately 40 mM NaCl) at 25 °C and has an exchangeable sodium of 15%. The yield of most crop plants is reduced at this EC (Munns 2005; Jamil et al. 2011; Shrivastava and Kumar 2015). It has been estimated that worldwide 20% of total cultivated land and 33% of irrigated agricultural lands are afflicted by high salinity (Machado and Serralheiro 2017). Furthermore, the salinized areas are increasing at a rate of 10% annually for various reasons, including low precipitation, high surface evaporation, weathering of native rocks, irrigation with saline water, and poor cultural practices. It has been estimated that more than 50% of the arable land would be salinized by the year 2050 (Jamil et al. 2011; Shrivastava and Kumar 2015; Machado and Serralheiro 2017).

The presence of high levels of NaCl salts, $NaCO^{3-}$, $NaSO_4{}^{-2}$, or magnesium salts in soils of arid and semiarid environments may induce three types of stresses: osmotic, ionic, and oxidative. This drastically affects the major processes of plant development: growth and productivity (Juniper and Abbott 1993; Feng et al. 2002; Saxena et al. 2017). Osmotic stress leads to altered water potential, thereby reducing the water use efficiency of plant due to induction of physiological drought conditions. Whereas ionic stress causes disruption of ion homoeostasis at both cellular and whole-plant levels, oxidative stress elicits release of reactive oxygen species, which inhibit cell growth and plant metabolism (Saxena et al. 2017). In addition, salt stress affects directly nucleic acids, plasma membrane, and the disruption of enzyme structure (Ramoliya et al. 2004; Evelin et al. 2009; Péréz-Tienda et al. 2011; Porcel et al. 2012).

Physiologically, the adverse impact of salinity stress results in the combination of an indirect effect on the water potential decreasing resulting in a reduction in the availability of water for the plants (Porcel et al. 2012) and also in the toxicity and disturbances of water mineral nutrition induced by excess of Na^+ and Cl^- (Plenchette 1991; Feng et al. 2002; Benkhaled et al. 2007; Estrada et al. 2013). Indeed, in the presence of salt stress, plants tend to select specific ions. This selectivity is generally related to the mechanisms of exclusion of Na^+ ions. High K^+/Na^+ selectivity has been demonstrated in the plasmalemma of the root cortex and in the xylem of the most NaCl-tolerant plants (Taleisnik and Grunberg 1994; Juniper and Abbott 2006; Porcel et al. 2012). Maintaining this high K^+/Na^+ ratio is a mechanism that helps the horizontal metabolic processes of the plant (Plenchette 1991; Juniper and Abbott 1993; Feng et al. 2002; Ramoliya et al. 2004; Porcel et al. 2012; Wu et al. 2013). The extracellular stress signal is first perceived by the receptor membrane, and then extensive and complex signaling is activated intracellular. The activated signals are the result of the expression of multiple stress-sensitive genes (Porcel et al. 2012) (Fig. 16.1). The voltage signal is first perceived at the membrane level by a specific receptor, resulting in the generation of secondary signaling molecules, such as Ca^{+2}, inositol phosphates, reactive oxygen (RO), and abscisic

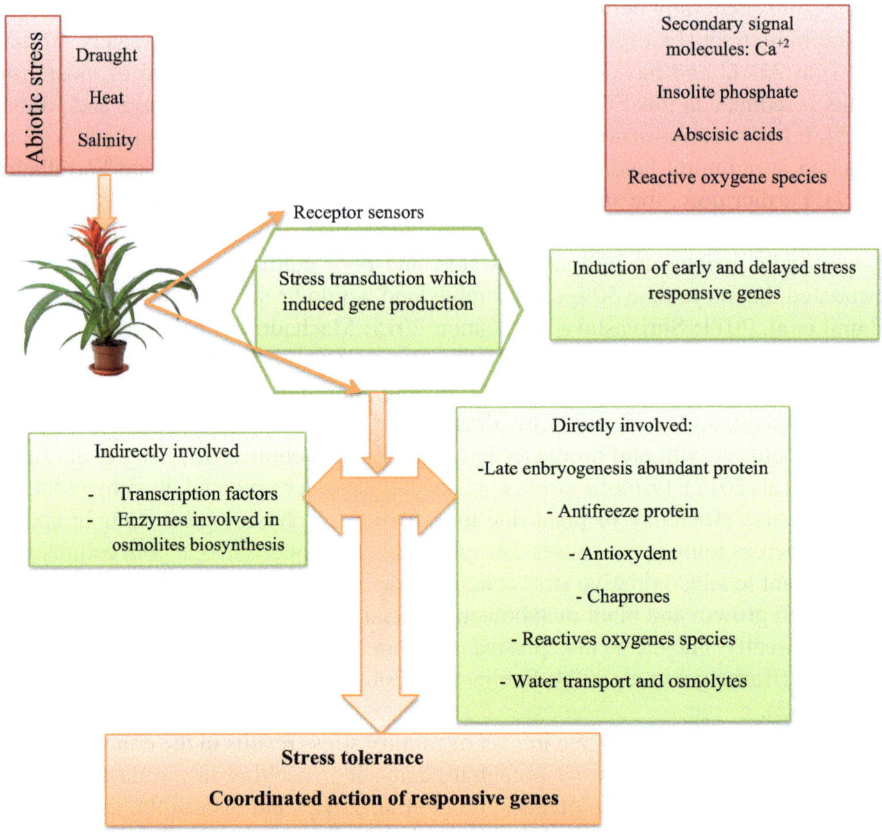

Fig. 16.1 Generic pathway for plant salinity stress response. The extracellular stress signal is first perceived by the membrane receptor and then activates a large and complex signaling cascade intracellularly. The signal cascade results in the expression of multiple stress responsive genes, the products of which can provide the stress tolerance directly or indirectly. Overall, the stress response could be a coordinated action of many genes (Author's design. Adapted from Porcel et al. 2012)

acid (ABA). The stress signal is then translated inside the nucleus whose products "tolerance genes" finally lead to the adaptation of plants to the reported stress. Early genes are induced within minutes of stress perception, prompting transcription factors to activate delayed gene expression: "RD" (responsive dehydration), "KIN" (cold induced), and "COR" (cold responsive). These genes are either directly involved in cellular protection against stress (e.g., the end of embryogenesis by the abundant production of proteins: antifreeze proteins, antioxidants, and detoxifying enzymes) or indirectly by the induction of factors of transcription and enzymes of phosphatidylinositol metabolism (Tuteja 2007; Porcel et al. 2012).

16.3 Alleviation of Salt Stress in Plant by Symbiotic Microorganisms

As discussed earlier in this chapter, high salt concentration is a threat to soil fertility and crop production as well, which necessitates increase in the food production for the growing world population (Shrivastava and Kumar 2015). Saxena et al. (2017) indicated that the exploitation and development of salt-tolerant plant species is the one strategy to overcome the problem of salt stress and food insecurity. Nevertheless, the biotic approach "plant-microbe interaction" to overcome salinity problems has recently received a considerable attention throughout the world. The halophilic microbes have potential for bioremediation of salt-dominant soils (Arora and Vanza 2017); but the mechanism is still not clear (Basu et al. 2018). Indeed, symbiotic microorganisms play a major role including mobilization of N and P from organic polymers, release of nutrients from mineral particles or rock surfaces via weathering, effects on carbon cycling, interactions with myco-heterotrophic plants, and mediation of plant responses to stress factors such as drought, salinity, and plant pathogens (Basu et al. 2018). Many studies indicate that bacterial strains belong to the plant growth-promoting rhizobacteria group (PGPB), and AMF can have significant agricultural applications to increase plant productivity and reduce the use of synthetic fertilizers and pesticides under saline conditions in arid and semiarid areas (Bacera et al. 2014; ElHindi et al. 2017; Diagne et al. 2018). According to Powell and Rilling (2018), soil richness in symbiotic microorganisms and their global or individual action facilitate ecosystem functioning (Fig. 16.2). For example, it was proven that inoculation with the efficient plant growth-promoting microorganism isolates from saline areas to maize and wheat seeds mitigates salt stress (Arora and Vanza 2017). The authors explained that maize yield increased about 10–12% with symbiotic microorganisms, while wheat yield increased about 6%. This result was obtained in soils with 6% of NaCl concentration as compared to control at 2% of NaCl concentration (Arora and Vanza 2017). It was also proven that inoculation of talh tree (*Acacia gerrardii* Benth.) with AMF increased tolerance to the damage caused by salinity stress by upregulating the antioxidant enzymes as well as by modulating the level of osmo-protectants like glycine, betaine, proline, and phenols (Hashem et al. 2016).

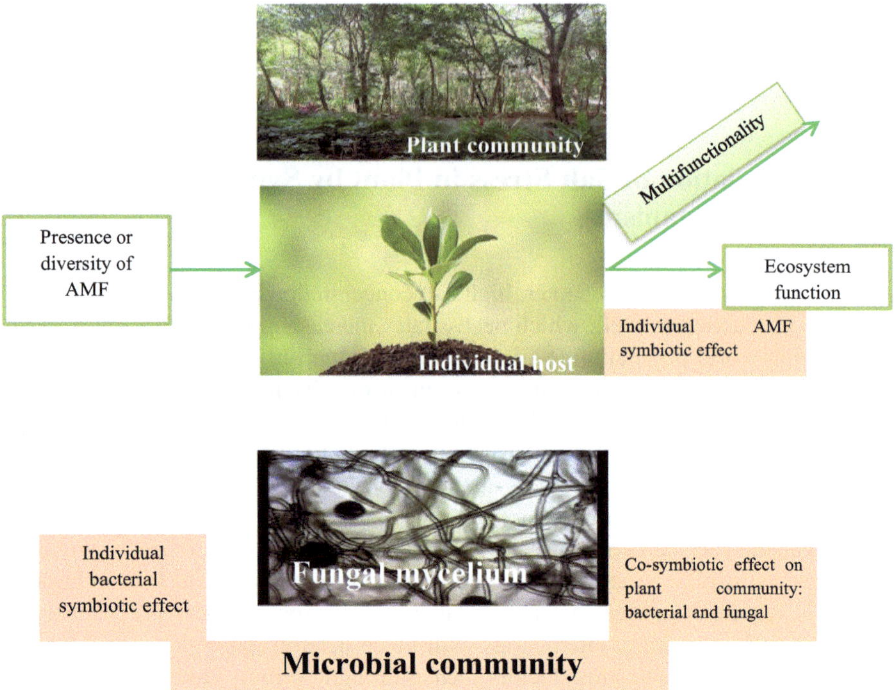

Fig. 16.2 Presence, diversity and effectiveness of microbial community in saline arid and semi-arid areas. High biodiversity is recorded on microorganisms in saline soils. According to host plants and to rhizospheric exudates, symbiosis is formed between plant/bacteria; plant/AMF; plant/AMF + bacteria to alleviate salt stress and to enhanced plant development and productivity. Microbial community presence and diversity can affect ecosystem functions at different hierarchical levels. If AMF and/or bacteria diversity effects on an ecosystem function are to be examined at the level of the plant community, all lower levels are also relevant. This is because AMF/bacteria can influence the composition of the plant community, with ripple on effects to the lower levels in this hierarchy (Adapted from Powell and Rilling 2018)

16.4 AMF in Saline Soils of Arid and Semiarid Areas

16.4.1 Natural Biodiversity of AMF in Saline Soils

According to Juniper and Abbott (1993), high soil salinity has adverse effects on plant colonization by AMF. However, there are reports in the literature from all over the world that plants of saline areas can be colonized by AMF (Table 16.1) (Juniper and Abbott 1993, 2006; Hildebrandt et al. 2001; Landwehr et al. 2002; Wang et al.

Table 16.1 Examples of osmo-tolerant AMF species

Order (number of species)	AMF species	References
Glomerales	*Funneliformis geosporum*	Hildebrandt et al. (2001)
Glomerales	*Funneliformis geosporum*	Carvalho et al. (2003)
Glomerales (19), *Archeosporales* (1), *Pacisporaceae* (1)	*Funneliformis geosporum*	Wilde et al. (2009)
Glomerales (15), *Diversisporales* (2)	*Funneliformis geosporum* *Funneliformis caledonium* *Rhizophagus irregularis*	Sonjak et al. (2009)
Glomerales (22)	*Rhizophagus irregularis* *Rhizophagus fasciculatus*	Guo and Gong (2014)
Glomerales (16), *Diversisporales* (5), *Gigasporales* (3), *Archeosporales* (3), *Pacisporaceae* (2), *Paraglomerales* (1)	*Claroideoglomus constricum* *Paraglomus occultum*	Estrada et al. (2013)
Glomerales (7), *Diversisporales* (1)	*Funneliformis caledonium* *Funneliformis mosseae* *Rhizophagus prolifer*	Estrada et al. (2013)
Glomales (7)	*Septoglomus constrictum* *Funneliformis mosseae* *Funneliformis geosporum* *Rhizophagus fasciculatus* *Funneliformis coronatum* *Gigaspora gigantea* *Funneliformis* sp1	Bencherif et al. (2015)

All species are listed according to the recent classification

2004; Benkhaled et al. 2007; Smith and Read 2008; Evelin et al. 2009; Porcel et al. 2012; Estrada et al. 2013; Barin et al. 2013; Bencherif et al. 2015). The AMF root colonization level is often shown to decrease with soil salinity (Juniper and Abbott 1993; Benkhaled et al. 2007; Evelin et al. 2009), while in some cases, it appeared to increase especially in natural saline soils (Aliasgharzadeh et al. 2001; Evelin et al. 2009; Bencherif et al. 2015). Many of the natural ecotypes of AMF are believed to contain 10% NaCl, which classified them as osmo-tolerant fungi "osmo-tolerant groups" (Smith and Read 2008; Evelin et al. 2009). On the other hand, soil salinity often promotes AMF sporulation levels (Aliasgharzadeh et al. 2001; Karaarslan and Uyanoz 2011), and these species belong mostly to the genus *Glomus* (Juniper and Abbott 2006; Sanchez-Castro et al. 2012).

16.4.2 Impact of Soil Salinity on AMF Biodiversity

Salinity does not only negatively affect plant growth but also AMF development (Juniper and Abbott 2006; Sheng et al. 2009; Porcel et al. 2012; Bencherif et al. 2015). It has been well established that soil factors, especially salinity, could negatively influence AMF spore germination and subsequent development of germination hyphae (Juniper and Abbott 1993; Barin et al. 2013; Badda et al. 2014; Muthukumar et al. 2017). Bowen (1987) showed that under saline conditions, the pre-symbiotic phases of AMF development proceeded normally. It is much more the symbiotic phase which is affected by salinity (Juniper and Abbott 1993). Recent study shows that the germ tube emergence of *Rhizophagus irregularis* and hyphal growth were the most affected phases by salinity (Muthukumar et al. 2017). However, other morphological features like the lateral brunching of the hyphal or the hyphal diameter was not affected. This was also proven by Juniper and Abbott (1993) on *Scutellospora calospora*. Nevertheless, Estrada et al. (2013) explained that brunched absorbing structure (ABS) was more negatively influenced by salt concentration during early stage than at later stages of culture. Table 16.2 summarizes the developmental stages of some species of AMF affected by salinity. On the other hand, it was reported that AMF isolates originating from saline soils were better adapted to salinity compared to those from nonsaline soils (Muthukumar et al. 2017). This support studies where AMF originating from saline soils promoted plant growth better than those originating from nonsaline soils (Estrada et al. 2013).

16.5 Role of AMF Biofertilizers in Mitigating Salinity Stress in Arid and Semiarid Areas

Worldwide arid and semiarid areas are submitted to extreme environmental abiotic stresses, mainly drought, extreme temperatures, and salinity. In reaction to these harmful conditions, arbuscular symbiosis helps vegetation to increase plant resistance (Porcel et al. 2016) and contributes to the improvement of soil structure (Garbaye 2013). It is fundamentally a nutritional symbiosis, in which nutrients (phosphorus and others) are traded with plant hosts for carbon (sugar and lipids). This very exchange, with all its physiological, biochemical, molecular, and ultrastructural intricacies, has captured the imagination of many researchers (Porcel et al. 2012; Wu et al. 2013; Estrada et al. 2013; Alqarawi et al. 2014; ElHindi et al. 2017; Powell and Rilling 2018). However, the devices put in place by the AMF are to be elucidated (Saxena et al. 2017; Elhindi et al. 2017; Powell and Rilling 2018).

Table 16.2 Impact of soil salinity on AMF development

Stress type	Tested AMF	AMF culture	Stage of development of AMF affected by salinity					References
Salinity stress			Spore germination	Germination hyphal elongation	Roots colonization	Elongation of extraradical hyphae	Sporulation	
	Rhizophagus irregularis	In vitro	No germination	No germination	No germination	Positive elongation	Positive sporulation	Estrada et al. (2013) Jahromi et al. (2008) Benkhaled et al. (2007)
	Funneliformis mosseae, *Rhizophagus irregularis*, *Diversispora versiforme*, *Rhizophagus fasciculatus* *Acaulospora laevis*, *Archeospora trappei*, *Gigaspora decipiens*, *Scutellospora calospora*	In pots culture	Positive germination	No germination	Positive germination	Positive elongation	Positive sporulation	Juniper and Abbott (2006) Kohler et al. (2009) Sheng et al. (2009) Fan et al. (2011) Aroca et al. (2013)

16.5.1 Impact of AMF Biofertilization on Soil Structure: Glomalin Production

The structure of the saline soils is destabilized by the excess of Na^+ ions which destroy the links between the different particles of the soil mostly in clay relations. This is where the AMF inoculation comes in. Correlative field studies and pot experiments with single plants indicate that AMF promote soil aggregation and soil structure stabilization by binding and involving soil particles together into bigger aggregates (Muthukumar et al. 2017). Further, AMF inoculation involved in the stabilization of aggregates is based on the fixation of soil particles by hyphae and roots and on the exudation of polysaccharides (Bearden and Petersen 2000). Indeed, it was proven that AMF extracellular mycelia colonizes the soil at a concentration of 50 g of hyphae/g of stable aggregates, secreting in abundance a glycoprotein: "glomalin" (Badda et al. 2014), the role of which in improving the structural stability of saline soils has been repeatedly demonstrated by previous studies (Bearden and Petersen 2000; Garbaye 2013). Glomalin is hydrophobic and thermotolerant, which makes it more slowly biodegradable by soil microorganisms, and when it binds to clay particles to which it gives a strong cohesion, it is even more recalcitrant to biological decomposition, thus ensuring strong structural stability of soils (Garbaye 2013; Badda et al. 2014). This property is used mainly in dune fixation (Garbaye 2013). However, it was demonstrated that NaCl stress strongly increased glomalin production in hyphae of *R. irregularis* under in vitro condition. This suggested that some physicochemical property of NaCl like the direct toxicity of ions may be involved in the induction of glomalin production (Muthukumar et al. 2017).

16.5.2 AMF Inoculation Effect in Plant Salt Tolerance

AMF interacts with their host plants enabling them to implement tolerance mechanisms (Ruiz-Lozano et al. 2012). AMF have been shown to promote salinity tolerance by employing various mechanisms, such as enhancing nutrient acquisition, producing plant growth hormone, improving rhizospheric and soil condition, altering the physiological and biochemical properties of the host, and defending roots against soilborne pathogens (Juniper and Abbott 1993; Ruiz-Lozano et al. 2012; Porcel et al. 2012; Badda et al. 2014). AMF adjust all physiological and biochemical properties of the host plant (Plenchette 1991) (Fig. 16.3). This can be manifested by:

- Increase and/or better selection in nutrient acquisition (Porcel et al. 2012)
- Accumulation of osmoregulatory compounds (ElHindi et al. 2017)
- An important stomatal conductance (Feng et al. 2002)
- Increase in photosynthetic activity or a limitation of leaf dehydration (Porcel et al. 2012; Plouznikoof et al. 2016)

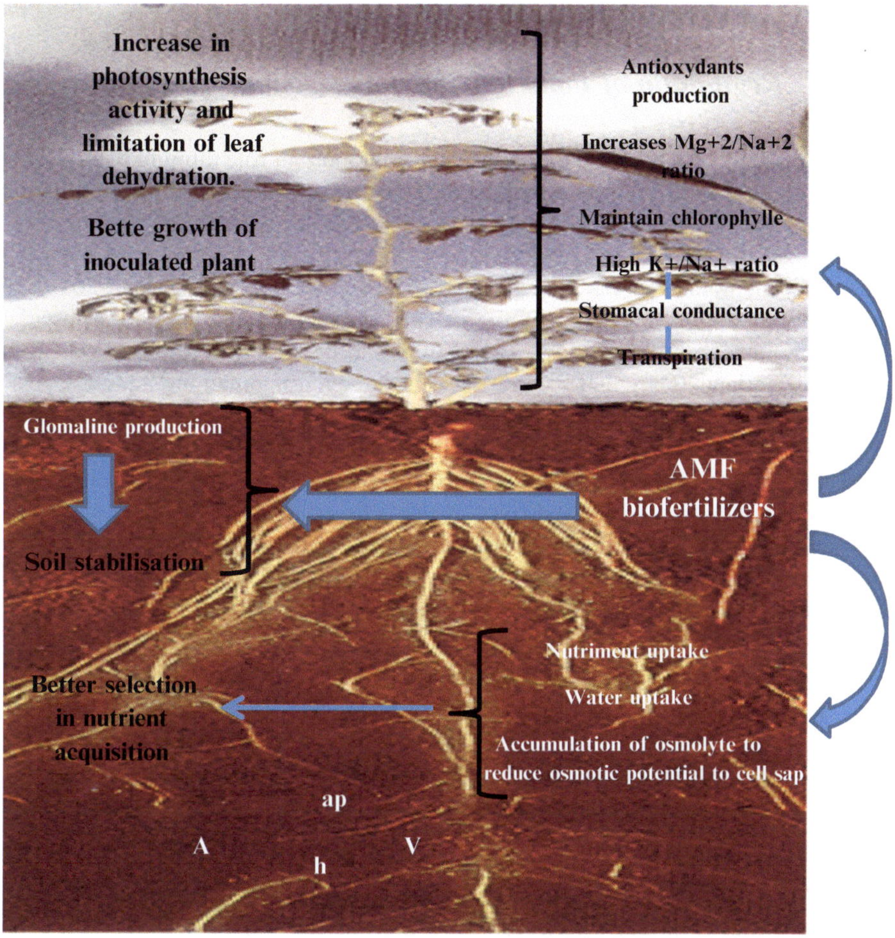

Fig. 16.3 The complex functioning of AMF biofertilizers in salt stress alleviation in host plants (Adapted from Evelin et al. 2009). The fungus forms an appressorium (ap) on the surface of the root and penetrates into the cortex of the root by extending its hyphae (h). Formulation of arbuscules (a) and vesicles (v) in the cortex. Salinity deprives plants of the basic requirements of water and nutrients, causing a water deficit and a decrease in osmotic potential accompanied by a decrease in nutrients leading to weak and unproductive plants. AMF increase the absorption of water and nutrients: a decrease in osmotic potential is counteracted by the increase and accumulation of osmolytes, photosynthesis and the production of antioxidants to trap (ROS) is more effective in plants under salt stress in the presence of AMF biofertilizers

16.6 Modification of Biochemical Properties

16.6.1 Regulation of Potassium (K^+) and Sodium (Na^+) Ions

It has recently been suggested that AMF play an important role through the mycelial membrane, in the selection of K^+ and Ca^{+2} ions, by attenuating saline stress in the plant through a preselection of nutrients and toxic ion prevention such as Na^+ (ElHindi et al. 2017). On the other hand, it has been suggested that the peri-arbuscular membrane serves as a selective barrier to all ions (Hajiboland 2013). In this case, this double selection (by the mycelial membrane and the peri-arbuscular membrane) acts to reduce the Na^+ acquisition and thus increase the K^+ content in mycorrhizal plants (Garg and Pandey 2016). Moreover, maintaining a high K^+/Na^+ ratio makes many enzymatic processes more or less stable (Porcel et al. 2012; Aroca et al. 2013). Indeed, K^+ activates a range of enzymes and plays an important role in stomatal movements. High concentrations of K^+ are required in protein synthesis and in the liaison between ARNt and ribosome (Evelin et al. 2009; Aroca et al. 2013). ElHindi et al. (2017) observed an enhancement in the uptake of K^+, P, and Ca^{+2} in the sweet basil plants inoculated with *Glomus deserticola*, with reduction of Na^+ and Cl^- uptake limiting their toxic effect while supporting a better growth of inoculated plants.

16.6.2 Impact of AMF Biofertilization on Proline and Other Osmolytes

In the case of salt stress, the plant tends to accumulate compounds such as proline, betaine, and carbohydrates (glucose and sucrose) to overcome the negative effect of salinity and maintain the osmotic balance at its stability level (Porcel et al. 2012; Plouznikoof et al. 2016; Saxena et al. 2017). The proline plays a role in primary metabolism as a constituent of proteins. In addition, it represents one of the most commonly accumulated compatible solutes in response to various environmental constraints. It plays an important role in the tolerance of plants to salt stress (Aroca et al. 2013). Indeed, several studies have found a lower proline content in plants inoculated with AMF biofertilizers compared to non-inoculated plants at different salinity levels. This accumulation has palliated stress symptoms in mycorrhizal plants (Benkhaled et al. 2007; Aroca et al. 2013; ElHindi et al. 2017). Sweet basil plants, for example, while under salt stress conditions, accumulated a proline lowest with AMF inoculation than with non-inoculated plant (ElHindi et al. 2017). Likewise, in Pearl millet shoots inoculated with *Rhizophagus fasciculatus*, proline accumulation was about 16–32% lower than non-inoculated plants after 90 days of culture under saline stress (Muthukumar et al. 2017). Reduction of proline amount in AMF-inoculated plants improved plant tolerance to salinity by maintaining the osmotic balance and reducing the free radical damage induced by osmotic stress and gives a clear indicator that these plants were less stressed (ElHindi et al. 2017).

Betaine's role is to stabilize the structures and activities of enzymes and protein complexes and maintain the integrity of cell membranes against the adverse effects of excessive salinity (Porcel et al. 2012). In these cases, the role of AMF biofertilizer is manifested by the increase in plant betaine content reducing salinity stress effect on plant development (Saxena et al. 2017). Indeed, wheat inoculated with a mixture of *Glomus* spp. accumulated more glycinebetaine than non-inoculated one at varying level of soil salinity (0.1–9.4 ds m^{-1}) (Muthukumar et al. 2017).

16.6.3 AMF Effect on Abscisic Acid Content and Antioxidant System

Abscisic acid (ABA) is a phytohormone that plays an important role in the development and growth of the plant. It is synthesized at the level of the parenchyma, mature roots, and leaves. This phytohormone is a very unstable molecule, so a high salinity level can lead to its conjugation with the different sugar. With AMF inoculation, the level of ABA can be adjusted in plants, which reduce signs of stress in the plant in saline environments (Evelin et al. 2009; Estrada et al. 2013; Rivero et al. 2014).

Salinity causes oxidative stress in plants; it is manifested by an increase in the cellular concentration of reactive oxygen species (ROS) (Rivero et al. 2014). However, an excess of ROS can lead to the oxidative damage of proteins, nucleic acids, and lipids, resulting in severe alterations of their structure and function causing sometimes lethal damage to cells (Evelin et al. 2009; ElHindi et al. 2017; Saxena et al. 2017). In fact, high levels of ROS under saline conditions induce a production of variety of reactive oxygen species such as superoxide anion radical (O^{-2}), hydrogen peroxide (H_2O_2), and hydroxyl radical (·OH). The alteration of lipids caused by a peroxidation in presence of ROS has been determined by quantification of malondialdehyde (MDA), which is a secondary end product of polyunsaturated fatty acid oxidation (Evelin et al. 2012). The accumulation of ROS at any point depends on the balance between the ROS production and ROS scavenging (Muthukumar et al. 2017). To counteract the effect of salt stress, literature from the last 10 years explained that AMF interfere in plant resistance by stimulating antioxidant systems (Badda et al. 2014), including antioxidant enzymes such as "superoxide dismutase" (SOD) and "ascorbate peroxidases" (APX) (Evelin et al. 2009; Plouznikoof et al. 2016), catalase (CAT), glutathione peroxidase (GPX), glutathione reductase (GR) (Alqarawi et al. 2014), dehydroascorbate reductase (Hashem et al. 2016), monodehydroascorbate reductase (Alqarawi et al. 2014), guaiacol peroxidase, and oxidized glutathione (Alqarawi et al. 2014; Hashem et al. 2016) and the enzymes involved in the ascorbate-glutathione cycle (Lee et al. 2015). These enzymes allow the degradation of oxygen reactive species (ROS) and

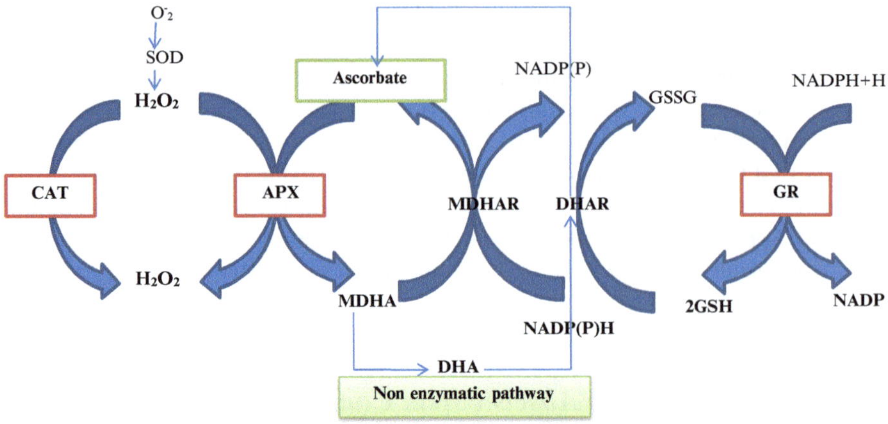

Fig. 16.4 Ascorbate-gluthatione pathway. *SOD* superoxide dismutase, *CAT* catalase, *APX* ascorbate peroxidase, *MDHAR* monodehydroascorbate reductase, *DHAR* dehydroascorbate reductase, *GR* gluthatione reductase, *MDHA* monodehydroascorbate, GSSG oxidized, GSH reduced gluthatione (Adapted from Porcel et al. 2012)

therefore limit the induction of membrane damage (Plouznikoof et al. 2016; Saxena et al. 2017) (Fig. 16.4). Inoculation with AMF has shown higher activities of SOD, peroxidase, and APOX in mycorrhizal *Ephedra aphylla* (Alqarawi et al. 2014) and in inoculated *Acacia gerrardii* (Hashem et al. 2016). In addition, some nonenzymatic antioxidant molecules such as ascorbic acid (AsA) and glutathione (GSH) (Miller et al. 2010), carotenoids, and tocopherols also scavenge activated oxygen species (Evelin et al. 2009; Wu et al. 2013; Hashem et al. 2016) (Fig. 16.4). It was demonstrated that AMF inoculation of *Brassica juncea* plants exhibit higher ROS scavenging capacity that could be attributed to the increased activities of antioxidant enzymes. Indeed, mycorrhizal fungi possess various oxidative stress-related genes (SOD genes), which upon upregulation enable mycorrhizal plants to tolerate salinity stresses (Ruiz-Lozano et al. 2001; Wu et al. 2013; Saxena et al. 2017).

16.7 Physiological Aspect

Salt stress inhibits photosynthetic capacity in plants and leads to water imbalance (Porcel et al. 2012) and to chlorophyll reduction (ElHindi et al. 2017). Several studies show that AMF inoculation avoid leaf dehydration caused by salinity and inhibit photosynthesis and particularly leaf photochemical properties under salt stress (Sheng et al. 2009; ElHindi et al. 2017). AMF inoculation leads to an increase in the water conductivity of host plant roots (Khalloufi et al. 2017; Saxena et al. 2017). Inoculated plants have the ability to fix more CO_2, resulting in improved growth (Aroca et al. 2013). Indeed, AMF biofertilization induces a higher contents of chlorophyll in leaves as compared to non-inoculated plants; this higher chlorophyll content means that the photosynthesis and carbon fixation processes are in a higher rate which enable the plant to afford mycorrhizal association as the cost for this relation is carbon (ElHindi et al. 2017; Basu et al. 2018). Khalloufi et al. (2017) demonstrated that AMF inoculation improved gibberellic acid concentration in *Solanum lycopersicum* under salinity stress. They suggest that there is an interactive positive effect between growth hormones and AMF which alleviates growth impairment under salinity conditions by modifying the hormonal balance of the plant.

16.8 Molecular Changes

The benefit of AMF inoculation is not only manifested in the growth and water potential of the plant. A beneficial effect on the expression of genes intervening in response to salt stress has been recorded (Porcel et al. 2012). Figure 16.5 summarizes gene regulation during plant and mycorrhiza interaction; it indicates how the plants perceive mycorrhizal signal through the involvement of various proteins like NOD factors, SYMRK, calcium signaling proteins, etc. This model is a summary of the information done on different plants combined into one to give detailed information of the complexity of the signaling network in plant mycorrhizal interaction (Basu et al. 2018).

16.8.1 Expression of the 1-Pyrroline-5-Carboxylate Synthetase (P5CS) Gene

Proline acts as a stabilizer of macromolecular proteins and complexes, free radical scavenger, and cellular redox potential regulator (Sheng et al. 2011). Yoshiba et al. (1995) demonstrate that in the case of environmental stresses, the accumulation of proline in plants, taking *Arabidopsis thaliana* as a model, necessarily follows the transcription induced by the "P5CS" gene. This gene interferes in the functioning of the two enzymes: "γ-glutamil-kinase" and "glutamic-γ-semialdehyde dehydrogenase" (Fig. 16.6) (Porcel et al. 2012). The "P5CS" gene codes for multifunctional proteins imperatively induced by saline and drought stress (Ábrahám et al. 2003) and necessarily triggers proline biosynthesis in plants from glutamic acids (Ábrahám et al. 2003; Saxena et al. 2017). AMF biofertilization reduce the "P5CS" gene transcription, indicating that host plant is not stressed.

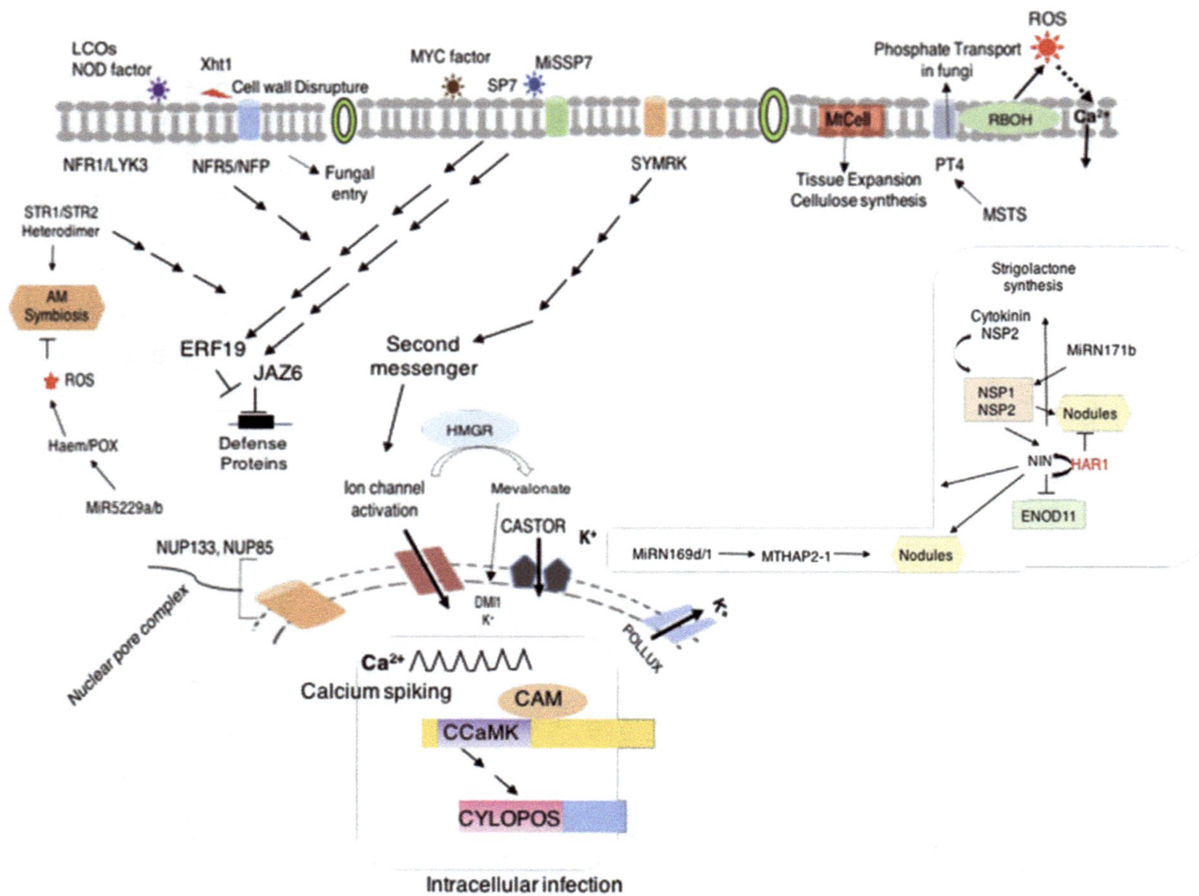

Fig. 16.5 Model of gene regulation during plant and mycorrhiza interaction. A model of gene regulation during plant and mycorrhiza interaction. The figure summarizes how the plants sense and perceive mycorrhizal signal through the involvement of various proteins like NOD factors, SYMRK, calcium signaling proteins, etc. This model is a summary of the information done on different plants combined into one to give detailed information of the complexity of the signaling network in plant mycorrhizal interaction (Basu et al. 2018)

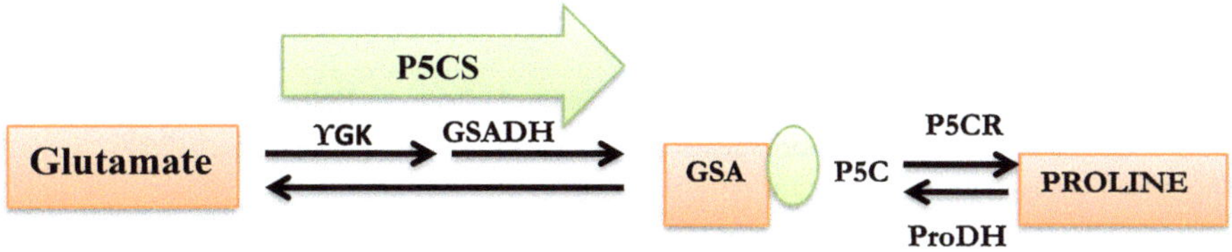

Fig. 16.6 Proline biosynthetic pathway. *GSA* glutamic-γ-semi-aldehydehyde, P5C Δ1-pyrroline-5-carboxylate, P5CS Δ1-pyrroline-5-carboxylate synthetase, *P5CR* P5C reductase, *γGK* γ-glutamyl kynase, *GSADH* glutamyl-γ-semi-aldehyde dehydrogenase, *P5CDH* P5C dehydrogenase, *proDH* proline dehydrogenase (Adapted from Porcel et al. 2012)

16.8.2 Aquaporin

Negative water potential in saline soils requires plants to acquire a sufficient amount of water (Yoshiba et al. 1995), which involves the functioning of water channel protein: aquaporin (Saxena et al. 2017). These proteins belong to the family of major intrinsic protein (MIP) and have a role in passive movement of water molecules following a water potential gradient and maintain the cellular osmoregulatory (Plouznikoof et al. 2016; Saxena et al. 2017). This protein chain is divided into four groups based on the homogeneity between their sequences. The four groups are known as "intrinsic proteins of the plasma membrane," "PIPs"; "intrinsic protein tonoplasts," "TIPs"; "intrinsic nitrogen-fixing nodule proteins," "NIPs"; and "small and basic intrinsic proteins," "SIPs" (Ábrahám et al. 2003; Porcel et al. 2012). In model plants, wild arabica, maize, and rice, more than 30 different aquaporin genes are identified (Maurel et al. 2009; Porcel et al. 2012). In addition, each group of aquaporin is different in its ability to carry water and small neutral solutes and in its subcellular location (Maurel et al. 2009).

Analysis of aquaporin gene expression revealed contrasting results (Porcel et al. 2012). Aroca et al. (2013) observed a higher expression of three PIP genes in *Beta vulgaris* under salt stress, especially in mycorrhizal plants. On the other hand, Benkhaled et al. (2007) and Jahromi et al. (2008) reported that in the presence of salt stress, the PIP1 and PIP2 genes are inhibited by mycorrhizal biofertilization in *Lactuca sativa*, while in mycorrhizal plants under salt stress, the expression of the LsPIP2 gene is maintained. A differential effect of arbuscular symbiosis on aquaporin isoforms under salinity conditions has also been described by Ouziad et al. (2006).

The study of Ouziad et al. (2006), Benkhaled et al. (2007), Jahromi et al. (2008), and Aroca et al. (2013) highlights the complex regulation of aquaporin genes in response to arbuscular symbiosis under salt stress (Table 16.3).

Table 16.3 Comparison between the expressions of the aquaporin (PIP1, TIP, and PIP2 genes) under salt stress in different studies

	Salinity effect	References
AMF inoculation effect on aquaporin genes expression d'aquaporines	PIP1 diminution TIP diminution PIP2 no effect	Ouziad et al. (2006)
	PIP1 augmentation PIP2 aucun changement	Jahromi et al. (2008), Benkhaled et al. (2007)
	PvPIP1.1 augmentation PvPIP1.2 no effect PvPIP1.3 augmentation PvPIP2.1 augmentation	Aroca et al. (2013)

16.8.3 Homeostasis Ion Genes Regulators

As previously explained, the plants inoculated with the AMF show a significant increase in K^+ with a reduction in Na^+, accompanied by an accumulation of the K^+/Na^+ ratio in their tissues. It has recently been shown that the increase in the K^+/Na^+ ratio is correlated with the expression of the genes involved in the Na^+ explosion and the acquisition of K^+. These genes are named "ZmAKT2," "ZmSOS1," and "ZmSKOR." The gene "ZmAKT2" codes for the expression of K^+ in phloem (Estrada et al. 2013). The gene "ZmSOS1" plays a double role in the plant. It codes for the restriction of the absorption of Na^+ ions by the roots in parallel with the control of their transport by xylem (Porcel et al. 2016). The "ZmSKOR" gene encodes the regulation of K^+ transport by xylem to leaves (Estrada et al. 2013; ElHindi et al. 2017). Estrada et al. (2013) evaluated the expression of these genes in corn roots under salt stress. These authors observed a regulation of ZmAKT2 and ZmSKOR gene expression exposed to 100 mM NaCl. This regulation contributes to the retention of K^+ in plant tissues and to the increase of the K^+/Na^+ ratio. Recently, Porcel et al. (2016) showed that rice plants inoculated with AMF *Claroideoglomus etunicatum* showed a lower distribution of Na^+ from root to shoot tissues. This lower root to shoot distribution of Na^+ ions could be due to increased regulation of rice transporter genes OsNHX3, OsSOS1, OsHKT2;1, and OsHKT1;5 involved in maintaining ion homeostasis by mycorrhizal fungi. They further suggested that the expression of these genes triggers Na^+ efflux from the cytoplasm and xylem and promotes Na^+ sequestration into the vacuole.

In addition, the members of the cyclic nucleotide gate ion channel "CNGC" gene cluster play an important role in the survival and adaptation of plants to stress and in

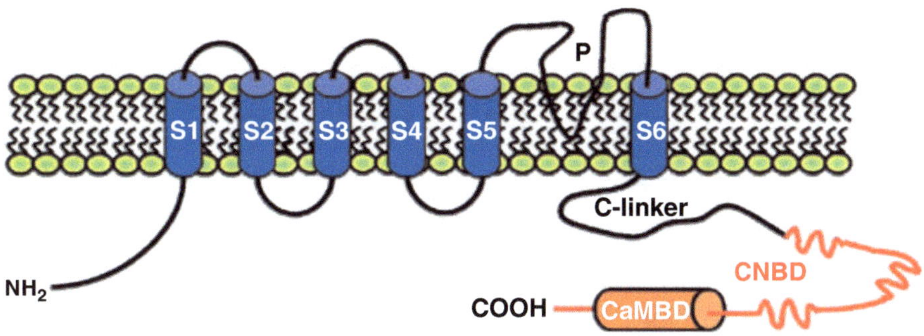

Fig. 16.7 Structure of cyclic-dependent nucleic acid plant channels (CNGC) (Porcel et al. 2012). CNGCs consist of six trans-membrane channels (S) and a pore region (P) between the fifth and sixth channel. The C-terminal end has a C-linker link followed by the cyclic nucleotide (CNBD) with mandatory overlap with calmodulin (CAMBD)

the regulation of different pathways of development and physiological functioning (Fig. 16.7) (Porcel et al. 2012). Indeed, the genes "AtCNGC19" and "AtCNGC20" encode the activation of the Ca^{2+} and K^+ ion transporter channels in the leaves (Chin et al. 2013). They are highly regulated in response to high salinity (Ruiz-Lozano et al. 2012; Porcel et al. 2012) and are involved in increasing the K^+/Na^+ ratio in plant tissues (Chin et al. 2013). Although it is evident from the literature that under saline condition inoculated plants with AMF maintain more favorable K^+/Na^+ ratio than non-AM plant, the molecular mechanisms involved in maintaining favorable ionic balance are not yet well documented. Therefore, attention is required to identify possible regulation of genes determining HKT, AKT, SOS, and NHX ion transporters and most likely CNGCs in AMF-colonized plant growing under salt stress (Ruiz-Lozano et al. 2012). Table 16.4 summarizes the most important impact of AMF biofertilization on salt stress alleviation in crops reported over the last decade.

16.9 Conclusion and Perspectives for Future Studies

The present review clearly emphasizes the important role played by AMF to the benefit of the terrestrial ecosystems. It also highlights the fact that the majority of vascular plants are capable of symbiotic association enabling survival under saline adverse atmospheric conditions. Biological improvement techniques based on AMF inoculation of plants to restore arid and semiarid areas have not only ecological benefits but also economic gains. Indeed, these biological models based on the assessment of plant symbiosis/mycorrhizal fungi have undeniably demonstrated the major role that this symbiotic process could play in ensuring the success of sustainable soil management. The valorization of AMF inoculum production

Table 16.4 Impact of AMF biofertilization on salinity stress alleviation in crops reported over the last decade

Salt stress	AMF species	Host plants	Growth conditions	Improved parameters					Remarks	References
				Growth	Nutrition	Antioxidant system	Osmotic balance	Photosynthesis		
50 mM NaCl	*R. irregularis* (MUCLA3194)	*Lactuca sativa*	Pot culture glasshouse	+		+	+		+RWC (shoot), -proline, -ABA, Lsp5cs, -LsLea, -L3nced	Jahromi et al. (2008)
100 mM NaCl		*L. cv. Romana*		+		+	+		+RWC (shoot), -proline, -ABA, =Lsp5cs, -LsLea, =Lsnced	
0.5, 1.0, 1.5, and 2 g NaCl/kg dry substrate	*F. mosseae*	*Zea mays*	Pot culture green house	=			+	+	-Intercellular O_2, +chlorophyll, +photosynthesis rate, +Fv/Fm, +qPSII, +qP, +RWC (leaves), +stomatal conductance	Sheng et al. (2008)
2 and 4 g NaCl/kg soil	*F. mosseae*	*Lactuca sativa* *L. cv. Tafalla*	Mesocosm	+					+Na (soil), +glomalin, -related soil protein, +aggregate stability	Kohler et al. (2009)

EC = 5 ds/m and 10 ds/m	*R. intraradices*	*Solanum lycopersicum L. cultivars Behta and Piazar*	Pot culture greenhouse	+	+P, Ca, K, +Ca/Na, +K/Na	+	+	+	+Electron transport rate, +APX, CAT, POD, SOD, proteins, -H_2O_2, MDA, +proline, +stomatal conductance to water vapor, +transpiration, +Fv/Fm, F/Fo, PSII+photosynthetic rate, +activity of ROS scavenging enzymes	Hajiboland (2013)
25 mM NaCl	*C. etunicatum*	*Zea mays*	Pot culture greenhouse	+	+N, P, Ca, Mg				-Proline	Lee et al. (2015)
EC = 0.64 ds/m and 5 ds/m and 10 ds/m	*Glomus deserticola*	*Ocimum basilicum*	Greenhouse	+	+P, Ca, K				+K/Na, -proline, +chlorophyll, +water efficiency	ElHindi et al. (2017)

originating from indigenous local strains well adapted to hostile environmental conditions (high salinity, temperature, degraded and deficient soil, etc.) and applying them to saline soils should allow preserving vegetation and exploiting more efficiently areas of arid and semiarid zones. Overall, these investigations should open new research lines aimed at obtaining maximum benefit from the AMF symbiosis under salinity or other osmotic stress conditions.

References

Ábrahám E, Rigó G, Székely G, Nagy R, Koncz C, Szabados L (2003) Light-dependent induction of proline biosynthesis by abscisic acid and salt stress is inhibited by brassinosteroid in Arabidopsis. Plant Mol Biol 51:363–372

Ahmad P, Hashem A, Abd-Allah EF, Alqarawi AA, John R, Egamberdieva D (2015) Role of *Trichoderma harzianum* in mitigating NaCl stress in Indian mustard (*Brassica juncea* L) through antioxidative defense system. Front Plant Sci 6:868

Aliasgharzadeh N, Rastin SN, Towfighi H et al (2001) Occurrence of arbuscular mycorrhizal fungi in saline soils of the Tabriz Plain of Iran in relation to some physical and chemical properties of soil. Mycorrhiza 11:119–122

Allen EB, Cunningham GL (1983) Effects of vesicular-arbuscular mycorrhizae on *Distichlis spicata* under three salinity levels. New Phytol 93:227–236

Alqarawi AA, Abd Allah EF, Hashem A (2014) Alleviation of salt-induced adverse impact via mycorrhizal fungi in *Ephedra aphylla* Forssk. J Plant Interact 9:802–810

Aroca R, Ruiz-Lozano JM, Zamarreño AM et al (2013) Arbuscular mycorrhizal symbiosis influences strigolactone production under salinity and alleviates salt stress in lettuce plants. J Plant Physiol 170:47–55

Arora S, Vanza M (2017) Microbial approach for bioremediation of saline and sodic soils. In: Singh (ed) Bioremediation of salt affected soils: an Indian perspective. Springer, Berlin, pp 87–100

Bacera A, Bartoloni N, Cofré N, Soteras F, Cabello M (2014) Arbuscular mycorrhizal fungi in saline soils: vertical distribution at different soil depth. Braz J Microbiol 45:585–594

Badda N, Aggarwal A, Kadian N, Sharma N (2014) Influence of arbuscular mycorrhizal fungi and different salinity levels on growth enhancement and nutrient uptake of *Gossypium arboreum* L. Kavaka 43:14–21

Barin M, Alisgharzad N, Olsson PA, Rasouli-Sadaghiani M (2013) Abundance of arbuscular mycorrhizal fungi in relation to soil salinity around Urmia northern Iran analyzed by use of lipid biomarkers and microscopy. Pedobiologia 56:225–232

Basu S, Rabara RC, Negiet S (2018) AMF: the future prospect for sustainable agriculture. Physiol Mol Plant Pathol 102:36–45

Bearden BN, Petersen L (2000) Influence of arbuscular mycorrhizal fungi on soil structure and aggregate stability of vertisols. Plant Soil 218:173–183

Bencherif K, Boutekrabta A, Fontaine J, Laruelle F, Dalpé Y, Lounès-Hadj Sahraoui A (2015) Impact of soil salinity on arbuscular mycorrhizal fungi biodiversity and microflora biomass associated with *Tamarix articulata* Vahll rhizosphere in arid and semi-arid Algerian areas. Sci Total Environ 533:488–494

Benkhaled L, Gomez AM, Ouarraqi E, Oihabi A (2007) Réponses physiologiques et biochimiques du tréfle (*Trifolium alexandrinun* L) à la double association Mycorhizes-Rhizobium sous une contrainte saline. Agronomie 23:571–580

Bowen GD (1987) The biology and physiology of infection and its development. In: Safir GR (ed) Ecophysiology of VA mycorrhizal plants. CRC, Boca Raton, pp 27–57

Carillo P, Annunziata MG, Pontecorvo G et al (2011) Salinity stress and salt tolerance. In: Shanker AK, Venkateswarlu B (eds) Abiotic stress plants-mechanisms and adaptations. InTech, Rijeka. https://doi.org/10.5772/22331

Carvalho LM, Correia PM, Caçador I, Martins-Louçao MA (2003) Effects of salinity and flooding on the infectivity of salt marsh arbuscular mycorrhizal fungi in *Aster tripolium* L. Biol Fertil Soils 38:137–143

Chin K, DeFalco TA, Moeder W, Yoshioka K (2013) The Arabidopsis cyclic nucleotide-gated ion channels AtCNGC2 and AtCNGC4 work in the same signaling pathway to regulate pathogen defense and floral transition. Plant Physiol 163:611–624

Diagne N, Baudin E, Svistoonoff S, Ouattara C, Diouf D, Kan A, Ndiaye C, Noba K, Bogusz D, Franche C, Duponnois R (2018) Effect of native and allochthonous arbuscular mycorrhizal fungi on *Casuarina equisetifolia* growth and its root bacterial community. Arid Land Res Manag. https://doi.org/10.1080/15324982.2017.1406413

Egamberdieva D, Davranov K, Wirth S, Hashem A, Abd-Allah E (2017) Impact of soil salinity on the plant-growth – promoting and biological control abilities of root associated bacteria. Saudi J Biol Sci 24:1601–1608

ElHindi KM, Sharaf El-Din A, ElGorban AM (2017) The impact of arbuscular mycorrhizal fungi in mitigating salt-induced adverse effects in sweet basil (*Ocimum basilicum* L.). Saudi J Biol Sci 24:170–179

Estrada B, Aroca MFJM et al (2013) Arbuscular mycorrhizal fungi native from a Mediterranean saline area enhance maize tolerance to salinity through improved ion homeostasis. Plant Cell Environ 36:1771–1782

Evelin H, Kapoor R, Giri B (2009) Arbuscular mycorrhizal fungi in alleviation of salt stress: a review. Ann Bot 104:1263–1280

Evelin H, Giri B, Kapoor R (2012) Ultrastructural evidence for AMF mediated salt stress mitigation in *Trigonella foenum-graecum*. Mycorrhiza 23(1):71–86

Fan L, Dalpé Y, Fang C, Dubé C, Khanizadeh S (2011) Influence of arbuscular mycorrhizae on biomass and root morphology of selected strawberry cultivars under salt stress. Botany 89:397–403

Feng G, Zhang FS, Xl L et al (2002) Improved tolerance of maize plants to salt stress by arbuscular mycorrhiza is related to higher accumulation of soluble sugars in roots. Mycorrhiza 12:185–190

Garbaye J (2013) La symbiose mycorhizienne: Une association entre les plantes et les champignons. Collection Synthèse. QUAE editions, p 280

Garg N, Pandey R (2016) High effectiveness of exotic arbuscular mycorrhizal fungi is reflected in improved rhizobial symbiosis and trehalose turnover in *Cajanus cajan* genotypes grown under salinity stress. Fungal Ecology 21:57–67

Giri B, Kapoor R, Mukerji KG (2007) Improved tolerance of *Acacia nilotica* to salt stress by arbuscular mycorrhiza, *Glomus fasciculatum* may be partly related to elevated K/Na ratios in root and shoot tissues. Microb Ecol 54:753–760

Guo X, Gong J (2014) Differential effects of abiotic factors and host plant traits on diversity and community composition of root-colonizing arbuscular mycorrhizal fungi in a salt-stressed ecosystem. Mycorrhiza 24:79–94

Hajiboland R (2013) Role of arbuscular mycorrhiza in amelioration of salinity. In: Ahmed P, Azooz MM, Prasad MNV (eds) Salt stress in plants: signaling, omics and adaptations. Springer, New York, pp 301–354

Hashem A, Abd_Allah EF, Alqaraw AA, Al-Huqail AA, Wirth S, Egamberdieva D (2016) The interaction between arbuscular mycorrhizal fungi and endophytic bacteria enhances plant growth of *Acacia gerrardii* under Salt Stress. Front Microbiol 7:1089

Hildebrandt U, Janetta K, Ouziad F, Renne B, Nawrath K, Bothe H (2001) Arbuscular mycorrhizal colonization of halophytes in Central European salt marshes. Mycorrhiza 10:175–183

Jahromi F, Aroca R, Porcel R et al (2008) Influence of salinity on the in vitro development of *Glomus intraradices* and on the in vivo physiological and molecular responses of mycorrhizal lettuce plants. Microb Ecol 55:45–53

Jamil A, Riaz S, Ashraf M, Foolad MR (2011) Gene expression profiling of plants under salt stress. Crit Rev Plant Sci 30(5):435–458

Juniper S, Abbott L (1993) VAM and soil salinity. Mycorrhiza 4:45–57

Juniper S, Abbott LK (2006) Soil salinity delays germination and limits growth of hyphae from propagules of arbuscular mycorrhizal fungi. Mycorrhiza 16:371–379

Karaarslan E, Uyanoz R (2011) Occurrence of arbuscular mycorrhizal fungi in some native plants grown on saline soils around the lake Tuz in Turkey and its relations with some physical and chemical properties of soil. Sci Res Essays 6:4238–4245

Khalloufi M, Andújar CM, Lachaâl M, Karray-Bouraoui N, Pérez-Alfocea F, Albacete A (2017) The interaction between foliar GA3 application and arbuscular mycorrhizal fungi inoculation improves growth in salinized tomato (*Solanum lycopersicum* L.) plants by modifying the hormonal balance. J Plant Physiol 214:134–144. https://doi.org/10.1016/j.jplph.2017.04.012

Kohler J, Hernandez JA, Caravaca F, Roldan A (2009) Induction of antioxidant enzymes is involved in the great effectiveness of a PGPR versus AM fungi with respect to increasing the tolerance of lettuce to sever salt stress. Environ Exp Bot 65:245–252

Landwehr M, Hildebrandt U, Wilde P, Nawrath K, Toth T, Biro B, Bothe H (2002) The arbuscular mycorrhizal fungus *Glomus geosporum* in European saline sodic and gypsum soils. Mycorrhiza 12:199–211

Lee Y, Krishnamoorthy R, Selvakumar G et al (2015) Alleviation of salt stress in maize plant by co-inoculation of arbuscular mycorrhizal fungi and *Methylobacterium oryzae* CBMB20. J Korean Soc Appl Biol 58:533–540

Machado RMA, Serralheiro RP (2017) Soil salinity: effect on vegetable crop growth. Management practices to prevent and mitigate soil salinization. Horticultirae 3:30. https://doi.org/10.3390/horticulturae3020030

Mathur N, Singh J, Bohra S et al (2007) Arbuscular mycorrhizal status of medicinal halophytes in saline areas of Indian Thar desert. Int J Soil Sci 2:119–127

Maurel C, Santoni V, Luu DT, Wudick MM, Verdoucq L (2009) The cellular dynamics of plant aquaporin expression and functions. Curr Opin Plant Biol 12:690–698. https://doi.org/10.1016/j.pbi.2009.09.002

Mayak S, Tirosh T, Glick BR (2004) Plant growth-promoting bacteria confer resistance in tomato plants to salt stress. Plant Physiol Biochem 42:565–572

Miller G, Suzuki N, Ciftci-Yilmaz S et al (2010) Reactive oxygen species homeostasis and signaling during drought and salinity stresses. Plant Cell Environ 33(4):453–467

Munns R (2005) Genes and salt tolerance: bringing them together. New Phytol 167:645–663

Muthukumar T, Bagyarag DJ, Ashwin R (2017) Arbuscular mycorrhizal fungi: Role in alleviating salt stress in crop plants. In: Bagyarag DJ, Jamaluddin (eds) Microbes for plant stress management, vol 1. New India, New Delhi, pp 221–243

Ouziad F, Wilde P, Schmelzer E et al (2006) Analysis of expression of aquaporins and Na+/H+ transporters in tomato colonized by arbuscular mycorrhizal fungi and affected by salt stress. Environ Exp Bot 57:177–186

Péréz-Tienda J, Testillanob PS, Balestrinic R, Fiorillic V, Azcón-Aguilara C, Ferrol N (2011) GintAMT2, a new member of the ammonium transporter family in the arbuscular mycorrhizal fungus *Glomus intraradices*. Fungal Genet Biol 48:144–1055

Plenchette C (1991) Utilisation des mycorhizes en agriculture et horticulture. In: Strullu DG, Garbaye J, Perrin R, Plenchette C (eds) Les mycorhizes des arbres et des plantes cultivées. Lavoisier, Paris, pp 131–196

Plouznikoof K, Declerck S, Calonne-Salmon M (2016) Mitigating abiotic stresses in crop plants by arbuscular mycorrhizal fungi. In: Vos CMF, Kazan K (eds) Belowground defence strategies in plants, signaling and communication in plants. Springer, Cham. https://doi.org/10.1007/978-3-319-42319-7_1

Pond EC, Menge JA, Jarrell WM (1984) Improved growth of tomato in salinized soil by vesicular arbuscular mycorrhizal fungi collected from saline soils. Mycologia 76:74–84
Porcel R, Aroca R, Ruíz-Lozano JM (2012) Salinity stress alleviation using arbuscular mycorrhizal fungi – a review. Agron Sustain Dev 32:181–200
Porcel R, Aroca R, Azcon R et al (2016) Regulation of cation transporter genes by the arbuscular mycorrhizal symbiosis in rice plants subjected to salinity suggests improved salt tolerance due to reduced Na+ root-to-shoot distribution. Mycorrhiza 26:673
Powell JR, Rilling MC (2018) Biodiversity of arbuscular mycorrhizal fungi and ecosystem function. New Phytol Trans. https://doi.org/10.1111/nph.15119
Ramoliya PJ, Patel HM, Pandey AN (2004) Effect of salinization of soil on growth and macro- and micro-nutrient accumulation in seedlings of *Salvadora persica* (Salvadoraceae). For Ecol Manag 202:181–193
Rivero RM, Mestre TC, Mittler R, Rubio F, Garcia-Sanchez F, Martinez V (2014) The combined effect of salinity and heat reveals a specific physiological biochemical and molecular response in tomato plants. Plant Cell Environ 37:1059–1073
Rozema J, Arp W, Van Diggelen J et al (1986) Occurrence and ecological significance of vesicular-arbuscular mycorrhiza in the salt marsh environment. Acta Bot Neerl 35:45
Ruiz-Lozano JM, Collados C, Barea JM et al (2001) Arbuscular mycorrhizal symbiosis can alleviate drought-induced nodule senescence in soybean plants. New Phytol 151:493–502
Ruiz-Lozano JM, Porcel R, Azcón C, Aroca R (2012) Regulation by arbuscular mycorrhizae of the integrated physiological response to salinity in plants: new challenges in physiological and molecular studies. J Exp Bot 63(11):4033–4044
Sabir P, Ashraf M, Hussain M, Jamil A (2009) Relationship of photosynthetic pigments and water relations with salt tolerance of proso millet (*Panicum Miliaceum* L.) accessions. Pak J Bot 41:2957–2964
Sanchez-Castro I, Ferrol N, Barea JM (2012) Analyzing the community composition of arbuscular mycorrhizal fungi colonizing the roots of representative shrubland species in a Mediterranean ecosystem. J Arid Environ 80:1–9
Saxena B, Shukla K, Giri B (2017) Arbuscular mycorrhizal fungi and tolerance of salt stress in plants. In: Wu QS (ed) Arbuscular mycorrhizas and stress tolerance of plants. Springer, Singapore, pp 76–106
Sheng M, Tang M, Chen H, Yang B, Zhang F, Huang Y (2008) Influence of arbuscular mycorrhizae on photosynthesis and water status of maize plants under salt stress. Mycorrhiza 18 (6–7):287–296
Sheng M, Tang M, Chen H, Yang B, Zhang F, Huanga Y (2009) Influence of arbuscular mycorrhizae on the root system of maize plants under salt stress. Rev Can Microbiol 55:879–886
Sheng M, Tang M, Zhang FF et al (2011) Influence of arbuscular mycorrhiza on organic solutes in maize leaves under salt stress. Mycorrhiza 21:423–430
Shrivastava P, Kumar R (2015) Soil salinity: a serious environmental issue and plant growth promoting bacteria as one of the tools for its alleviation. Saudi J Biol Sci 22:123–131
Smith SE, Read DJ (2008) Mycorrhizal symbiosis. Academic, New York
Sonjak S, Beguiristain T, Leyval C, Regvar M (2009) Temporal temperature gradient gel electrophoresis (TTGE) analysis of arbuscular mycorrhizal fungi associated with selected plants from saline and metal polluted environments. Plant Soil 314:25–34
Taleisnik EL, Grunberg K (1994) Ion balance in tomato cultivars differing in salt tolerance. Sodium and potassium accumulation and fluxes under moderate salinity. Physiol Plant 92:528–534
Tuteja N (2007) Mechanisms of high salinity tolerance in plants. Methods Enzymol 428:419–438
Wang FY, Liu RJ, Lin XJ, Zhou JM (2004) Arbuscular mycorrhizal status of wild plants in saline-alkaline soils of the Yellow River Delta. Mycorrhiza 14:133–137

Wilde P, Manal A, Stodden M, Sieverding E, Hildebrandt U, Bothe H (2009) Biodiversity of arbuscular mycorrhizal fungi in roots and soils of two salt marshes. Environ Microbiol 11:1548–1561

Wu Z, Sawada T, Shiba K, Liu S, Kanao T, Takahashi R, Hattori N, Imai Y, Lu B (2013) Tricornered/NDR kinase signaling mediates PINK1-directed mitochondrial quality control and tissue maintenance. Genes Dev 27:157–162

Yoshiba Y, Kiyosue T, Katagiri T, Ueda H, Mizoguchi T, Y amaguchishinozaki K, Wada K, Harada Y, Shinozaki K (1995) Correlation between the induction of a gene for delta (1)-pyrroline-5-carboxylate synthetase and the accumulation of proline in Arabidopsis thaliana under osmotic stress. Plant J 7:751–760

Yun P, Xu L, Wang SS, Shabala L, Shabala S, Zhang WY (2018) *Piriformospora indica* improves salinity stress tolerance in *Zea mays* L. plants by regulating Na+ and K+ loading in root and allocating K+ in shoot. Plant Growth Regul. https://doi.org/10.1007/s10725-018-0431-3

Chapter 17
Salinity Stress-Dependent Coordination of Metabolic Networks in Relation to Salt Tolerance in Plants

Ashutosh Dubey and A. K. Verma

Abstract The major environmental stress, salinity, results in other secondary stresses such as osmotic stress, oxidative stress, and ionic stress leading to physiological, biochemical, and molecular alterations in plants. Many plants develop mechanisms, either to exclude salt from their cells or to tolerate its presence within the cells. Salinity stress signal initiates multiple secondary signals in plants leading the activation of several physiological and biochemical strategies for gradual adaption of plants against salt stresses. The essential processes leading to plant adaption to salt stress include metabolic adjustment, toxic ion homeostasis, and osmotic adjustment. During the salinity stress, active accumulation of osmotic solutes such as soluble carbohydrates, proteins, and free amino acids is one of effective stress tolerance mechanisms as accumulation of these osmotic solutes lowers osmotic potential within cells and tissues. Environmental constrains have ultimate impact on plant growth, although these can be overcome by production of various metabolites through appropriate genes of the cells. In this chapter, salt tolerance and subsequent metabolic alterations in plants are under the following aspects: impact and response of plants for abiotic stress specifically to salinity stress; the morphological, biochemical, and metabolic adjustments of plants during salinity stress; role of rhizosphere microbes in plant stress tolerance; and expression of salt tolerance genes. This chapter will help to understand the mechanism of plant salt tolerance along with biochemical and metabolic changes required to adapt plants for salinity stress.

Keywords Plant stress tolerance · Abiotic environment · Salinity · Salt stress · Salt tolerance

---

A. Dubey (✉) · A. K. Verma
Department of Biochemistry, College of Basic Sciences and Humanities, G. B. Pant University of Agriculture and Technology, Pantnagar, India

B. Giri, A. Varma (eds.), *Microorganisms in Saline Environments: Strategies and Functions*, Soil Biology 56, https://doi.org/10.1007/978-3-030-18975-4_17

17.1 Introduction

The adverse surroundings not only affect plant growth but also strongly modify metabolic reactions, responses, and regulations in plants to counter these challenges. The metabolic diversity ascended from constant evolution of plants has been manifested in the form of plant secondary metabolites amid miscellaneous functions. The biosynthesis, concentration, transport, and storage of not only primary but also secondary metabolites are affected by abiotic stresses. Among abiotic stresses, soil salinity is one of the prominent environmental condition that limits plant productivity (Ashraf and Harris 2004; Bui 2013; Hussain et al. 2009). Major factors contributing to high level of soluble salts/salinity are use of saline water for irrigating cultivated lands, poor drainage during irrigation, low precipitation, high temperature, and high transpiration rate (Shrivastava and Kumar 2015). A soil is considered to be saline or salt affected when the electric conductivity (EC) of the soil solution becomes 4 dS m^{-1} (equivalent to 40 mM NaCl), which generates about 0.2 MPa osmotic pressure and causes significant reduction in the yields of most crops (Munns and Tester 2008; Numan et al. 2018). Salinity has been the foremost ecological stress among others that results in generation of other secondary stresses such as osmotic stress, oxidative stress, and ionic stress leading to physiological, biochemical, and molecular alterations in plants resulting in substantial loss in plant productivity. Plants are grouped as either glycophytes or euhalophytes, on the basis of their ability to grow in saline conditions. Most crop plants are glycophytes which cannot survive at 100–200 nM NaCl concentration because plant growth is completely inhibited at such high salt concentrations (Munns and Termaat 1986) even with short-term exposure (Hernández and Almansa 2002). Contrary to glycophytes, halophytes can survive high NaCl concentrations (300–500 mM) due to suitable salt resistance mechanisms (Flowers and Colmer 2015; Parida and Das 2005). Salt-tolerant plants are able to grow and complete their life cycle in high concentrations of soluble salt present in soil. Halophytes are additionally assorted as obligate and facultative on the basis of their proficiency to tolerate the salt concentration. Obligate halophytes have growth rate increasing up to 50% seawater but have low morphological and taxonomical diversity, while facultative halophytes are found in less saline habitats, i.e., along the border between saline and nonsaline upland, and have broader physiological diversity which facilitate survival with saline as well as nonsaline conditions (Inan et al. 2004). Euhalophytes can accumulate that much high salt concentration in cytoplasm so that the osmotic potential of soil solutions can be higher than the cellular osmotic potential. Although low osmotic cell sap potential in plant cells can also be attributed by soluble carbohydrates accumulated along with salts. The proper ion compartmentalization, primarily in the vacuoles within the plant cell, may also enable the cells to endure high salt concentrations.

Soil salinity exerts abiotic stress to plant as high intracellular salt concentrations can be harmful for cells and water extraction by roots becomes tedious due to high salt concentrations of soil (Munns and Tester 2008). Na^+ accumulation in plants occurs due excess salt cause's not only ion toxicity-dependent chlorosis and necrosis but also interferes with several physiological processes of plants (Munns 2002).

Thus, for the acclimation and survival of plants under abiotic stresses, adjustments and activation of suitable metabolic responses reestablish chemical and energetic balances which are disturbed due to the stress.

17.2 Impact of Biotic Stresses on Plants

Worldwide agricultural practices are facing difficulties due to abiotic constraints generated by salinity in soils but also in irrigation water due to tremendous increased irrigation for arid and semiarid regions (Acosta-Motos et al. 2017b; Cirillo et al. 2016; Colla et al. 2010; De Pascale et al. 2011; Munns and Gilliham 2015). High salt concentration dependent stress is incipiently sensed by the plant roots which are further manifested to affect the overall growth of plant. High salt concentrations primarily cause decreased water availability to induce osmotic stress in plants and further cause cytosolic imbalance of nutrients to create ion toxicity. Osmotic stress and ion toxicity together felicitate added impact on plant growth reduction (Munns 2005). In plants, elemental nutrient deficiencies or imbalances can also be induced by salinity-dependent composition of external solution (Grattan and Grieve 1994). Studies had revealed that saline environments lead to reduction in uptake of nutrient element by plants and the degree of uptake reduction depends upon nutrient element and composition of external environment. The increased salinity decreases the nutrient activity for elements other than the salinizing salts, e.g., Ca^{2+}, Mg^{2+}, or SO_4^{2-} (Hasan and Miyake 2017; Marosz 2004; Patel et al. 2010; Talei et al. 2012). Under saline conditions nutrient uptake and accumulation by plants are often reduced due to competition amid the nutrients and salt species present in soil as saline soils reduce nutrient availability not only by ionic strength effect but also by affecting sorption and precipitation processes of minerals (Sharpley et al. 1992). Salinity not only affects nutrient ion activities but also produces extreme ion ratios (e.g., Na/Ca^{2+}, Na/K, Cl^-/NO^{3-}) in the soil solution.

On any plant, the salinity inflicts two main threatening consequences: (a) induces osmotic stress and (b) stimulates excessive Cl^- and Na^+ uptake-dependent ionic toxicity that not only leads to Ca^{2+} and K^+ deficiency but also causes imbalances to other nutrients (Marschner 1995). Soil salinity affects nutrient availability, the nutrient uptake along distribution within the plant system, and also increases the internal nutrient element requirement due to various physiological inactivation; all these factors build nutrient imbalances in the plant (Grattan and Grieve 1999). Cl^- salts can reduce nongrowth-limiting NO^{3-} uptake and accumulation in plants (Munns and Termaat 1986). Na-dependent K deficiencies can be observed in crops having salinity stress due to Na salts (Janzen and Chang 1987). Awad et al. (1990) had reported that the internal requirement of plant for phosphorus increases as physiological inactivation of phosphorus occurs due to the salinity as it was observed that foliar symptoms of phosphorus deficiency increased with increased NaCl salinity.

Reactive oxygen species are also formed as a result of the oxidative stress caused by salinity. Such salinity-based retaliations collectively add toward harmful consequences for plants (Hernández et al. 1993, 1995, 2001, 2003; Mittova et al. 2003, 2004). Among different plant growth, repression level and critical death concentrations of salts vary significantly (Wenbo and Qing 2017). According to Tang et al. (2015) environmental circumstances, viz., soil, light, and habitat, along with species of plant under consideration exert influence on the detrimental consequence originated by salinity. Salt stress affects almost all physiological and biochemical processes, i.e., plant growth, photosynthetic rate, and metabolism of proteins and phospholipids. Almost all plants show growth inhibition with salt stress, and lethal effect of salt stress induces decline in productivity or even plant death. Thus, to sustain under salinity stress, plants turn on diverse physiological and biochemical mechanisms. These survival mechanisms include anatomy and morphology change, alteration in water relations, efficiency and efficacy of photosynthesis, the phytohormonal profile, distribution of toxic ions, and antioxidative metabolism response like biochemical adjustments (Acosta-Motos et al. 2015a, b; Ashraf and Harris 2013; Hernández et al. 2001; Parida and Das 2005).

17.3 Plant Responses to Salinity Stress

Most plants are sensitive to salinity in the soil; thus every aspect of plant development, viz., germination, vegetative growth, and reproductive growth, is affected by salt stress (Wenbo and Qing 2017; Acosta-Motos et al. 2016, 2017a). Soil salinity affects seed germination, seedling emergence, and eventually the survival of plantlets (Baldwin et al. 1996; Mariko et al. 1992) as soil substrate salinity can create high osmotic potential and ion toxicity (Grieve and Suarez 1997). Hence according to Zhu (2001) to resist the environmental stress, plants acclimatize biochemical, physiological, morphological, and even developmental changes. Glycophytes are intolerant toward salinity which may cause physiological and biochemical disorders. These physiological and biochemical changes either prevent or delay the germination or can produce anomalous seedlings (Rehman et al. 1996; Ungar 1996) in these pants. Halophyte plants are able to grow in saline habitats, cultivate diverse resistance mechanisms (Larcher 2003) to succeed consequences of salt stress, and regulate their salt content by (a) salt exclusion by thwarting the salts entry within the vascular system; (b) by process of salt elimination with use of salt-secreting glands and hairs for active removal of salts which maintains salt concentration within endurable margins; (c) by salt succulence, i.e., gradual increase in salt storage capacity with increased salt uptake by the plant; and (d) by redistribution of salt components, viz., Na^+ and Cl^- through translocation all throughout the plant by phloem system. Sodium is an essential microelement for plant growth as low salt concentration stimulates the plant growth and increase the plant biomass but high concentration of sodium in plants inhibit plant growth, reduce the fresh weight and dry weight of root, stem, and leaf (Ali Dinar et al. 1999; Chartzoulakis and Klapaki 2000). It had been reported that high concentration of NaCl causes biomass decrease in the root, shot, and leaf of cotton (Meloni et al. 2001). It had also been reported that salt stress can repress the growth of plant tissues and organs, shorten the flowering phase, and accelerate the aging, death, and falloff of mature leaves (Parida and Das 2005).

17.4 Morphological Adaptations of Plants to Salinity

For growth during salt stress, maintenance of ion homeostasis and compartmentalization of ions is essential (Hasegawa 2013; Niu et al. 1995; Serrano et al. 1999; Flowers and Yeo 1986, 1988; Koyro 2006). Glycophytes as well as halophytes cannot tolerate high salt concentration in their respective cytoplasms; hence, extra salt is either sequestered and then excreted out or transported to vacuoles for storage so that cytoplasm of plant cells has balanced salt concentrations (Reddy et al. 1992; Zhu 2003).

Plants have early and delayed responses as reflex to salinity exposure. The primitive initial responses are expressed within minutes which are ion-independent and stay for days, while the lagging second phase depends on ions generated due to salinity and becomes effective only after days to weeks' time. Enhanced salinity influence plant-water relations hence stimulate stomata closure and leaf expansion inhibition as initial salinity stress response (Munns and Termaat 1986). The early responses are reported to be associated with nutrient sensing and subsequent signaling (Gilroy et al. 2014, 2016; Roy et al. 2014). According to Munns and Tester (2008), the dawdling next stage of salinity response occurs only after the formation of lethal concentrations of ions in old leaves leading to premature leaves senescence that reduces the plant yield and causes untimely plant demise. Munns and Tester (2008) also proposed basic mechanisms applied by plant to tolerate salinity, as toxic ions exclusion, use of selected subcellular organelles for toxic ions compartmentalization within cells, and Na^+ accumulation to maintain plant-water relation. Over the period various other physiological processes are reported which help the plants to tolerate high concentrations of the salt such as early seedling growth (Kingsbury and Epstein 1984), protection during germination of seed (Foolad and Lin 1997), antioxidant production (Ashraf 2009), maintenance of transpiration use efficiency (Barbieri et al. 2012; Harris et al. 2010; This et al. 2010), and conservation of leaf area (Maggio et al. 2007).

17.5 Biochemical and Metabolic Adjustments of Plants During Salinity Stress

Through stress conditions suitable metabolic alterations are vital for plant survival as plants not only have intricate interconnected metabolic reaction networks but also have subcellular compartmentalization with large number of metabolites present which are influenced by plant developmental stages. Thus, any metabolic response is not only dynamic but also complex that involves more than one metabolite modification. Along with specific compound accumulation for tolerance against any stress, metabolite flux regulation is also necessary to confer the stress tolerance in plant (Allen et al. 2009; Stepien and Johnson 2009). Plants synthesize a vast array of metabolites categorized as primary metabolites which are essential for life functions, e.g., carbohydrates, proteins, nucleic acids, and lipids, while secondary metabolites facilitate interactions between the plant and its environment (Croteau et al. 2000). The secondary metabolites are not commonly found in plants; hence they do not have any specific function or provide any assistance to plant cells producing such metabolites. Although accumulation of these secondary metabolites increases competence of plants to counter biotic or abiotic stresses and thereby increases adaptive capability of plants under adverse environmental conditions (van der Fits and Memelink 2000), e.g., toxic nicotine, plant secondary metabolite is a natural plant protectant against herbivore animals and microflora. Many halophytes and glycophytes under salt stress conditions synthesize these secondary metabolite compounds to protect cellular functions (Pichersky and Gang 2000). Phenols and flavonoids are the major secondary plant metabolites which facilitate plants to acquire tolerance against environmental stresses (Ali and Abbas 2003). These phenolic and flavonoid secondary metabolites increase antioxidant activities to control the lipid peroxidation, i.e., major oxidative damage sign of cell membranes under stress conditions, by scavenging the reactive oxygen species (ROS) (Baque et al. 2012). The secondary plant metabolites generated in response of biotic/abiotic stress also serve as osmolytes and osmoprotectant which safeguard plants under extreme abiotic stresses (Shulaev et al. 2008). Thus, under stressed circumstances plants accumulate compatible solutes as a common self-protection maneuver. These low molecular weight organic molecules have diverse chemical nature, viz., amino acids (asparagine, proline, serine), amines (polyamines and glycine betaine), and γ-amino-*N*-butyric acid (GABA) along with sugars, e.g., fructose, sucrose, trehalose, raffinose, and polyols, e.g., myo-inositol, D-pinitol (Krasensky and Jonak 2012; Banu et al. 2010), and natural organic antioxidants, e.g., glutathione (GSH) and ascorbate (Phang et al. 2008; El-Shabrawi et al. 2010). All these diverse solutes have some common characteristics as (1) high solubility in cytoplasm and (2) inability for enzyme inhibition even at high concentrations. As a defense mechanism for stress response, compatible solute accumulation is an evolutionarily conserved trait as it is observed not only in plants but also in animals, marine algae, and bacterial cells (Grant 2004; Empadinhas and da Costa 2008). These days abiotic stress tolerance in plants is probed with the targeted metabolite analysis and profiling based on metabolic fingerprinting (Renberg et al. 2010; Shulaev et al. 2008).

17.5.1 Osmolyte Accumulation

Elevation in dissolved salt ions leads to osmotic stress and subsequent depletion in water osmotic potential of plant cells, and this osmotic stress is adjusted by accumulation of osmolytes, i.e., inorganic salt ions and hydrophilic compounds along with hydrophilic proteins having osmoprotective function. A diverse group of organic compounds, e.g., proline, glycine betaine, sugar, and polyols also known as compatible osmolytes, are water soluble, polar but uncharged, and inert even at high concentration for the cellular metabolism (Kerepesi and Galiba 2000; Ashraf and Foolad 2007; Ahmad et al. 2010; Saxena et al. 2013) and are synthesized and accumulated in varying amounts among different plant species. These osmolytes create constant water influx that maintain osmotic balance inside the cell and guard the cell structure (Hasegawa et al. 2000). Since concentration of these osmolytes is relative to the outside osmolarity, the constant intracellular osmolytes concentration is maintained by precise formation and degeneration (Gupta and Huang 2014). Under the salt stress accumulation of osmolytes, viz., proline, glycine betaine occurs (Pang et al. 2010) due to overexpression of enzymes from proline biosynthesis (Askari et al. 2006; Veeranagamallaiah et al. 2008) along with decreased quantity of proline hydrolyzing enzyme (Pang et al. 2010). Hydrophilic proteins, e.g., proteins from LEA superfamily including dehydrins with osmoprotective function, are reported to be elevated (Battaglia et al. 2008; Hundertmark and Hincha 2008; Kosová et al. 2010).

17.5.1.1 Proline

Proline accumulation occurs in taxonomically diverse sets of plants (Saxena et al. 2013) as accumulation of proline is a major method adopted by plants to mitigate the salinity stress (Matysik et al. 2002; Saxena et al. 2013) since accumulated intracellular proline is responsible for tolerance toward stress but also assists as an organic nitrogen reserve during stress recovery (Gupta and Huang 2014). Proline is O_2 quencher, i.e., antioxidant molecule (Matysik et al. 2002), which is synthesized either from proteinaceous amino acid glutamate or nonprotein amino acid ornithine. It has been observed that proline supplements enhanced salt tolerance by increasing photosynthetic activity along with some antioxidative enzyme activities and support plant growth by preservating suitable plant-water status under salinity conditions in *Olea europaea* (Ben Ahmed et al. 2010). Salt tolerance in *Nicotiana tabacum* has been reported to improve due to increased antioxidant enzyme activity by proline (Hoque et al. 2007a, b, 2008, 2010). Proline-pretreated rice seedlings had exhibited growth improvement during salt stress (Deivanai et al. 2011).

17.5.1.2 Glycine Betaine

Glycine betaine is a water-soluble amphoteric quaternary ammonium compound, which is electrically neutral over a range of pH and ubiquitously present in higher plants, animals, and microorganisms. As this compound is amphoteric in nature, it can interact with hydrophobic and hydrophilic domains of proteins and enzymes. Glycine betaine is a nontoxic cellular osmolyte that reduce abiotic stress influence to protect cell by raising cellular osmolarity, i.e., osmotic adjustment (Gadallah 1999), proteins stabilization (Makela et al. 2000), photosynthetic apparatus protection from stress (Cha-Um and Kirdmanee 2010), and removal of reactive oxygen species generated during salt stress (Ashraf and Foolad 2007; Saxena et al. 2013). Glycine betaine pretreatment of *Oryza sativa* seedlings prevents salt stress induced thylakoids swelling, grana and intergranal lamellae disintegration in choloroplast along with mitochondrial disruption (Rahman et al. 2002). Foliar spray of glycine betaine on stressed plants had resulted in increase in photosynthetic rate with growth revival and pigment stabilization (Cha-Um and Kirdmanee 2010; Ahmad et al. 2013).

17.5.1.3 Polyols and Sugars

Polyols are multiple hydroxyl groups containing organic compounds, e.g., sugar alcohols; those are compatible solutes with low molecular weight which act as chaperones and ROS-scavenging compounds (Ashraf and Foolad 2007). Polyols are primarily grouped as cyclic (e.g., pinitol) and acyclic (e.g., mannitol) compounds. Accumulation of cyclic, acyclic, and their derivative polyols in many species of microbes, plants, and animals is interrelated to drought and/or salinity tolerance (Bohnert and Sheveleva 1998). Besides polyols; sugars; e.g., glucose, fructose, fructans, and trehalose; and starch are reported to accumulate under salt stress (Parida et al. 2004) which help in stress mitigation by osmoprotection and ROS scavenging. Kerepesi and Galiba (2000) had reported increase in reducing sugars in the cells due to salt stress in various plant species. These reducing sugars play an osmoprotective role in physiological responses along with being sugar reserve (Ahmad et al. 2013). Alteration in concentrations of various sugars are also reported in plants due to salinity stress; increased sucrose content in *Solanum lycopersicum* Gao et al. (1998), decrease in sugar content within rice genotypes roots (Alamgir and Ali 1999), decreased starch but increased reducing and nonreducing sugar contents in the leaves of *Bruguiera parviflora* (Parida et al. 2004).

17.5.2 Antioxidative Metabolism

Under abiotic stresses plants have interruption/deregulation of mitochondrial as well as chloroplastal electron transport chains (ETC), because under abiotic stress the molecular oxygen O_2, the terminal electron acceptor, produces and accumulates reactive oxygen species, viz., singlet oxygen ($_1O^2$), the hydroxyl radical (OH^-), the superoxide radical (O^{-2}), and hydrogen peroxide (H_2O_2). Under normal growth conditions also, low amounts of ROS are metabolic byproducts of plant cells (Cai-Hong et al. 2005), essentially from photosynthesis, photorespiration, and respiration. Additional sources include NADPH oxidases, amine oxidases, and cell wall-bound peroxidases (Mittler 2002). All these ROS molecules are strong oxidizing agents thus highly reactive and extremely harmful for cell integrity (Grob et al.

2013) as these ROS molecules cause oxidative damage to membrane lipids and proteins along with cellular nucleic acids and also cause lipid peroxidation, protein denaturation, and DNA mutation (Imlay 2003). The oxidative damage-related plant injuries are interrelated at the cellular level (Foyer and Noctor 2003), and surplus ROS production leads to membrane damage that ultimately causes the cell death, i.e., in the event of stress, ROS production may overcome the antioxidative scavenging mechanisms, and oxidative stress will be imparted to cell leading to impact on vital cellular biomolecules and cellular metabolism causing eventual cell death (Sakihama et al. 2002). Thus, imbalance of reactive oxygen species (ROS) production and antioxidant defense in any cell compartment leads to oxidative stress which causes severe damage to cell integrity (Mittler 2002).

Therefore, during abiotic stresses, reactive oxygen species (ROS) scavenging for restoration of redox metabolism along with compatible osmoprotectants mediated protection/stabilization of proteins, and cellular structures with cellular turgor preservation are primary and necessary cellular responses (Rathinasabapathi 2000; Yancey 2005; Mittler 2006). Plants have developed a complex antioxidant mechanisms for ROS scavenging, and these mechanisms are grouped as enzymatic and nonenzymatic (Demiral and Turkan 2005; Reddy et al. 2004) systems. Hernandez et al. (2000) had reported that different subcellular organelles contain different components of antioxidant defense system. Plant secondary metabolites like carotenoids, ascorbate, glutathione, and tocopherols constitute the primary components of antioxidant system along with different antioxidant enzymes, e.g., catalase (CAT), superoxide dismutase (SOD), glutathione peroxidase (GPX), ascorbate peroxidase (APX), and glutathione reductase (GR) are deployed by plant defense machinery (Asada 1999; Gupta et al. 2005). It had been reported that plant salinity tolerance can be facilitated by helicase proteins (e.g., DESD-box helicase and OsSUV3 dual helicase), as these proteins improve photosynthetic and antioxidative machinery (Gill et al. 2013; Tuteja et al. 2013).

17.5.2.1 Enzymatic Response to Oxidative Stress Response

Various defensive mechanisms are established in plants to either reduce or eliminate ROS at different ranks of stress (Beak and Skinner 2003). According to Scandalios (1993) superoxide dismutase (SOD: EC 1.15.1.1) is one of the protective enzyme found in different cellular organelles which remove the disproportion of two $\cdot O_2^-$ radicals to H_2O_2 and O_2. As ROS increases in the system, superoxide dismutase (SOD) gets activated and converts the O_2^- radicals to molecular O_2 and H_2O_2 (Meloni et al. 2003) in the chloroplast, mitochondrion, cytoplasm, apoplast, and peroxisome (Bowler et al. 1992). Although under low stress concentrations, H_2O_2 is reported to have a signaling role (Apel and Hirt 2004; Hancock et al. 2006); therefore, the H_2O_2-detoxifying enzymes activities are tightly regulated for maintenance of appropriate concentration of H_2O_2 important for cellular signaling.

Hydrogen peroxide is detoxified as it's converted to water by two major antioxidant enzymes, viz., catalases (CAT: EC 1.11.1.6) (Kono and Fridovich 1983;

Scandalios 1993) and peroxidases (POX: EC 1.11.1.7) (De Gara et al. 2003; Jablonski and Anderson 1982). The generated H_2O_2 can also be detoxified by using it to oxidize ascorbate with ascorbate peroxidase (APX) and re-reduced glutathione with the help of glutathione reductase (GR: EC 1.6.4.2) enzyme (Noctor and Foyer 1998). This detoxification of H_2O_2 is collectively known as the ascorbate-glutathione cycle (Asada 1999; Mittler 2002). Other important enzymes of ROS-scavenging system are monodehydro ascorbate reductase (MDHAR: EC 1.6.5.4) and dehydroascorbate reductase (DHAR: EC 1.8.5.1) (Candan and Tarhan 2003; Yoshimura et al. 2000).

17.5.2.2 Nonenzymatic Response to Oxidative Stress Response

Among the important nonenzymatic ROS-scavenging antioxidants are ascorbate (which acts as a reductant for peroxidases), glutathione, tocopherol, and carotenoids. Ascorbate is most abundant, water-soluble robust antioxidant present in plants to counter the effect of ROS. GSH is a sulfhydryl group containing water-soluble tripeptide that acts as a foundation for dehydroascorbate reductase in the ascorbate-glutathione (AsA–GSH) cycle. GSH directly scavenges OH· and $^1O_2$ and can protect enzyme thiol groups (Gill and Tuteja 2010). The presence of ascorbate and glutathione under the salt stress are more affected in sensitive plants in comparison to the tolerant plants (Acosta-Motos et al. 2017b). Although the organic osmolytes, proline and glycine betaine, are important and efficient compatible solutes (Tang et al. 2015), proline provide tolerance to plant against the environmental stresses, as it is not only a osmoprotectant with antioxidant properties but also act as a molecular chaperone for biological macromolecules structural defense during dehydration (Ashraf and Foolad 2007; Tang et al. 2015). Kim et al. (2014) showed that silicon (Si) application to rice root zone influenced the hormonal and antioxidant responses under salinity stress as Si treatments reduced the sodium accumulation resulting in low electrolytic leakage and lipid peroxidation compared to control plants under salinity stress. Enzymatic antioxidant (catalase, peroxidase, and polyphenol oxidase) responses were more pronounced in control plants than in Si-treated plants under salinity stress.

17.6 Role of Rhizosphere Microbes in Plant Stress Tolerance

The root-associated microbiota, rhizobacteria, mainly consisted of bacteria and fungi having symbiotic association with host plants (Smith et al. 2015a, b, 2017) which get protection through rhizodeposition along with carbon the major growth requirement. The composition and activities of rhizospheric microorganisms are also influenced by plants found in the surroundings (Qin et al. 2016). In the past few decades, studies on bacteria belonging to different genera including *Rhizobium*, *Bacillus*, *Pseudomonas*, *Paenibacillus*, *Achromobacter*, *Azospirillum*, *Microbacterium*, *Methylobacterium*, *Variovorax*, *Enterobacter*, etc. have been reported to provide tolerance to host plants against different abiotic stresses (Yang et al. 2009; Hayat et al. 2010). *Klebsiella* sp. SBP-8 has been reported to improve plant growth under temperature and salt stress (Singh et al. 2015), while *Pseudomonas* strains can also be helpful in counteracting the effect of aluminum acidity along with the salinity (Zerrouk et al. 2016). The halophilic bacteria were reported to alleviate salt stress of up to 200 mM NaCl in *Triticum aestivum* (Orhan 2016). Various bacterial secondary metabolic products and pathways, viz., ACCD activity, exopolysaccharides, siderophores, volatiles, IAA production, phosphate solubilization etc. also actively participate in stress tolerance (Kumari et al. 2015; Nadeem et al. 2016).

17.6.1 Phytohormonal Modulations by Microbes for Plant Stress Tolerance

In plants, ethylene is important for plant growth, development, and fruit-ripening process, but under the stress, viz., drought, salinity, or pathogen attack, undue ethylene production is ensured by plants, and ethylene overproduction might reduce seed germination and root growth (Abeles et al. 1992). Therefore, bacteria possessing ACCD activity have been regarded as prime candidates for salt stress alleviation (Qin et al. 2016). ACC deaminase (ACCd) enzyme of PGPR origin reduces the quantity of stress-endued ethylene because this enzyme metabolizes 1-aminocyclopropane-1-carboxylate (ACC), the immediate precursor of ethylene production in higher plants, as alpha-ketobutyrate and ammonia. These metabolites of ACC benefit the allied PGPR organism as energy and nitrogen source (Glick et al. 2007; Glick 2014). These PGPR bacteria improve morphology and size of roots, thus enhance nitrogen metabolism with improved nutrient uptake and regulate osmotic balance culminating in plant growth promotion (Glick 2014). Habib et al. (2016) demonstrated that the application of PGPR, *Enterobacter* sp. as a soil enhancer, resulted in improved seed germination, chlorophyll content, and growth of okra plants under salinity stress by maintaining low stress ethylene levels through ACCD activity and increasing the ROS-scavenging enzymes. Endogenous fungi *Fusarium equiseti* of *Bambusa balcooa* is reported to have PGPR activity (Tyagi et al. 2017).

17.6.2 Antioxidant Defense by Microbes for Plant Stress Tolerance

Salinity stress also generates reactive oxygen species (ROS), such as H_2O_2, O^{2-}, and OH^- that damages DNA, RNA, and proteins. This ROS compound also causes chlorophyll impairment and hampers the root meristem activity. Antioxidant enzymes such as superoxide dismutase (SOD), catalase (CAT), and ascorbate peroxidase (APX) have potential for ROS scavenging and maintain them at low levels (Habib et al. 2016). An increased plant salt stress tolerance by enhanced activities of peroxidase, catalase, superoxide dismutase, and glutathione reductase in *Glycyrrhiza uralensis Fisch* inoculated with *P. extremorientalis* and *Mesorhizobium* sp. was reported (Egamberdieva et al. 2017).

17.6.3 Ion Homeostasis by Microbes for Plant Stress Tolerance

As it has been established that salinity inhibits plant growth due to increased Na^+ concentration and high Na^+/K^+ ratio in plant, it has been also reported that PGPR inoculation to plants avoided the over accumulation of Na^+ and maintained the ion homeostasis under salinity stress conditions. *Bacillus subtilis* has been reported to lower Na^+/K^+ ratio in leaves of wheat plant growing under saline condition with up to 23% reduction in sodium content (Upadhyay and Singh 2015). *Bacillus amyloliquefaciens* SQR9 strain had reported to provide protection to maize plants in salt stress-prone area, as the bacteria protected the plant by regulating Na^+ homeostasis along with enhancing peroxidase/catalase activity as well as glutathione content for scavenging reactive oxygen species (Chen et al. 2016). Wang et al. (2016) had reported improved growth and photosynthesis rate *Pisum sativum* by limiting Na^+ accumulation in plants.

17.7 Stress Genes Expression for Salt Tolerance in Plants

Salt tolerance, a complex mechanism, involves multitudinous interacting properties which can be identified and understood at molecular level by genomic, transcriptomic, proteomic, and metabolomic studies. At the molecular level, abiotic stress may induce or repress genes from stress gene networks with precise regulation (Shinozaki and Yamaguchi-Shinozaki 2007; Delano-Frier et al. 2011; Grativol et al. 2012; Li and Li 2017) for products required in stress response and tolerance at the cellular level. As initial stress protection response, osmoprotectant compounds, detoxification enzymes, transporter proteins, and chaperones biosynthesizing pathways are activated along with stimulation for regulatory proteins (e.g., transcription factors) and enzymes, e.g., kinases and phosphatases. The signaling molecules essential for signal transduction regulation and stress-responsive gene expression are also synthesized (Wang et al. 2009; Krasensky and Jonak 2012). Functional genomic tool, i.e., metabolomics, can be applied to understand plant metabolic reaction toward salt stress (Lu et al. 2013) because high-throughput metabolomics technique can provide an insight to understand salt stress-induced changes in gene expression for various proteins and subsequent metabolites production in plants (Urano et al. 2010; Wu et al. 2013). Reestablishment of lost equilibrium of energy as well as redox and osmotic imbalances created because of stress is the first priority of plant cells; thus under the stress plant response, mechanisms initially prevent or alleviate cellular damages and reestablish homeostatic conditions for growth continuation (Peleg et al. 2011). To reestablish homeostatic conditions in plants under salt stress, the genes liable for salt sensors and SOS systems including diverse signaling pathways are triggered. All these genes synergistically stimulate/initiate other correlated reactions to counter the salinity stress, such as assembly of salt gland or bladders to change in influx and efflux of ions with production of diverse antioxidants and osmoprotectants (Cheeseman 2014; Shabala et al. 2014; Slama et al. 2015; Himabindu et al. 2016). In *Arabidopsis thaliana*, it had been reported that complex of nuclear-localized protein AtRSA1 (with calcium-binding domain) along with AtRITF1 is responsible for regulation of genes associated with salt stress-dependent reactive oxygen species detoxification (Guan et al. 2013). Another saline-responsive gene studied is AtWRKY8 (At5g46350), expressed mainly in roots. It was also determined that AtWRKY8 only specifically interacted with AtVQ9 (At1g78310) and the interaction decreased the DNA-binding activity of AtWRKY8 thus indicating that AtVQ9 acts antagonistically with AtWRKY8 (Hu et al. 2013). Sukweenadhi et al. (2015) further found that after inoculation with *Paenibacillus yonginensis* DCY84T in *Arabidopsis thaliana*, the transcription of AtRSA1 and AtWRKY8 genes were increased, while AtWRKY8 antagonist, i.e., AtVQ9 gene, was decreased in *Arabidopsis thaliana*. These observations further confirmed that under the salt stress conditions, plant receptiveness for salt stress had increased.

Salinity stress-specific response mechanisms along with general adaptive responses provide salinity stress tolerance responses in plants to provide strategic advantages against salinity stress. General response mechanisms include pathways for energy maintenance and calcium signal cascades (Reddy et al. 2011; Pan et al. 2012), scavenging of ROS (Ahmad et al. 2010; Loiacono and De Tullio 2012), and stress signaling (Baena-Gonzalez and Sheen 2008). Not only salinity but different types of stress lead to induction of these central pathways for plant acclimatization towards environmental stresses. There is a genome structure level difference between tolerant and susceptible plants as certain distinctive stress-receptive genes are found within tolerant plants which are absent in susceptible plants. Stress-responsive genes for salt tolerance are basically grouped in three classes as (1) genes responsible for salt uptake and transport, (2) genes with osmoprotective function, and (3) genes responsible for fast plant growth under saline conditions (Munns 2005).

17.8 Conclusions

All life forms on earth either eukaryotes or prokaryotes are dependent on salts as it is essential in cellular signal transduction and also aids for plant osmosis, blood flow, and pressure regulation in animals. But, it becomes detrimental if concentration/quantity in system increases than that of the required quantity. In the case of plants, salt creates one of the primary abiotic stresses, i.e., salinity that imposes serious impacts on plant growth and development. With identification of multiple signaling pathways, certain elements and several common salt tolerance determinants are identified (Wenbo and Qing 2017). It is now well established that ROS sequestering is important in plants for abiotic stress adaptation (Karpinski et al. 1999; Mittler 2002). The precise understanding of each ROS-scavenging enzyme and antioxidants osmolytes for stress adaptation followed by accurate phenotypic characterization of complex stress tolerance is indispensable to develop ultimate stress-tolerant plants. Still, only facts of antioxidative mechanism are not enough for the same. The salt stress-dependent resistance genes of plants could be better understood with advancements in molecular biology techniques in plant genome research. These biotechnological approach-dependent molecular strategies such as marker-assisted breeding can be used to develop lines of stress-resistant lines/genotypes. Advance studies on many more aspects and components of salinity-dependent signaling pathways such as noncoding RNAs and epigenetics regulation are needed to fully understand to aptly appreciate the plant salinity stress tolerance mechanisms. Last but not the least for emphatic use of these approaches with full potential, knowledge of applied plant ecology is also essential to make significant contribution in agricultural yield with environmental protection.

References

Abeles FB, Morgan PW, Saltveit ME Jr (1992) Ethylene in plant biology, 2nd edn. Academic, New York

Acosta-Motos JR, Díaz-Vivancos P, Álvarez S, Fernández-García N, Sánchez-Blanco MJ, Hernández JA (2015a) NaCl-induced physiological and biochemical adaptative mechanism in the ornamental *Myrtus communis* L. plants. J Plant Physiol 183:41–51

Acosta-Motos JR, Díaz-Vivancos P, Álvarez S, Fernández-García N, Sánchez-Blanco MJ, Hernández JA (2015b) Physiological and biochemical mechanisms of the ornamental *Eugenia myrtifolia* L. plants for coping with NaCl stress and recovery. Planta 242:829–846

Acosta-Motos JR, Ortuño MF, Álvarez S, López-Climent MF, Gómez-Cadenas A, Sánchez-Blanco MJ (2016) Changes in growth, physiological parameters and the hormonal status of *Myrtus communis* L. plants irrigated with water with different chemical compositions. J Plant Physiol 191:12–21

Acosta-Motos JR, Ortuño MF, Bernal-Vicente A, Diaz-Vivancos P, Sanchez-Blanco MJ, Hernandez JA (2017a) Plant responses to salt stress: adaptive mechanisms. Agronomy 7:18. https://doi.org/10.3390/agronomy7010018

Acosta-Motos JR, Hernández JA, Álvarez S, Barba-Espín G, Sánchez-Blanco MJ (2017b) Long-term resistance mechanisms and irrigation critical threshold showed by *Eugenia myrtifolia* plants in response to saline reclaimed water and relief capacity. Plant Physiol Biochem 111: 244–256

Ahmad P, Jaleel CA, Salem MA, Nabi G, Sharma S (2010) Roles of enzymatic and nonenzymatic antioxidants in plants during abiotic stress. Crit Rev Biotechnol 30(3):161–175

Ahmad R, Lim CJ, Kwon SY (2013) Glycine betaine: a versatile compound with great potential for gene pyramiding to improve crop plant performance against environmental stresses. Plant Biotechnol Rep 7:49–57

Alamgir ANM, Ali MY (1999) Effect of salinity on leaf pigments, sugar and protein concentrations and chloroplast ATPase activity of rice (*Oryza sativa* L.). Bangladesh J Bot 28(2):145–149

Ali RM, Abbas HM (2003) Response of salt stressed barley seedlings to phenylurea. Plant Soil Environ 49:158–162

Ali Dinar HM, Ebert G, Ludders P (1999) Growth, chlorophyll content, photosynthesis and water relations in guava (*Psidium guajava* L.) under salinity and different nitrogen supply. Garten Bauwissens Chaft 64(2):54–59

Allen DK, Liboure IG, Shachar-Hill Y (2009) Metabolic flux analysis in plants: coping with complexity. Plant Cell Environ 32(9):1241–1257

Apel K, Hirt H (2004) Reactive oxygen species: metabolism, oxidative stress and signal transduction. Annu Rev Plant Biol 55:373–399

Asada K (1999) The water-water cycle in chloroplasts: scavenging of active oxygen and dissipation of excess photons. Annu Rev Plant Physiol Plant Mol Biol 50:601–639

Ashraf M (2009) Biotechnological approach of improving plant salt tolerance using antioxidants as markers. Biotechnol Adv 27:84–89

Ashraf M, Foolad MR (2007) Roles of glycine betaine and proline in improving plant abiotic stress resistance. Environ Exp Bot 59(2):206–216

Ashraf M, Harris JC (2004) Potential biochemical indicators of salinity tolerance in plants. Plant Sci 166:3–16

Ashraf M, Harris PJC (2013) Photosynthesis under stressful environments: an overview. Photosynthetica 51:163–190

Askari H, Edqvist J, Hajheidari M, Kafi M, Salekdeh GH (2006) Effects of salinity levels on proteome of *Suaeda aegyptiaca* leaves. Proteomics 6:2542–2554

Awad AS, Edwards DG, Campbell LC (1990) Phosphorus enhancement of salt tolerance of tomato. Crop Sci 30:123–128

Baena-Gonzalez E, Sheen J (2008) Convergent energy and stress signaling. Trends Plant Sci 13(9): 474–482

Baldwin AH, McKee KL, Mendelssohn IA (1996) The influence of vegetation salinity and inundation on seed banks of oligohaline coastal marshes. Am J Bot 83:470–479

Banu MN, Hoque MA, Watanabe-Sugimoto M, Islam MM, Uraji M, Matsuoka K, Nakamura Y, Murata Y (2010) Proline and glycinebetaine ameliorated NaCl stress via scavenging of hydrogen peroxide and methylglyoxal but not superoxide or nitric oxide in tobacco cultured cells. Biosci Biotechnol Biochem 74(10):2043–2049

Baque MA, Elgirban A, Lee E, Paek K (2012) Sucrose regulated enhanced induction of anthraquinone, phenolics, flavonoids biosynthesis and activities of antioxidant enzymes in adventitious root suspension cultures of *Morinda citrifolia* (L.). Acta Physiol Plant 34:405–415

Barbieri G, Vallone S, Orsini F, Paradiso R, De Pascale S, Negre-Zakharov F, Maggio A (2012) Stomatal density and metabolic determinants mediate salt stress adaptation and water use efficiency in basil (*Ocimum basilicum* L.). J Plant Physiol 169:1737–1746

Battaglia M, Olvera-Carillo Y, Garciarrubio A, Covarrubias AA (2008) The enigmatic LEA proteins and other hydrophilins. Plant Physiol 148:6–24

Beak KH, Skinner DZ (2003) Alteration of antioxidant enzyme gene expression during cold acclimation of near-isogenic wheat lines. Plant Sci 165:1221–1227

Ben Ahmed C, Ben Rouina B, Sensoy S, Boukhriss M, Ben Abdullah F (2010) Exogenous proline effects on photosynthetic performance and antioxidant defense system of young olive tree. J Agric Food Chem 58(7):4216–4222

Bohnert HJ, Sheveleva E (1998) Plant stress adaptations-making metabolism move. Curr Opin Plant Biol 1:267–274

Bowler C, Montagu MV, Inze D (1992) Superoxide dismutases and stress tolerance. Annu Rev Plant Physiol Plant Mol Biol 43:83–116

Bui EN (2013) Soil salinity: a neglected factor in plant ecology and biogeography. J Arid Environ 92:14–12

Cai-Hong P, Jun ZS, Zhong GZ, Bao-Shan WW (2005) NaCl treatment markedly enhances H_2O_2 scavenging system in leaves of halophyte Suaeda salsa. Physiol Plant 125:490–499

Candan N, Tarhan L (2003) The correlation between antioxidant enzyme activities and lipid peroxidation levels in *Mentha pulegium* organs grown in Ca^{2+}, Mg^{2+}, Cu^{2+}, Zn^{2+} and Mn^{2+} stress conditions. Plant Sci 163:769–779

Chartzoulakis K, Klapaki G (2000) Response of two greenhouse pepper hybrids to NaCl salinity during different growth stages. Sci Hortic 86(3):247–260

Cha-Um S, Kirdmanee C (2010) Effect of glycinebetaine on proline, water use, and photosynthetic efficiencies, and growth of rice seedlings under salt stress. Turk J Agric For 34:517–527

Cheeseman JM (2014) The evolution of halophytes, glycophytes and crops, and its implications for food security under saline conditions. New Phytol 206:1–14

Chen L, Liu Y, Wu G, Veronican Njeri K, Shen Q, Zhang N, Zhang R (2016) Induced maize salt tolerance by rhizosphere inoculation of *Bacillus amyloliquefaciens* SQR9. Physiol Plant 158(1): 34–44

Cirillo C, Rouphael Y, Caputo R, Raimondi G, Sifola MI, De Pascale S (2016) Efects of high salinity and the exogenous of an osmolyte on growth, phosynthesis and mineral composition in two ornamental shrubs. J Hortic Sci Biotechnol 91:14–22

Colla G, Rouphael Y, Leonardi C, Bie Z (2010) Role of grafting in vegetable crops grown under saline conditions. Sci Hortic 127:147–155

Croteau R, Kutchan TM, Lewis NG (2000) Natural products (secondary metabolites). In: Buchanan BW et al (eds) Biochemistry and molecular biology of plants. American Society of Plant Biologists, Rockville, pp 1250–1268

De Gara L, de Pinto MC, Tommasi F (2003) The antioxidant systems vis-a-vis reactive oxygen species during plant-pathogen interaction. Plant Physiol Biochem 41:863–870

De Pascale S, Dalla-Costa L, Vallone S, Barbieri G, Maggio A (2011) Increasing water use efficiency in vegetable crop production: from plant to irrigation system efficiency. Hortic Technol 21:301–308

Deivanai S, Xavier R, Vinod V, Timalata K, Lim OF (2011) Role of exogenous proline in ameliorating salt stress at early stage in two rice cultivars. J Stress Physiol Biochem 7(4):157–174

Delano-Frier JP, Aviles-Arnaut H, Casarrubias-Castillo K, Casique-Arroyo G, Castrillon-Arbelaez PA, Herrera-Estrella L, Massange-Sánchez J, Martínez-Gallardo NA, Parra-Cota FI, Vargas-Ortiz E, Estrada-Hernández MG (2011) Transcriptomic analysis of grain amaranth (*Amaranthus hypochondriacus*) using 454 pyrosequencing: comparison with *A. tuberculatus*, expression profiling in stems and in response to biotic and abiotic stress. BMC Genomics 12:363. https://doi.org/10.1186/1471-2164-12-363

Demiral T, Turkan I (2005) Comparative lipid peroxidation, antioxidant defense systems and proline content in roots of two rice cultivars differing in salt tolerance. Environ Exp Bot 53: 247–257

Egamberdieva D, Wirth S, Li L, Abd-Allah EF, Lindström K (2017) Microbial cooperation in the rhizosphere improves liquorice growth under salt stress. Bioengineered 8(4):433–438

El-Shabrawi H, Kumar B, Kaul T, Reddy MK, Singla-Pareek SL, Sopory SK (2010) Redox homeostasis, antioxidant defense, and methylglyoxal detoxification as markers for salt tolerance in Pokkali rice. Protoplasma 245(1–4):85–96

Empadinhas N, da Costa MS (2008) Osmoadaptation mechanisms in prokaryotes: distribution of compatible solutes. Int Microbiol 11:151–161

Flowers TJ, Colmer TD (2015) Plant salt tolerance: adaptations in halophytes. Ann Bot 115: 327–331

Flowers TJ, Yeo AR (1986) Ion relations of plants under drought and salinity. Aust J Plant Physiol 13:75–91

Flowers TJ, Yeo AR (1988) Ion relation of salt tolerance. In: Baker DA, Hall JL (eds) Solute transport in plant cells and tissues. Longman Scientific and Technical, Harlow, pp 392–413

Foolad MR, Lin GY (1997) Genetic potential for salt tolerance during germination in *Lycopersicon* species. Hortscience 32:296–300

Foyer CH, Noctor G (2003) Redox sensing and signalling associated with reactive oxygen in chloroplasts, peroxisomes and mitochondria. Physiol Plant 119:355–364

Gadallah MAA (1999) Effects of proline and glycinebetaine on *Vicia faba* responses to salt stress. Biol Plant 42(2):249–257

Gao Z, Sagi M, Lips SH (1998) Carbohydrate metabolism in leaves and assimilate partitioning in fruits of tomato (*Lycopersicon esculentum* L.) as affected by salinity. Plant Sci 135:149–159

Gill SS, Tuteja N (2010) Reactive oxygen species and antioxidant machinery in abiotic stress tolerance in crop plants. Plant Physiol Biochem 48:909–930

Gill SS, Tajrishi M, Madan M, Tuteja N (2013) A DESD box helicase functions in salinity stress tolerance by improving photosynthesis and antioxidant machinery in rice (*Oryza sativa* L. cv. PB1). Plant Mol Biol 82(1–2):1–22

Gilroy S, Suzuki N, Miller G, Choi WG, Toyota M, Devireddy AR, Mittler R (2014) A tidal wave of signals: calcium and ROS at the forefront of rapid systemic signaling. Trends Plant Sci 19: 623–630

Gilroy S, Białasek M, Suzuki N, Górecka M, Devireddy AR, Karpinski S, Mittler R (2016) ROS, calcium, and electric signals: key mediators of rapid systemic signaling in plants. Plant Physiol 171:1606–1615

Glick BR (2014) Bacteria with ACC deaminase can promote plant growth and help to feed the world. Microbiol Res 169(1):30–39

Glick BR, Todorovic B, Czarny J, Cheng Z, Duan J, McConkey B (2007) Promotion of plant growth by bacterial ACC deaminase. Crit Rev Plant Sci 26:227–242

Grant WD (2004) Life at low water activity. Philos Trans R Soc Lond B Biol Sci 359(1448): 1249–1267

Grativol C, Hemerly AS, Ferreira PC (2012) Genetic and epigenetic regulation of stress responses in natural plant populations. Biochim Biophys Acta 1819(2):176–185

Grattan SR, Grieve CM (1994) Mineral nutrient acquisition and response by plants grown in saline environments. In: Pessarakli M (ed) Handbook of plant and crop stress. Marcel Dekker, New York, pp 203–226

Grattan SR, Grieve CM (1999) Salinity-mineral nutrient relations in horticultural crops. Sci Hortic 78:127–157

Grieve C, Suarez D (1997) Purslane (*Portulaca oleracea* L.): a halophytic crop for drainage water reuse systems. Plant Soil 192:277–283

Grob F, Durner J, Gaupels F (2013) Nitric oxide, antioxidants and prooxidants in plant defence responses. Front Plant Sci 4:419

Guan Q, Wu J, Yue X, Zhang Y, Zhu J (2013) A nuclear calcium-sensing pathway is critical for gene regulation and salt stress tolerance in Arabidopsis. PLoS Genet 9(8):e1003755

Gupta B, Huang B (2014) Mechanism of salinity tolerance in plants: physiological, biochemical, and molecular characterization. Int J Genomics (online):701596. https://doi.org/10.1155/2014/701596

Gupta KJ, Stoimenova M, Kaiser WM (2005) In higher plants, only root mitochondria, but not leaf mitochondria reduce nitrite to NO, in vitro and in situ. J Exp Bot 56(420):2601–2609

Habib SH, Kausar H, Saud HM (2016) Plant growth-promoting rhizobacteria enhance salinity stress tolerance in okra through ROS-scavenging enzymes. BioMed Res Int 6284547. https://doi.org/10.1155/2016/6284547

Hancock J, Desikan R, Harrison J, Bright J, Hooley R, Neill S (2006) Doing the unexpected: proteins involved in hydrogen peroxide perception. J Exp Bot 57:1711–1718

Harris BN, Sadras VO, Tester M (2010) A water-centred framework to assess the effects of salinity on the growth and yield of wheat and barley. Plant Soil 336:377–389

Hasan R, Miyake H (2017) Salinity stress alters nutrient uptake and causes the damage of root and leaf anatomy in maize. In: ICBS conference proceedings, international conference on biological science, 2015. KnE Life Sciences, pp 219–225

Hasegawa PM (2013) Sodium (Na+) homeostasis and salt tolerance of plants. Environ Exp Bot 92: 19–31

Hasegawa PM, Bressan RA, Zhu JK, Bohnert HJ (2000) Plant cellular and molecular responses to high salinity. Annu Rev Plant Physiol Plant Mol Biol 51:463–499

Hayat R, Ali S, Amara U, Khalid R, Ahmed I (2010) Soil beneficial bacteria and their role in plant growth promotion: a review. Ann Microbiol 60(4):579–598

Hernández JA, Almansa MS (2002) Short-term effects of salt stress on antioxidant systems and leaf water relations of pea plants. Physiol Plant 115:251–257

Hernández JA, Corpas FJ, Gómez M, del Río LA, Sevilla F (1993) Salt induced oxidative stress mediated by activated oxygen species in pea leaf mitochondria. Physiol Plant 89:103–110

Hernández JA, Olmos E, Corpas FJ, Sevilla F, del Río LA (1995) Salt-induced oxidative stress in chloroplast of pea plants. Plant Sci 105:151–167

Hernandez JA, Jimenez A, Mullineaux P, Sevilia F (2000) Tolerance of pea (*Pisum sativum* L.) to long-term salt stress is associated with induction of antioxidant defences. Plant Cell Environ 23:853–862

Hernández JA, Ferrer MA, Jiménez A, Ros-Barceló A, Sevilla F (2001) Antioxidant systems and O_2/H_2O_2 production in the apoplast of *Pisum sativum* L. leaves: it's relation with NaCl-induced necrotic lesions in minor veins. Plant Physiol 127:817–831

Hernández JA, Aguilar A, Portillo B, López-Gómez E, Mataix-Beneyto J, García-Legaz MF (2003) The effect of calcium on the antioxidant enzymes from salt-treated loquat and anger plants. Funct Plant Biol 30:1127–1137

Himabindu Y, Chakradhar T, Reddy MC, Kanygin A, Redding KE, Chandrasekhar T (2016) Salt-tolerant genes from halophytes are potential key players of salt tolerance in glycophytes. Environ Exp Bot 124:39–63

Hoque MA, Okuma E, Banu MNA, Nakamura Y, Shimoishi Y, Murata Y (2007a) Exogenous proline mitigates the detrimental effects of salt stress more than exogenous betaine by increasing antioxidant enzyme activities. J Plant Physiol 164:553–561

Hoque MA, Banu MNA, Okuma E, Amako K, Nakamura Y, Shimoish Y, Murata Y (2007b) Exogenous proline and glycinebetaine ingresses NaCl-induced ascorbate glutathione cycle enzyme activities and proline improves salt tolerance more than glycinebetaine in tobacco Bright yellow-2 suspension-cultured cells. J Plant Physiol 164:553–561

Hoque MA, Banu MNA, Nakamura Y, Shimoishi Y, Murata Y (2008) Proline and glycinebetaine enhance antioxidant defense and methylglyoxal detoxification systems and reduce NaCl-induced damage in cultured tobacco cells. J Plant Physiol 165:813–824

Hoque MA, Uraji M, Banu MNA, Mori C, Namamura Y, Murata Y (2010) The effects of methylglyoxal on glutathione S-tansferase from *Nicotiana tabacum*. Biosei Biotechnol Biochem 74: 2124–2126

Hu Y, Chen L, Wang H, Zhang L, Wang F, Yu D (2013) Arabidopsis transcription factor WRKY 8 functions antagonistically with its interacting partner VQ 9 to modulate salinity stress tolerance. Plant J 74(5):730–745

Hundertmark M, Hincha DK (2008) LEA (late embryogenesis abundant) proteins and their encoding genes in *Arabidopsis thaliana*. BMC Genomics 9:118–139

Hussain K, Majeed A, Nawaz K, Khizar HB, Nisar MF (2009) Effect of different levels of salinity on growth and ion contents of black seeds (*Nigella sativa* L.). Curr Res J Biol Sci 1:135–138

Imlay JA (2003) Pathways of oxidative damage. Annu Rev Microbiol 57:395–418

Inan G, Zhang Q, Li P, Wang Z, Cao Z, Zhang H, Zhang C, Quist TM, Goodwin SM, Zhu J, Shi H, Damsz B, Charbaji T, Gong Q, Ma S, Fredricksen M, Galbraith DW, Jenks MA, Rhodes D, Hasegawa PM, Bohnert HJ, Joly RJ, Bressan RA, Zhu JK (2004) Salt cress. A halophyte and

cryophyte *Arabidopsis* relative model system and its applicability to molecular genetic analyses of growth and development of extremophiles. Plant Physiol 135:1718–1737
Jablonski PP, Anderson JW (1982) Light-dependent reduction of hydrogen peroxide by ruptured pea chloroplasts. Plant Physiol 69:1403–1407
Janzen HH, Chang C (1987) Cation nutrition of barley as influenced by soil solution composition in a saline soil. Can J Soil Sci 67:619–629
Karpinski S, Reynolds H, Karpinska B, Wingsle G, Creissen G, Mullineaux P (1999) Systemic signaling and acclimation in response to excess excitation energy in Arabidopsis. Science 284:654–657
Kerepesi I, Galiba G (2000) Osmotic and salt stress-induced alteration in soluble carbohydrate content in wheat seedlings. Crop Sci 40:482–487
Kim YH, Khan AL, Waqas M, Shim JK, Kim DH, Lee KY, Lee IJ (2014) Silicon application to rice root zone influenced the phytohormonal and antioxidant responses under salinity stress. J Plant Growth Regul 33:137–149. https://doi.org/10.1007/s00344-013-9356-2
Kingsbury RW, Epstein E (1984) Selection for salt-resistant spring wheat. Crop Sci 24:310–315
Kono Y, Fridovich I (1983) Inhibition and reactivation of Mn-catalase. Implications for valence changes at the active site manganese. J Biol Chem 258:13646–13668
Kosová K, Vítámvás P, Prášil IT (2010) The role of dehydrins in plant stress response. In: Pessarakli M (ed) Handbook of plant and crop stress. CRC/Taylor and Francis, Boca Raton, pp 239–285
Koyro HW (2006) Effect of salinity of growth, photosynthesis, water relations and solute composition of the potential cash crop halophyte *Plantago coronopus* (L.). Environ Exp Bot 56: 136–146
Krasensky J, Jonak C (2012) Drought, salt, and temperature stress-induced metabolic rearrangements and regulatory networks. J Exp Bot 63(4):1593–1608
Kumari S, Vaishnav A, Jain S, Varma A, Choudhary DK (2015) Bacterial-mediated induction of systemic tolerance to salinity with expression of stress alleviating enzymes in soybean (*Glycine max* L. Merrill). J Plant Growth Regul 34(3):558–573
Larcher W (2003) Physiological plant ecology, 4th edn. Springer, Heidelberg
Li W, Li Q (2017) Effect of environmental salt stress on plants and the molecular mechanism of salt stress tolerance. Int J Environ Sci Nat Res 7(3):555714. https://doi.org/10.19080/IJESNR.2017.07.555714
Loiacono FV, De Tullio MC (2012) Why we should stop inferring simple correlations between antioxidants and plant stress resistance: towards the antioxidomic era. Omics 64:160–167
Lu Y, Lam H, Pi E, Zhan Q, Tsai S, Wang C, Kwan Y, Ngai S (2013) Comparative metabolomics in *Glycine max* and *Glycine soja* under salt stress to reveal the phenotypes of their offspring. J Agric Food Chem 61:8711–8721
Maggio A, Raimondi G, Martino A, DePascale S (2007) Salt stress response in tomato beyond the salinity tolerance threshold. Environ Exp Bot 59:276–282
Makela P, Karkkainen J, Somersalo S (2000) Effect of glycinebetaine on chloroplast ultrastructure, chlorophyll and protein content, and RuBPCO activities in tomato grown under drought or salinity. Biol Plant 43(3):471–475
Mariko S, Kachi N, Ishikawa S, Furukawa A (1992) Germination ecology of coastal plants in relation to salt environment. Ecol Res 7:225–233
Marosz A (2004) Effect of soil salinity on nutrient uptake, growth, and decorative value of four ground cover shrubs. J Plant Nutr 27(6):977–989
Marschner H (1995) Mineral nutrition of higher plants, 2nd edn. Academic, London, pp 596–680
Matysik J, Alai Bhalu B, Mohanty P (2002) Molecular mechanisms of quenching of reactive oxygen species by proline under stress in plants. Curr Sci 82:525–532
Meloni DA, Oliva MA, Ruiz HA, Martinez CA (2001) Contribution of proline and inorganic solutes to osmotic adjustment in cotton under salt stress. J Plant Nutr 24(3):599–612

Meloni DA, Oliva MA, Martinez CA, Cambraia J (2003) Photosynthesis and activity of superoxide dismutase, peroxidase and glutathione reductase in cotton under salt stress. Environ Exp Bot 49:69–76
Mittler R (2002) Oxidative stress, antioxidants and stress tolerance. Trends Plant Sci 7:405–410
Mittler R (2006) Abiotic stress, the field environment and stress combination. Trends Plant Sci 11(1):15–19
Mittova V, Tal M, Volokita M, Guy M (2003) Up-regulation of the leaf mitochondrial and peroxisomal antioxidative systems in response to salt-induced oxidative stress in the wild salt-tolerant tomato species *Lycopersicon pennellii*. Plant Cell Environ 26:845–856
Mittova V, Guy M, Tal M, Volokita M (2004) Salinity up-regulates the antioxidative system in root mitocondria and peroxisomes of the wild salt-tolerant tomato species *Lycopersicon pennellii*. J Exp Bot 399:1105–1113
Munns R (2002) Comparative physiology of salt and water stress. Plant Cell Environ 25:239–250
Munns R (2005) Genes and salt tolerance: bringing them together. Plant Physiol 167:645–663
Munns R, Gilliham M (2015) Salinity tolerance of crops—what is the cost? New Phytol 208: 668–673
Munns R, Termaat A (1986) Whole-plant responses to salinity. Aust J Plant Physiol 13:143–160
Munns R, Tester M (2008) Mechanisms of salinity tolerance. Annu Rev Plant Biol 59:651–681
Nadeem SM, Ahmad M, Naveed M, Imran M, Zahir ZA, Crowley DE (2016) Relationship between in vitro characterization and comparative efficacy of plant growth-promoting rhizobacteria for improving cucumber salt tolerance. Arch Microbiol 198(4):379–387
Niu XNX, Bressan RA, Hasegawa PM, Pardo JM (1995) Ion homeostasis in NaCl stress environments. Plant Physiol 109(3):735–742
Noctor G, Foyer CH (1998) Ascorbate and glutathione: keeping active oxygen under control. Annu Rev Plant Physiol Mol Biol 49:249–279
Numan M, Bashir S, Khan Y, Mumtaz R, Shinwari ZK, Khan AL, Khan A, Ahmed AH (2018) Plant growth promoting bacteria as an alternative strategy for salt tolerance in plants: a review. Microbiol Res 209:21–32
Orhan F (2016) Alleviation of salt stress by halotolerant and halophilic plant growth-promoting bacteria in wheat (*Triticum aestivum*). Braz J Microbiol 47(3):621–627
Pan Z, Zhao Y, Zheng Y, Liu J, Jiang X, Guo Y (2012) A high-throughput method for screening Arabidopsis mutants with disordered abiotic stress induced calcium signal. J Genet Genomics 39(5):225–235
Pang Q, Chen S, Dai S, Wang Y, Chen Y, Yan X (2010) Comparative proteomics of salt tolerance in *Arabidopsis thaliana* and *Thellungiella halophila*. J Proteome Res 9:2584–2599
Parida AK, Das AB (2005) Salt tolerance and salinity effects on plants: a review. Eco Toxicol Environ Saf 60(3):324–349
Parida AK, Das AB, Mohanty P (2004) Investigations on the antioxidative defence responses to NaCl stress in a mangrove, *Bruguiera parviflora*: differential regulations of isoforms of some antioxidative enzymes. Plant Growth Regul 42(3):213–226
Patel PR, Kajal SS, Patel VR, Patel VJ, Khristi SM (2010) Impact of salt stress on nutrient uptake and growth of cowpea. Braz J Plant Physiol 22(1):43–48
Peleg Z, Apse MP, Blumwald E (2011) Engineering salinity and water stress tolerance in crop plants: getting closer to the field. Adv Bot Res 57:405–443
Phang TH, Shao G, Lam HM (2008) Salt tolerance in soybean. J Integr Plant Biol 50(10): 1196–1212
Pichersky E, Gang DR (2000) Genetics and biochemistry of secondary metabolites in plants: an evolutionary perspective. Trends Plant Sci 10:439–445
Qin Y, Druzhinina IS, Pan X, Yuan Z (2016) Microbially-mediated plant salt tolerance and microbiome-based solutions for saline agriculture. Biotechnol Adv 34(7):1245–1259
Rahman S, Miyake H, Takeoka Y (2002) Effects of exogenous glycinebetaine on growth and ultrastructure of salt-stressed rice seedlings (*Oryza sativa* L.). Plant Product Sci 5(1):33–44

Rathinasabapathi B (2000) Metabolic engineering for stress tolerance: installing osmoprotectant synthesis pathways. Ann Bot 86(4):709–716
Reddy MP, Sanish S, Iyengar ERR (1992) Photosynthetic studies and compartmentation of ions in different tissues of *Salicornia brachiata* Roxb. under saline conditions. Photosynthetica 26: 173–179
Reddy AR, Chaitanya KV, Vivekanandan M (2004) Drought-induced responses of photosynthesis and antioxidant metabolism in higher plants. J Plant Physiol 161:1189–1202
Reddy AS, Ali GS, Celesnik H, Day IS (2011) Coping with stresses: roles of calcium- and calcium/calmodulin-regulated gene expression. Plant Cell 23(6):2010–2032
Rehman S, Harris PJC, Bourne WF, Wilkin J (1996) The effect of sodium chloride on germination and the potassium and calcium contents of acacia seeds. Seed Sci Technol 25:45–57
Renberg L, Johansson A, Shutova T, Stenlund H, Aksmann A, Raven JA, Gardeström P, Moritz T, Samuelsson G (2010) A metabolomic approach to study major metabolite changes during acclimation to limiting co2 in *chlamydomonas reinhardtii*. Plant Physiol 154:187–196
Roy SJ, Negrao S, Tester M (2014) Salt resistant crop plants. Curr Opin Biotechnol 26:115–124
Sakihama YM, Cohen F, Grace SC, Yamasaki H (2002) Plant phenolic antioxidant and prooxidant activities: phenolics-induced oxidative damage mediated by metals in plants. Toxicology 177: 67–80
Saxena SC, Kaur H, Verma P (2013) Osmoprotectants: potential for crop improvement under adverse conditions. In: Tuteja N, Sarvajeet SG (eds) Plant acclimation to environmental stress. Springer, New York, pp 197–232
Scandalios JG (1993) Oxygen stress and superoxide dismutases. Plant Physiol 101:7–12
Serrano R, Mulet JMG, Rios Marquez JA, de Larrinoa IF, Leube MP, Mendizabal I, Pascual-Ahuir A, Proft M, Ros R, Montesinos C (1999) A glimpse of the mechanisms of ion homeostasis during salt stress. J Exp Bot 50:1023–1036
Shabala S, Bose J, Hedrich R (2014) Salt bladders: do they matter? Trends Plant Sci 19(11): 687–691
Sharpley AN, Meisinger JJ, Power JF, Suarez DL (1992) Root extraction of nutrients associated with long-term soil management. In: Hatfield JI, Stewart BA (eds) Advances in soil science, limitations to plant root growth. Springer, New York, pp 19151–19217
Shinozaki K, Yamaguchi-Shinozaki K (2007) Gene networks involved in drought stress response and tolerance. J Exp Bot 58(2):221–227
Shrivastava P, Kumar R (2015) Soil salinity: a serious environmental issue and plant growth promoting bacteria as one of the tools for its alleviation. Saudi J Biol Sci 22(2):123–131
Shulaev V, Cortes D, Miller G, Mittler R (2008) Metabolomics for plant stress response. Physiol Plant 132:199–208
Singh RP, Jha P, Jha PN (2015) The plant-growth-promoting bacterium *Klebsiella* sp. SBP-8 confers induced systemic tolerance in wheat (*Triticum aestivum*) under salt stress. J Plant Physiol 184:57–67
Slama I, Abdelly C, Bouchereau A, Flowers T, Savoure A (2015) Diversity, distribution and roles of osmoprotective compounds accumulated in halophytes under abiotic stress. Ann Bot 115(3): 433–447
Smith DL, Praslickova D, Ilangumaran G (2015a) Inter-organismal signaling and management of the phytomicrobiome. Front Plant Sci 6:722. https://doi.org/10.3389/fpls.2015.00722
Smith DL, Subramanian S, Lamont JR, Bywater-Ekegard M (2015b) Signaling in the phytomicrobiome: breadth and potential. Front Plant Sci 6:709. https://doi.org/10.3389/fpls.2015.00709
Smith DL, Gravel V, Yergeau E (2017) Editorial: signaling in the phytomicrobiome. Front Plant Sci 8:611. https://doi.org/10.3389/fpls.2017.00611
Stepien P, Johnson GN (2009) Contrasting responses of photosynthesis to salt stress in the glycophyte Arabidopsis and the halophyte Thellungiella: role of the plastid terminal oxidase as an alternative electron sink. Plant Physiol 149:1154–1165

Sukweenadhi J, Kim YJ, Choi ES, Koh SC, Lee SW, Kim YJ, Yang DC (2015) *Paenibacillus yonginensis* DCY84T induces changes in *Arabidopsis thaliana* gene expression against aluminum, drought, and salt stress. Microbiol Res 172:7–15

Talei D, Kadir MA, Yusop MK, Valdiani A, Abdullah MP (2012) Salinity effects on macro and micronutrients uptake in medicinal plant King of Bitters (*Andrographis paniculata* Nees.). POJ 53:271–278

Tang X, Mu X, Shao H, Wang H, Brestic M (2015) Global plant-responding mechanisms to salt stress: physiological and molecular levels and implications in biotechnology. Crit Rev Biotechnol 35:425–437

This D, Comstock J, Courtois B, Xu Y, Ahmadi N, Vonhof WM, Fleet C, Setter T, McCouch S (2010) Genetic analysis of water use efficiency in rice (*Oryza sativa* L.) at the leaf level. Rice 3:72–86

Tuteja N, Sahoo RK, Garg B, Tuteja R (2013) *OsSUV3* dual helicase functions in salinity stress tolerance by maintaining photosynthesis and antioxidant machinery in rice (*Oryza sativa* L. cv. IR64). Plant J 76(1):115–127

Tyagi B, Tewari S, Dubey A (2017) Biochemical characterization of fungus isolated during In vitro propagation of *Bambusa balcooa*. Pharmacogn Mag 13:775–779

Ungar IA (1996) Effect of salinity on seed germination, growth, and ion accumulation of *Atriplex patula* (Chenopodiaceae). Am J Bot 83:604–607

Upadhyay SK, Singh DP (2015) Effect of salt-tolerant plant growth-promoting rhizobacteria on wheat plants and soil health in a saline environment. Plant Biol 17:288–293

Urano K, Kurihara Y, Seki M, Shinozaki K (2010) Omics' analyses of regulatory networks in plant abiotic stress responses. Curr Opin Plant Biol 13:132–138

van der Fits L, Memelink J (2000) ORCA3, a jasmonate-responsive transcriptional regulator of plant primary and secondary metabolism. Science 289:295–297

Veeranagamallaiah G, Jyothsnakumari G, Thippeswamy M, Reddy PCO, Surabhi GK, Sriranganayakulu G, Mahesh Y, Rajasekhar B, Madhurarekha C, Sudhakar C (2008) Proteomic analysis of salt stress responses in foxtail millet (*Setaria italica* L. cv. Prasad) seedlings. Plant Sci 175:631–641

Wang XQ, Yang PF, Liu Z, Liu WZ, Hu Y, Chen H, Kuang TY, Pei ZM, Shen SH, He YK (2009) Exploring the mechanism of Physcomitrella patens desiccation tolerance through a proteomic strategy. Plant Physiol 149:1739–1750

Wang Q, Dodd IC, Belimov AA, Jiang F (2016) Rhizosphere bacteria containing 1-aminocyclopropane-1-carboxylate deaminase increase growth and photosynthesis of pea plants under salt stress by limiting Na+ accumulation. Funct Plant Biol 43:161–172

Wenbo L, Qing L (2017) Effect of environmental salt stress on plants and the molecular mechanism of salt stress tolerance. Int J Environ Sci Nat Res 7:555714

Wu D, Cai S, Chen M, Ye L, Chen Z, Zhang H, Dai F, Wu F, Zhang G (2013) Tissue metabolic responses to salt stress in wild and cultivated barley. PLoS One 8:e55431

Yancey PH (2005) Organic osmolytes as compatible, metabolic and counteracting cytoprotectants in high osmolarity and other stresses. J Exp Biol 208:2819–2830

Yang J, Kloepper JW, Ryu CM (2009) Rhizosphere bacteria help plants tolerate abiotic stress. Trends Plant Sci 14:1–4

Yoshimura K, Yabute Y, Ishikawa T, Shigeoka S (2000) Expression of spinach ascorbate peroxidase isoenzymes in response to oxidative stresses. Plant Physiol 123:223–233

Zerrouk IZ, Benchabane M, Khelifi L, Yokawa K, Ludwig-Müller J, Baluska F (2016) A *Pseudomonas* strain isolated from date-palm rhizospheres improves root growth and promotes root formation in maize exposed to salt and aluminum stress. J Plant Physiol 191:111–119

Zhu JK (2001) Plant salt tolerance. Trends Plant Sci 6:66–71

Zhu JK (2003) Regulation of ion homeostasis under salt stress. Curr Opin Plant Biol 6:441–445

Chapter 18
Effect of Salt Stress on Plants and Role of Microbes in Promoting Plant Growth Under Salt Stress

Manik Prabhu Narsing Rao, Zhou-Yan Dong, Min Xiao, and Wen-Jun Li

Abstract Salinity has become a severe issue for plant growth, reducing crop yield worldwide. Salt stress in plants causes various physiological and metabolic changes such as nutritional imbalance, inhibition of water uptake, seed germination, photosynthesis, and decrease in growth. It is important to combat soil pollution to feed the world's growing population. Microorganisms colonize plants in the natural environment. Plant root-associated microbes, including endophytes, closely cooperate with plants and mediate important physiological and metabolic processes, thereby increasing plant's tolerance to salinity stress. In the present chapter, we discussed the effect of salt stress on plants and the role of microbes in promoting plant growth under salt stress.

Keywords Soil salinity · Salt stress on plant growth · Osmotic stress · Ionic stress · Salt stress tolerance mechanisms in plants · Plant growth-promoting microbes

18.1 Introduction

The term soil salinity is used to describe the salt content like NaCl, Na_2SO_4, $MgSO_4$, $MgCl_2$, $CaSO_4$, KCl, and Na_2CO_3 within the soil (Tavakkoli et al. 2010), and the process of increase in their content is known as salinization (Clarke et al. 2015). Generally, saline soil is defined as one in which the electrical conductivity (EC) of the saturation extract (EC_e) in the root zone exceeds 4 dS m^{-1} (approximately

M. P. Narsing Rao · Z.-Y. Dong · M. Xiao
State Key Laboratory of Biocontrol and Guangdong Key, Laboratory of Plant Resources, School of Life Sciences, Sun Yat-Sen University, Guangzhou, PR China

W.-J. Li (✉)
State Key Laboratory of Biocontrol and Guangdong Key, Laboratory of Plant Resources, School of Life Sciences, Sun Yat-Sen University, Guangzhou, PR China

Key Laboratory of Biogeography and Bioresource in Arid Land, Xinjiang Institute of Ecology and Geography, Chinese Academy of Sciences, Ürümqi, PR China
e-mail: liwenjun3@mail.sysu.edu.cn

B. Giri, A. Varma (eds.), *Microorganisms in Saline Environments: Strategies and Functions*, Soil Biology 56, https://doi.org/10.1007/978-3-030-18975-4_18

40 mM NaCl) at 25 °C and has exchangeable sodium of 15% (Shrivastava and Kumar 2015). Indeed, NaCl is the most prevalent salt in the soils and has been a matter of great concern for the researchers (Tavakkoli et al. 2010; Clarke et al. 2015). Salt stress causes many adverse effects in plants, among them the significant impact is on plant growth and development (Liang et al. 2018). Salt stress in plants causes changes in various physiological and metabolic processes (Gupta and Huang 2014). The adverse effect is noted at different stages like germination, vegetative growth, and flowering (Kaleem et al. 2018). The severity of salt stress varies depending on the stage of plant growth, growth conditions, the degree of plant tolerance, environmental factors, and irrigation management practices (Acosta-Motos et al. 2017). About 50% of irrigated land in the world is facing a salinity problem and keep increasing at the rate of 1% (approximately) yearly (Kumar et al. 2017). To overcome this problem, many researches developed transgenic plants which reduced plants salt stress (Nongpiur et al. 2016), but these approaches are expensive and time-consuming (Bianco and Defez 2012).

Microorganisms colonize plants in the natural environment. Microbes associated with plants and roots mediate important physiological processes and help plants to tolerate various abiotic stresses (Egamberdieva et al. 2017). Several microorganisms, especially plant growth-promoting rhizobacteria, have been reported to provide salt tolerance to host plants (Bano and Fatima 2009; Waller et al. 2005). In this chapter, we discuss the effect of salt stress on plants and the role of microbes in promoting plant growth under salt stress.

18.2 Effect of Salt Stress on Plants

In the present scenario, soil salinity has become a severe issue for plants, reducing crop yield worldwide. It is essential to know how plants react to salt stress (Fig. 18.1) to avoid further damage. Munns (1988) put forward a model in which plants under salt stress experience osmotic and ionic stress. Osmotic stress experienced immediately when the plant roots were exposed to salts causing cell expansion and water uptake inhibition (Munns et al. 1995). Ionic stresses develop when toxic ions (e.g., Na^+) accumulate in cells causing chlorosis and necrosis and decrease in the activity of cellular metabolism including photosynthesis (Panuccio et al. 2014). As a consequence of osmotic and ionic stress, secondary stresses such as oxidative stress often occur (Perez-Lopez et al. 2009) causing damage to membrane lipids, proteins, and nucleic acids (Perez-Lopez et al. 2009).

Roots act as a connecting link between plants and the soil (Acosta-Motos et al. 2017). Plants uptake water because they have a higher water pressure than the soil solution. When salt stress occurs, the osmotic pressure of the soil solution will be higher than that in plant cells, and thus, the plant cannot get enough water (Kader and Lindberg 2010). Under salt stress, root system, namely, root length, root diameter, root number, number of root hairs, the diameter of xylem vessels, the width of root cortex, and the suberin deposition get effected (Acosta-Motos et al. 2017).

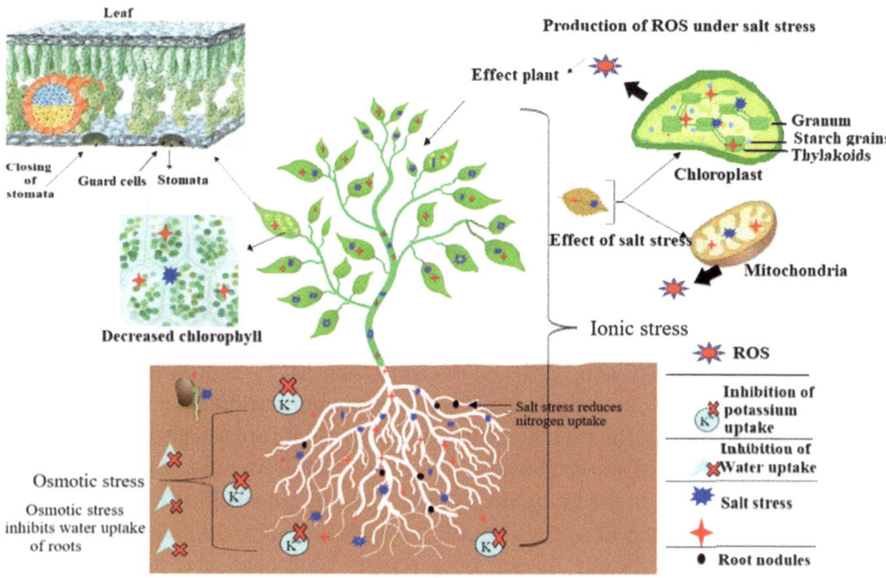

Fig. 18.1 Illustration of physiological and biochemical changes occurring during salt stress

Salt enters into the roots along with water and introduces to the vascular part either through symplastic or apoplastic route. In apoplastic pathway, salt along with the water delivers the in xylem through the intracellular region, whereas in the symplastic pathway, salt along with water enters in the roots through epidermal plasma membranes and then travels cell to cell through plasmodesmata until discharging to the xylem (Das et al. 2015). Apical region of roots grown under salinity shows extensive vacuolization and lacks typical apical tissue organization. Under salt stress, epidermis and parenchyma cells of cortex and pith show shrinkage (Carillo et al. 2011; Munns and Rawson 1999). Salt stress prevents root growth by reducing cyclin level and suppressing cell division and elongation (Duan et al. 2015).

Mineral nutrition of plants dramatically affects their ability to adapt to stress conditions (Paul and Lade 2014). Nutritional imbalance hampers plant growth, development, and also the yield (Paul and Lade 2014). The cellular cation of plants such as potassium plays a vital role in preserving membrane potential and enzyme activities. Excessive Na^+ around the root exterior disturbs potassium uptake because of the identical chemical properties of Na^+ and K^+. After entering the cytosol, Na^+ inhibits the activity of many enzymes and proteins, causing damage to the cells (Das et al. 2015; Xiong and Zhu 2002).

Seed germination is a crucial stage in plants' life history, and salt stress during this stage is critical for the establishment of plants. Salt stress during this stage can prolong seed germination and also inhibit seed germination (Xue et al. 2004). High salt concentrations severely affect radicle elongation (Li 2008) due to the reduction in the turgor of the radicle cells (Bewley and Black 1994).

Soil salinity also affects photosynthesis through decrease in leaf area. Old leaves start to develop chlorosis and fall with prolonged salt stress. If the rate of dying leaves exceeds than the rate at which they produced, then the photosynthetic capacity of the plant will not supply the required carbohydrate to young leaves, which further reduces their growth rate (Carillo et al. 2011). Salt stress affects photosynthesis both in the short and long term. Short- term salt stress is rapid and occurs within a short period of salt exposure affecting photosynthesis by stomatal limitations, leading to decrease carbon assimilation (Parida and Das 2005). Long-term salt stress affects photosynthesis due to salt accumulation in young leaves, decreasing chlorophyll and carotenoid content. The decrease in chlorophyll content may be due to the changes in the lipid-protein ratio of pigment-protein complexes or increased chlorophyllase activity (Saravanavel et al. 2011). Salt stress also affects stomata size and density, causing decrease in stomatal conductance (Hanin et al. 2016). Under salt stress, the thylakoid structure of the chloroplasts gets disorganized. The number and size of plastoglobuli increase, and their starch content decreases. The number and depth of the grana stacks reduce and cause swelling of the thylakoid (Parida and Das 2005). Salt stress also affects shoot and reproductive development. Young leaves emerge slowly, lateral buds remain quiescent, and flowering starts earlier (Ilangumaran and Smith 2017).

Plants' lipid content plays an important role in cell permeability; under salt stress the total lipids content is found to decrease (Keshtehgar et al. 2013). Under salt stress, increases in compatible solutes such as proline, glycine betaine, sugars, and polyols have been noticed (Singh et al. 2000; Khan et al. 2000). Salt stress also increases ROS (reactive oxygen species) production in chloroplast, mitochondria, and peroxisomes. Antioxidative enzyme activities such as ascorbate peroxidase, glutathione reductase, monodehydroascorbate reductase, and dehydroascorbate reductase have significantly increased under salt stress (Parida and Das 2005).

Nitric oxide regulates plant growth and various developmental processes like respiration, flowering, root growth, stomata closure, seed germination, and cell death. In the expression of various redox-regulated genes, many antioxidant enzymes were directly or indirectly activated by the nitric oxide (Numan et al. 2018). Under salt stress, reduced nitrate reductase activity was noticed due to the presence of Cl^- salts in the external medium. Salt stress also inhibits nitrogen fixation by reducing nodulation, leghemoglobin content, and nitrogenase activity (Parida and Das 2005).

18.3 Role of Microorganisms in Enhancing Plant Tolerance to Salt Stress

Millions of microorganisms form a complex ecological community with plants root system and influence plant growth with their interactions (Vurukonda et al. 2016). Microorganisms can survive in diverse environments and therefore have enormous

Table 18.1 List of microbes and their role in enhancing plant growth under salt stress

Microbes	Plant	Salt tolerance strategy	References
Pseudomonas simiae	*Glycine max*	4-nitroguaiacol and quinoline promote seed germination	Vaishnav et al. (2016)
Azospirillum brasilense	*Phaseolus vulgaris*	Stimulation of persistent exudation of flavonoids	Dardanelli et al. (2008)
Pseudomonas koreensis AK-1	*Glycine max*	Reduction in Na^+ level and increase in K^+ level	Kasotia et al. (2015)
Co-inoculation of *Azospirillum brasilense* and *Pantoea dispersa*	*Capsicum annuum*	High stomatal conductance and photosynthesis	de Zelicourt et al. (2013)
Pseudomonas fluorescens	*Arachis hypogea*	Enhanced ACC deaminase activity	Saravanakumar and Samiyappan (2007)
Azospirillum	*Zea mays*	Restricted Na + uptake and increased K^+ and Ca^{2+} uptake along with increased nitrate reductase and nitrogenase activity	Hamdia et al. (2004)
Pseudomonas pseudoalcaligenes and *Bacillus pumilus*	*Oryza sativa*	Increased concentration of compatible solute	Jha et al. (2011)
Frankia	*Casuarina glauca* and *Casuarina equisetifolia*	Plant growth, shoot, root, and total dry weight, proline, and chlorophyll contents	Ngom et al. (2016)
Endophytic diazotrophic *Enterobacter* sp. EN-21	Sugarcane	ACC deaminase activity	Kruasuwan and Thamchaipenet (2018)
Arbuscular mycorrhizal fungi	Citrus seedlings	Enhance growth, photosynthesis, root morphology, and ionic balance	Wu et al. (2010)
Arbuscular mycorrhizal fungi	*Panicum turgidum*	Alter photosynthetic and antioxidant pathways	Hashem et al. (2015)

capabilities to mitigate stresses (Meena et al. 2017). Microbes associated with plants and roots system help to alleviate plant stress by a variety of mechanisms (Table 18.1). Bacteria which are tolerant to high salt concentration can grow and colonize better to root rhizospheres and external spaces of saline soil. These bacteria expose themselves to high salinity and protect the plants against salt stress (Vejan et al. 2016; Vurukonda et al. 2016). *Azospirillum brasilense* improved salt tolerance of jojoba plant during the rooting stage. It was found that *Azospirillum brasilense* reduced the undesirable effects of saline conditions during jojoba rooting (Gonzalez et al. 2015).

The preliminary impact that plants face during salt stress is water uptake. A wide variety of microorganisms have been reported which help plants to uptake/maintain water under salt stress. The water absorbing ability of maize roots under salt stress was increased when they were inoculated with *Bacillus megaterium* strain (Marulanda et al. 2010). Similarly, the water absorbing ability of maize root under salt stress was increased by *Pantoea agglomerans* (Gond et al. 2015).

Formation of biofilm and exopolysaccharides (EPS) is a significant strategy of microbes to survive under salt stress (Kasim et al. 2016; Qurashi and Sabri 2012). Biofilm- forming *Bacillus amyloliquefaciens* strain showed ameliorative effects on the growth of barley plants growing under salt stress (Kasim et al. 2016). Exopolysaccharide and biofilm- producing strains *Halomonas variabilis* and *Planococcus rifietoensis*, under salt stress, stimulate chickpea growth and soil aggregation (Qurashi and Sabri 2012). The biofilm produced by these strains maintains sufficient moisture around the seeds and helps plants to survive and germinate under salt stress (Qurashi and Sabri 2012). Microbial EPS showed better water holding capacity (Qurashi and Sabri 2012) due to its cation binding potential. It is believed that EPS-producing bacteria in the root zone decrease the Na^+ content available for plants and thus help plants to alleviate salt stress (Ashraf et al. 2004). EPS-producing bacteria belonging to *Aeromonas* and *Bacillus* found to restrict Na^+ uptake by wheat roots (Ashraf et al. 2004). EPS-producing *Azotobacter chroococcum* significantly increased root and shoot dry weights and also chlorophyll a and b and carotenoid contents (Awad et al. 2012).

Microbes can also alter plant's toxic ion uptake by changing host physiology or by directly reducing foliar accumulation of toxic ions and improving the nutritional status (Paul and Lade 2014). *Bacillus subtilis* GB03 in *Arabidopsis* decreased whole plant Na^+ content by downregulating *HKT1* expression in roots and upregulating *HKT1* expression in shoots to enhance shoot-to-root Na^+ recirculation (Zhang et al. 2008). In white clover, shoot and root Na^+ accumulation was reduced by *Bacillus subtilis* GB03 (Han et al. 2014). Colonization of arbuscular mycorrhizal fungi reduced Cl ions uptake by translocating Na^+ to shoot tissues (Evelin et al. 2009). *Glomus intraradices* prevented Na^+ accumulation in *Lotus glaber* by enhancing K^+ concentrations in roots (Sannazzaro et al. 2006). *Pseudomonas putida* Rs-198 increased salt stress tolerance of cotton by increasing the absorption of the Mg^{2+}, K^+, and Ca^{2+} and decreased the uptake of Na^{2+} (Yao et al. 2010).

Phosphorous is considered one of the significant and essential macronutrients of plants. Under salt stress, phosphorous becomes unavailable for plants (Numan et al. 2018). However, phosphate-solubilizing microbes help to solubilize inorganic phosphate (Alori et al. 2017). A phosphate-solubilizing bacterium, *Pseudomonas fluorescens* Ms-01, when inoculated with wheat plants, increased the plant height and weight under saline conditions (Kadmiri et al. 2018). *Bacillus cereus* Pb25 having phosphorous solubilization ability increased *Vigna radiata* salt stress tolerance (Islam et al. 2016).

Microbes with unique attributes when associated with plants are found to increase plant's growth under salt stress (Enebe and Babalola 2018). *Brevibacterium epidermidis* RS15, *Micrococcus yunnanensis* RS222, and *Bacillus aryabhattai* RS341, with 1-aminocyclopropane-1-carboxylate (ACC) deaminase activity, found to increase root elongation (40%) and dry weight in canola (Siddikee et al. 2010).

Enterobacter sp. has been found to promote rice seedling growth under salt stress by ACC deaminase activity and production of siderophore, indole-3-acetic acid (IAA), and phosphate solubilization activity (Sarkar et al. 2017). *Achromobacter piechaudii* having (ACC) deaminase activity increased tomato seedling. The bacterium increased plants' water, phosphorous, and potassium uptake and reduced ethylene production (Glick 2014).

Volatile organic compounds (VOCs) from microbes play a crucial role in enhancing plant salt stress. VOCs from microorganisms have been reported to play a significant role in the morphogenetic process of plants (Numan et al. 2018). VOCs regulate auxin homeostasis and cell expansion (Zhang et al. 2007). VOCs from *Bacillus subtilis* GB03 showed salt tolerance in *Arabidopsis thaliana* by downregulating the expression of high-affinity K^+ transporter (HKT1) in roots and on the other hand upregulating in shoots which results in lower accumulation of Na^+ inside the plant (Zhang et al. 2008).

Excessive production of ROS under salt stress causes oxidative damage to plants' membrane lipids, proteins, and nucleic acids (Perez-Lopez et al. 2009). Microbes producing antioxidants significantly increased plant tolerance to salt stress. *Bacillus cereus* Pb25, when inoculated with *Vigna radiata*, increased superoxide dismutase, catalase, and peroxidase activity (Islam et al. 2016). Similarly, *Pseudomonas* sp. DW1 decreased superoxide dismutase activities and increased peroxidase activities in eggplant (Fu et al. 2010). *Bacillus pumilus* and *Pseudomonas pseudoalcaligenes*, when inoculated with paddy plant, increased catalase, peroxidase, and superoxide dismutase activity. These enzymes helped plant in eliminating H_2O_2 from salt-stressed environments (Jha and Subramanian 2013). Salt tolerance of barley was increased by root endophyte *Piriformospora indica* which produced antioxidants (Baltruschat et al. 2008).

Nitrogen in saline habitats is inadequate; therefore, nitrogen input in these environments is essential (Wang et al. 2018). *Funneliformis mosseae* and *Diversispora versiformis*, alone and in combination, helped *Chrysanthemum morifolium* to uptake nitrogen under salt stress. The growth was enhanced by the increase in root length and shoot and root dry weight (Wang et al. 2018). Microbial N_2-fixation under saline conditions has been proven to improve plant growth (Hanin et al. 2016). *Rhizobium* strain inoculated with *Acacia ampliceps* improved biological N_2 fixation under saline conditions (Zou et al. 1995).

Rhizosphere is relatively nutrient-rich environment containing amino acids, sugars, fatty acids, and other organic compounds. Microbes utilize these nutrients, and in return, they synthesize biologically active compounds, including phytohormones (auxins, cytokinins, gibberellins, etc.), antifungal compounds, and compatible solutes. These biological compounds not only help plant growth but also increase plant tolerance to various stresses (Ilangumaran and Smith 2017). *Bacillus amyloliquefaciens* SQR9 enhanced salt stress tolerance of maize seedlings by increasing chlorophyll, total soluble sugar, and glutathione content. Improved peroxidase, catalase activity, and enhanced K^+/Na^+ ratio were also noticed. Besides, the ABA level which was increased due to salt stress was counteracted by *Bacillus amyloliquefaciens* SQR9. *Bacillus amyloliquefaciens* SQR9 decreased salt stress by upregulation of genes *RBCS*, *RBCL* (encoding RuBisCo subunits), *H(+)-Ppase* (encoding H^+ pumping pyrophosphatase), *HKT1*, *NHX1*, *NHX2*, and *NHX3* and also the downregulation of *NCED* expression (encoding 9-*cis*-epoxycarotenoid dioxygenase) in inoculated seedlings (Chen et al. 2016). *Pseudomonas putida* UW4 has been found to increase tomato growth under salt stress. Increased shoot growth and *Toc GTPase* expression (gene responsible for chloroplast protein import) have been upregulated which may facilitate the import of proteins involved as a part of the stress response (Yan et al. 2014).

Streptomyces rochei SM3 increased the salt tolerance of chickpea by increasing the activity of phenylalanine ammonia lyase. Further, investigation at a genetic level showed that *Streptomyces rochei* SM3 triggered the ethylene-responsive transcription factor (CaTF2). The ET-mediated defense pathway and activation of the phenylpropanoid pathway alleviated salt stresses in chickpea (Srivastava et al. 2015). *Azotobacter chroococcum* and *Azotobacter vinelandii* facilitate plant growth in saline soils by the production of auxins (IAA) (Allam et al. 2018). *Streptomyces* sp. strain PGPA39 alleviated salt stress in micro tom tomato plants by increasing chlorophyll content and plant biomass and reducing leaf proline content (Palaniyandi et al. 2014). *Bacillus drentensis* and *Enterobacter cloacae* reported for phosphate solubilization, IAA production, and exopolysaccharide activity. These strains help to grow mung bean under salt stress (Mahmood et al. 2016).

Co-culture techniques find multiple applications in biology (Goers et al. 2014). Co-culture techniques have been used to increase the salt tolerance of plants. Co-inoculation of *Rhizobium* and *Pseudomonas* enhanced corn plants' salt stress tolerance by selective uptake of K^+, maintaining leaf water content, decreased electrolyte leakage, and increased proline production (Bano and Fatima 2009). Co-culture of *Rhizophagus intraradices* and *Massilia* sp. RK4 significantly increased shoot total nitrogen in maize (Krishnamoorthy et al. 2016). Under field trials, co-culture of *Rhizobium* and *Pseudomonas* strains improved physiology, ionic balance, and quality of mung bean grown under salt-affected conditions (Ahmad et al. 2013). *Azospirillum brasilense* co-cultured with *Rhizobium* increased root development, nitrogen fixation, production of flavonoid signals, nod-gene transcription, and Nod factor patterns (Dardanelli et al. 2008). Dual microbial inoculation of arbuscular mycorrhizal fungi (AMF) and *Sinorhizobium terangae* enhanced nodulation and improved salinity tolerance of *Acacia saligna* (Soliman et al. 2012).

18.4 Conclusion and Outlook

In the past few decades, soil salinity increased dramatically caused naturally or due to human activities. About 50% of irrigated land in the world is facing salinity, and the problem keeps increasing at the rate of 1% (approximately) yearly. It is important not only to increase soil fertility but also to find appropriate methods, which can help plants to overcome salt stress. Microorganisms can survive in diverse environments and have enormous capabilities to mitigate various stresses. Microorganisms not only promote plant growth but also enhance salt stress tolerance. Bacteria which grow under salt conditions can better colonize root rhizospheres and external spaces under saline conditions. These bacteria expose themselves to high salinity and protect plants against salt stress. Microorganisms have been reported to help plants to overcome salt stress by increasing water and nutrients uptake. Phytohormones, exopolysaccharides, biofilm, volatile organic compounds, antioxidants, and osmolytes from microbes helped plants to overcome salt stress by removing toxic substances, maintaining water level, and increasing photosynthesis, germination, root growth, and fresh and dry weight. Though extensive studies have been carried out in understanding the role of microorganisms in enhancing plant's salt stress tolerance, still many microorganisms are hidden in the soil. Extensive studies have to be initiated to uncover the hidden microbes and understand their role in enhancing plant growth under salt stress.

Acknowledgments This research was supported by the Xinjiang Uygur Autonomous Region Regional Coordinated Innovation Project (Shanghai Cooperation Organization Science and Technology Partnership Program; No. 2017E01031). W. -J. L. was also supported by Guangdong Province Higher Vocational Colleges and Schools Pearl River Scholar-Funded Scheme (2014).

References

Acosta-Motos JR, Ortuno MF, Bernal-Vicente A, Diaz-Vivancos P, Sanchez-Blanco MJ, Hernandez JA (2017) Plant responses to salt stress: adaptive mechanisms. Agronomy 7:18. https://doi.org/10.3390/agronomy7010018

Ahmad M, Zahir ZA, Khalid M, Nazli F, Arshad M (2013) Efficacy of *Rhizobium* and *Pseudomonas* strains to improve physiology, ionic balance and quality of mung bean under salt-affected conditions on farmer's fields. Plant Phys Biochem 63:170–176

Allam NG, Kinany R, El-Refai E, Ali WY (2018) Potential use of beneficial salt tolerant bacteria for improving wheat productivity grown in salinized soil. J Microbiol Res 8:43–53

Alori ET, Glick BR, Babalola OO (2017) Microbial phosphorus solubilization and its potential for use in sustainable agriculture. Front Microbiol 8:971. https://doi.org/10.3389/fmicb.2017.00971

Ashraf M, Hasnain S, Berge O, Mahmood T (2004) Inoculating wheat seedlings with exopolysaccharide-producing bacteria restricts sodium uptake and stimulates plant growth under salt stress. Biol Fertil Soils 40:157–162

Awad NM, Turky AS, Abdelhamid MT, Attia M (2012) Ameliorate of environmental salt stress on the growth of *Zea mays* L. plants by exopolysaccharides producing bacteria. JASR 8:2033–2044

Baltruschat H, Fodor J, Harrach BD, Niemczyk E, Barna B, Gullner G, Janeczko A, Kogel KH, Schafer P, Schwarczinger I, Zuccaro A, Skoczowski A (2008) Salt tolerance of barley induced by the root endophyte *Piriformospora indica* is associated with a strong increase in antioxidants. New Phytol 180:501–510

Bano A, Fatima M (2009) Salt tolerance in *Zea mays* (L.) following inoculation with *Rhizobium* and *Pseudomonas*. Biol Fert Soils 45:405–413

Bewley JD, Black M (1994) Seeds: physiology of development and germination. Plenum Press, New York

Bianco C, Defez R (2012) Soil bacteria support and protect plants against abiotic stresses. In: Shanker A, Venkateswarlu B (eds) Abiotic stress in plants, Mechanisms and adaptations. IntechOpen, London, pp 143–170

Carillo P, Annunziata MG, Pontecorvo G, Fuggi A, Woodrow P (2011) Salinity stress and salt tolerance. In: Shanker A, Venkateswarlu B (eds) Abiotic stress in plants, Mechanisms and adaptations. IntechOpen, London, pp 21–38

Chen L, Liu Y, Wu G, Veronican Njeri K, Shen Q, Zhang N, Zhang R (2016) Induced maize salt tolerance by rhizosphere inoculation of *Bacillus amyloliquefaciens* SQR9. Physiol Plant 158:34–44

Clarke D, Williams S, Jahiruddin M, Parks K, Salehin M (2015) Projections of on-farm salinity in coastal Bangladesh. Environ Sci Process Impacts 17:1127–1136

Dardanelli MS, Fernandez de Cordoba FJ, Espuny MR, Carvajal MAR, Diaz MES, Serrano AMG, Okon Y, Megias M (2008) Effect of *Azospirillum brasilense* coinoculated with rhizobium on *Phaseolus vulgaris* flavonoids and nod factor production under salt stress. Soil Biol Biochem 40:2713–2721

Das P, Nutan KK, Singla-Pareek SL, Pareek A (2015) Understanding salinity responses and adopting 'omics-based' approaches to generate salinity tolerant cultivars of rice. Front Plant Sci 6:712. https://doi.org/10.3389/fpls.2015.00712

de Zelicourt A, Al-Yousif M, Hirt H (2013) Rhizosphere microbes as essential partners for plant stress tolerance. Mol Plant 6:242–245

Duan L, Sebastian J, Dinneny JR (2015) Salt-stress regulation of root system growth and architecture in *Arabidopsis* seedlings. Methods Mol Biol 1242:105–122

Egamberdieva D, Davranov K, Wirth S, Hashem A, Abd_Allah EF (2017) Impact of soil salinity on the plant-growth-promoting and biological control abilities of root associated bacteria. Saudi J Biol Sci 24:1601–1608

Enebe MC, Babalola OO (2018) The influence of plant growth-promoting rhizobacteria in plant tolerance to abiotic stress: a survival strategy. Appl Microbiol Biotechnol 102:7821–7835

Evelin H, Kapoor R, Giri B (2009) Arbuscular mycorrhizal fungi in alleviation of salt stress: a review. Ann Bot 104:1263–1280

Fu QL, Liu C, Ding NF, Lin YC, Guo B (2010) Ameliorative effects of inoculation with the plant growth-promoting rhizobacterium *Pseudomonas* sp. DW1 on growth of eggplant (*Solanum melongena* L.) seedlings under salt stress. Agr Water Manage 97:1994–2000

Glick BR (2014) Bacteria with ACC deaminase can promote plant growth and help to feed the world. Microbiol Res 169:30–39

Goers L, Freemont P, Polizzi KM (2014) Co-culture systems and technologies: taking synthetic biology to the next level. J R Soc Interface 1:20140065. https://doi.org/10.1098/rsif.2014.0065

Gond SK, Bergen MS, Torres MS, White JF, Kharwar RN (2015) Effect of bacterial endophyte on expression of defense genes in Indian popcorn against *Fusarium moniliforme*. Symbiobosis 66:133–140

Gonzalez AJ, Larraburu EE, Llorente BE (2015) *Azospirillum brasilense* increased salt tolerance of jojoba during in vitro rooting. Ind Crop Prod 76:41–48

Gupta B, Huang B (2014) Mechanism of salinity tolerance in plants: physiological, biochemical, and molecular characterization. Int J Genomics 701596:1. https://doi.org/10.1155/2014/701596

Hamdia MA, Shaddad MAK, Doaa MM (2004) Mechanism of salt tolerance and interactive effect of *Azospirillum bransilense* inoculation on maize cultivars grown under salt stress conditions. Plant Growth Regul 44:165–174

Han QQ, Lu XP, Bai JP, Qiao Y, Pare PW, Wang SM, Zhang JL, Wu YN, Pang XP, Xu WB, Wang ZL (2014) Beneficial soil bacterium *Bacillus subtilis* (GB03) augments salt tolerance of white clover. Front Plant Sci 5:525. https://doi.org/10.3389/fpls.2014.00525

Hanin M, Ebel C, Ngom M, Laplaze L, Masmoudi K (2016) New insights on plant salt tolerance mechanisms and their potential use for breeding. Front Plant Sci 7:1787. https://doi.org/10.3389/fpls.2016.01787

Hashem A, Abd_Allah EF, Alqarawi AA, Aldebasi A, Egamberdieva D (2015) Arbuscular mycorrhizal fungi enhance salinity tolerance of *Panicum turgidum* Forssk by altering photosynthetic and antioxidant pathways. J Plant Inter 10:230–242

Ilangumaran G, Smith DL (2017) Plant growth promoting rhizobacteria in amelioration of salinity stress: a systems biology perspective. Front Plant Sci 8:1768. https://doi.org/10.3389/fpls.2017.01768

Islam F, Yasmeen T, Arif MS, Ali S, Ali B, Hameed S, Zhou W (2016) Plant growth promoting bacteria confer salt tolerance in *Vigna radiata* by up-regulating antioxidant defense and biological soil fertility. Plant Growth Regul 80:23–36

Jha Y, Subramanian RB (2013) Paddy plants inoculated with PGPR show better growth physiology and nutrient content under saline conditions. Chil J Agric Res 73:213–219

Jha Y, Subramanian RB, Patel S (2011) Combination of endophytic and rhizospheric plant growth promoting rhizobacteria in *Oryza sativa* shows higher accumulation of osmoprotectant against saline stress. Acta Physiol Plant 33:797–802

Kader MA, Lindberg S (2010) Cytosolic calcium and pH signaling in plants under salinity stress. Plant Signal Behav 5:233–238

Kadmiri IM, Chaouqui L, Azaroual SE, Sijilmassi B, Yaakoubi K, Wahby I (2018) Phosphate-solubilizing and auxin-producing Rhizobacteria promote plant growth under saline conditions. Arab J Sci Eng 43:3403–3415

Kaleem F, Shabir G, Aslam K, Rasul S, Manzoor H, Shah SM, Khan AR (2018) An overview of the genetics of plant response to salt stress: present status and the way forward. Appl Biochem Biotechnol 186:306–334

Kasim WA, Gaafar RM, Abou-Ali RM, Omar MN, Hewait HM (2016) Effect of biofilm forming plant growth promoting rhizobacteria on salinity tolerance in barley. Ann Agric Sci 61:217–227

Kasotia A, Varma A, Choudhary DK (2015) *Pseudomonas* mediated mitigation of salt stress and growth promotion in *Glycine max*. Agric Res 4:31–41

Keshtehgar A, Rigi K, Vazirimehr MR (2013) Effects of salt stress in crop plants. Int J Agric Crop Sci 5:2863–2867

Khan MA, Ungar IA, Showalter AM (2000) Effects of sodium chloride treatments on growth and ion accumulation of the halophyte *Haloxylon recurvum*. Commun Soil Sci Plant Anal 31:2763–2774

Krishnamoorthy R, Kim K, Subramanian P, Senthilkumar M, Anandham R, Sa T (2016) Arbuscular mycorrhizal fungi and associated bacteria isolated from salt-affected soil enhances the tolerance of maize to salinity in coastal reclamation soil. Agric Ecosyst Environ 231:233–239

Kruasuwan W, Thamchaipenet A (2018) 1-Aminocyclopropane-1-carboxylate (ACC) deaminase-producing modulates salt–stress response in sugarcane. J Plant Growth Regul 37:849–858

Kumar J, Singh S, Singha M, Srivastava PK, Mishra RK, Singh VP, Prasad SM (2017) Transcriptional regulation of salinity stress in plants: a short review. Plant Gene 11:160–169

Li Y (2008) Effect of salt stress on seed germination and seedling growth of three salinity plants. Pak J Biol Sci 11:1268–1272

Liang W, Ma X, Wan P, Liu L (2018) Plant salt-tolerance mechanism: a review. Biochem Biophys Res Commun 495:286–291

Mahmood S, Daur I, Al-Solaimani SG, Ahmad S, Madkour MH, Yasir M, Hirt H, Ali S, Ali Z (2016) Plant growth promoting rhizobacteria and silicon synergistically enhance salinity tolerance of mung bean. Front Plant Sci 7:876. https://doi.org/10.3389/fpls.2016.00876

Marulanda A, Azcon R, Chaumont F, Ruiz-Lozano JM, Aroca R (2010) Regulation of plasma membrane aquaporins by inoculation with a *Bacillus megaterium* strain in maize (*Zea mays* L.) plants under unstressed and salt-stressed conditions. Planta 232:533–543

Meena KK, Sorty AM, Bitla UM, Choudhary K, Gupta P, Pareek A, Singh DP, Prabha R, Sahu PK, Gupta VK, Singh HB (2017) Abiotic stress responses and microbe mediated mitigation in plants: the omics strategies. Front Plant Sci 8:1–25. https://doi.org/10.3389/fpls.2017.00172

Munns R (1988) Effect of high external NaCl concentrations on ion transport within the shoot of *Lupinus albus*. I. Ions in xylem sap. Plant Cell Environ 11:283–289

Munns R, Rawson HM (1999) Effect of salinity on salt accumulation and reproductive development in the apical meristem of wheat and barley. Aust J Plant Physiol 26:459–464

Munns R, Schachtman DP, Condon AG (1995) The significance of a two-phase growth response to salinity in wheat and barley. Aust J Plant Physiol 22:561–569

Ngom M, Gray K, Diagne N, Oshone R, Fardoux J, Gherbi H, Hocher V, Svistoonoff S, Laplaze L, Tisa LS, Sy MO, Champion A (2016) Symbiotic performance of diverse *Frankia* strains on salt-stressed *Casuarina glauca* and *Casuarina equisetifolia* plants. Front Plant Sci 7:1331. https://doi.org/10.3389/ fpls.2016.01331

Nongpiur RC, Singla Pareek SL, Pareek A (2016) Genomics approaches for improving salinity stress tolerance in crop plants. Curr Genomics 17:343–357

Numan M, Bashir S, Khan Y, Mumtaz R, Shinwari ZK, Khan AL, Khan A, Al-Harrasi A (2018) Plant growth promoting bacteria as an alternative strategy for salt tolerance in plants: a review. Microbiol Res 209:21–32

Palaniyandi SA, Damodharan K, Yang SH, Suh JW (2014) *Streptomyces* sp. strain PGPA39 alleviates salt stress and promotes growth of 'micro tom' tomato plants. J Appl Microbiol 1117:766–773

Panuccio MR, Jacobsen SE, Akhtar SS, Muscolo A (2014) Effect of saline water on seed germination and early seedling growth of the halophyte quinoa. AoB Plants 6:plu047. https://doi.org/10.1093/aobpla/plu047

Parida AK, Das AB (2005) Salt tolerance and salinity effects on plants: a review. Ecotoxicol Environ Saf 60:324–349

Paul D, Lade H (2014) Plant-growth-promoting rhizobacteria to improve crop growth in saline soils: a review. Agron Sustain Dev 34:737–752

Perez-Lopez U, Robredo A, Lacuesta M, Sgherri C, Munoz-Rueda A, Navari-Izzo F, Mena-Petite A (2009) The oxidative stress caused by salinity in two barley cultivars is mitigated by elevated CO_2. Physiol Plant 135:29–42

Qurashi AW, Sabri AN (2012) Bacterial exopolysaccharide and biofilm formation stimulate chickpea growth and soil aggregation under salt stress. Braz J Microbiol 43:1183–1191

Sannazzaro AI, Ruiz OA, Alberto EO, Menendez AB (2006) Alleviation of salt stress in *Lotus glaber* by *Glomus intraradices*. Plant Soil 285:279–287

Saravanakumar D, Samiyappan R (2007) ACC deaminase from *Pseudomonas fluorescens* mediated saline resistance in groundnut (*Arachis hypogaea*) plants. J Appl Microbiol 102:1283–1292

Saravanavel R, Ranganathan R, Anantharaman P (2011) Effect of sodium chloride on photosynthetic pigments and photosynthetic characteristics of *Avicennia officinalis* seedlings. Recent Res Sci Technol 3:177–180

Sarkar A, Ghosh PK, Pramanik K, Mitra S, Soren T, Pandey S, Mondal MH, Maiti TK (2017) A halotolerant *Enterobacter* sp. displaying ACC deaminase activity promotes rice seedling growth under salt stress. Res Microbiol 169:20–32

Shrivastava P, Kumar R (2015) Soil salinity: a serious environmental issue and plant growth promoting bacteria as one of the tools for its alleviation. Saudi J Biol Sci 22:123–131

Siddikee M, Chauhan P, Anandham R, Han GH, Sa T (2010) Isolation, characterization, and use for plant growth promotion under salt stress, of ACC deaminase-producing halotolerant bacteria derived from coastal soil. J Microbiol Biotechnol 20:1577–1584

Singh SK, Sharma HC, Goswami AM, Datta SP, Singh SP (2000) In vitro growth and leaf composition of grapevine cultivars as affected by sodium chloride. Biol Plant 43:283–286

Soliman AS, Shanan NT, Massoud ON, Swelim DM (2012) Improving salinity tolerance of *Acacia saligna* (Labill.) plant by arbuscular mycorrhizal fungi and *Rhizobium* inoculation. Afr J Biotechnol 11:1259–1266

Srivastava S, Patel JS, Singh HB, Sinha A, Sarma BK (2015) *Streptomyces rochei* SM3 induces stress tolerance in chickpea against *Sclerotinia sclerotiorum* and NaCl. J Phytopathol 163:583–592

Tavakkoli E, Rengasamy P, McDonald GK (2010) High concentrations of Na+ and Cl– ions in soil solution have simultaneous detrimental effects on growth of faba bean under salinity stress. J Expt Bot 61:4449–4459

Vaishnav A, Kumari S, Jain S, Verma A, Tuteja N, Choudhary DK (2016) PGPR-mediated expression of salt tolerance gene in soybean through volatiles under sodium nitroprusside. J Basic Microbiol 56:1274–1288

Vejan P, Abdullah R, Khadiran T, Ismail S, Nasrulhaq Boyce A (2016) Role of plant growth promoting Rhizobacteria in agricultural sustainability-a review. Molecules 21:573. https://doi.org/10.3390/molecules21050573

Vurukonda SS, Vardharajula S, Shrivastava M, SkZ A (2016) Enhancement of drought stress tolerance in crops by plant growth promoting rhizobacteria. Microbiol Res 184:13–24

Waller F, Achatz B, Baltruschat H, Fodor J, Becker K, Fischer M, Heier T, Huckelhoven R, Neumann C, Von Wettstein D, Franken P, Kogel KH (2005) The endophytic fungus *Piriformis indica* reprograms barley to salt-stress tolerance, disease resistance, and higher yield. Proc Natl Acad Sci 102:13386–13391

Wang Y, Wang M, Li Y, Wu A, Huang J (2018) Effects of arbuscular mycorrhizal fungi on growth and nitrogen uptake of *Chrysanthemum morifolium* under salt stress. PLoS One 13(4): e0196408. https://doi.org/10.1371/journal. pone.0196408

Wu QS, Zou YN, He XH (2010) Contributions of arbuscular mycorrhizal fungi to growth, photosynthesis, root morphology and ionic balance of citrus seedlings under salt stress. Acta Physiol Plant 32:297–304

Xiong L, Zhu JK (2002) Molecular and genetic aspects of plant responses to osmotic stress. Plant Cell Environ 25:131–139

Xue ZY, Zhi DY, Xue GP, Zhang H, Zhao YX, Xia GM (2004) Enhanced salt tolerance of transgenic wheat (*Triticum aestivum* L.) expressing a vacuolar Na^+/H^+antiporter gene with improved grain yields in saline soils in the field and a reduced level of leaf Na^+. Plant Sci 167:849–859

Yan JM, Smith MD, Glick BR, Liang Y (2014) Effects of ACC deaminase containing rhizobacteria on plant growth and expression of toc GTPases in tomato (*Solanum lycopersicum*) under salt stress. Botany 92:775–781

Yao LX, Wu ZS, Zheng YY, Kaleem I, Li C (2010) Growth promotion and protection against salt stress by *Pseudomonas putida* Rs-198 on cotton. Eur J Soil Biol 46:49–54

Zhang H, Kim MS, Krishnamachari V, Payton P, Sun Y, Grimson M, Farag MA, Rayu CM, Allen R, Melo IS, Pare PW (2007) Rhizobacterial volatile emissions regulate auxin homeostasis and cell expansion in *Arabidopsis*. Planta 226:839–851

Zhang H, Kim MS, Sun Y, Dowd SE, Shi H, Pare PW (2008) Soil bacteria confer plant salt tolerance by tissue-specific regulation of the sodium transporter *HKT1*. Mol Plant-Microbe Interact 21:737–744

Zou N, Dart PJ, Marcar NE (1995) Interaction of salinity and rhizobial strain on growth and N_2-fixation by *Acacia ampliceps*. Soil Biol Biochem 27:409–413

Part IV
Strategies for Mitigation and Reclamation of Saline Soils

Chapter 19
Strategies for Reclamation of Saline Soils

Vinay Shankar and Heikham Evelin

Abstract Salinization of soil is one of the major constrains to food security. Salt deposition in the soil is a natural process; however, anthropogenic activities such as improper agricultural practices, poor drainage system, and inefficient water management have hastened the process of salinization. The increase in areas under salinity is coupled with the exponentially growing human population posing a huge challenge to mankind to ensure food security. It is therefore the need of the hour to reclaim these saline areas as well as improve the plant's ability to resist soil salinity. Many methods have been developed to desalinize the soils, such as leaching, growing halophytes, subsurface drainage, biomimicry, etc. This chapter deals with various strategies that are used for desalinization and also identifies the challenges in the application of these techniques.

Keywords Biomimicry · Desalinization · Halophytes · Leaching · Salinity

19.1 Introduction

Soil salinity is a major impediment to modern agriculture as it can adversely affect the growth, development, and yield of plants. The problem of salinity is enormous as it occurs in more than 100 countries covering approximately 1 billion hectares (ha) (FAO report 2015). Salinity problems are long-lasting and manifold—(1) it is rapidly increasing, with an estimated 0.3- 1.5 million ha of farmland succumbing to salinity per year; (2) it causes a decline in crop productivity by 20%; (3) it decreases the production capacity of another 20–46 million ha; (4) it reduces diversity of flora; and (5) it increases soil erosion (Porcel et al. 2012; FAO 2015; Swallow and O'Sullivan 2019). Then again, the exponential increase of human population adding

V. Shankar
Department of Botany, Deshbandhu College, University of Delhi, New Delhi, India

H. Evelin (✉)
Department of Botany, Rajiv Gandhi University, Itanagar, India

B. Giri, A. Varma (eds.), *Microorganisms in Saline Environments: Strategies and Functions*, Soil Biology 56, https://doi.org/10.1007/978-3-030-18975-4_19

83 million people every year exerts an added pressure on earth (UN 2018). Thus, modern agriculture is tasked with ensuring food security to sustain the mounting population from the ever-declining cultivable land.

Soluble salts are an invariable component of soils; however, certain environmental conditions accelerate accumulation of salts in the soil (Sharma and Singh 2017). Salt deposition in the soil occurs either by primary (natural) or secondary (anthropogenic) processes. Weathering of rocks, infiltration of sea water, scanty rainfall coupled with high temperature, and atmospheric deposition constitute the primary processes of salt deposition (Biswas and Biswas 2014; Hanay et al. 2014; FAO 2015). The secondary processes include inappropriate irrigation practices, poor drainage system, and preferring annual plants against perennials leading to rise in saline groundwater level (Biswas and Biswas 2014; Hanay et al. 2014; FAO 2015). The perennial plants having deep roots and higher water usage can keep saline groundwater at bay. Excess deposition of salt can impair the structure of soil by changing some of its physical and chemical properties (Qadir et al. 2000). High Na^+ concentration in the soil solution can cause flocculation of soil colloids resulting in decreased stability. As a result, saline soils suffer from poor aeration and subsequent diminished hydraulic conductivity and increased condensation (Hanay et al. 2014; Amini et al. 2015). These conditions are hostile for the establishment and growth of plants by subjecting them to osmotic stress, ionic imbalance, and oxidative stress (Evelin et al. 2009). This subsequently diminishes the yield of the plants or may even result into crop failure (Qadir et al. 2000). Reclamation of saline soil is thus the need of the hour. This chapter describes saline soils and the various strategies used to reclaim them. It also highlights challenges in remediation of these soils.

19.2 Features of Saline Soil

Saline soils are those possessing excessive amounts of soluble salts in their profile (Hanay et al. 2015; Amini et al. 2015). The primary constituents of dissolved salts include highly soluble Cl^-, NO_3^-, and SO_4^{2-} as well as low soluble salts, $CaCO_3$ and $CaSO_4$ (Richards 1954; Qadir et al. 2000; Hanay et al. 2014). The salinity of soil is expressed in terms of electrical conductivity (EC_e) of saturated soil paste, pH, and exchangeable sodium percentage (ESP). Saline soils are characterized by an $EC_e > 4$ dS m^{-1} at 25 °C, pH < 8.2, and ESP < 15. However, many fruit and ornamental plants are susceptible to salt effects even in soil EC_e range of 2–4 dS m^{-1}. Therefore, it has been recommended to lower the EC_e limit for saline soils from 4 to 2 dS m^{-1} (Sparks 2003). High salt concentration in soils causes instability in soil structure, low infiltration rate, and reduced water-holding capacity (Richards 1954). These soils have poor biodiversity, low organic matter, and stressed microbial community (Rengasamy 2002; Rietz and Haynes 2003). Plants growing in saline areas are subjected to osmotic stress, nutrient deficiencies, as well as toxicities causing ionic imbalances. These phenomena result in the development of oxidative stress (Evelin et al. 2009; Porcel et al. 2012). These effects hamper growth, development, and reproduction in plants subsequently leading to decrease in crop yield.

19.3 Reclamation of Saline Soil

Salt-induced crop loss amounts to about $27 billion per year (Gies 2017). If salinity problem is not managed, the saline areas will become non-cultivable (FAO 2017). Therefore, providing solutions to this problem is the need of the hour. To overcome this problem and enhance crop yield, saline soils may be reclaimed to favor plant growth, or the plant is modified to adjust and survive in the saline environment (Maas 1993). Saline soils can be reclaimed by minimizing salt concentration, decreasing pH value and improving soil structure (Yu et al. 2010). Soil desalinization can be brought about by harvesting aboveground parts of salt-tolerant plants or allowing leaching of salts from the upper layers of the soil below the root zone (Swallow and O'Sullivan 2019). A reclaimed soil should (1) be able to control salt concentration, (2) maintain porosity, and (3) provide a conducive environment for water transport in the soil and promote growth and development of plant roots (Qadir et al. 2000).

Reclamation of soil can be accomplished by using various strategies. For successful implementation of each method, a major point of consideration is to provide adequate soil drainage system (Qadir et al. 2000). However, the suitability of each method depends on many factors:

1. Kinds and concentrations of salts present
2. Availability of good-quality water (low electrolyte concentration) for leaching
3. Texture of subsoil
4. The level of groundwater and its quality
5. Volume of soil that should be reclaimed
6. The landscape of the salt-affected areas
7. The type of plant(s) to be cultivated after reclamation
8. The climatic conditions
9. Availability of time for reclamation
10. Cost-effectiveness (Qadir et al. 2000; Mahanta et al. 2015)

The methods to reclaim a saline soil are discussed subsequently. These methods may be implemented individually or in conjunction with other methods for a successful reclamation program.

19.3.1 Flushing

In this method water is passed over the soil surface to flush out the salts. It is possible only when the salt forms a layer of crust on the soil surface. However, it is an inefficient process and can remove very little amount of salt (Qadir et al. 2000; Abdel-Fattah 2018).

19.3.2 Scrapping

Accumulated surface salts can also be removed by scrapping off the surface. However, like flushing, it is also of less practical significance (Qadir et al. 2000; Abdel-Fattah 2018).

19.3.3 Leaching

It is a method by which salts from the upper layer of soil (top 45–60 cm) are forced to move downward, much below the root zone (Keren and Miyamoto 1990; Qadir et al. 2000). This method requires good-quality water with low electrolyte concentration that percolates through the soil profile to facilitate dissolution of salts and their downward movement to prevent buildup of excess salt in the root zone. The volume of water required to reclaim a soil is determined by the texture of soil, salts present in it, volume of soil to be reclaimed, and desired salt level in the rhizosphere as well as plants to be grown after reclamation (Qadir et al. 2000; Biswas and Biswas 2014). Leaching is preferable when the soil has low moisture content and the ground water table is deep (FAO).

Leaching can be carried out in different ways, such as continuous ponding, intermittent ponding, and sprinkling (Qadir et al. 2000). Continuous ponding is more appropriate for medium-textured soil and can achieve leaching in a short duration of time. In fact, it is estimated that upon continuous ponding with a depth of water equivalent to the depth of soil to be reclaimed, more than 70% of soluble salts present in the soil before leaching are removed (Hoffman 1980). Intermittent ponding can accomplish the same level of leaching with about one-third less water used in continuous ponding; however, it may take longer duration of time (Hoffman 1980; Qadir et al. 2000). It is more suitable for fine-textured soils, especially in fields with shallow water table or a tile drainage system. It enhances efficiency of leaching by lowering the water table (Talsma 1967). In situations, where a surface seal develops on soil surface, intermittent ponding can induce formation of cracks to allow water infiltration (Abdel-Fattah 2018). Sprinkling is an energy- and cost-intensive method suitable for fields that are not prepared for ponded leaching.

19.3.4 Subsurface Drainage

It is a process by which excess salts from the root zone are removed through leaching the salts into a collection system (Gajja et al. 2002). The salty water can be collected in deep open drains. Else, percolated pipe drains can be buried in the soil to collect leached salty water and conveyed to a collector drain. This technique not only has the ability to lower salinity by as much as 50%; it can also bring down the water table (Raju et al. 2016). Changes in cropping pattern and intensity can enhance the benefits of subsurface drainage system and improve crop yield and income of farmers (Raju et al. 2016). The prerequisites of subsurface drainage technology are availability of technically skilled manpower and development of infrastructure. However, for its successful implementation, active participation by the government as well as farmers is essential (Raju et al. 2016). This method has been proven to be technically viable, economically feasible, and socially acceptable by farmers (Chinnappa and Nagraj 2007; Tripathi 2011; Raju et al. 2016).

19.3.5 Biomimicry

It is a novel technique which attempts to mimic the capillary action of vascular plants to allow harvesting effloresced salt on the soil surface (Swallow and O'Sullivan 2019). Using this technique for 30 days, the researchers were able to lower soil EC_e as well as salt concentration by almost tenfold (from 120 mS cm^{-1} to 14 mS cm^{-1} and the salt concentration from 8% to 0.8%). Crystal inhibitors, such as ferrocyanides, are applied on soil surface to prevent crystal formation and optimize evapotranspiration (Gupta et al. 2014) while promoting formation of dendritic crystal structures (Klaustermeier et al. 2017). Dendritic crystals create capillary pump in the soil to allow rapid uphill movement of soil solution containing the dissolved salts (Sghaier and Prat 2009). On the soil surface, the salts effloresced due to evapotranspiration and can be harvested. This method is at nascent stage and yet to be tested in fields, for it to be used as an effective tool for desalinization to reduce soil salinity (Swallow and O'Sullivan 2019).

19.3.6 Phyto-desalinization

It is a technique by which plants and their parts are used to reclaim a saline soil. Excess salts from the root zone can be removed by growing salt-tolerant crops (halophytes) on the saline area or applying organic materials, such as farmyard manure, straw mulch, etc. It is a clean, efficient, cost-effective, and eco-friendly method (Devi et al. 2015). Halophytes are the flora of saline environments that can survive and produce fertile offsprings in saline soil with EC_e as high as 20 dS m^{-1}

(approximately 200 mM NaCl) (Flowers and Colmer 2008, 2015). These plants are widely dispersed across the families of angiosperms and can be annuals or perennials, dicotyledonous or monocotyledonous species, shrubs, and some trees (Rozema and Flowers 2008; Flowers and Colmer 2008, 2015). Halophytes have a wide range of morphological, biochemical, and physiological adaptation mechanisms, such as vacuolar sequestration of Na^+ ions, synthesis of osmolytes to overcome osmotic stress, as well as induction of antioxidant systems in such plants, which vary widely in their level of salt tolerance (Glenn et al. 1999; Breckle 2002; Shevyakova et al. 2003; Flowers and Colmer 2008, 2015). These plants are endowed with the ability to modulate the uptake and intrinsic concentrations of Na^+ and Cl^- while maintaining adequate cytoplasmic K^+ and Mg^{2+} concentrations for cellular functions (Flowers and Colmer 2015). Thus, these plants can exclude, excrete, or accumulate salts and are classified into three types—(1) salt excluding, (2) salt secreting, and (3) salt accumulating (Walter 1961). Salt-excluding halophytes can exclude salts in the root system through anatomical modifications. Examples include *Bruguiera gymnorrhiza*, *Ceriops candolleana*, *Kandelia candel*, and *Rhizophora mucronata*. Some halophytes, such as *Avicennia alba*, *A. marina*, *A. officinalis*, and *Aegiceras corniculatum*, can excrete salts through their foliar glands. Salt-accumulating halophytes develop succulence to overcome salt toxicity by accumulating high concentrations of salt in their cells and tissues. Examples include *Excoecaria agallocha*, *Lumnitzera racemosa*, *Pentatropis sianshoides*, *Salvadora persica*, *Sonneratia apetala*, *S. acida*, *Sesuvium portulacastrum*, and *Suaeda nudiflora*.

Halophytes have been investigated for their potential to reclaim saline soil (Zaharan et al. 1982; Qadir et al. 1996; Ravindran et al. 2007; Rabhi et al. 2009; Muchate et al. 2016). Some studies in which halophytes were successful in lowering the soil EC_e are shown in Table 19.1. Halophytes help in saline soil reclamation by acquiring Na^+ and Cl^- from the soil and accumulating them in their tissues (Manousaki and Kalogerakis 2011; Shelef et al. 2012; Walker et al. 2013). In addition, they also enhance leaching and reduce pH of the soil (Rasouli et al. 2013; Walker et al. 2013). The decrease in soil pH increases the dissolution of $CaCO_3$ which makes more Ca^{2+} available for exchange with Na^+ ions. Ravindran et al. (2007) reported that salt-accumulating halophytes, *Suaeda maritima* plants, were able to lower EC_e of saline soil from 4.9 to 1.4 dS m^{-1}, while *Sesuvium portulacastrum* brought down the soil EC_e to 2.5 dS m^{-1} after 120 days of cultivation in the saline soil. Correspondingly, plant sample EC_e increased from 4.5 to 18.8 dS m^{-1} in *S. maritima* and from 4.3 to 15.3 dS m^{-1} in *S. portulacastrum* (Ravindran et al. 2007). It was also projected that these two halophytes could remove NaCl from saline areas at the rate of 504 and 474 kg NaCl per hectare, respectively, in 120 days. Muchate et al. (2016) also reported that growing *S. portulacastrum* for 90 days could lower soil EC_e from 7.1 to 4.9 dS m^{-1}. These plants absorb NaCl from the soil and compartmentalize in their vacuoles as cheap osmolytes (Ravindran et al. 2007).

Organic amendments, such as straw, farmyard manure, etc., have been shown to promote desalinization of soil. Straw mulching has been found to enhance leaching

Table 19.1 Soil EC_e before and after plantation with halophytes in some phyto-desalinization studies

Plant	Soil EC_e (dS m^{-1}) before treatment	Soil EC_e (dS m^{-1}) after treatment	Reference
Suaeda maritime	4.9	1.4	Ravindran et al. (2007)
Sesuvium portulacastrum	4.9	2.5	
Clerodendrum inerme	4.8	2.66	
Heliotropium curassavicum	4.8	3.56	
Excoecaria agallocha	4.8	2.45	
Ipomoea pes-caprae	4.7	3.08	
Arthrocnemum indicum	19.6	10.1	Rabhi et al. (2009)
Suaeda fruticosa	19.6	12	
Sesuvium portulacastrum	19.6	9.1	
Glycine max	16.4–18	1.5–2.1	Boonsaner and Hawker (2012)
Typha angustifolia	16.4–18	2.5–2.8	
Acanthus ebracteatus	16.4–18	2.1–2.6	
Sulla carnosa	8.4	5.8	Jlassi et al. (2013)
Sesuvium portulacastrum	7.1	4.9	Muchate et al. (2016)

efficiency after rainfall (Cui et al. 2017). Straw mulch reduces soil water evaporation and limits the upward movement of soil solution to the root zone or above, increases water intake and storage, improves infiltration of water, and decreases velocity and amount of surface runoff (Cook et al. 2006; Mulumba and Lal 2008; Jordán et al. 2010; Ram et al. 2013; Gholami et al. 2014; Prosdocimi et al. 2016; Leogrande and Vitti 2018). In another study, Sadegh-Zadeh et al. (2009) reported that layered mulch consisting of three different layers developed using farmyard manure, common reed (Sadegh-Zadeh and She-Bardan 2004, 2005), was successful in lowering soil EC_e till the depth of 56 cm. Organic amendments not only lower salinity but also improve vegetation in saline areas. Tejada et al. (2006) reported that cotton gin crushed compost (CGCC) and poultry manure (PM) improved the physical, chemical, and biological properties of saline soil. They reported that after a year of treatment of saline soil with these organic wastes, there was growth of spontaneous vegetation, and after a period of 5 years, 80% of land was covered by vegetation as compared to 0.8% in control plots. These findings suggest that promotion of organic farming can enhance desalinization of saline areas. Figure 19.1 depicts the various methods of reclamation of saline soil.

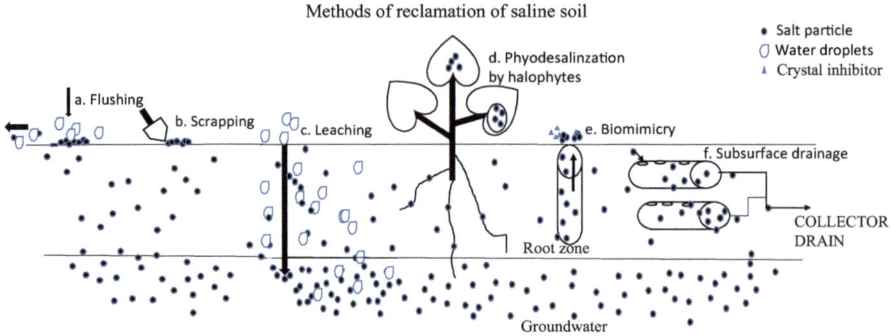

Fig. 19.1 The figure shows various methods of reclamation of saline soil. (**a, b**) Flushing and scrapping are done when salt forms a crust on the surface of the soil. In flushing, surface salts are flushed out using good-quality water, while in scrapping, surface salts are physically removed; (**c**) in leaching, good-quality water is supplied to forcefully leach the salts from root zone toward deeper level; (**d**) halophytes remove excess salts from the root zone by accumulating them in their aboveground parts; (**e**) in biomimicry the capillary action of vascular plants is mimicked to allow upward movements of salts that can crystallize on the soil surface; and (**f**) in subsurface drainage, excess salts from the root zone are collected in a collector drain through conveyor systems

19.4 Conclusions

Soil salinity problems are enormous and reclaiming these soils is the need of the hour. A major constraint to desalinization strategies is the availability of good-quality water as salinity problems are mostly encountered in arid and semiarid parts of the world where the water availability is naturally scarce. In addition, the conventional techniques are cost and labor intensive and may deter the reclamation process. For example, leaching is cost intensive and dependent on the availability of water and proper drainage system as well as it reduces the total nitrogen-total organic carbon, microbial activity, and overall soil fertility (Laudicina et al. 2009). In this regard, phyto-desalinization is a promising alternative. Halophytes offer a significant interest as they are naturally found in saline environments and have wide applicability. They can be used as food, fodder, and oil producing and pharmaceutically important plants. Halophytes have been reported to form symbiotic union with arbuscular mycorrhizal fungi (AMF) as well as plant growth-promoting rhizobacteria (PGPR) by many authors. As AMF and PGPR are well known to enhance salinity tolerance in many plants, these symbioses may be explored for reclaiming saline areas. Organic wastes, such as farmyard manure, straw, etc., are the primary components of organic farming, and they have also been shown to improve leaching during desalinization. Therefore, the possibility of organic farming in saline areas may also be explored as a reclamation strategy. Phyto-desalinization is a slow process; however, it is the most economical, eco-friendly, and sustainable method of all the reclamation techniques. Future research should be directed toward transferring these laboratory or glasshouse findings to the field conditions as well as explore the salt removal capacity of more halophytes and organic wastes.

References

Abdel-Fattah MK (2018) Reclamation of saline-sodic soils for sustainable agriculture in Egypt. In: Negm AM, Abuhashim M (eds) Sustainability of agricultural environment in Egypt: Part II - Soil-water-plant nexus. Springer, Cham. https://doi.org/10.1007/698-2018-310

Amini S, Ghadiri H, Chen C, Marschner P (2015) Salt-affected soils, reclamation, carbon dynamics, and biochar: a review. J Soils Sediments 16:939. https://doi.org/10.1007/s11368-015-1293-1

Biswas A, Biswas A (2014) Comprehensive approaches in rehabilitating salt affected soils: a review on Indian perspective. Open Trans Geosci 1(1):13–24

Breckle SW (2002) Salinity, halophytes and salt affected natural ecosystems. In: Salinity: environment - plants - molecules. Kluwer Academic, Dordrecht

Boonsaner M, Hawker DW (2012) Remediation of saline soil from shrimp farms by three different plants including soybean (*Glycine max* (L.) Merr.). J Environ Sci Health A Tox Hazard Subst Environ Eng 47(4):558–564

Chinnappa B, Nagaraj N (2007) An economic analysis of public interventions for amelioration of irrigation induced soil degradation. Agric Econ Res Rev 20:375–384

Cook HF, Valdes GS, Lee HC (2006) Mulch effects on rainfall interception, soil physical characteristics and temperature under *Zea mays* L. Soil Till Res 91:227–235

Cui S, Zhang J, Sun M, Chen H, Feng Z (2017) Leaching effectiveness of desalinization by rainfall combined with wheat straw mulching on heavy saline soil. Arch Agron Soil Sci 64:891. https://doi.org/10.1080/03650340.2017.1407871

Devi S, Nandwal AS, Angrish R, Arya SS, Kumar N, Sharma SK (2015) Phytoremediation potential of some halophytic species for soil salinity. Int J Phytoremediation 18:693. https://doi.org/10.1080/15226514.2015.1131229

Evelin H, Kapoor R, Giri B (2009) Arbuscular mycorrhizal fungi in alleviation of salt stress: a review. Ann Bot 104:1263–1280

Flowers TJ, Colmer TD (2008) Salinity tolerance in halophytes. New Phytol 179(4):945–963

Flowers TJ, Colmer TD (2015) Plant salt tolerance: adaptations in halophytes. Ann Bot 115:327–331

Food and Agriculture Organization of the United Nations (2015) Status of the world's soil resources, main report. http://www.fao.org/3/a-i5199e.pdf

Food and Agriculture Organization of the United Nations (2017) Report of the 25th session of the committee on agriculture, Rome, 21–30 Sept 2016. http://www.fao.org/3/a-mr949e.pdf

Gajja BL, Tiwari JC, Prasad R (2002) Management of natural resources in sustainable surface, 12th edn. ISCO conference, Beijing

Gholami L, Banasik K, Sadeghi SH, Khaledi Darvishan A, Hejduk L (2014) Effectiveness of straw mulch on infiltration, splash erosion, runoff and sediment in laboratory conditions. J Land Water Dev 22:51–60

Gies E (2017) Soil salinization threatens California farms. Food and Farm Discussion Lab. http://fafdl.org/blog/2017/06/29soil-salinity-threatens-california-farms/

Glenn EP, Brown JJ, Blumwald E (1999) Salt tolerance and crop potential of halophytes. Crit Rev Plant Sci 18:227

Gupta S, Huinink HP, Pel L, Kopinga K (2014) How ferrocyanide influences NaCl crystallization under different humidity conditions. Cryst Growth Des 14(4):1591–1599

Hanay A, Büyüksönmez F, Kiziloglu FM, Canbolat MY (2014) Reclamation of saline-sodic soils with gypsum and MSW compost. Compost Sci Util 12(2):175–179

Hoffman GJ (1980) Guidelines for the reclamation of salt affected soils. In: Proceedings of inter-American conference on salinity and water management technology, vol. 198, Juarez, 11–12 Dec, pp 49–64

Jlassi A, Zorrig W, El K a, Lakhdar A, Smaoui A, Abdelly C, Rabhi M (2013) Phytodesalinization of a moderately salt-affected soil by *Sulla carnosa*. Int J Phytoremediation 15(4):398–404

Jordán A, Zavala LM, Gill J (2010) Effects of mulching on soil physical properties and runoff under semi-arid conditions in southern Spain. Catena 81:77–85

Keren R, Miyamoto S (1990) Reclamation of saline sodic and boron-affected soils. In: Tanji KK (ed) Agricultural salinity assessment and management. American Society of Civil Engineers, New York

Klaustermeier AW, Daigh AL, Limb RF, Sedivec K (2017) Crystallization inhibitors and their remediation potential on brine-contaminated soils. Vadose Zone J 16(4)

Laudicina V, Hurtado M, Badalucco L, Delgado A, Palazzolo E, Panno M (2009) Soil chemical and biochemical properties of a salt-marsh alluvial Spanish area after long-term reclamation. Biol Fertil Soils 45:691–700

Leogrande R, Vitti C (2018) Use of organic-amendments to reclaim saline and sodic soils: a review. Arid Land Res Manag 33:1. https://doi.org/10.1080/15324982.2018.1498038

Maas EV (1993) Salinity and citriculture. Tree Physiol 12(2):195–216

Mahanta KK, Kansal ML, Mishra GC (2015) Managing salt affected soils: issues and strategies. Sky J Soil Sci Environ Manag 4(1):1–9

Manousaki E, Kalogerakis N (2011) Halophytes—an emerging trend in phytoremediation. Int J Phytoremediation 13:959–969

Muchate NS, Nikalje GC, Rajurkar NS, Suprasanna P, Nikam TD (2016) Physiological responses of the halophyte *Sesuvium portulacastrum* to salt stress and their relevance for saline soil bio-reclamation. Flora 224:96–105

Mulumba LN, Lal R (2008) Mulching effects on selected soil physical properties. Soil Till Res 98:106–111

Porcel R, Aroca R, Ruiz-Lozano JM (2012) Salinity stress alleviation using arbuscular mycorrhizal fungi. A review. Agron Sustain Dev 32:181–200

Prosdocimi M, Jordán A, Tarolli P, Keesstra S, CerdáA NA (2016) The immediate effectiveness of barley straw mulch in reducing soil erodibility and surface runoff generation in Mediterranean vineyards. Sci Total Environ 547:323–333

Qadir M, Qureshi RH, Ahmad N, Ilyas M (1996) Salt tolerant forage cultivation on a saline sodic field for biomass production and soil reclamation. Land Degrad Dev 7:11–18

Qadir M, Ghafoor A, Murtaza G (2000) Amelioration strategies for saline soils: a review. Land Degrad Devel 11:501–521

Rabhi M, Hafsi C, Lakhdar A, Hajji S, Barhoumi Z, Hamrouni MH, Abdelly C, Smaoui A (2009) Evaluation of the capacity of three halophytes to desalinize their rhizosphere as grown on saline soils under non leaching conditions. Afr J Ecol 47:463–468

Raju R, Thimmappa K, Kumar P, Kumar S, Tripathi RS (2016) Reclamation of saline soils through subsurface drainage technology in Haryana – an economic impact analysis. J Soil Salin Water Qual 8(2):194–201

Ram H, Dadhwal V, Vashist KK, Kaur H (2013) Grain yield and water use efficiency of wheat (*Triticum aestivum* L.) in relation to irrigation levels and rice straw mulching in North West India. Agric Water Manage 128:92–101

Rasouli F, Kiani Pouya A, Karimian N (2013) Wheat yield and physico-chemical properties of a sodic soil from semi-arid area of Iran as affected by applied gypsum. Geoderma 193–194:246–255

Ravindran KC, Venkatesan K, Balakrishnan V, Chellappan KP, Balasubramaniam T (2007) Restoration of saline land by halophytes for Indian soils. Soil Biol Biochem 39:2661–2664

Rengasamy P (2002) Transient salinity and subsoil constraints to dryland farming in Australian sodic soils: an overview. Aust J Exp Agric 42(3):351–361

Richards LA (1954) Diagnosis and improvement of saline and alkali soils. Soil Sci 78(2):154

Rietz DN, Haynes RJ (2003) effects of irrigation-induced salinity and sodicity on soil microbial activity. Soil Biol Biochem 35:845–854

Rozema J, Flowers T (2008) Ecology. Crops for a salinized world. Science 322(5907):1478–1480

Sadegh-Zadeh F, Seh-Bardan BJ (2004) An investigation effects of mulch with various layers (layered mulch) on soil temperature, pp 84–86. In: First scientific conference of plant residues management. Ministry of Agriculture, Tehran, Iran

Sadegh-Zadeh F, Seh-Bardan BJ (2005) Decreasing of evaporation by means of layered mulch, pp 39–41. In: Ninth iranian soil science conference. Tehran, Iran
Sadegh-Zadeh F, She-Bardan BJ, Samsuri AW, Mohammadi A, Chorom M, Yazdani GA (2009) Saline soil reclamation by means of layered mulch. Arid Land Res Manage 23:127–136
Sghaier N, Prat M (2009) Effect of efflorescence formation on drying kinetics of porous media. Transp Porous Media 80(3):441
Sharma DK, Singh A (2017) Reclamation and management strategies under salty soils. In: 5th National seminar on climate resilient saline agriculture: sustaining livelihood security
Shelef O, Gross A, Rachmilevitch S (2012) The use of *Bassia indica* for salt phytoremediation in constructed wetlands. Water Res 46:3967–1259
Shevyakova NI, Netronina IA, Aronova EE, Kuznetsov VV (2003) Compartmentation of cadmium and iron in *Mesembryanthemum crystallinum* plants during the adaptation to cadmium stress. Russ J Plant Physiol 50:678
Sparks D (2003) Environmental soil chemistry, 2nd edn. Academic Press, Cambridge, MA
Swallow MJB, O'Sullivan G (2019) Biomimicry of vascular plants as a means of saline soil remediation. Sci Total Environ 655:84–91
Talsma T (1967) Leaching of tile-drained saline soils. Aust J Soil Res 5:37–46
Tejada M, Garcia C, Gonzalez JL, Hernandezb MT (2006) Use of organic amendment as a strategy for saline soil remediation: influence on the physical, chemical and biological properties of soil. Soil Biol Biochem 38(6):1413–1421
Tripathi RS (2011) Socio-economic impact of reclaiming salt affected lands in India. J Soil Salinity Water Qual 3(2):110–126
UN (2018). https://www.un.org/en/development/desa/population/publications/trends/population-prospects.asp
Walker DJ, Lutts S, Sánchez-García M, Correal E (2013) *Atriplex halimus* L.: its biology and uses. J Arid Environ 100–101:111–121. https://doi.org/10.1016/j.jaridenv.2013.09.004
Walter H (1961) Salinity problems in the acid zones. The adaptations of plants to saline soils. Arid Zone Res 14:65–68
Yu J, Chen X, Sun Z, Xie W, Mao P, Wu C, Dong H, Mu X, Li Y, Guan B (2010) The spatial distribution characteristics of soil nutrients in new-born coastal wetland in Yellow River delta. Acta Sci Circum 30(4):855–861
Zaharan MA, Amal A, Wahid A (1982) Halophytes and human welfare. In: Sen DM, Rajpurohit KS (eds) Contributions to the ecology of halophytes, Tasks of vegetation science, vol 2. D.W. Junk Publishers, The Hauge

Chapter 20
Biochar for Reclamation of Saline Soils

Sharma Mona, Rachna Bhateria, Bansal Deepak, Bala Kiran, and Rani Nisha

Abstract Salinity is one of the important environmental factors, which influence fertility and quality of the soil. In semiarid and arid regions of the world, salinity is the main constraint that is affecting productivity and yield of the crop plants. Many crop plants are very sensitive to high concentration of salinity. A recent concept of biochar application to soil is enhancing various physicochemical properties under acidic and saline conditions. Biochar is a carbonaceous compound and can be utilized as a better soil ameliorant in comparison to other organic amendments for improving the fertility of soil for an elongated time period. The present review chapter is based on the results of many research papers, showing that the application of biochar to soil is potentially suitable for highly weathered soils as it enhances physicochemical properties and reduces soil loss. Although the information on biochar for enhancing the soil chemical properties is well-known, information on physical properties' enhancement is scarce. Therefore, the main aim of the current chapter is to provide all the available literature and information on the role of biochar for the reclamation of soil by enhancing their physicochemical properties under saline conditions. Further, the present review examines and deliberates the investigations on the effect of biochar in reclamation and enhancing the properties of salt-affected soils.

S. Mona (✉)
Department of Environmental Science & Engineering, Guru Jambheshwar University of Science & Technology, Hisar, Haryana, India

R. Bhateria
Department of Environmental Sciences, Maharshi Dayanand University, Rohtak, Haryana, India

B. Deepak
JBM Group, Gurugram, Haryana, India

B. Kiran
Discipline of Biosciences & Biomedical Engineering, Indian Institute of Technology, Indore, India

R. Nisha
Department of Biosciences, Himachal Pradesh University, Shimla, India

B. Giri, A. Varma (eds.), *Microorganisms in Saline Environments: Strategies and Functions*, Soil Biology 56, https://doi.org/10.1007/978-3-030-18975-4_20

Keywords Biochar · Soil physical properties · Chemical properties · Organic carbon · Amelioration

20.1 Introduction

In recent years, growing problem of soil salinity is becoming a serious concern because it affects soil fertility, which is an important factor for crop yield and production (Szabolcs 1992; Ghassemi et al. 1995; Barrow 2011; Abbas et al. 2018). There are a number of reasons which are affecting land fertility and causing their degradation due to increasing level of salinity, poor water holding capacity, pH, etc. A very old practice of utilization of agricultural waste is transforming the waste into a possible good material, which can act as an enhancer for improving the soil properties. Recently, climate change and sustainable management of soil are emerging as two major global issues, and therefore, the application of biochar to soil could be an important tool for resolving these problems (Chan et al. 2007; Abbas et al. 2018). Degradation of soil resulting to intensive agricultural consumption and changing climatic conditions jeopardizes global security of food, necessitating the development of new and environment-benign technologies that can improve soil health and resilience (Gisladottir and Stocking 2005; She et al. 2018). A knowledgeable emergent group illustrates that application of biochar efficiently improves various chemical, physical, and biological properties of saline soils (Saifullah et al. 2018; Abbas et al. 2018).

Biochar is a carbonaceous compound or char produced from thermal decomposition (pyrolysis) of organic content of plants and animal residues (manure, crop residues, algae, wood, leaves, etc.) in a closed container under low-oxygen and high-temperature conditions (Jien and Wang 2013; Ali et al. 2017). This procedure generates a highly porous, finely grained char that helps soil to maintain nutrients and water and known as biochar. The demand of biochar has been increasing for its unique role in the compartmentalization of soil through various activities like increasing water holding capacity, ion exchange capacity, and soil quality amendment. Biochar is a stable carbon-rich solid, which can sustain for a longer period of time in soil or even for thousands of years. Therefore, this approach is also used for carbon sequestration, and thus it has potential to mitigate the problem of greenhouse which ultimately leads to global warming (Woolf et al. 2010; Yousaf et al. 2016). The biochar is potentially an important tool for reclamation of saline soils by improving food security, binding of carbon for a longer period, increasing biodiversity of soil, and also decreasing the rate of deforestation in areas where extremely depleted soils with low organic resource and inadequate supply of water with chemical fertilizers are available (Lu et al. 2015; Kim et al. 2016). Biochar is having a number of unique properties and acts as an enhancer of physical and chemical properties of soil (Van Zwieten et al. 2010; Deenik et al. 2011; She et al. 2018). Biochar also enhances quality and quantity of nutrients through retention and agro-based chemicals in soil for crop use. More nutrients get retained in the soil instead leaching into groundwater and spread pollution. Biochar has many vital characteristics like the ability for:

1. CO_2 sequestration
2. Enhancing physicochemical properties of soil

Biochar can be utilized as a better ameliorant for soil as compared to other organic amendments for improving the physicochemical properties or fertility of the soil in sustainable manner. Biochar is also recommended as a potential biosorbent having unique characteristics owing its highly porous surface and carbon content. Hence, addition of small amount of biochar to soil could result in higher capacity of adsorption and, simultaneously, decrease the contaminants' availability to communities of microbes, earthworms, plants, and other organisms available in the soil. Based on the results obtained by many research papers, biochar application is considered as most suitable for weathered soil as it improves the chemical and physical properties of soil efficiently and reduces the soil loss. Although information on biochar for enhancing the soil chemical and physical properties is well-known, information on soil property enhancement under saline conditions is limited.

Hence, the major aim of this review chapter is to make available all the information on the role of biochar for soil amelioration by enhancing physicochemical properties under saline conditions. Moreover, some key features of environmental and agricultural applications of biochar are also discussed.

20.2 Production of Biochar

Different feedstocks like algae, wood, leaves, sticks, crop residues, and twigs can be used to produce biochar for enhancing the fertility of soil. Biochar can be produced by feeding residue into the reactors for the process of pyrolysis under limited supply of oxygen. Feedstocks used for the production of biochar through pyrolysis are playing a very important role in determining the quality and quantity of the biochar. A feedstock affects various biochar properties, and these properties were studied by Indren and Cheruvu (2017) as well.

20.2.1 Pyrolysis Reactor: Laboratory Scale

The design and construction of pyrolysis reactor is very important for the production of various biochar. A variety of reactors have been utilized for biochar synthesis. Essential parts of typical reactors are shown in a schematic diagram (Fig. 20.1) for the batch studies. The different parts of a well-designed and mechanically constructed reactor have major ten parts which are shown in Fig. 20.1. This reactor is used to produce the biochar from various biomaterials, and further this biochar is

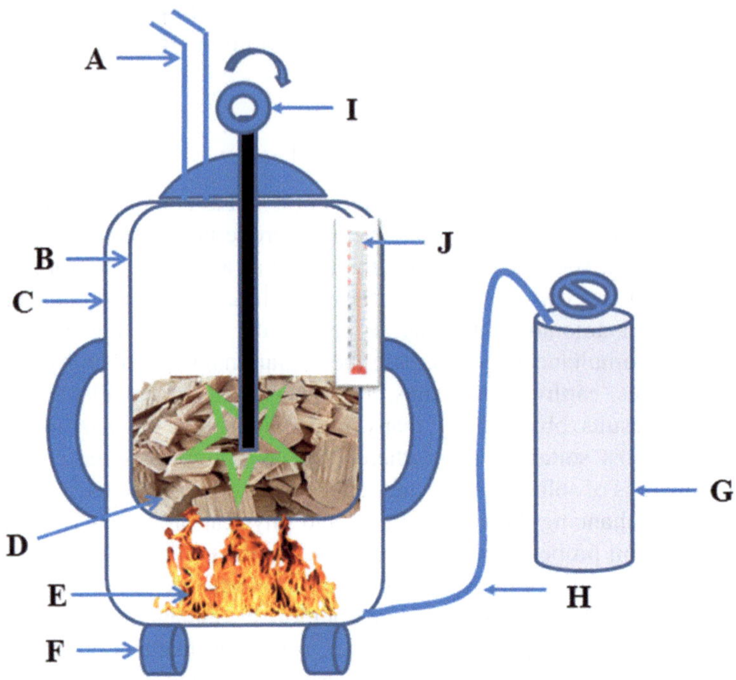

Fig. 20.1 Schematic design of a biochar producing pyrolysis reactor from waste biomass. Description: (a) chimney, (b) inner stainless steel chamber, (c) outer stainless steel chamber, (d) wood feedstock, (e) fire flame, (f) wheels, (g) gas cylinder, (h) gas tube, (i) mixing arm (j) temperature probe

used for their characterization also. The capacity of this system is varying and able to pyrolyze many kilograms of biomass (algae, wood, wheat bran, rice husk, and many others) at every run. Generally, the reactor's main vessels (horizontal and vertical tube) are made up of stainless steel, which can tolerate the temperatures between 300 and 1000 °C (Xie et al. 2015), and time for production is also varying from 50 minutes to 4 hours, by using a gas tank for providing fire feed. Temperature of the container is monitored time to time by using a temperature-sensing device made up of iron. This iron device is inserted into the open chamber of the container for measuring the accurate temperature of the system. The gas vapors generated during pyrolysis were collected and evacuated through a pipe from the reactor. After each running cycle, the left charred biomass remains inside the container until it obtained its ambient temperature to avoid further oxidation with oxygen (present in air). Initially, smoke is released from the process which indicates the content of moisture evaporation from the biomass, and after some time when the process of pyrolysis starts, the smoke is converted and becomes clear, and then the air vents are fastened as the heat from the feedstock was used to endure the process through to finishing point. The only method which is available for controlling the combustion rate for generation of heat against pyrolysis is closing and opening of air holes for the

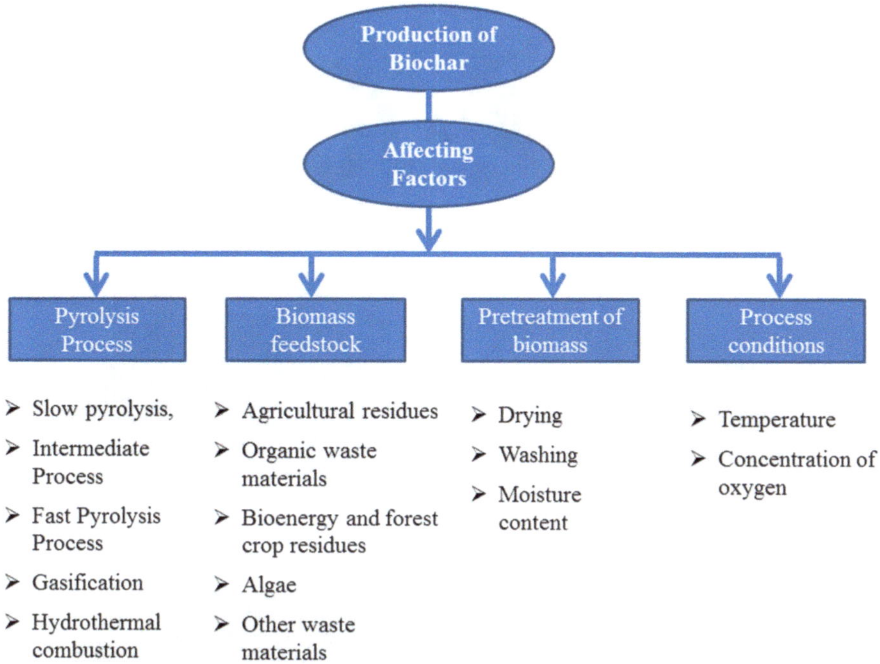

Fig. 20.2 Schematic diagram showing the list of factors affecting the production of biochar (Adopted from Nartey and Zhao 2014)

production of biochar. The complete process of pyrolysis is managed by observing the type of smoke discharging from the chimney.

There are a number of factors which do affect the production of biochar (Nartey and Zhao 2014). A list of factors is summarized in Fig. 20.2. The major factors which influence the quality and quantity of the biochar productions are as follows:

1. Pyrolysis process (types of pyrolysis like slow, intermediate, fast, hydrothermal, etc.)
2. Oxygen concentration (low or no concentration of oxygen)
3. Feedstock of biomass (like agricultural wastes, organic waste materials, algae, etc.)
4. Pretreatment of biomass (drying, washing, etc.)
5. Process conditions (like temperature, oxygen concentration, etc.)

Biochar is a steadied, stabilized, and carbon-containing compound, produced through heating of biomass under low or preferably no oxygen (Xie et al. 2015). Biochar can also be produced from a variety of biomaterial feedstock like algae, agricultural residues, wheat bran, rice husk, wood chips, organic waste materials, etc. by means of different methods of thermal combustions (Fig. 20.3). Process conditions like temperature and oxygen concentration are also important factors in

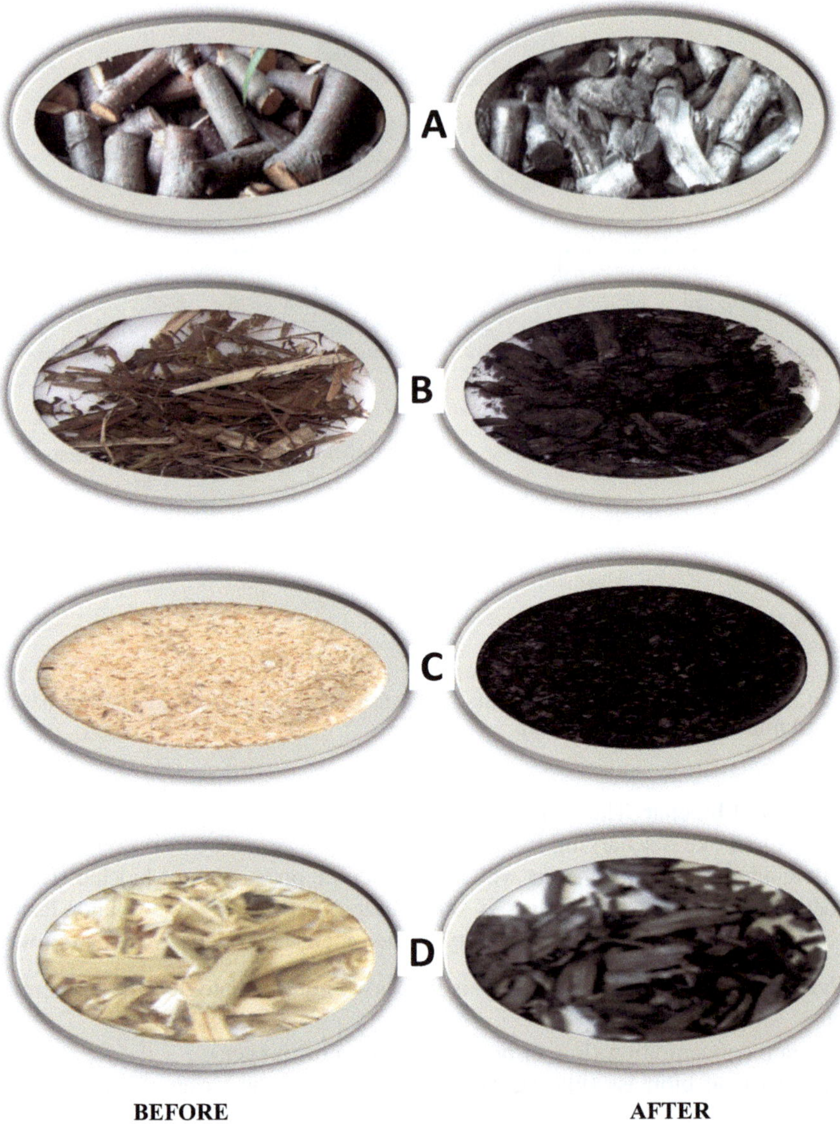

Fig. 20.3 Photographs of feedstock before and after the production of biochar. (**a**) Wood chips, (**b**) wood residues, (**c**) husk, (**d**) leaves

controlling or determining the quality and quantity of biochar. Biochar has highly movable and customizable surface chemistry, and hence, it suggests that biochar has a great potential in handling the various emerging environmental applications (Fig. 20.4).

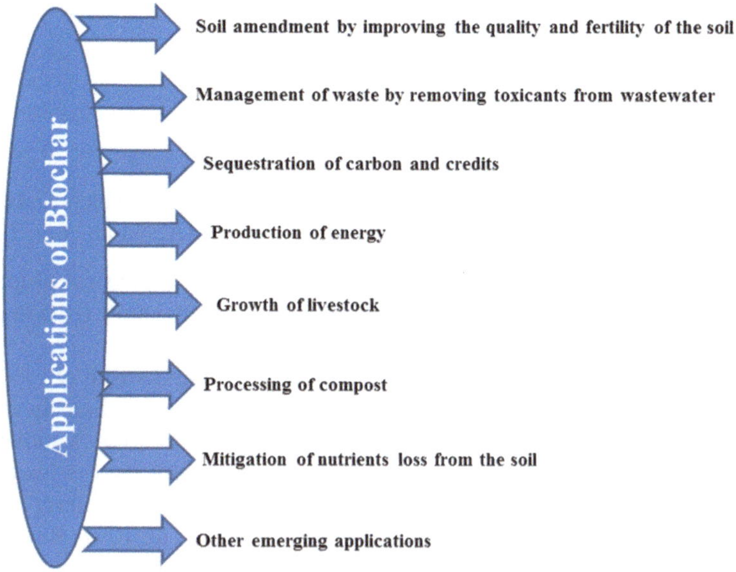

Fig. 20.4 Flow diagram showing various important applications of biochar produced from various kind of feedstock

20.3 Effects of Biochar on Soil Properties

Utilization of biochar as an amendment of soil is provoked with the challenge that it essentially benefits health of the soil. It cannot be separated from soils after their addition. Still very little literature is available which is mostly based on short-term batch studies. Biochar application to soil has been described to have many positive impacts on soil properties benefiting the biology of the soil like enhancing fixation of N_2, controlling soilborne pathogens, enhancing physicochemical properties of the soil, decreasing nitrate, leaching, and nitrous oxide emission for the remediation of contaminated soils (Carter et al. 2013). Biochar also has the potential to reduce the amount of carbon emission into the atmosphere, and it is eco-friendly in nature. A number of activation methods for biochar preparation also enhance chemical and physical properties of the soil (Verheijen et al. 2009; Sohi 2010; Haefele et al. 2011; Jeffery et al. 2011).

Physicochemical conditions of the soil have direct and indirect effects on the productivity of the soil for the production of crop by checking aeration, H_2O holding capacity, ion exchange capacity, and increasing soil strength for root activity (Benjamin et al. 2003; Verheijen et al. 2009; Sohi 2010; Haefele et al. 2011; Jeffery et al. 2011). A good soil has many qualities like:

1. Highly porous structure
2. Good bulk density
3. Increased water retention
4. More conductivity
5. High root strength
6. Better medium for the growth of beneficial microfauna
7. Easy movement of sap (water + minerals) into the soil profile which leads to more root growth resulting in high production of crop yield (Abdallah et al. 1998)

Another important factor is organic matter, which affects physical properties of soil by improving structure of the soil through soil aggregation (Bowman et al. 1990).

But, due to rapid growth in human population and rising demand of food, fibers, and other fresh materials to fulfill the needs of human beings, there is an urgent requirement to support agriculture for the food supply security (Middleton and Thomas 1992; Reynolds and Stafford 2002). Moreover, this excess load of food supply on agriculture results in degradation of soil due to anthropogenic or man-made activities (Middleton and Thomas 1992; Díaz et al. 1997; Reynolds and Stafford 2002). Hence, for improving the soil quality and fertility, there is a need to enhance the soil's physical properties through addition of organic carbon content.

Various important properties of biochar like having highly porous structure are ultimately helpful in increasing the water retention or holding capacity of soil. It also alters distribution of soil pore size, water holding, flow pathways, and patterns of percolation (Major et al. 2009). A detailed literature review is discussed here which shows biochar effects on the physicochemical properties of the soil under saline conditions.

20.3.1 Effect of Biochar on Chemical Properties of Soil

20.3.1.1 Effect of Biochar on pH, Electrical Conductivity (EC), and Cation Exchange Capacity (CEC)

Application of biochar on the various chemical properties of saline and acidic soils such as pH of the soil, CEC, and electrical conductivity (EC) were determined by many researchers which is helpful in determining the liming possible factor of biochar application (Lehmann et al. 2011; Zhang et al. 2012, 2013; Novak et al. 2014; Xu et al. 2014; Edenborn et al. 2015). Biochar application on soil significantly affects electrical conductivity, cation exchange capacity (CEC), and pH (Agbna et al. 2017). Biochar has the ability to alter chemical properties like pH, electrical conductivity, cation exchange capacity (CEC), and organic carbon content (Chan et al. 2007; Yuan and Xu 2012; Abbasi and Anwar 2015). However, the effect of biochar is also dependent upon many factors like feedstock used, type of pyrolysis, temperature, rates of application, and types of soil. The potential of biochar in enhancing the pH of the acidic soil has been advocated by a number of studies (Lehmann et al. 2011; Zhang et al. 2012, 2013; Novak et al. 2014; Xu et al. 2014; Edenborn et al. 2015); hence researchers have recommended amelioration effects of biochar application to acidic soils (Briggs et al. 2005; Dugan et al. 2010; Brockhoff et al. 2010; Laird et al. 2010a, b; Yuan and Xu 2011; Karhu et al. 2011; Usman et al. 2015).

However, a few reports showed no effect of biochar application in alkaline soils (Zhang et al. 2012, 2015; Usman et al. 2016). One of the reasons behind this is buffering capacity of alkaline soils that hinder the biochar's alkaline effects. El-Mahrouky et al. (2015) reported that pH of calcareous sandy soil increased after application of biochar at high rate of 50 g kg^{-1}. The results of Yuan and Xu's (2012) study established the effect of biochar produced by using ten different types of crop residues prepared by pyrolysis (under 350 °C temperature) into acidic soils. Biochar produced from ten residues enhancing the pH and CEC of soil and also decreased soil exchangeable acidity and resultant decreasing the overall acidity of the soil (Yuan and Xu 2012). Similar findings also showed that application of biochar produced through combustion of residual crops applies to highly acidic soils, which significantly increased the pH and cation exchange capacity of the soil (Yuan and Xu 2011; Wang et al. 2014).

Moreover, Liu et al. (2012) also investigated the biochar effect on pH of alkaline soils and found that there is no pH change in the alkaline soil after adding biochar or rather pH decreases with high concentration of biochar. Biochar application on alkaline soils has been examined by many scientists and showed that very less or almost no effect of biochar occurred on the alkaline and saline soil (Zhang et al. 2012, 2015; El-Mahrouky et al. 2015; Usman et al. 2016), but its effect depends upon the factors like feedstock, pyrolysis, rate of application, and temperature. Agbna et al. (2017) demonstrated that the biochar applications reduced pH and electrical conductivity of the soil under saline condition. This report also showed decreased soil pH (27%) after treatments of biochar into soil. The electrical conductivity of soil indicated that increase in the rate of biochar application could decrease EC (Shamim et al. 2015).

20.3.1.2 Effect of Biochar on Organic Carbon Content

Soil amendment with biochar positively alters the content of soil organic carbon (Agbna et al. 2017). Organic carbon content of soil increased with varying concentration of biochar application in soils (Jeffery et al. 2011). Many researchers proved that the organic carbon of the soil increased significantly with the application of biochar (Briggs et al. 2005; Verheijen et al. 2009; Brockhoff et al. 2010; Dugan et al. 2010; Laird et al. 2010a, b; Karhu et al. 2011; Konboon et al. 2011), and it is often recommended as an important factor in describing improved yield of crops (Sohi et al. 2009).

20.3.2 Biochar Effects on Physical Properties of the Soil

The addition of biochar in soils can improve production and yield of crop by enhancing root growth, water holding capacity, porosity, soluble nutrients, surface area, distribution in pore size, texture and structure of soil, decreased uptake of Na^+, and increased uptake of K^+ in plants facing salt stress (Sohi et al. 2009; Downie et al. 2009; Ali et al. 2017). All these biochar-mediated positive changes ultimately influence conductance of stomata, phytohormones, mineral accumulation, growth and production of the plants as the root depth, and the availability of water and air in the root zone (Ali et al. 2017). However, the abovementioned characteristics of the soil are significant factors, which determined the good or bad physical condition of the soil (Downie et al. 2009). A few other effects of biochar on the physical properties of soil could be as follows.

20.3.2.1 Effect of Biochar on Soil Porosity

Biochar can be implemented as a tool for enhancing the various properties of the soil for an extended period of time as compared to other amendments (organic in nature). Biochar can act as a good ameliorator and increase the porosity. After reviewing literature, it can be concluded that application of biochar can decrease the bulk density and increase the porosity of the soil which leads to increase infiltration rate through increasing porosity of soil. Furthermore, biochar may also contribute in enhancing the capacity of H_2O retention and availability to plants through its internal porosity (Uzoma et al. 2011; Bruun et al. 2014; Hansen et al. 2016). Several researches have reported their findings on the biochar effects on plant growth (Sohi et al. 2010; Waters et al. 2011).

20.3.2.2 Biochar Effect on Bulk Density, Water Droplet Penetration Time, and Water Holding Capacity

Bulk density of the soil is an important constraint required for the enhanced crop yield. Initially, the bulk density of the soil (saline and acidic) increased with the application of biochar, which significantly decreased with the higher concentration of the biochar during each growing period (Agbna et al. 2017; Saifullah et al. 2018). This may be due to the biochar porosity, which is very high, but when applied with the soil, it declines bulk density significantly by enhancing the volume of the pores (Lehmann et al. 2011). Moreover, this change in the bulk density and of the soil (decrease) by the addition of biochar can positively influence growth and development of the root system of crop (Atkinson et al. 2010; Laird et al. 2010a, b).

The amount of water held and retained by the soil is known as the water holding capacity. The water holding capacity is an important factor with respect to plant growth. If a soil has high water holding capacity, then it decreases the frequency of irrigation for the growth of plants. Many reports are available which showing the application of biochar boosts water retention up to 97% and the saturated water contents are around 56% and 15% (Laird et al. 2010a, b; Uzoma et al. 2011; Herath et al. 2013).

A report also found that treatment of soil with addition of biochar can improve soil field capacity as compared to non-biochar soil results (Agbna et al. 2017; Saifullah et al. 2018). The permanent wilting point value of crops is also decreased significantly with the application of biochar in the soil (Agbna et al. 2017). This may happen due to the biochar addition, which can increase the pore size of soil and then enhance moisture of the soil (Novak et al. 2009; Agbna et al. 2017; Saifullah et al. 2018).

Biochar showed increase in water repellency of soil with the addition of biochar (5%), but water repellency is decreased with increased concentration of biochar through decreasing hydrophobicity of the soil (Doerr et al. 2009; Agbna et al. 2017). This may be due to soil particles being coated by hydrophobic substances. Soils having coarse texture are extra prone to hydrophobicity due to lower surface area (Scott 2000). Biochar produced through low temperature is commonly hydrophobic in nature (Major et al. 2009; Sohi et al. 2009); hence, it may be assumed that hydrophobicity of the biochar would increase. Biochar addition also increases water release measurements in the soil by maintaining soil water content of a matric potential (Gaskin et al. 2007; Agbna et al. 2017). Another report also showed a significant improvement of water holding after the addition of biochar, but this was the result of only one treatment (biochar application) out of six (Gaskin et al. 2007).

20.3.2.3 Effect of Biochar on Soil Aggregation

Application of biochar also encouraged the development of macroaggregates in the sandy loam soil during the period of incubation. Soil aggregation is a very important property for the good structure of soil because a well-aggregated soil has better structure and provides good medium for the movement of water and nutrient (sap) for plant uptake (Borselli et al. 1996). Another study discussed that some microbes secrete polysaccharides which could improve the adherence in colloidal particles of soil (Dorioz et al. 1993). Biochar also provides good microfauna and also prevents microorganism from predators and desiccation (Dorioz et al. 1993).

The macroaggregates get changed with the application of biochar. Initially, macroaggregates increased and then reached their peak, and then at later stage, they decreased. Biochar addition increased the formation and stabilization of the soil macroaggregates, typically in the sandy loam soil. The stabilization and formation of aggregates are affected by many factors like the type of organic materials, quality and concentration of residue used, and amount of organic materials, which include the synthesis of microfauna (Bossuyt et al. 2001; Six et al. 2004).

However, few studies also demonstrated that the sodicity and salinity of the soil increase with increasing concentration of biochar (Saifullah et al. 2018). Moreover, cost of biochar production and high rates of its application are also a major challenge for its widespread application in the regions, which are affected by salinity and sodicity (Saifullah et al. 2018). Furthermore, very little information is available on the consistent performance of biochar on soils affected by excessive concentration of soluble salts (Saifullah et al. 2018).

20.4 Conclusion and Future Perspectives

The production and yield of crops are poorly decreased by salt stress. After the available research articles suggest that biochar application could reclaim and enhance the quality and fertility of soil through improving their physical and chemical properties under different conditions (saline and acidic). Moreover, biochar application also enhances efficiency for utilizing irrigation water and hence could be an efficient and effective option to improve plant growth under saline conditions. The quality of the biochar is also important for improving the fertility of the soil. Biochar not only improves biological and chemical properties of the soil but also increases soil cation exchange capacity, pH, bulk density, and activity of microbes. So, it is pertinent to state that biochar application efficiently enhances soil health of highly weathered soil and also reduces loss of soil. These factors showed that the biochar, as an ameliorator, may be applied to reclaim saline soils and help crop production in arid and semiarid regions. However, it is too early to declare application of biochar is not feasible for the soil affected by salts. Indeed, biochar can be a successful organic amendment having potential to improve soil fertility and quality and also decline the salinity level of the soil.

This review chapter tried to contribute for better understanding of the biochar-mediated changes in the soil which enhance their tolerance by modifying various mechanisms in plants under saline conditions. The following experiments can be performed further to improve the understanding of biochar application to saline soils:

1. Effects of different kinds of biochar with additional plant species under saline and different environmental conditions.
2. More intense and depth research studies are required to determine the biochar role under saline conditions.
3. The studies also showed that the application of biochar with microbes could additionally improve the tolerance of plant against various stresses.
4. Pilot and large-scale studies are also required for the generation and production of new models for the improved recommendation of biochar dose according to the soil condition, types of crop, and various environmental conditions.
5. One of the most important is to investigate the economic feasibility of various types of feedstock used for the production of biochar.

Thus, many more detailed and focused time-course experiments and research are required to explore the properties and mechanisms of various biochar-mediated plant production under saline conditions.

References

Abbas T, Rizwan M, Ali S, Adrees M, Zia-ur-Rehman M, Qayyum MF, Ok YS, Murtaza G (2018) Effect of biochar on alleviation of cadmium toxicity in wheat (*Triticum aestivum* L.) grown on Cd-contaminated saline soil. Environ Sci Pollut Res 25:25668–25680

Abbasi MK, Anwar AA (2015) Ameliorating effects of biochar derived from poultry manure and white clover residues on soil nutrient status and plant growth promotion - greenhouse experiments. PLoS One 10(6):e0131592

Abdallah IHM, Alawneh A, Abu-safaqan O (1998) Effects of organic matter on the physical and the physicochemical properties of an illitic soil. J Appl Clay Sci 14(5–6):257–278

Agbna HD, Ali G, Albashir E, Mohammed M, Bakir M, Osman A, Elshaikh A (2017) Effect of biochar on some soil properties and tomato growth under saline water conditions. Int J Sci Eng Res 8:24–28

Ali S, Rizwan M, Qayyum MF, Ok YS, Ibrahim M, Riaz M, Arif MS, Hafeez F, Al-Wabel MI, Shahzad AN (2017) Biochar soil amendment on alleviation of drought and salt stress in plants: a critical review. Environ Sci Pollut Res 24:12700–12712

Atkinson CJ, Fitzgerald JD, Hipps NA (2010) Potential mechanisms for achieving agricultural benefits from biochar application to temperature soils: a review. Plant Soil 337:1–18

Barrow CJ (2011) Biochar: potential for countering land degradation and for improving agriculture. Appl Geogr 34:21–28

Benjamin JG, Nielsen DC, Vigil MF (2003) Quantifying effects of soil conditions on plant growth and crop production. Geoderma 116:137–148

Borselli L, Carnicelli S, Ferrari GA, Pagliai M, Lucamante G (1996) Effect of gypsum on hydrological, mechanical and porosity properties of a kaolinitic crusting soil. Soil Technol 9:39–54

Bossuyt H, Denef K, Six J, Frey SD, Merckx R, Paustian K (2001) Influence of microbial populations and residue quality on aggregate stability. Appl Soil Ecol 16:195–208

Bowman RA, Reeder JD, Lober RW (1990) Changes in soil properties in a central plains rangeland soil after 3, 20 and 60 years of cultivation. Soil Sci 150:851–857

Briggs CM, Breiner J, Graham RC (2005) Contributions of *Pinus ponderosa* charcoal to soil chemical and physical properties. In: The ASA–CSSA–SSSA international annual meetings, Salt Lake City, 6–10 Nov 2005

Brockhoff SR, Christians NE, Killorn RJ, Horton R, Davis DD (2010) Physical and mineral-nutrition properties of sand-based Turfgrass root zones amended with biochar. Agron J 102: 1627–1631

Bruun EW, Petersen CT, Hansen E, Holm JK, Hauggaard-Nielsen H (2014) Biochar amendment to coarse sandy subsoil improves root growth and increases water retention. Soil Use Manag 30:109–118

Carter S, Shackley S, Sohi S, Suy TB, Haefele S (2013) The impact of biochar application on soil properties and plant growth of pot grown lettuce (*Lactuca sativa*) and cabbage (*Brassica chinensis*). Agronomy 3:404–418

Chan KY, Van Zwieten L, Meszaros I, Downie A, Joseph S (2007) Agronomic values of green waste biochar as a soil amendment. Aust J Soil Res 45:629–634

Deenik J, Diarra A, Uehara G, Campbell S, Sumiyoshi Y, Antal M Jr (2011) Charcoal ash and volatile matter effects on soil properties and plant growth in an acid Ultisol. Soil Sci 176:336–345

Díaz RA, Magrin GO, Travasso MI, Rodríguez RO (1997) Climate change and its impact on the properties of agricultural soils in the Argentinean Rolling Pampas. Clim Res 9:25–30

Doerr SH, Woods SW, Martin DA, Casimiro M (2009) 'Natural background' soil water repellency in conifer forests of the North-Western USA: its prediction and relationship to wildfire occurrence. J Hydrol 371: 12–21

Dorioz JM, Robert M, Chenu C (1993) The role of roots, fungi and bacteria in clay particle organisation: an experimental approach. Geoderma 56:179–194

Downie A, Crosky A, Munroe P (2009) Physical properties of biochar. In: Lehmann JL, Joseph S (eds) Biochar for environmental management: science and technology. London: Earthscan 13–32

Dugan E, Verhoef A, Robinson S, Sohi S (2010) Bio-char from sawdust, maize stover and charcoal: impact on water holding capacities (WHC) of three soils from Ghana. In: 19th world congress of soil science, soil solutions for a changing world 11, Brisbane, published on DVD

Edenborn SL, Edenborn HM, Krynock RM, Haug KLZ (2015) Influence of biochar application methods on the phytostabilization of a hydrophobic soil contaminated with lead and acid tar. J Environ Manage 150:226–234

El-Mahrouky M, El-Naggar AH, Usman AR, Al-Wabel M (2015) Dynamics of CO_2 emission and biochemical properties of a sandy calcareous soil amended with *Conocarpus* waste and biochar. Pedosphere 25:46–56

Gaskin JW, Speir A, Morris LM, Ogden L, Harris K, Lee D, Das KC (2007) Potential for pyrolysis char to affect soil moisture and nutrient status of a loamy-sand soil. In: Proceedings of the Georgia water resources conference. University of Georgia, Athens, p 15

Ghassemi F, Jakeman AJ, Nix HA (1995) Salinisation of land and water resources: human causes, extent management and case studies, UNSW Press/CAB International, Sydney/Wallingford, 526p

Gisladottir G, Stocking M (2005) Land degradation control and its global environmental benefits. Land Degrad Dev 16:99–112

Haefele SM, Konboon Y, Wongboon W, Amarante S, Maarifat AA, Pfeiffer EM, Knoblauch C (2011) Effects and fate of biochar from rice residues in rice-based systems. Field Crop Res 121:430–440

Hansen V, Hauggaard-Nielsen H, Petersen CT, Mikkelsen TN, Müller-Stöver D (2016) Effects of gasification biochar on plant-available water capacity and plant growth in two contrasting soil types. Soil Tillage Res 161:1–9

Herath HMSK, Camps-Arbestain M, Hedley M (2013) Effect of biochar on soil physical properties in two contrasting soils: an Alfisol and an Andisol. Geodema 209–210:188–197

Indren M, Cheruvu N (2017) Biochar production and characterisation - a field study. 978-1-5090-6046-7/17/$31.00 IEEE 22-37

Jeffery S, Verheijen FGA, van der Velde M, Bastos AC (2011) A quantitative review of the effects of biochar application to soils on crop productivity using meta-analysis. Agric Ecosyst Environ 144:175–187

Jien SH, Wang CS (2013) Effects of biochar on soil properties and erosion potential in a highly weathered soil. Catena 110:225–233

Karhu K, Mattila T, Bergström I, Regina K (2011) Biochar addition to agricultural soil increased CH_4 uptake and water holding capacity results from a short-term pilot field study. Agric Ecosyst Environ 140:309–313

Kim H-S, Kim K-R, Jae E, Yang OYS, Owens G, Nehls T, Wessolek G, Kim K-H (2016) Effect of biochar on reclaimed tidal land soil properties and maize (*Zea mays* L.) response. Chemosphere 142:153–159

Konboon S, Wongboon Y, Amarante W, Maarifat S, Pfeiffer A, Knoblauch EC (2011) Effects and fate of biochar from rice residues in rice-based systems. Field Crops Res 121:430–441

Laird DA, Fleming P, Wang B, Horton R, Karlen D (2010a) Biochar impact on nutrient leaching from a midwestern agricultural soil. J Geod 158(3–4):436–442

Laird D, Fleming P, Wang B, Horton R, Karlen D (2010b) Impact of biochar amendments on the quality of a typical Midwestern agricultural soil. Geoderma 158: 443–449

Lehmann J, Rillig MC, Thies J, Masiello CA, Hockaday WC, Crowley D (2011) Biochar effects on soil biota–a review. Soil Biol Biochem 43:1812–1836

Liu XH, Han FP, Zhang XC (2012) Effects of biochar on soil aggregates in the loess plateau: results from incubation experiments. Int J Agr and Bio 14:975–979

Lu H, Lashari MS, Liu X, Ji H, Li L, Zheng J, Kibue GW, Joseph S, Pan G (2015) Changes in soil microbial community structure and enzyme activity with amendment of biochar-manure compost and pyroligneous solution in a saline soil from Central China. European J Soil Bio 70:67–76

Major J, Steiner C, Downie A, Lehmann J (2009) Biochar effects on nutrient leaching. In: Lehmann JL, Joseph S (eds) Biochar for environmental management: science and technology. Earthscan, London, pp 271–288

Middleton NJ, Thomas DSG (1992) World atlas of desertification. Edward Arnold/UNEP, Seven Oaks, p 182

Nartey OD, Zhao B (2014) Biochar preparation, characterization, and adsorptive capacity and its effect on bioavailability of contaminants: an overview. Adv Mater Sci Eng 2014:1–12

Novak JM, Lima I, Xing B, Gaskin JW, Steiner C, Das KC, Ahmedna M, Rehrah D, Watts DW, Busscher WJ, Schomberg H (2009) Characterization of designer biochar produced at different temperatures and their effects on a loamy sand. Ann Environ Sci 2:195–206

Novak JM, Cantrell KB, Watts DW, Busscher WJ, Johnson MG (2014) Designing relevant biochars as soil amendments using lignocellulosic-based and manure-based feedstocks. J Soils Sediments 14(2):330–343

Reynolds JF, Stafford DMS (2002) Do human cause deserts? In: Reynolds JF, Stafford Smith DMS (eds) Global desertification: do human cause deserts? JF Dahlem University Press, Berlin, pp 2042–2048

Saifullah DS, Rengel NA, Naidu RZ (2018) Biochar application for the remediation of salt-affected soils: challenges and opportunities. Sci Total Environ 625:320–335

Scott HD (2000) Soil physics: agriculture and environmental applications, 1st edn. Iowa State University Press, Ames, pp 35–87

Shamim M, Dijkstra FA, Abuyusf M, Hossain AI (2015) Synergistic effects of biochar and NPK fertilizer on soybean yield in an alkaline soil. Pedosphere 25:713–719

She D, Sun X, Agbna HD, Gamareldawla NEA, Hu W, Edith K, Yu S (2018) Benefits of soil biochar amendments to tomato growth under saline water irrigation. Nat Sci Rep 8:14743

Six J, Bossuyt H, Degryze S, Denef K (2004) A history of research on the link between (micro) aggregates, soil biota, and soil organic matter dynamics. Soil Tillage Res 79:7–31

Sohi SP (2010) Appendix 1. Analysis of scientific studies published on the function of char, its quantification, and its stability in soil. In: Shackley S, Sohi SP (eds) An assessment of the benefits and issues associated with the application of biochar to soil. A report to the Department for Environment, Food and Rural Affairs and the Department of Energy and Climate Change, London, pp 1–4

Sohi S, Lopez-Capel E, Krull E, Bol R (2009) Biochar, climate change and soil: a review to guide future research. CSIRO Land and Water Science report 05/09. CSIRO, Melbourne, pp.1–64

Sohi SP, Krull E, Lopez-Capel E, Bol R (2010) A review of biochar and its use and function in soil. In: Donald LS (ed) Advances in agronomy, vol 105. Academic Press, Burlington, pp 47–82

Szabolcs I (1992) Salinization of soils and water and its relation to desertification. Desert Control Bull 21:32–37

Usman ARA, Abduljabbar A, Vithanage M, Ok YS, Ahmad M, Ahmad M (2015) Biochar production from date palm waste: charring temperature induced changes in composition and surface chemistry. J Anal Appl Pyrolysis 115:392–400

Usman ARA, Al-Wabel MI, Ok YS, Al-Harbi A, Wahb-Allah M, El-Naggar AH, Al-Omran A (2016) Conocarpus biochar induces changes in soil nutrient availability and tomato growth under saline irrigation. Pedosphere 26:27–38

Uzoma KC, Inoue M, Andry H, Zahoor A, Nishihara E (2011) Influence of biochar application on sandy soil hydraulic properties and nutrient retention. J Food Agric Environ 9:1137–1143

Van Zwieten L, Kimber S, Morris S, Chan KY, Downie A, Rust J, Joseph S, Cowie A (2010) Effects of biochar from slow pyrolysis of paper mill waste on agronomic performance and soil fertility. Plant Soil 327:235–246

Verheijen FGA, Jefferey S, Bastos AC, van der Velde M, Diafas I (2009) Biochar application to soils–a critical scientific review of effects on soil properties, processes and functions. JRC scientific and technical report EUR 24099 EN: European Communities, pp 1–162
Wang L, Butterfly CR, Wang Y, Herath H, Xi YG, Xiao XJ (2014) Effect of crop residue biochar on soil acidity amelioration in strongly acidic tea garden soils. Soil Use Manag 30:119–128
Waters D, van Zwieten L, Singh BP, Downie A, Cowie AL, Lehmann J (2011) Biochar in soil for climate change mitigation and adaptation. Soil Health Clim Change 29:345–368
Woolf D, James E, Amonette F, Street-Perrott A, Lehmann J, Joseph S (2010) Sustainable biochar to mitigate global climate change. Nat Commun 1:56
Xie T, Krishna R, Wang RC, Yargicoglu E, Spokas K (2015) Characteristics and applications of biochar for environmental remediation: a review. Crit Rev Environ Sci Technol 45:939–969
Xu G, Sun J, Shao H, Chang SX (2014) Biochar had effects on phosphorus sorption and desorption in three soils with differing acidity. Ecol Eng 62:54–60
Yousaf B, Liu G, Wang R, Abbas Q, Imtiaz M, Liu R (2016) Investigating the biochar effects on C-mineralization and sequestration of carbon in soil compared with conventional amendments using stable isotope (δ13C) approach. GCB Bioenergy 9:1085–1091
Yuan JH, Xu RK (2011) The amelioration effects of low temperature biochar generated from nine crop residues on an acidic Ultisol. Soil Use Manag 27:110–115
Yuan JH, Xu RK (2012) Effects of biochars generated from crop residues on chemical properties of acid soils from tropical and subtropical China. Soil Res 50:570–578
Zhang A, Bian R, Pan G, Cui L, Hussain Q, Li L, Yu X (2012) Effects of biochar amendment on soil quality, crop yield and greenhouse gas emission in a Chinese rice paddy: a field study of 2 consecutive rice growing cycles. Field Crop Res 127:153–160
Zhang A, Bian R, Hussain Q, Li L, Pan G, Zheng J (2013) Change in net global warming potential of a rice–wheat cropping system with biochar soil amendment in a rice paddy from China. Agric Ecosyst Environ 173:37–45
Zhang H, Voroney RP, Price GW (2015) Effects of temperature and processing conditions on biochar chemical properties and their influence on soil C and N transformations. Soil Biol Biochem 83:19–28

MIX
Papier aus verantwortungsvollen Quellen
Paper from responsible sources
FSC® C105338

If you have any concerns about our products,
you can contact us on
ProductSafety@springernature.com

In case Publisher is established outside the EU,
the EU authorized representative is:
Springer Nature Customer Service Center GmbH
Europaplatz 3, 69115 Heidelberg, Germany

Printed by Libri Plureos GmbH
in Hamburg, Germany